Neurodegenerative Diseases – Laboratory and Clinical Research Series

PROGRESS IN NEURODEGENERATION: THE ROLE OF METALS

NEURODEGENERATIVE DISEASES – LABORATORY AND CLINICAL RESEARCH SERIES

Progress in Neurodegeneration: The Role of Metals
Maria Rosa Avila-Costa and Veronica Anaya Martinez (Editors)
2009. ISBN: 978-1-60741-317-2

Neurodegenerative Diseases – Laboratory and Clinical Research Series

PROGRESS IN NEURODEGENERATION: THE ROLE OF METALS

MARIA ROSA AVILA-COSTA
AND
VERONICA ANAYA-MARTINEZ
EDITORS

Nova Science Publishers, Inc.
New York

For permission to use material from this book please contact us:
Telephone 631-231-7269; Fax 631-231-8175
Web Site: http://www.novapublishers.com

Library of Congress Cataloging-in-Publication Data

Avila-Costa, Maria Rosa.
Progress in neurodegeneration : the role of metals / Maria Rosa Avila-Costa and Veronica Anaya Martinez.
p. cm.
Includes bibliographical references and index.
ISBN 978-1-60741-317-2 (hardcover)
1. Nervous system--Degeneration. 2. Metals--Toxicology. 3. Neurotoxicology. I. Martinez, Veronica Anaya. II. Title.
[DNLM: 1. Neurodegenerative Diseases--chemically induced. 2. Metals--adverse effects. 3. Metals--toxicity. WL 359 A958p 2009]
RC365.A95 2009
616.8'0471--dc22

2009006398

Published by Nova Science Publishers, Inc. ✝ New York

Dedication

We dedicate this book humbly and reverently to the people who have suffered, is suffering, or will suffer from any neurodegenerative disease. We have high hope that the etiology of these mysterious diseases will be discovered soon and a cure and prevention will be found in the very near future.

Contents

Acknowledgments

I would like to thank all the authors for their contributions. This book would not exist without you. This was an exciting journey. We made it!

Thanks for reading chapters and for many excellent suggestions to Professor Maria del Consuelo Susuki Costa Arriaga.

We also thank Alicia Cruz Resendiz who performed expert secretarial assistance.

This Book would not have been completed without the technical help of several persons, particularly Jesús Espinosa Villanueva and Patricia Aley Medina, who participated in this venture with great skill and enthusiasm.

Special thanks go to my son, Leo, for his unwavering love and patience, especially during the times when my mind was distracted by the development of this book.

This project was partially financed by CONACyT 102031 and 104017, PAPCA 07-08 and PAPIIT 215708.

Maria Rosa Avila-Costa

Preface

In the past 50 years there have been major changes on health: since the discovering and use of antibiotics, infectious diseases are not the main causes of death and because of that, we have witnessed a rapid increase of life expectancy, so to does the incidence of such age-related neurodegenerative disorders as Parkinson, Alzheimer and Huntington diseases. Rapid advances of the molecular genetics and environmental factors that either cause or increase risk for age-related neurodegenerative disorders have been made in the past decade.

The genetic and environmental factors that promote neurodegeneration may differ among diseases; shared biochemical features that will ultimately lead to the death of neurons have been identified. These features imply oxyradical production, aberrant regulation of cellular ion homeostasis and activation of a stereotyped sequence of events involving mitochondrial dysfunction and activation of specific proteases.

As we age the number of reactive oxygen species (ROS) that are produced increases. The central nervous system (CNS) seems to be particularly susceptible to damage by ROS. A number of issues contribute to the relatively high vulnerability of the CNS to oxidative damage including low levels of the natural antioxidant glutathione in neurons, membranes containing a high proportion of polyunsaturated fatty acids, and a relatively increased requirement for oxygen because of the high metabolic activities of the brain. Several lines of evidence implicate redox-active metals, as intermediaries of oxidative stress and ROS production in neurodegenerative diseases. Data are now rapidly accumulating to demonstrate that metallochemical reactions that result in formation of ROS might be the common denominator underlying Parkinson disease, Alzheimer disease, amyotrophic lateral sclerosis, prion disease and Friedreich ataxia. The association between metals, oxidative stress and neuronal death is complex, but there is support for a causative role of oxidative stress induced neurotoxicity in neurodegenerative diseases. Copper, manganese, iron, and other trace redoxactive transition metals are essential in most biological reactions, e.g., in the synthesis of DNA, RNA, and proteins, and as cofactors of numerous enzymes, particularly those involved in respiration. Thus their deficiency can lead to alterations in CNS. However, accumulation of redox-active transition metals in tissues in excess of the capacity of the cellular complement of metalloproteins can be cytotoxic as a consequence of their participation in an array of cellular disturbances characterized by oxidative stress and increased free radical production.

The purpose of the book is to provide valuable knowledge about the most recent studies about the role of metals in neurodegenerative disorders giving special attention to the synaptic interactions of the structures which are involved in the different pathologies and the physiological, anatomical, histological and structural mechanisms related with the cell death, highlighting the role of oxidative stress-inducing cell alterations. The way in which is to be achieved: first a general description of the CNS and its related cells, the description of the different neurodegenerative diseases highlighting the most recent studies. Metals, Toxicity and Neurodegeneration and its relation with oxidative stress, common mechanisms of neurotoxicity, the role of neuroglia in the inflammation process and the activation of signaling pathways; and finally the description of different metals (Aluminum, Manganese, Cadmium, Zinc, Vanadium, Lead, Iron, Mercury and Copper) and its relation with the different pathologies.

Maria Rosa Avila-Costa

Introduction

This book highlights the role of some metals which induce oxidative stress and imbalances in the neurodegenerative diseases such as Alzheimer disease, Parkinson disease, Amyotrophic Lateral Sclerosis, Huntington Disease, and other dementias. The chemistry and biochemistry of metals induced-oxidative stress, protein damage is first described, followed by the evidence for a pathological role of oxidative stress in these disease states. It is tempting to speculate that free radical oxygen chemistry contributes to pathogenesis in all these conditions, though it is as yet undetermined what types of oxidative changes occur early in the disease, and what types are secondary manifestations of neuronal degeneration. Finally we will review different metals to describe their specific role in the different pathologies.

In: Progress in Neurodegeneration
Editors: M.R. Avila-Costa and V.A. Martinez
ISBN 978-1-60741-317-2

Chapter 1

The Nervous System and Brain Damage

***Maria Rosa Avila-Costa*[1], *Leonardo Reynoso-Erazo*[2], *Ana Luisa Gutierrez-Valdez*[1] *and Jose Luis Ordoñez-Librado*[1]**
1) Dept. of Neuroscience, Neuromorphology Lab
2) Research Division, Health Education Team. UNAM, Mexico

1. The Nervous System

In order to understand the ultrastructural mechanisms of brain damage, initially we need to describe in a general way some morphological aspects of the Central Nervous System.

This introductory chapter describes in a general form the subdivisions of the nervous system, and then focuses on the cellular elements found within it and some anatomical specializations that adapt these cellular elements to their respective functions, focusing these functions through the brain damage.

1.1. Central Nervous System

The Central Nervous System (CNS) is the collective name for the brain and the spinal cord. It contrasts with the Peripheral Nervous System, which passes nerve signals between the CNS and rest of the body.

Despite the large size and widespread distribution of the nervous system, it contains only two principal categories of cells, nerve cells or neurons, which are the information-processing and signaling elements, and glial cells, which play a variety of supporting roles.

The CNS contains billions of neurons, which do all the processing and communication, and trillions of glial cells, which assist its functioning, repair damage and isolate it from the rest of the body.

The Nervous System is the body's information gatherer, storage center and control system. Its overall function is to collect information about the external conditions in relation to the body's internal state, to analyze this information, and to initiate appropriate responses to satisfy certain needs.

1.1.1. Neurons

Neurons are nerve cells that transmit signals to and from the brain at up to 200 mph. In most neurons three distinct regions can be recognized: a cell body; a number of ramifying dendrites; and a single, smooth and relatively straight axon that extends farther away from the cell body than the dendrites and may acquire a myelin sheath (Figure 1).

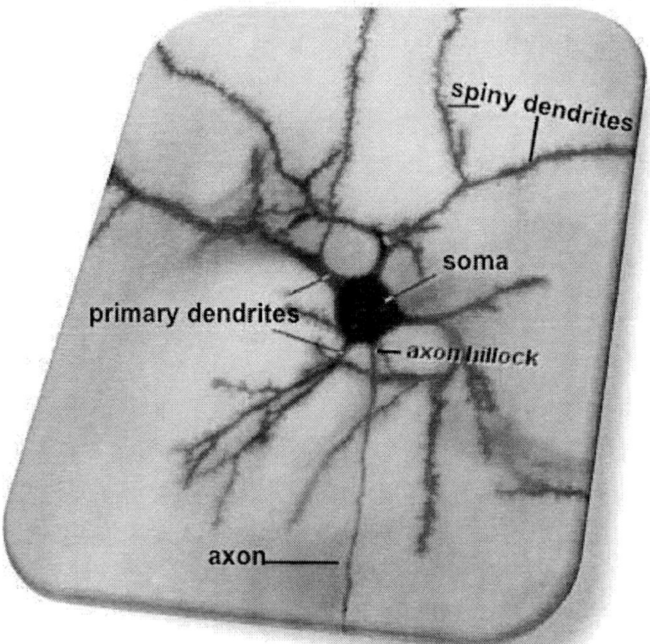

Figure 1. Pyramidal neuron from the cerebral cortex. Golgi method (400 X).

The cell body, soma or perikarion (Figure 2) contains the neuron's nucleus. This structure supports the chemical processing of the neuron, the most important of which is the production of neurotransmitters. In addition to the nucleus, the soma contains other cellular organelles, structures with distinctive structure and function that are found within all living animal cells. Organelles include ribosomes, mitochondria, endoplasmic reticulum, the Golgi complex, and lysosomes. The soma is, therefore, the site of major metabolic activity in the neuron. The size of neuronal somas range widely from 0.005 mm to 0.1 mm in mammals. A typical neuron receives about 1,000 to 10,000 synapses. Dendrites branch from the cell body and receive synaptic contacts.

A dendrite is usually one of several similar processes that issue from the soma of a neuron (Figures 1, 2). Most neurons give rise to a number of dendrites and for this reason are said to be multipolar. Others have only two main trunks, normally extending from opposite poles of the cell body. The dendrites seem to be a structure for vastly increasing the surface area and therefore, the receptor area of the cell. But the dendrites have a more subtle function than merely increasing the surface area. The form of the dendrites and the pattern of their arborizations permit the neuron to make specific contacts with certain axons.

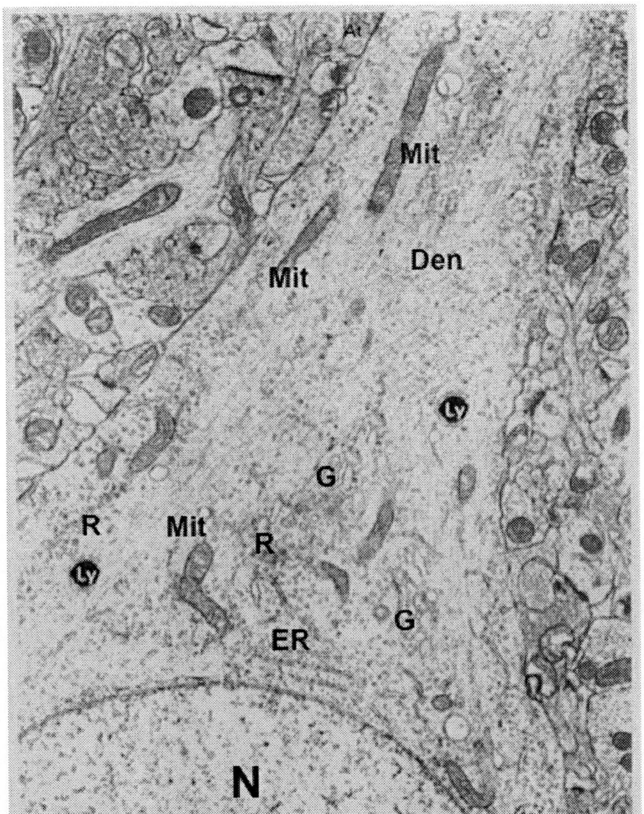

Figure 2. This low magnification electron micrograph of a cerebral cortex pyramidal cell shows the basic features of a neuronal cell body. In the lower extreme is the nucleus (N). In the surrounding perikaryal cytoplasm, the most prominent organelles are rough endoplasmic reticulum (ER), ribosome (R) Golgi apparatus (G), mitochondria (Mit) and a few lysosomes (Ly). The process extending from the perikaryon is a dendrite (Den). Bar= 0.65 μm

The dendrites appear to be reaching out toward the axonal terminals and, reciprocally, the axons appear to ramify in such a way as to terminate in synapse with certain dendrites and on specific parts of dendrites. As the efficacy of synapses must vary with, among other factors, their distance from the initial segment of the axon [axon hillock (Figure 1)], the form of the dendritic tree provides a topographic map of the world as seen by particular cell. The shape of the cell body and particularly the dendritic tree is an expression of the receptive field of the neuron.

Neurons have the capability of forming spiny outgrowths on dendrites (Figure 1) that are associated with neuroplasticity. Stimulation, especially during post-natal development can lead to brain activation, referred to as Long Term Potentiation (LTP) associated with the growth of spines. These dendritic spines, which can number thousands to a single neuron, can have synaptic heads, and more than 90 percent of synapses in the brain occur on these structures (Shepherd, 1996). Through experimentation it has been found spine's glutamate receptors, calcium concentrations, and actin can affect its shape, length, and even the presence on a dendrite. It has been proposed that many conditions lead to decreased number of dendritic spines (Fiala et al. 2002), i.e. mental retardation, epilepsy, alcoholism, and neurodegenerative disorders.

The Axon of a neuron is a singular fiber that carries information away from the soma to the synaptic sites of other neurons (dendrites and somas), muscles, or glands. The axon is considerably thicker and longer than the dendrites. Larger neurons have a markedly expanded region at the initial end of the axon. This axon hillock is the site of temporal and spatial summation for incoming information. At any given moment, the collective influence of all neurons that conduct impulses to a given neuron will determine whether or not an action potential will be initiated at the axon hillock and propagated along the axon. In vertebrates, the axons of many neurons are sheathed in myelin (Figure 3).

The myelin sheath of a neuron consists of fat-containing cells that isolate the axon from electrical activity. This insulation acts to increase the rate of transmission of signals. A gap exists between each myelin sheath cell along the axon. Since fat inhibits the propagation of electricity, the signals jump from one gap to the next.

The neuropil (Figure 3) is a term given to the portions of the CNS that contain a feltwork of intermingled and interconnected processes of neurons. It is in the neuropil that most of the synaptic interactions occur. At first sight the intermingling of the axons and dendrites appears to be random, but more careful study shows that the functional connections between these neuronal processes are very specific and precise.

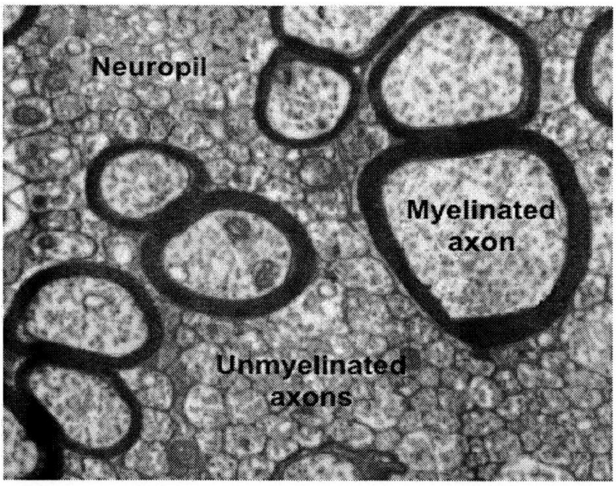

Figure 3. The figure shows complete transverse sections of myelinated and unmyelinated axons within the neuropil. 25,000X.

1.1.2. Synapse

Sir Charles Sherrington first used the term synapse in 1897. It was derived from Greek words meaning "to fasten together", and was used to signify a site where the axon terminal of one neuron comes into functional contact with a second neuron. It follows from the neuron doctrine that a synapse is not a site of cytoplasmic confluence between neurons but an interface at which they are functionally related (Peters et al. 1991). In the light microscope, synapses are usually revealed by means of silver-staining techniques, but virtually all that these methods show the profile of the expanded terminal bouton of the axon attached to the surface of a perikarion or dendrite. Apart from the mitochondria and, sometimes, neurofibrils, the light microscope shows nothing of the organelles within the terminal bouton or of the junctional interface itself. It remained for the electron microscope to reveal the detailed morphology of synapses and to show that there is a gap between the membranes of the synaptic partners (Peters et al. 1991).

A typical presynaptic bouton is a specialized portion of the axon. It is characterized by an active zone, a region where the presynaptic plasma membrane comes in close contact with the postsynaptic membrane and an associated cluster of vesicles (De Camilli et al. 2001). Vesicle exocytosis occurs at the active zone; subsequent endocytic retrieval of vesicular components may occur both at the active zone and in a peri-active zone area (Roos and Kelly, 1999).

Synapses are typically surrounded by glial cells, which might perform auxiliary roles such as secretion of neuromodulatory factors, regulation of extracellular ion concentration, and uptake of neurotransmitters (Haydon, 2001; Yang and Rothstein 2009).

Ultrastructurally the synapse can be seen as a postsynaptic electron density (PSD) in direct opposition to a presynaptic profile associated with synaptic vesicles that is referred to as an active zone. The synaptic cleft is a tight intercellular junction that is resistant to biochemical disruption. The intervening space at the synapse is termed the synaptic cleft (10 to 20 nm wide) (Figure 4).

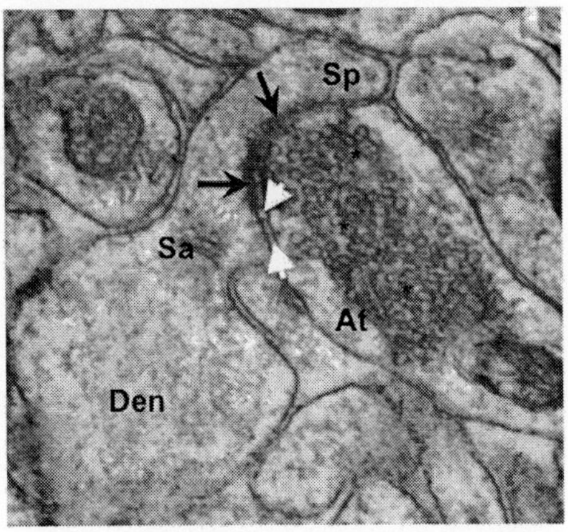

Figure 4. Electron micrograph of an axon terminal (At) that forms asymmetric synapse with a dendritic spine (Sp). This axon terminal contains spherical vesicles (*). Note the membrane asymmetry (black arrows) called

the postsynaptic electron density and the synaptic cleft (white arrowheads). Sa- spine apparatus; Den- dendrite. 40,000X.

Since the late 1950s, the ultrastructural features of individual synapses have been studied extensively via electron microscopy. Gray (1959) classified two types of synapses within the brain based on the ultrastructural characteristics of the presynaptic (vesicle-bearing) and postsynaptic partners (length of apposed membrane, membrane thickenings and synaptic cleft):

- Gray type 1- were found on dendritic spines and dendrite shafts
- Gray type 2- occurred primarily on dendrite shafts and neuronal cell bodies.

Virtually synonymous with Gray's nomenclature are the terms asymmetric or symmetric synapses described by Colonnier (1964):

- Asymmetric synapses- includes axons that contain predominantly round or spherical vesicles and form synapses that are distinguished by a thickened postsynaptic density (Figure 4).
- Symmetric synapses- involves axons that contain clusters of vesicles that are predominantly flattened or elongate in their appearance. The pre-and postsynaptic membranes are more parallel than the surrounding nonsynaptic membrane, and the synapse does not contain a prominent postsynaptic density (figure 5).

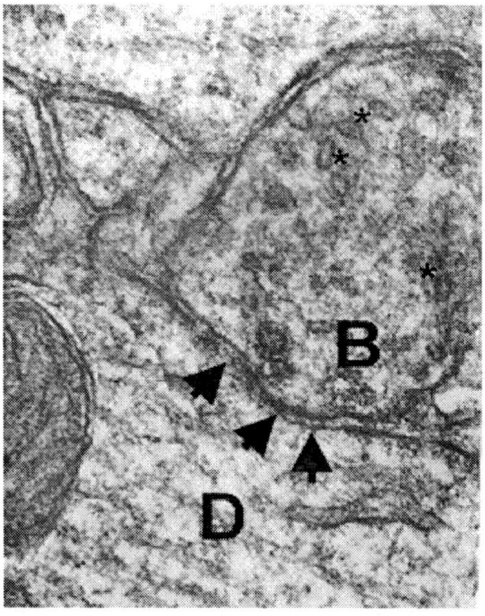

Figure 5. This electron micrograph shows a presynaptic bouton (B) forming a symmetric synapse with a dendrite (D). This axon terminal contains flattened synaptic vesicles (*). The arrows show the postsynaptic electron density. 40,000 X.

The stereotypical and most abundant synapse in the central nervous system is the asymmetric synapse occurring between an axon and a dendritic spine (figure 4). Other synaptic relationships exist and involve different parts of the neuron. For instance, axo-axonic, axo-somatic, dendro-somatic, axo-dendritic, and dendro-dendritic synapses can occur and provide alternate mechanisms for functional communication.

As we mentioned above, the principal neurons of most brain regions are covered with small protrusions known as dendritic spines. Spines are extremely numerous on many kinds of dendrites; in fact they account for the majority of postsynaptic sites in the vertebrate brain. They are especially prominent in the cerebral cortex, basal ganglia, hippocampus and cerebellar cortex.

The dendritic spines mediate most excitatory contacts in the mammalian nervous system, so they are likely to be crucial for brain function. At ultrastructural level, spines can be distinguished from other dendritic elements in the neuropil by the presence of a characteristic spine apparatus (figure 4) composed of a calcium-binding protein, and it has been an important marker for spines in quantitative electron microscopic studies. Within cerebral cortex, for example, about 79% of all excitatory synapses are made onto spines and the rest directly onto dendrites, whereas 31% of all inhibitory synapses are made onto spines. A spine with an inhibitory synapse always carries an excitatory synapse as well (Beaulieu and Colonnier, 1983). Given the dominance of excitatory synapses, about 15% of all dendritic spines carry both excitatory and inhibitory synaptic profiles.

As principal sites of synaptic input, spines play a key role in connectivity in the brain. Alterations in spine morphology that accompany disease are expected to have a significant impact on brain function. In addition, understanding spine plasticity in the extreme of pathological conditions will help elucidate the limits of normal synaptic plasticity during development, and during learning and memory in the mature brain (Chang and Greenough, 1984).

The existence of distinct morphological subtypes of axospinous synapses is consistent with this idea and suggests a hypothetical model of synapse restructuring that may account for synaptic plasticity associated with the induction phase of LTP. According to this model, LTP induction initiates a sequence of structural synaptic modifications, which initiates with the enlargement of typical small-nonperforated synapses and their conversion into atypically large nonperforated ones. This is followed by the consecutive formation of perforated synapses that have initially a focal spine partition with a fenestrated post synaptic density, and finally a complete partition(s) with a segmented post synaptic density. Synapses of the latter subtype have multiple transmission zones instead of only a single one, and an increase in their number after LTP induction, may result in an augmentation of synaptic transmission.

Perforated synapses (Figure 6) are characterized by a perforation (fenestration) extended through the paramembranous densities, from the presynaptic terminal across the cleft and into the postsynaptic process. This type of synapses has been reported to increase in number during development (Itarat and Jones, 1992) and under various experimental conditions such as complex environments (Sirevaag and Greenough, 1985), behavioral training (Vrensen and Nunez, 1981), repetitive electrical stimulation (Geinisman et al. 1992), and hormonal alterations (Hatton and Ellisman, 1982). It has also been suggested that a preserved complement of hippocampal perforated synapses is required for the maintenance of good spatial memory during aging (Geinisman et al. 1986).

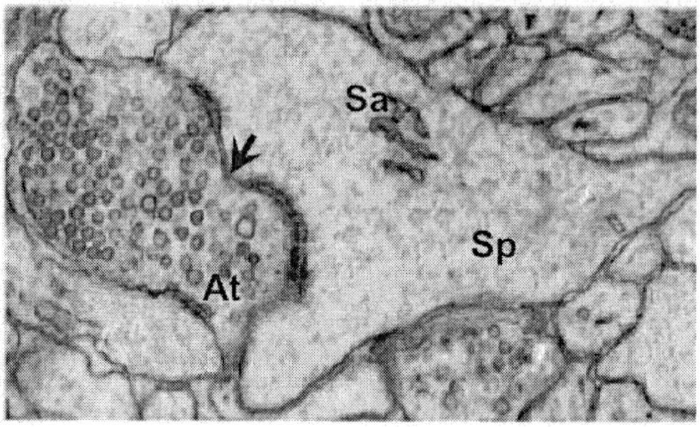

Figure 6. The typical perforated synapses can be distinguished with a rupture toward the presynaptic terminal membrane (arrow), and the presynaptic density is in association to the spine apparatus (Sa); Sp = dendritic spine; At = pre synaptic button. 40,000X.

1.1.3. Neuroglial Cells

Glial cells regulate the internal environment of the brain, especially the fluid surrounding neurons and their synapses, and provide nutrition to nerve cells. These cells have important developmental roles, guiding migration of neurons in early development. Recent findings have indicated that glia are also active participants in synaptic transmission, regulating clearance of neurotransmitter from the synaptic cleft, releasing factors such as ATP which modulate presynaptic function, and even releasing neurotransmitters themselves (Yang and Rothstein 2009). Four main types of glia exist, namely astrocytes, oligodendrocytes, ependymal cells and microglia.

Traditionally glia cells had been thought to lack certain features of neurons. For example, it was thought that glia cells were not believed to have synapses, nor were they believed to generate action potentials or release neurotransmitters. They were considered to be passive bystanders of neural transmission. However, recent studies disproved this. For example, astrocytes (Figure 7) are crucial in neurotransmitter clearance within the synaptic cleft, which temporally and spatially restricts neurotransmission and limit the toxicity of certain neurotransmitters such as glutamate. And, at least *in vitro* astrocytes can release neurotransmitter glutamate in response to certain stimulation (Yang and Rothstein 2009).

Neuronal–astrocyte interactions typically require chemical signaling through the extracellular space (Parpura et al. 1994; Bezzi et al. 1998). For example, exogenous application or synaptically released neurotransmitters activate receptors on glia to affect their membrane potential, evoke second messenger signals, and induce calcium waves (Dani et al. 1992; Chen et al. 1997; Pasti et al. 1997). In turn, the astrocytes remove neurotransmitters from the extracellular space and maintain proper ionic balance (Jendelova and Sykova, 1991; Bergles et al. 1999). Moreover, glutamate released from the astrocytes has been shown to modulate evoked and spontaneous synaptic transmission in neurons (Araque et al. 1998; Yang and Rothstein 2009). Together, these bidirectional interactions mediate neuronal–glial communication and indicate that astrocytes play an active role in influencing neuronal activity.

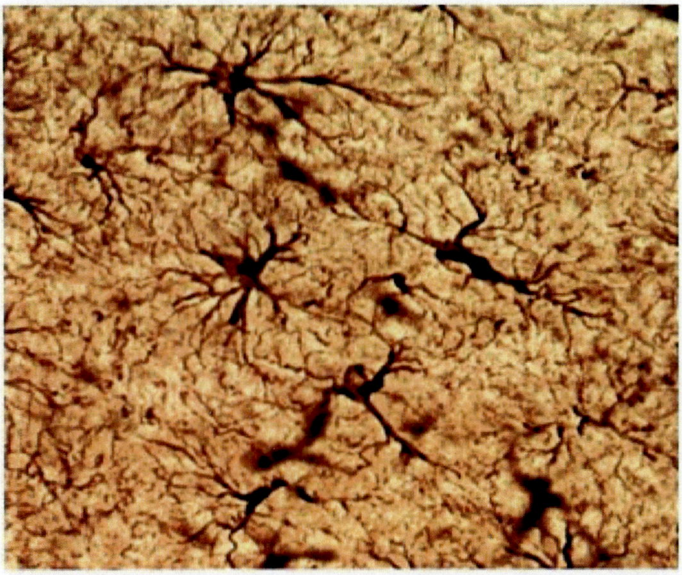

Figure 7. Astrocyte of stellate morphology in rat striatum stained with antibody to GFAP. 400X.

Oligodendrocytes comprise several short processes, which wrap themselves around neurons present in the CNS. Oligodendrocytes are responsible for axonal regulation and the generation and maintenance of the myelin sheath that surrounds axons. As we mentioned above, the main role of myelin is to mediate rapid saltatory propagation of action potential between Nodes of Ranvier, thereby facilitating neurotransmission (Ransom and Sontheimer 1992; Edgar and Garbern, 2004). Oligodendrocytes also secrete a number of neurotrophins including nerve growth factor (NGF), brain derived neurotrophic factor (BDNF) and neurotrophin-3 (Dai et al. 2003), which provide local trophic support for neurons. Peripheral axons are myelinated by Schwann cells and these cells in contrast to oligodendrocytes faciliate neuronal regeneration following injury (Torigoe et al. 1996).

The choroid plexus and the oligodendrocytes are the only two cell types in the CNS capable of producing transferrin (Zheng et al. 2003; Gonzalez-Reyes et al. 2007).

Ependymal cells (Figure 8) line the ventricles and the central canal of the spinal cord (Del Bigio, 1995). The functions of ependymal cells remain largely speculative. However, ependymal cells possess microvilli, which beat in a co-ordinated manner; therefore these cells are believed to be involved in the directional movement of cerebral spinal fluid (CSF), disturbances of which lead to hydrocephaly. The directional flow of CSF is thought to facilitate the transport of nutrients into the brain and the removal of toxic metabolites. Ependymal cells have also been suggested to serve as an axonal guidance system during early development.

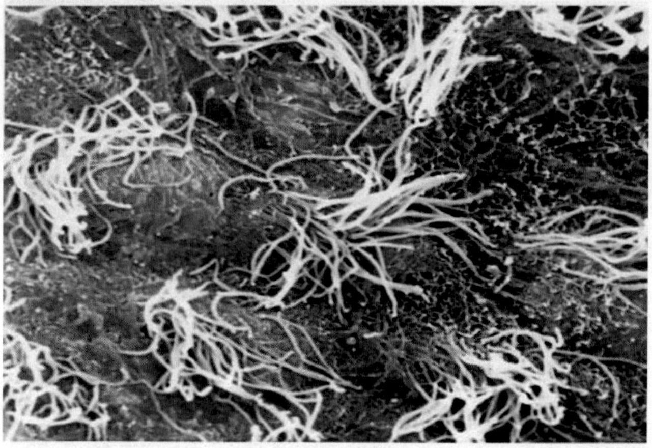

Figure 8. Scanning electron microscopy micrograph of the floor of the fourth ventricle. Note the ependymal cell cilia. 5000 X.

Microglia are the immune effector cells of the CNS and are present in abundance in the brain parenchyma and constitute approximately 10-20% of the total population of glial cells in the adult (Banati, 2003). These cells are derived from myeloid progenitor cells of the bone marrow, which migrate into the CNS during early development. Microglia are small round cells that comprise numerous branching processes and possess little cytoplasm. Classically, microglia were considered to be inactive under physiological conditions, however, it is now known that microglia exhibit pinocytotic activity and localized motility (Glenn et al. 1991) particularly of their ramified protrusions (Nimmerjahn et al. 2005). Microglial processes directly contact neuronal cell bodies, astrocytes and blood vessels (Nimmerjahn et al. 2005), therefore it seems likely that microglia monitor the well being of the brain and also function to cleanse the extracellular fluid in order to maintain central homeostasis (Fetler and Amigorena 2005). Furthermore, the presence of neurotransmitter receptors on microglia means that these cells can respond to released neurotransmitters (Boucsein et al. 2003; Light et al. 2006; Taylor et al. 2003, 2005).

In response to injury or pathogen, invasion microglia transform into active phagocytic macrophages in an attempt to combat disease (Stence et al. 2001). Following a damaging event, reactive microglia accumulates at the site of injury (Eugenin et al. 2001) where they play a neuroprotective role phagocytosing damaged cells and debris. In acute lesions the peak of microglial activation occurs 2-3 days post insult, but if the pathological stimulus persists microglial activation continues (Banati, 2003). In addition to their roles in the adult, microglia play a pivotal role during development by removing inappropriate axons (Marin-Teva et al. 2004) and through the promotion of axonal migration and growth (Polazzi and Contestabile, 2002).

1.2. Brain Damage

The brain is a highly specialized tissue. Due to its complexity, even the slightest damage can have extreme consequences.

The brain can be damaged in a variety of ways, and depending on the areas and the severity of the damage, it can prove relatively harmless to fatal. One of the possible causes of brain damage is the presence of oxidative stress, which is the consequence of reactive oxygen species (ROS).

ROS are byproducts generated by cellular oxidative metabolism. However, enhanced production of ROS, exceeding the intrinsic antioxidant scavenging capacity, leads to the development of several pathophysiological conditions. Brain tissue may be especially vulnerable to ROS damage because of high oxygen consumption, moderate antioxidant capacity, and membranes rich in easily oxidized polyunsaturated lipids. Mounting evidence has implicated the role of ROS in the pathophysiology of neurodegenerative conditions such as Parkinson disease, Huntington disease, Alzheimer disease, and in normal aging (for review, see Beal, 1995; Simonian and Coyle, 1996). An essential component of ROS-induced neurotoxicity may involve the modulation of various ion transport proteins and receptor-gated ion channels either directly via protein oxidation or indirectly via peroxidation of membrane phospholipids (Kourie, 1998).

Neurodegeneration is a progressive loss of structure or function of neurons, which leads to cell death. Acute neurodegeneration which includes stroke, head trauma and subarachnoid hemorrhage share common mechanisms of neuronal loss with chronic forms of neurodegeneration, e.g., Alzheimer and Parkinson diseases. There is relationship between excitotoxicity, cellular calcium overload, metabolic failure, and oxidative stress with emphasis on the role that mitochondrial dysfunction plays in both necrotic and apoptotic cell death (Figure 9). Animal models are used both as tools to test mechanistic hypotheses and as a means to develop clinically useful neuroprotective interventions.

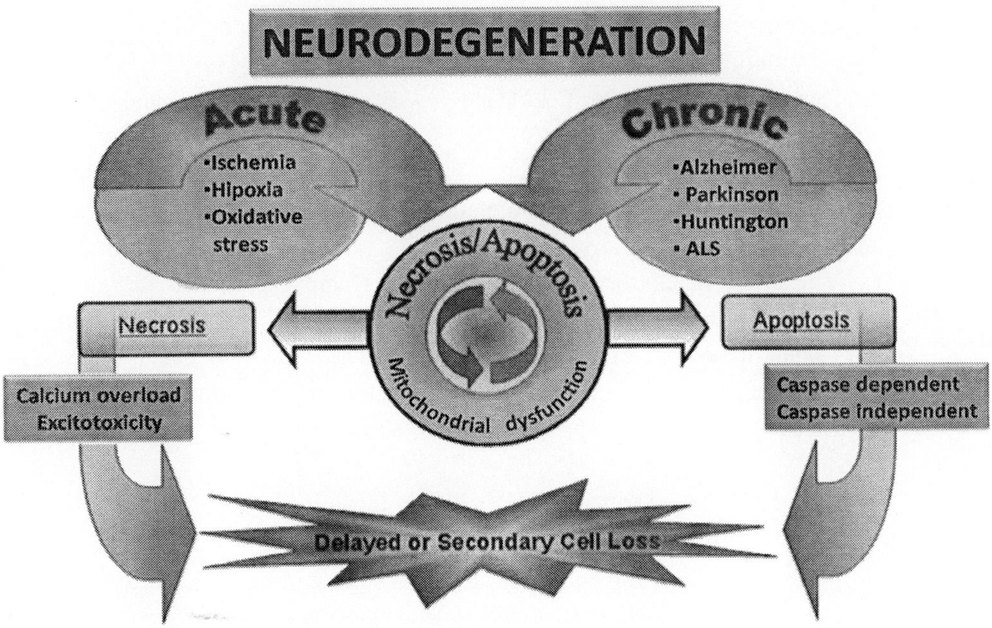

Figure 9. Schematic representation of neurodegeneration.

Neurodegenerative diseases result from the gradual and progressive loss of neural cells, leading to nervous system dysfunction. These diseases represent a large group of neurological disorders with diverse clinical and pathological indicators affecting specific subsets of neurons in certain functional anatomic systems; they arise for unknown reasons and progress in a continual manner. On the contrary, edema, hemorrhage, neoplasm, and trauma of the nervous system, which are not primary neuronal diseases, are not considered neurodegenerative disorders. Diseases of the nervous system that involve not neurons per se but rather their features, such as the myelin sheath as seen in multiple sclerosis, are not neurodegenerative disorders either, nor are pathologies in which neurons die as the result of a known cause such as hypoxia, poison, metabolic defects, or infections.

In chapter 2 we will try to provide an overview of the complexity of the field of neurodegeneration, classification of the neurodegenerative diseases and a brief explanation of each disease.

References

Araque, A., Sanzgiri, R.P., Parpura, V. and Haydon, P.G. (1998) Calcium elevation in astrocytes causes an NMDA receptor-dependent increase in the frequency of miniature synaptic currents in cultured hippocampal neurons. *J Neurosci.* 18:6822– 6829.

Banati, R. (2003) Neuropathological imaging: in vivo detection of glial activation as a measure of disease and adaptive change in the brain. *Brit. Med. Bul.* 65, 121-131.

Beal, MF. (1995) Aging, energy and oxidative stress in neurodegenerative diseases. *Ann Neurol.* 38:357-366.

Beaulieu, C. and Colonnier, M. (1983). The number of neurons in the different laminae of the binocular and monocular regions of area 17 in the cat. *J. Comp. Neurol.* 217:337-344.

Bergles, D.E., Diamond, J.S. and Jahr, C.E. (1999) Clearance of glutamate inside the synapse and beyond. *Curr Opin Neurobiol.* 9:293–298.

Bezzi, P., Carmignoto, G., Pasti, L., Vesce, S., Rossi, D., Rizzini, B.L., Pozzan, T. and Volterra, A (1998) Prostaglandins stimulate calcium-dependent glutamate release in astrocytes. *Nature,* 391:281–285.

Boucsein, C., Zacharias, R., Farber, K., Pavlovic, S., Haqnisch, U. and Kettenmann, H. (2003) Purinergic receptors on microglial cells: Functional expression in acute brain slices and modulation of microglial activation in vitro. *Euro. J. Neurosci.* 17, 2267-2276.

Chang, F.L.F. and Greenough, W.T. (1984). Transient and enduring morphological correlates of synaptic activity and efficacy change in the rat hippocampal slice. *Brain Res.* 309: 35-46.

Chen, J., Backus, K.H. and Deitmer, J.W. (1997) Intracellular calcium transients and potassium current oscillations evoked by glutamate in cultured rat astrocytes. *J Neurosci.* 17:7278 –7287.

Colonnier, M. (1964). The tangential organization of the visual cortex. *J. Anat.* 98: 327-344.

Dai, X., Lercher, L.D., Clinton, P.M., Du, Y., Livingston, D.L., Viera, C., Yang, L., Shen, M.M. and Dreyfus, C. F. (2003) The trophic role of oligodendrocytes in the basal forebrain. *J. Neurosci.* 23, 5846-5853.

Dani, J.W., Chernjavsky, A. and Smith, S.J (1992) Neuronal activity triggers calcium waves in hippocampal astrocyte networks. *Neuron.* 8:429–440.

De Camilli, P., Haucke, V., Takei, K. and Mugnaini, E. 2001. The structure of synapses. In WM Cowan, TC Sudhof, CF Stevens (eds.). Synapses, (pp. 89–133). Baltimore, MD: Johns Hopkins Univ. Press.

Del Bigio, M. (1995) The ependyma: a protective barrier between brain and cerebrospinal fluid. *Glia.* 14, 1-13.

Edgar, J.M. and Garbern, J. (2004) The myelinated axon is dependent on the myelinating cell for support and maintenance: molecules involved. *J. Neurosci. Res.* 76, 593-598.

Eugenin, E.A., Eckardt, D., Theis, M., Willecke, K., Bennett, M.V. and Saez, J.C. (2001) Microglia at brain stab wounds express connexin 43 and in vitro form functional gap junctions after treatment with interferon gamma and tumour necrosis factor alpha. *Proc. Natl. Acad. Sci USA.* 98, 4190-4195.

Fetler, L. and Amigorena, S. (2005) Brain under surveillance the microglia patrol. *Science*, 309, 392-393.

Fiala, J.C., Spacek, J. and Harris, K.M. (2002) Dendritic Spine Pathology: Cause or Consequence of Neurological Disorders? *Brain Res Rev.* 39: 29–54.

Geinisman, Y., de Toledo Morrell, L. and Morrell, F. (1986). Aged rats need a preserved complement of perforated axospinous synapses per hippocampal neuron to maintain good spatial memory. *Brain Res.* 398: 266 275.

Geinisman, Y., Morrell, F. and de Toledo Morrell, L. (1992). Increase in the number of axospinous synapses with segmented postsynaptic densities following hippocampal kindling. *Brain Res.* 569: 341 347.

Glenn, J.A., Booth, P.L. and Thomas, W.E. (1991) Pinocytic activity in ramified microglia. *Neurosci. Lett.* 123, 27-31.

Gonzalez-Reyes, R.E., Gutierrez-Alvarez, A.M. and Moreno, C.B. (2007) Manganese and epilepsy: A systematic review of the literature. *Brain Res Rev.* 332–336.

Gray, E.G. (1959) Axo-somatic and axo-dendritic synapses of the cerebral cortex An electron microscope study. *J Anat.* 93: 420–433.

Hatton, J.D. and Ellisman M.H. (1982). A restructuring of hypothalamic synapses is associated with motherhood. *J. Neurosci.* 2: 704 707.

Haydon, P.G. 2001. Glia: listening and talking to the synapse. *Nat. Rev. Neurosci.* 2:185–93.

Itarat, W. and Jones, D.G. (1992). Perforated synapses are present during synaptogenesis in rat neocortex. *Synapse*, 11: 279–286.

Jendelova, P. and Sykova, E (1991) Role of glia in K1 and pH homeostasis in the neonatal rat spinal cord. *Glia*, 4:56–63.

Kourie, J.I. (1998) Interaction of reactive oxygen species with ion transport mechanisms. *Am J Physiol.* 275:C1-C24.

Light, A.R., Wu, Y., Hughen, R.W. and Guthrie, P.B. (2006) Purinergic receptors activating rapid intracellular Ca2+ increases in microglia. *Neuron Glia Biology*, 2:125-128.

Marin-Teva, J.L., Dusart, I., Colin, C., Gervais, A., van Rooijen, N. and Mallat, M. (2004) Microglia promote the death of developing purkinje cells. *Neuron.* 41, 535-547.

Nimmerjahn, A., Kirchhoff, F. and Helmchen, F. (2005) Resting microglial cells are highly dynamic surveillants of brain parenchyma in vivo. *Science*, 308, 1314-1318.

Parpura, V., Basarsky, T.A., Liu, F., Jeftinija, K., Jeftinija, S. and Haydon, P.G. (1994) Glutamate-mediated astrocyte-neuron signalling. *Nature*, 369:744 –747.

Pasti, L., Volterra, A., Pozzan, T. and Carmignoto, G. (1997) Intracellular calcium oscillations in astrocytes: a highly plastic, bidirectional form of communication between neurons and astrocytes in situ. *J Neurosci.* 17:7817–7830.

Peters, A., Palay, S. and Webster, H. (1991) Synapses. In: The fine Structure of the nervous system. 3rd edition. Oxford University Press, New York.

Polazzi, E. and Contestabile, A. (2002) Reciprocal interactions between microglia and neurons: from survival to neuropathology. *Rev. Neurosci.* 13, 221-242.

Ransom, B.R. and Sontheimer, H. (1992) The neurophysiology of glial cells. *J. Clin. Neurophysiol.* 9, 224-251.

Roos, J. and Kelly, R.B. (1999). The endocytic machinery in nerve terminals surrounds sites of exocytosis. *Curr. Biol.* 9:1411–14.

Shepherd, G.M. (1996) The dendritic spine: A multifunctional integrative unit. *J. Neurophysiol.* 75 (6): 2197-2210.

Simonian, N.A. and Coyle, J.T (1996) Oxidative stress in neurodegenerative disease. *Annu Rev Pharmacol Toxicol.* 36:83-106.

Sirevaag, A.M. and Greenough, W.T. (1985). Differential rearing effects on rat visual cortex synapses. II. Synaptic morphometry. *Dev. Brain Res.* 19: 215 226.

Stence, N., Waite, M. and Dailey, E. (2001) Dynamics of microglial-activation: a confocal time-lapse analysis in hippocampal slices. *Glia.* 33, 256-266.

Taylor, D.L., Diemel, L.T. and Pocock, J.M. (2003) Activation of microglial group III metabotropic glutamate receptors protects neurons against microglial neurotoxicity. *J Neurosci.* 15:2150-2160.

Taylor, D.L., Jones, F., Chen Seho Kubota, E.S.F., and Pocock, J.M. (2005) Stimulation of microglial metabotopic glutamate receptor, mGlu2 triggers TNFa-induced neurotoxicity in concert with microglial derived FasL. *J Neurosci.* 25:2952-2964.

Torigoe, K., Tanaka, H.F., Takahashi, A., Awaya, A. and Hashimoto, K. (1996) Basic behaviour of migratory schwann cells in peripheral nerve regeneration. *Exp. Neurol.* 137, 301-308.

Vrensen, G. and Nunez, J. (1981). Changes in size and shape of synaptic connections after visual training: an ultrastructural approach of synaptic plasticity. *Brain Res.* 218: 79 97.

Yang, Y. and Rothstein, J.D. (2009) Specialized Neurotransmitter Transporters in Astrocytes. In: Astrocytes in (Patho)Physiology of the Nervous System. V., Parpura and P.G., Haydon (Eds) Springer Science, NY. Pp. 69-106.

Zheng, Wei, Aschner, M. and Ghersi-Egeac, J.F (2003) Brain barrier systems: a new frontier in metal neurotoxicological research. *Toxicol and Applied Pharmacol.* 192 1–11.

In: Progress in Neurodegeneration
Editors: M.R. Avila-Costa and V.A. Martinez
ISBN 978-1-60741-317-2

Chapter 2

Neurodegenerative Diseases

***Maria Rosa Avila-Costa*[1]*, Leonardo Reynoso Erazo*[2] *and Jose Luis Ordoñez Librado*[1]**

1) Dept. of Neuroscience, Neuromorphology Lab
2) Research Division, Health Education Team. UNAM, Mexico

Neurodegeneration is a complex and multifaceted process leading to many chronic disease states. A conventional definition implies a progressive neuronal death, which usually affects a specific population of nerve cells, the vulnerability of which determines the clinical picture of a particular neurodegenerative disease. Owing to the prevalence, morbidity and mortality, as well as social, ethical and personal burden of neurodegenerative disorders, considerable effort has been directed towards the identification of a rational strategy to treat these devastating brain pathological conditions. A substantial challenge is to discern phenomena that may represent causes from those that may represent effects. The neurodegenerative process is associated with many events, of which each one may correspond to a typical feature of a specific disease.

The factors that underlie neurodegeneration can be classified as genetic, environmental (extrinsic neurotoxins, e.g. metals, infective damage), biological (aggregation, abnormal folding and accumulation of proteins), metabolic (oxidative stress), excitotoxic (intrinsic neurotoxins, e.g. metals and excitatory amino acids), autoimmunity and ageing. Each of these factors can play a crucial role in the etiology of neuronal degeneration, but with an influence that varies between the different pathologies (Gaeta and Hider, 2005).

Evidence is accumulating to suggest that such chronic neurodegenerative disorders are caused by a combination of events that impair normal neuronal function. Clinical signs are evident before frank neuronal loss. Therefore, new efforts have focused on identifying crucial changes, of genetic, epigenetic or environmental origin, that hamper normal neuronal function (Bossy-Wetzel et al. 2004).

The number of neurodegenerative diseases is currently estimated to be a few hundred, and, among these, many appear to overlap with one another clinically and pathologically, rendering their practical classification quite challenging. The issue is further complicated by the fact that, in diseases such as multisystem atrophy in which several areas of the brain are affected, different combinations of lesions can give rise to different clinical pictures (Burn and Jaros, 2001). Furthermore, the same neurodegenerative process, especially at the beginning, can affect different areas of the brain, making a given disease appear very different from a symptomatic standpoint. Despite these difficulties, the most popular categorization of neurodegenerative disorders is still based on the predominant clinical feature or the topography of the predominant lesion, or often on a combination of both. Accordingly, neurodegenerative disorders of the CNS may, for example, be first grouped into diseases of the cerebral cortex, the basal ganglia, the brainstem and cerebellum, or the spinal cord. Then, within each group, a given disease may be further classified based on its main clinical features. For instance, the group of diseases that predominantly affect the cerebral cortex may be divided into dementing (e.g., Alzheimer disease) and nondementing conditions. Of note, while Alzheimer disease is by far the most frequently cited cause of dementing cerebral cortex pathology (Sulkava et al. 1983), dementia can apparently be observed in at least 50 different diseases (Tomlinson et al. 1977). Moreover, dementia is not exclusively observed in neurodegenerative disorders; it is also frequently observed in ischemic, metabolic, toxic, infectious, and traumatic insults of the brain.

Diseases that predominantly involve the basal ganglia (a series of deep nuclei situated at the base of the forebrain, including the caudate nucleus putamen, globus pallidus, substantia nigra, subthalamic nucleus, and some thalamic and brainstem nuclei) are essentially characterized by abnormal movements. Yet, based on the phenomenology of the abnormal movements, diseases of the basal ganglia can be classified as hypokinetic or hyperkinetic. Hypokinetic basal ganglia disorders are epitomized by Parkinson disease, in which the amplitude and velocity of voluntary movements are diminished or, in extreme cases, even nonexistent, causing the patient to become a prisoner within his or her own body (Przedborski et al. 2003). Aside from Parkinson disease, parkinsonism — which refers to an association of at least two of the following clinical signs: resting tremor, slowness of movements, stiffness, and postural instability— is found in a variety of other diseases of the basal ganglia as well. In some (e.g., striatonigral degeneration), there is only parkinsonism, but in others, often called Parkinson-plus syndromes, there is parkinsonism plus signs of cerebellar ataxia (e.g., olivopontocerebellar atrophy), orthostatic hypotension (e.g., Shy-Drager syndrome), or paralysis of vertical eye movements (e.g., progressive supranuclear palsy). Because, early on, parkinsonism may be the only clinical expression of Parkinson-plus syndromes, it is difficult to reach an accurate diagnosis before the patient reaches a more advanced stage of the disease (Przedborski et al. 2003). This problem is well illustrated by the fact that more than 77% of patients with parkinsonism are diagnosed in life as having Parkinson disease (Stacy and Jankovic, 1992), but as much as a quarter of these are found at autopsy to have lesions incompatible with Parkinson disease (Hughes et al. 1992). At the other end of the spectrum are the hyperkinetic basal ganglia disorders, which are epitomized by Huntington disease and essential tremor. In these two conditions, excessive abnormal movements such as chorea or tremor are superimposed onto and interfere with normal voluntary movements. Although hyperkinetic basal ganglia disorders are probably as diverse as are hypokinetic basal ganglia disorders, their accurate classification, even during life, is less problematic, in part because specific disease markers such as gene mutations exist for several of these syndromes.

Classification of neurodegenerative diseases of the cerebellum and its connections is particularly challenging because of the striking overlap among the various pathological conditions. Indeed, some diseases of the cerebellum can readily be grouped into three main neuropathological types: cerebellar cortical atrophy (lesion confined to the Purkinje cells and the inferior olives), pontocerebellar atrophy (lesion affecting several cerebellar and brain structures), and Friedreich ataxia (lesion affecting the posterior column of the spinal cord, peripheral nerves, and the heart). However, several other diseases of the cerebellum and its connections cannot be situated in one of these categories such as dentatorubral degeneration, in which the most conspicuous lesions are in the dentate and red nuclei, and Machado-Joseph disease, in which degeneration involves the lower and upper motor neurons, the Substantia Nigra, and the dentate system (Przedborski et al. 2003).

Among the neurodegenerative diseases that predominantly affect the spinal cord are Amyotrophic Lateral Sclerosis and spinal muscular atrophy, in which the most severe lesions are found in the anterior part of the spinal cord, and the already cited Friedreich ataxia, in which the most severe lesions are found in the posterior part of the spinal cord. Finally, there is one group of neurological diseases that are often, but not always, considered neurodegenerative because of their chronic course and unknown etiopathogenesis but that, unlike those described above, show no apparent structural abnormalities. These include torsion dystonia, Tourette syndrome, essential tremor, and schizophrenia. Various brain-imaging studies and electrophysiological investigations have revealed significant functional abnormalities in all of these singular neurodegenerative disorders but have not yet enabled us to unravel their chemical neuroanatomical substrates (Przedborski et al. 2003).

In general, a clear classification can be achieved based on the principal neuropathological changes (Gaeta and Hider, 2005):

- Neurodegenerative disorders characterized by the presence of abnormal protein components (Butterfield and Kanski, 2001; Shastry, 2003), which accumulate in the brain leading to a selective loss of neurons in an age-dependent manner such as Alzheimer disease (Aβ-amyloid neuritic plaques [Figure 1A] and neurofibrillary tangles [Figure 1B]), Parkinson disease (α-synuclein, Lewy bodies [Figure 1C]), prion disease (amyloid plaque core surrounded by 'petals' of sponge-like tissue, spongiosis [Figure 1D]) (Prusiner, 2001), Pick disease (Pick bodies [Figure 1E]) and Huntington disease (huntingtin protein aggregates [Figure 1F]) (Wisniewsky et al. 1972; Brion et al. 1973).
- Neurodegenerative disorders resulting from dysfunction/degeneration of motor neurons, which can be characterized by different pathological hallmarks as exemplified in multiple sclerosis (demyelinisation), amyotrophic lateral sclerosis, Friedreich ataxia and progressive supranuclear palsy.

The major neurodegenerative disorders are the object of intense study since they are characterized by a comprehensive pattern of findings, the interpretation of which can be wide ranging. Striking similarities between different pathologies render investigation complicated, as it is difficult to associate every disorder with specific neuropathological characteristics. The neurodegenerative disorders discussed in detail in this chapter are Huntington disease (HD), Parkinson disease (PD), Alzheimer disease (AD) and Amyotrophic Lateral Sclerosis (ALS).

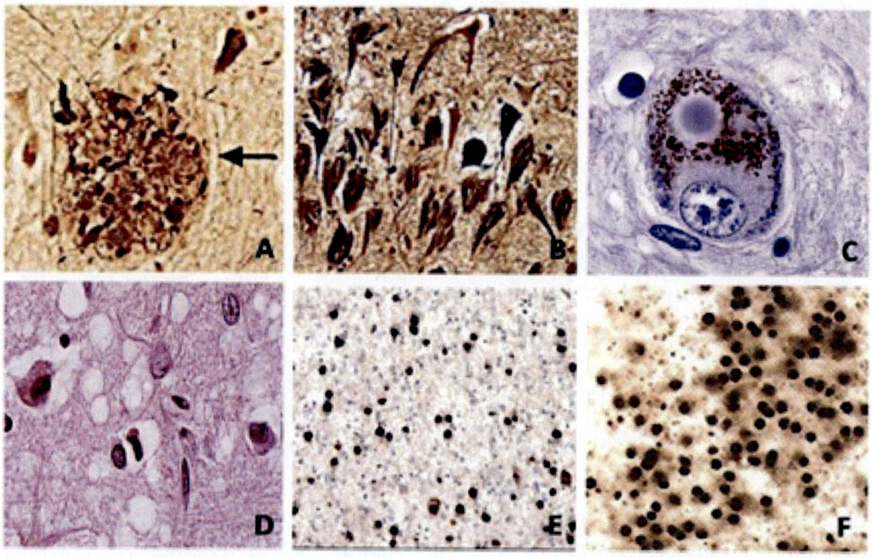

Figure 1. **A.** Senile plaques of the neuritic type usually contain an amyloid. The brown clump in the center of the picture above is amyloid stained by the Bielschowsky stain (arrow). **B**. Shows a dark black neurofibrillary tangle stained with Bielschowsky silver stain. **C**. Haematoxylin-eosin-stained section of substantia nigra with a pigmented nerve cell containing a Lewy body. **D**. Frontal cortex sample from the brain of a patient suffering from Creutzfeldt–Jakob disease (Prion disease). Brain sections were stained with haematoxylin-eosin. Neuronal loss and prominent spongiosis are visible. **E.** Pick bodies (tau-positive spherical cytoplasmic neuronal inclusions, composed of straight filaments), and Pick cells (ballooned neurons with dissolution of chromatin). **F.** Mouse Anti-Huntingtin. Brain sections containing the hipocampus from Huntington Disease transgenic mouse.

2.1. Huntington Disease

More than a century ago, in 1872, George Huntington first described the devastating movement disorder now carrying his name. HD is an autosomal dominant inherited slowly progressive neurodegenerative disorder with a prevalence of 4 to 8 per 100,000 persons (Harper, 1992). The relentless course, leading to death after 10 to 20 years, and the high penetrance in parental generations pose a burden on afflicted families and have challenged generations of physicians and scientists (Lanska, 1995). There is no treatment known so far to postpone disease progression.

This disease is characterized by degeneration of neurons in the striatum and cerebral cortex, which results in involuntary motor movements and cognitive deterioration.

People with HD may express a wide variety of symptoms, which are grouped into three categories: movement, cognitive, and psychiatric symptoms.

- Some of the movement symptoms of HD include muscle spasms, tics, rigidity, falling down, difficulty physically producing speech, and, in the later stages of the disease, difficulty swallowing (which can lead to significant weight loss). Uncontrollable movements such as twisting are also quite common symptoms of HD. These uncontrollable movements are known as "chorea" (from the Greek word for "dance").
- The most significant cognitive symptoms of HD are the altered organization and generally slowed processing of information in the brain. These symptoms can lead to difficulty learning, difficulty planning and prioritizing, impairment of one's perception of space, and difficulty "multitasking" (paying attention to several things at once). Individuals frequently adhere to common routines because these routines are the easiest for them to accomplish. Finally, because they have trouble organizing incoming and outgoing words in their brains, many people with HD experience difficulty communicating with others.
- Depression is the most common of the psychiatric symptoms of HD. Other symptoms include personality changes, apathy, anxiety, irritability and obsession, with certain activities (such as hand washing), delirium, and mania. Denial of having HD is also a common symptom of the disease.

The onset of HD usually happens in the fourth and fifth decades of life. Gross pathology of HD is limited to the brain, with atrophy most prominent in the caudate, putamen, and cerebral cortex. Brain weight may be reduced by as much as 25–30% in advanced cases resulting from neuronal cell death, with a direct correlation between brain atrophy and duration and severity of the disease (Vonsattel et al. 1985). Initial atrophy is particularly evident in the basal ganglia, with up to 60% loss of mass in the caudate nucleus, putamen, and globus pallidus. The neuronal loss in the striatum is largely confined to spiny striatal GABAergic type II neurons, which project to the globus pallidus and receive glutamatergic signals from cerebral cortex, and dopaminergic signals from the substantia nigra; neuronal loss is progressive and associated with astrogliosis (Ferrante et al. 1985; van Dellen et al. 2005). The basal ganglia atrophy is readily visible by magnetic resonance imaging scan and progresses over time (Aylward et al. 1997). Gross cortical atrophy is also readily detectable on magnetic resonance imaging, and increasingly sophisticated volumetric analysis has demonstrated early and progressive changes in the cortex (Rosas et al. 2002) in large neurons of layers III, IV, and V (Hedreen et al. 1991). An especially intriguing finding is the presence of intranuclear inclusion bodies, consisting of amyloid-like fibrils that contain mutant huntingtin, ubiquitin, synuclein, and other proteins (DiFiglia et al. 1997; Becher et al. 1998).

HD is a hereditary autosomal dominant disorder in which the underlying mutation is an expanded CAG repeat in exon 1 of the coding region of the HD gene located in the 4p16.3 region of the short arm of chromosome 4. The CAG triplet encodes the amino acid glutamine in the gene product called huntingtin (htt). The mutant htt has elongated polyglutamine (polyQ) stretch in the N-terminal region (>37 units) that affects its interaction with htt-binding proteins and makes them susceptible to aggregation (Harjes and Wanker, 2003). As is the case with other neurodegenerative disorders, polyQ repeat diseases are characterized by selective vulnerability of neurons; in the case of HD, it is primarily striatal neurons that degenerate, especially in the caudate nucleus, and to lesser extent affects neurons in the cerebral cortex and elsewhere.

In healthy subjects the CAG repeat length ranges between 10 and 35; usually an expression of more than 40 CAG repeats leads to phenotypic symptoms of HD. Normal htt is localized in the cytoplasm but mutant htt in addition to being found in the cytoplasm is also localized in the nucleus (DiFiglia et al. 1997; Becher et al. 1998). Experimental models carried out *in vitro* and *in vivo*, suggest that htt aggregates occur first in the cytoplasm and subsequently move to the nucleus. In fact, htt is a cytoplasmic protein and its cleavage occurs mostly in the cytoplasm before the translocation of cleaved products to the nucleus where they associate (Sieradzan and Mann, 2001). These expanded proteins are prone to structural alterations making them resistant to proteolysis by the ubiquitin-proteasome system, which is the cell pivot to clear misfolded proteins. The lack of efficacy of the ubiquitin-proteasome system in these neurological diseases is supposed to be the underlying metabolic alteration responsible for accumulating expanded proteins within affected cells, thereby leading to neuronal inclusions (Cummings et al. 1999).

Cooper et al. (1999) suggested that expanded polyQ repeats serve as glutamyl donor substrates and lead to a tissue transglutaminase catalyzed crosslinking to form aggregates. Saudou et al. (1998) have shown that inclusions are not necessary for cell death, while an aberrant decompartmentalization e.g., translocation of mutant htt into the nucleus appears to be a more critical pathogenic event, leading to mechanisms of apoptotic cell death. In this way, htt was shown to be cleaved during apoptosis by caspase-3, and increasing the length of the polyQ tract (Martindale et al. 1998; Wellington et al. 1998) enhances the rate of cleavage. Butterworth et al. (1998) provided evidence for apoptotic DNA fragmentation in post-mortem striatal tissue of HD patients.

Cha (2000) has proposed some mechanisms by which mutant htt induce neuronal alterations: (1) the increased polyQ stretch might confer a ‘gain of function’ property resulting in direct binding of the mutant protein to DNA. Thus, when abnormally bound to DNA, mutant htt might disrupt the normal pattern of transcription. (2) htt might bind aberrantly to specific transcription factor proteins, forming an inactive transcriptional complex that functions essentially as a repressor. (3) htt might form complexes with co-repressor proteins and therefore aberrantly de-repress transcription of normally silent genes. (4) htt might sequester transcription factors through aberrant interactions, depleting the levels of required factors within the cell.

htt is only inconsistently found in striatal projection neurons, a neuronal cell subtype that is supposed to die; surviving striatal cholinergic interneurons and cortical pyramidal neurons express htt abundantly (Fusco et al. 1999). A similar finding by Sapp et al. (1999) raises the question as to whether htt preferentially damages the corticostriatal neuronal pathway and via excessive synaptic release of glutamate in its terminal nerve endings leads to excitotoxic striatal cell death.

Mitochondrial dysfunction is also implicated in HD pathogenesis. The mitochondrial toxin 3-nitropropionate produces neuropathology in animals similar to that observed in human HD (Sipione and Cattaneo, 2001; Rubinsztein, 2002). Mitochondrial membrane potential in lymphocytes of HD patients is depolarized at lower calcium concentrations than normal (Panov et al. 2002), and brain mitochondria isolated from transgenic mice expressing full-length mutant htt show similar deficiencies in calcium handling. These findings suggest that mutant htt may make neurons susceptible to excitotoxicity, associated with increases in cytosolic calcium (Bossy-Wetzel et al. 2004).

The link between the genetic defect and circumscribed cell death is still missing. It is widely accepted that interactions between defective energy metabolism, free radical-induced oxidative stress, and excitotoxicity contribute to the pathogenesis of HD as well as to other neurodegenerative disorders such as ALS, AD, and PD (Grünewald and Beal, 1999).

The last several years have been marked by unprecedented progress in HD and other polyglutamine diseases. The genetic cause of HD has been identified; enormous progress has been made in the molecular properties of expanded htt; the pattern of neuronal death in HD has been better characterized; and transgenic animal models that recapitulate aspects of the illness more faithfully may soon replace the more traditional excitotoxic and metabolic toxic models. Despite these remarkable developments, there is a big gap of knowledge between the expansion of CAG in htt and the massive, as well as selective, death of neostriatal neurons and glial cells. Some clues begin to emerge from the combination of both molecular and systems analysis, but the evident complexity of the problem will require much more work across a number of disciplines in the years to come.

2.2. Parkinson Disease

PD was first described in England in 1817 by Dr. James Parkinson. PD is the most common neurodegenerative movement disorder. Approximately 1% of the population older than 65 years suffers from this slowly progressive neurodegenerative disease; 95% of PD cases are sporadic. PD is characterized by resting tremors, bradykinesia (slowness of voluntary movement), rigidity, and a loss of postural reflexes. Patients with PD typically have a flat, expressionless face and walk with a stooped gait characterized by small steps. Many patients also experience severe depression. The molecular pathogenesis of PD is still not understood. Two forms of PD exist: idiopathic and heritable (familial).

The lifetime risk for PD is estimated to be 2 and 1.3% for men and women, respectively, and between 3.7 and 4.4% for "parkinsonism," a term used to distinguish other clinical conditions characterized by akinesia and rigidity that do not meet clinical or pathologic criteria for idiopathic PD (Elbaz et al. 2002).

The diagnosis of PD continues to be based on presenting signs and symptoms. Tremor is the most obvious clinical symptom, often starts in one extremity, and worsens with precipitating factors such as stress, fatigue and cold weather. It may be confused with the more common essential tremor but can be differentiated by noting if the tremor occurs predominantly at rest (PD) or with action (essential tremor). Essential tremor typically occurs in both arms, whereas patients with PD usually have a unilateral tremor that may affect one arm or a leg. Bradykinesia is usually the most troublesome symptom. Patients report slowness in performing their activities of daily living, including dressing, walking and doing household chores. Writing may become micrographic, with a progressively smaller character size as the person continues to write. Watching a patient get up from a chair and walk is helpful. Patients with PD may need to push themselves up, take a longer time to get up or fall backward.

Reduced arm swing, flexed posture and a shuffling gait may be very early features of the disease. Rigidity of the muscles on passive movement is characteristic of PD but must be distinguished from the rigidity resulting from upper motor neuron lesions, for example, in patients with a stroke (Guttman et al. 2003).

There have been important recent advances of the etiology and pathogenesis of PD. The study of familial parkinsonism has led to the identification of at least four genes linked to PD, including α-synuclein (PARK1), parkin (PARK2), DJ-1 (PARK7), and PTEN (phosphatase and tensin homolog deleted on chromosome 10)-induced kinase 1 (PINK1, also known as PARK6) (Checkoway et al. 1998; Valente et al. 2004). Although a number of genetic polymorphisms are linked to PD, it is likely that the majority of cases of PD are not inherited but related to environmental factors. This is supported by a genetic study of twins by Tanner et al. (1999), who observed that monozygotic–dizygotic concordance rates are indistinguishable, implying a lack of genetic influence and a strong probability of an environmental influence (Brown et al. 2005). Rare hereditary forms of PD have provided insight into the molecular pathways of this disorder (Hardy et al. 2003). Most PD cases are idiopathic, with no known genetic component. The cause of idiopathic PD is unknown. Epidemiological studies reveal several risk factors for developing idiopathic PD in addition to aging, including exposure to pesticides, herbicides, and some industrial chemicals (Olanow and Tatton, 1999).

The symptoms of PD are caused by selective and progressive degeneration of pigmented dopaminergic neurons in the substantia nigra pars compacta (SNc). Current treatments, such as administration of L-DOPA to produce dopamine (DA), are only symptomatic and do not stop or delay the progressive loss of neurons. In fact, some studies have suggested that oxidative injury via DA may lead to further neuronal damage (Xu et al. 2002). The ability to overcome oxidative stress is essential for neuronal survival, and a decline in this function is a major cause of age-related neurodegenerative diseases (Prolla and Mattson, 2001). Because DA generates reactive oxygen species (ROS), DA neurons encounter higher levels of ROS than other neurons (Adams et al. 2001; Jenner, 2003). Oxidative stress induced by mitochondrial inhibition promotes cell death specific to DA neurons accompanied by Lewy body formation (Betarbet et al. 2000; Sherer et al. 2002).

In this way, oxidative stress has received the most attention in PD because of the potential of the oxidative metabolism of DA to yield hydrogen peroxide and other ROS (reviewed in Halliwell and Gutteridge 1985; Olanow, 1990, 1993) (Fig. 2). Oxidative stress and consequent cell death could develop in the SNc under circumstances in which there is (1) increased DA turnover, resulting in excess peroxide formation; (2) a deficiency in glutathione (GSH), thereby diminishing the brain's capacity to clear hydrogen peroxide; or (3) an increase in reactive iron, which can promote OH• formation. Indeed, postmortem studies in PD brains demonstrate increased iron, decreased GSH, and oxidative damage to lipids, proteins, and DNA, suggesting that the SNc is in a state of oxidative stress (reviewed in Jenner and Olanow 1996).

There are several deleterious outcomes from excess oxidative stress in nigral neurons. These include modification of macromolecules by ROS or oxidized catechols, depletion of intracellular thiols, and inhibition of mitochondrial function (Graham, 1978; Hastings et al. 1996; Ben et al. 1995). Indeed, some of these processes already have been shown to readily induce cell death in experimental systems (Li and Dryhurst, 1997; Gabby et al. 1996; Ankarcrona et al. 1995).

$$(1)\quad DA + O_2 + H_2O \xrightarrow{MAO} 3,4\ DHPA + NH_3 + H_2O_2$$

$$(2)\quad DA + O_2 \longrightarrow SQ^{\bullet} + {}^{\bullet}O_2^- + H^+$$

$$DA + {}^{\bullet}O_2^- + 2H^+ \longrightarrow SQ^{\bullet} + H_2O_2$$

$$(3)\quad H_2O_2 + 2GSH \longrightarrow GSSG + 2H_2O$$

$$(4)\quad H_2O_2 + Fe^{2+} \longrightarrow OH^{\bullet} + OH^- + Fe^{3+}$$

Figure 2. Both the enzymatic and the chemical metabolism of dopamine result in the formation of hydrogen peroxide (H_2O_2) (1 and 2). H_2O_2 is normally cleared by reduced glutathione (GSH) (3). However, an increase in the steady-state concentration of H_2O_2 can lead to a reaction with ferrous iron that generates the highly reactive and potentially cytotoxic hydroxyl radical ($OH^{\bullet}$) according to the Fenton reaction (4).

Although the cause of PD is still unknown, it is widely believed that most cases of idiopathic disease are caused by an interaction of environmental and genetic factors. As we mentioned above, the primary brain abnormality found in all affected patients is a degeneration of nigrostriatal DA neurons, with a loss of pigmented neurons in the SNc. The remaining neurons contain eosinophilic, cytoplasmic inclusions of fibrillar, misfolded proteins, termed Lewy bodies. The exact composition of Lewy bodies is unknown, but they contain ubiquinated α-synuclein, parkin, synphilin, neurofilaments and synaptic vesicle proteins (Bossy-Wetzel et al. 2004).

There is a moderate-to-severe loss of striatal (caudate and putamen) DA (Kish et al. 1988). Neurodegeneration and Lewy bodies can also be found in the locus ceruleus, nucleus basalis, hypothalamus, cerebral cortex, cranial nerve motor nuclei, and central and peripheral components of the autonomic nervous system (Olanow and Tatton, 1999).

Over the past 20 years, several pathological processes (e.g., oxidative stress, apoptosis and mitochondrial DNA defect) have been identified that might be involved in the pathway leading to the degeneration of nigrostriatal DA neurons; however, definitive proof that any one of these processes is critically involved is lacking. A number of exogenous toxins have been associated with the development of parkinsonism, including trace metals, cyanide, lacquer thinner, organic solvents, carbon monoxide, and carbon disulfide. There has also been interest in the possible role of endogenous toxins such as tetrahydroisoquinolines and beta-carbolines. However, no specific toxin has been found in the brain of PD patients, and in many instances, the parkinsonism seen in association with toxins is not that of typical Lewy body PD (Olanow and Tatton, 1999; Stoessl, 1999).

The most compelling evidence for an environmental factor in PD relates to the toxin 1,2,3,6-methyl-phenyl-tetrahydropyridine (MPTP). MPTP is a byproduct of the illicit manufacture of a synthetic meperidine derivative. Drug addicts who took MPTP developed a syndrome that strikingly resembled PD, both clinically and pathologically (Langston et al. 1983). MPTP induces toxicity through its conversion in astrocytes to the 1-methyl-4-phenyl pyridium (MPP+) in a reaction catalyzed by monooxidase type B (MAO-B) (Singer et al 1987). MPP+ is then taken up by dopamine neurons via the dopamine transporter and where it targets mitochondria, inhibits respiratory complex I, defect similar to that found in PD, and promotes ROS production (Nicklas et al. 1985).

As with AD, altered metal homeostasis may have a role in the etiology of PD. Metal chelation by clioquinol rescues DA neurons and improves balance and posture in MPTP-treated animals (Kaur et al. 2003). This observation supports the possibility that an environmental factor might cause PD; however, no MPTP-like factor has been identified in PD patients to date. Moreover, viral exposure-causing clusters of cases of PD and specific pesticides induce dopamine neuron loss in experimental animals. These clues may be helpful in defining the cause of PD but are unlikely to be the cause of the majority of sporadic cases (Guttman et al. 2003).

Additional support for mitochondrial dysfunction in PD pathogenesis comes from evidence that the insecticide rotenone induces a parkinsonian-like syndrome in animal models and probably humans, with protein deposits that resemble Lewy bodies (Betarbet et al. 2000). Rotenone also inhibits mitochondrial complex I, and DA neurons seem to be most severely affected.

There is increasing evidence that nigrostriatal cell death involves multiple processes that form a final pathway that may be common to many neurodegenerative diseases. Oxidative stress and free-radical generation occurs. Free-radical scavenger systems such as glutathione are suboptimal in the basal ganglia of PD patients, which enable an ever-increasing cycle of oxidative stress to occur. Mitochondrial function is further compromised, leading to decreased ATP production and subsequent cell death. Additional inflammatory changes occur, predominantly in the glia, which further accelerate these processes. Manipulation of these processes may lead to effective neuroprotective agents that can slow rate of neuronal loss.

The key question is what triggers this final common pathway in PD. Available data imply that multiple etiologies are more likely than a single common factor. This hypothesis for etiology in PD may also include a combination of genetic and environmental factors where the contribution of each may vary between affected patients.

Genetic susceptibility may be determined in part through impaired metabolism of free radicals or complex I activity, which may in turn be the product of nuclear or mitochondrial genomic defects. Environmental interactions may include exogenous toxins with uptake and conversion characteristic similar to MPTP, such that they are targeted to the substantia nigra, or endogenously generated neurotoxins such as tetrohydroisoquinolines or β-carbolines. Evidence is emerging that genetic factors may be more important in idiopatic PD that previously thought.

Based on current knowledge regarding the etiology, pathogenesis, and mechanism of cell death in PD, numerous neuroprotective strategies might be devised.

Eliminating a primary etiology is most desirable, but it is unlikely to be effective in view of the probability that different environmental and genetic factors likely contribute to the development of PD and that multiple causes may be operative even in an individual patient. Neuroprotection might be provided by agents that interfere with factors involved in pathogenesis. These could include antioxidants, bioenergetics, agents that interfere with excitotoxicity or prevent a rise in cytosolic free calcium, trophic factors, and anti-inflammatory drugs.

2.3. Alzheimer Disease

AD is a chronic progressive, degenerative disease, which causes thinking and memory to become seriously impaired. It is the most common form of dementia, perhaps is the prototypical degenerative disease affecting the central nervous system; involves areas of the brain related with cognition, starting in the hippocampus and eventually involving the amygdala, the cerebral cortex and other regions. The disease was first identified by Dr. Alois Alzheimer in 1906. He described the two hallmarks of the disease: "plaques" - numerous tiny dense deposits scattered throughout the brain which become toxic to brain cells at excessive levels and "tangles" which interfere with vital processes eventually "choking" off the living cells. As well, when brain cells degenerate and die, the brain markedly shrinks in some regions. It seems that this disease is the major cause of dementia among the elderly.

AD is characterized by memory loss and deficits in one or more of the following cognitive domains: aphasia (language disturbance), agnosia (failure to recognize people or objects in presence of intact sensory function), apraxia (inability to perform motor acts in presence of intact motor system), or executive function (plan, organize, sequence actions, or form abstractions). In addition, these deficits must be severe enough to interfere with daily life or work, and they must represent a significant decline from an earlier level of function. The prevalence rate is about 7% for individuals aged 65 or more, with the risk doubling every 5 years after age 65 (McCullagh et al. 2001; McDowell, 2001).

Although most cases of AD are thought to be sporadic, there are at least four well-known risk factors for AD: increasing age, familial association, Down syndrome, and the apolipoprotein E4 allele (Cedazo-Mínguez and Cowburn, 2001; McCullagh et al. 2001; Raber et al. 2004; Rubinsztein and Easton, 2000; Weisgraber and Mahley, 1996).

Mutations in three genes that are inherited in an autosomal dominant fashion have been linked to rare familial, early-onset forms of AD. These genes include those encoding amyloid precursor protein (APP), presenilin 1 (PS1) and presenilin 2 (PS2). Although these familiar forms account for only a few cases of AD, one common event in both familial and sporadic types of AD is the increased production and accumulation of the toxic β-amyloid (Aβ). This observation led to the 'amyloid cascade hypothesis' that excessive Aβ production is the primary cause of the disease (Hardy, 1997; Selkoe, 1999).

AD is characterized pathologically by the presence of two hallmark lesions in the brain, extracellular amyloid plaques, known as senile plaques (SP), and intraneuronal neurofibrillary tangles (NFT) as well as neuropil threads and a selective loss of neurons. SP contains mainly the Aβ peptide, whereas NFT are composed mainly of the microtubule-associated protein (MAP) tau present as paired helical filaments (PHF) (Khachaturian, 1985; National Institute on Aging, 1997). Amyloid plaques contain small, toxic cleavage products (denoted as Aβ40 and Aβ42) of the amyloid precursor protein (APP). The apolipoprotein E4 (apoE4) genotype is a powerful risk factor for developing AD, and it may possibly affect Aβ deposition and NTF (Roses, 1996).

Histological studies suggest that tau pathology is prominent in the medial temporal lobe early in the disease and progresses outward; amyloid has a broader cortical distribution that includes but is not especially prominent in medial temporal lobe (Braak and Braak, 1997; Price et al. 1991). Profound cell loss is also observed in the hippocampal formation (Hyman et al. 1984; Price et al. 1991; Gómez-Isla et al. 1996) and loss of presynaptic terminals and neuronal subpopulations and reactive gliosis (Price and Sisodia, 1998). Human studies have pointed to synaptic damage and loss as the key determinant of the degree of dementia observed in patients, correlating better with cognitive loss than either plaque number or neuronal loss (Terry et al. 1991).

Consistent with this pathology, *in vivo* magnetic resonance imaging (MRI) structural measures in early-stage AD demonstrate medial temporal atrophy (for review, see Jack and Petersen, 2000). While the temporal relationship between these changes in the human brain is difficult to demonstrate because of the necessity of using postmortem materials, the availability of transgenic mice that over produce Aβ, the major component of SP and the key toxic element in AD, has enabled temporal studies of the consequences of Aβ overproduction and deposition (Moolman et al. 2004).

On autopsy, the pathological signature of AD is a collection of abnormal deposits: NFT consisting of aggregates of hyperphosphorylated forms of the microtubule associated protein, tau, and waxy protein plaques consisting of fragments of the membrane protein APP. Neuronal death is regionally variable with heavy losses of neurons (up to 80%) found in hippocampus and other limbic regions of cortex, as well as in basal nucleus, locus ceruleus, and the dorsal raphe (Herrup et al. 2004).

Tau is one of the major and most studied microtubule-associated proteins (MAPs) in the central nervous system. It promotes assembly of tubulin both *in vitro* and *in vivo* into microtubules and maintains the microtubule structure (Weingarten et al. 1975). Tau is a phosphoprotein, and its biological activity is regulated by its degree of phosphorylation (Lindwall and Cole, 1984). In AD brain, tau is abnormally hyperphosphorylated and is present as cytosolic protein (Köpke et al. 1983) and as polymerized into PHF (Grundke-Iqbal et al. 1986). Unlike normal tau, which contains two or three phosphate groups, the cytosolic hyperphosphorylated tau from AD brain contains 5–9 mol of phosphate/mol of the protein (Köpke et al. 1993).

It seems clear that tau hyperphosphorylation is a critical factor underlying neurofibrillary pathology. Supporting evidence is the appearance of epitopes recognized by anti-PHF antibodies, as a result of kinase stimulation in neurons that precedes death of these cells (Nuydens et al. 1995). Not only does hyperphosphorylated tau fail in its normal function in stabilizing microtubules, but also it reflects a "gain of toxic function" due to its sequestering normal tau and other MAPs, resulting in the disruption of microtubules (Iqbal et al. 1994, 2005).

Much excitement in recent years has come from the discovery of a multiplicity of mutations in the genes encoding the Aβ precursor protein (AβPP) and/or the presenilins, that account for the bulk of the familial cases of AD on the basis of overproduction of AβPP and/or altered AβPP proteolytic processing, both leading to increased Aβ. Transgenic mice overexpressing AβPP or other human mutant AD-related proteins, as well as animals expressing more than one of these mutations, exhibit many of the neuropathologic and behavioral features of the human disease, including the development of SP and some neurotoxicity. In addition, NFT can be produced in mice expressing mutant tau protein, and tangle formation is further enhanced in animals that also express mutant AβPP (Gotz et al. 2004).

There is considerable debate as to whether neuronal loss in AD reflects appearance of either SP or NFT and if so, which one plays a more important role. A majority of workers in the field still favor the amyloid cascade hypothesis, specifying that the onset and progression of AD is initiated by aggregation of Aβ into toxic fibrillar deposits within the extracellular space of the brain, thereby disrupting neuronal and synaptic function and eventually leading to neuronal degeneration and dementia (Selkoe, 2005). However, Aβ, formed upon proteolytic processing of AβPP by β- and γ-secretases, has a widespread distribution through the brain and body in healthy individuals throughout life. Moreover, soluble Aβ appears to serve a variety of physiological functions, including modulation of synaptic function, facilitation of neuronal growth and survival, protection against oxidative stress, and surveillance against neuroactive compounds, toxins and pathogens (Bishop and Robinson, 2004). At the other extreme, the mature SP are apparently also non-toxic because neuronal loss in AD does not occur in the vicinity of SP deposition. In addition, Aβ deposition is inversely correlated with the levels of 8OHG since this oxidative marker is found physically distant from the Aβ deposits (Nunomura et al. 1999). Thus, recent focus on Aβ toxicity has been on fibrillar polymers that are toxic to cultured neurons, and especially on smaller oligomers (Kim et al. 2003), particularly those containing more Aβ(1-42) than Aβ(1- 40) (Kirkitadze et al. 2002). The mechanism of neurotoxicity of Aβ(1-42), wherein Met-35 appears to play a critical role, is usually considered to involve induction of oxidative stress and lipid peroxidation (Butterfield and Boyd-Kimball, 2004).

Despite the genetic evidence implicating a pathologic role of Aβ, only tau pathology and NFT have been found to correlate with symptom presentation in patients (King, 2005). As we have mentioned above, in normal adult brain, the microtubule-associated protein tau binds microtubules through its tubulin-binding domains, and this assembly depends partially upon the degree of phosphorylation. By regulating microtubule assembly, tau has a role in modulating the functional organization of the neuron, particularly in axonal morphology, growth, and polarity (Buee et al. 2000).

Regarding NFT, it seems that display resistance to proteolysis, and this may reflect in part the presence of transaminase-mediated γ-glutamyl-ε-lysine crosslinks (Miller and Johnson, 1995) among proteins identified in NFT (hsp27, SN, ubiquitin, parkin) (Nemes et al. 2004). It has been an important contention that the persistent insolubilization and permanency of NFT aggregates probably represents, at least in part, cementing of the aggregates by processes associated with oxidative stress. In particular, the finding that most of the covalent, including cross-linking modifications induced by products of oxidative stress are seen in apparently normal neurons in AD and at pre-NFT PHF-tau stages (Perry et al. 2000) suggests that these modifications play at least partially a causative rather than by-stander role in the neurofibrillary pathology in AD. Liu et al. (2005) recently showed that seven distinct antibodies raised against NFT that recognize unique epitopes of tau in AD, recognize phosphorylated tau more strongly after treatment with 4-hydroxy-2-nonenal, an α,β-unsaturated hydroxyalkenal which is produced by lipid peroxidation. These findings support the idea that 4-hydroxy-2-nonenal modifications of tau promote and contribute to the generation of the major conformational properties defining NFT (Sayre et al. 2005).

It is still unclear how Aβ does its damage, but several mechanisms have been proposed. One view suggests that Aβ protofibrils activate microglia, inciting an inflammatory response and release of neurotoxic cytokines. Nonsteroidal anti-inflammatory drugs (NSAIDs) including ibuprofen seem to delay the onset of AD (Stewart et al. 1997). Additionally, NSAIDs reduce the production of Aβ42 (Weggen et al. 2001). In a second view, Aβ protofibrils trigger excessive release of excitatory amino acids like glutamate from glial cells that may injure nearby neurons by excitotoxicity. Overactivation of glutamate receptors of the N-methyl-D-aspartate (NMDA) subtype results in increased intracellular Calcium (Ca^{2+}), which activates neuronal nitric oxide synthase and consequently generates nitric oxide (NO). When generated in excess, NO combines with superoxide anion ($O2^-$), forming the highly reactive and neurotoxic product peroxynitrite ($ONOO^-$), which leads to further oxidative and nitrosative stress in part via mitochondrial injury. In fact, positive phase III human trials of the uncompetitive NMDA receptor channel blocker, memantine, led to its recent approval for the treatment of AD (Lipton, 2004). A third view suggests that protofibrils and aggregates convey harmful effects to neurons by paralyzing axonal and dendritic transport. Both APP and PS may bind kinesin I and regulate vesicular traffic (Pigino et al. 2003). PS mutants increase glycogen synthase kinase-3β activity, which hampers kinesin-mediated, anterograde axonal transport (Pigino et al. 2003). In addition, Aβ deposits may act as nonspecific 'roadblocks', representing a physical transport barrier (Bossy-Wetzel et al. 2004).

An additional mechanism of Aβ injury is synaptic dysfunction and loss, which are early events in AD and occur before amyloid plaque formation (Selkoe, 2002). Cholinergic transmission and synaptic density are considerably decreased in AD patients. The mechanism for synaptic damage is unknown, but diffusible oligomeric forms of Aβ may be important. Synaptic dysfunction probably contributes to memory loss and cognitive deficits in AD. In fact, APP transgenic mice manifest cellular, biochemical and electrophysiological evidence of synaptic deficits before Aβ deposition, including reduced excitatory postsynaptic potentials and Long-Term Potentiation, regarded as a correlate of learning and memory (Chapman et al. 1999). Inhibition of gamma-secretase decreases oligomeric Aβ and Long-term potentiation deficits (Walsh et al. 2002). Microinjection of Aβ43 and Aβ40 peptides into the rat hippocampus disrupts synaptic transmission and short-term memory (Stephan et al. 2001).

Aβ may also mediate harmful effects by binding redox-reactive metals, which in turn release free radicals (Lovell et al. 1998; Bush, 2003; Dong et al. 2003, and others). Bossy-Wetzel et al. (2004) found that nitrosative stress mobilizes zinc from intracellular stores, resulting in mitochondrial dysfunction. Chelation of zinc and copper provides neuroprotective effects (Bush, 2002). For example, clioquinol, an antibiotic that also chelates zinc and copper and crosses the blood-brain barrier, decreases brain Aβ deposition and improves learning in mutant APP transgenic mice (Cherny et al. 2001).

Oxidative stress from mitochondrial dysfunction occurs early in AD, and Aβ may directly or indirectly injure mitochondria (Anandatheerthavarada et al. 2003). Aβ blocks respiratory complex I, thus producing a decline in ATP (Casley et al. 2002a). In isolated mitochondria, Aβ inhibits respiration and enzyme activity (α-ketoglutarate dehydrogenase and pyruvate dehydrogenase) (Casley et al. 2002b). The mechanism of Aβ translocation to mitochondria remains unknown. New data describe an interaction partner of Aβ in mitochondria, termed Aβ-binding alcohol dehydrogenase (ABAD) (Lustbader et al. 2004). ABAD is upregulated in neurons of AD patients. Expression of ABAD in concert with mutant APP enhances free-radical production and toxicity. Conversely, a peptide that blocks Aβ-ABAD interaction prevents free-radical formation and cell death. Aβ interacts with the LD loop of ABAD, close to its nicotinamide adenine dinucleotide (NAD) binding site, and may therefore induce a conformational change that prevents NAD binding. To date, however, details of the interaction are not apparent in the crystal structure (Lustbader et al. 2004).

In summary, mitochondrial dysfunction and resulting energy deficits may contribute to impaired clearance of protein aggregates and neuronal dysfunction, affecting ion channel and pump activity, neurotransmission, and axonal and dendritic transport (Bossy-Wetzel et al. 2004).

Studies in brain tissue and cerebrospinal fluid from AD patients have provided evidence for increased levels of oxidative stress, mitochondrial dysfunction, and impaired glucose uptake in vulnerable neuronal populations. Experimental cell-culture and animal models of AD indicate that increased levels of oxidative stress, impaired energy metabolism, and perturbed Ca^{2+} homeostasis are critical events underlying synaptic degeneration and neuronal death in AD. Abnormal production and aggregation of Aβ may play a key role in promoting oxidative stress, resulting in impaired function of membrane ion-motive ATPases, glucose and glutamate transporters. In this way oxidative and metabolic compromise, render neurons vulnerable to excitotoxicity and apoptosis. Studies of inherited forms of AD caused by mutations in APP and presenilins strongly support central roles for perturbed cellular Ca^{2+} homeostasis and aberrant proteolytic processing of APP in the pathogenesis of AD. Novel mediators of neuronal cell death in AD and related disorders are being identified and included Par-4 caspases. On the other hand, metabolic and signaling pathways that may increase neuronal resistance to aging and AD are being identified and include dietary restriction and neurotrophic factors. Although cognitive dysfunctions dominate the clinical presentation, recent findings from studies of AD patients and mouse models of AD suggest rather widespread alterations of energy metabolism, stress responses and immune function. Exciting new findings concerning mechanisms of neuronal degeneration and neuroprotection are leading to the development of new approaches aimed at preventing and treating AD patients.

2.4. Amyotrophic Lateral Sclerosis

ALS, also known as Lou Gehrig's disease, is a fatal neurodegenerative disorder that is characterized by the selective loss of motor neurons in the spinal cord, brain stem, and motor cortex. Described in 1869 by the French neurobiologist and physician Jean-Martin Charcot, the disease's primary hallmark is the selective dysfunction and death of the neurons in the motor pathways, which leads to spasticity, hyperreflexia (upper motor neurons), generalized weakness, muscle atrophy, and paralysis (lower motor neurons) (Mulder et al. 1986). Failure of the respiratory muscles is generally the fatal event, occurring within one to five years of onset. Selectivity of killing occurs even among motor neurons. The neurons that control the bladder (i.e., Onuf's nucleus in the sacral cord) and eye movements are relatively spared.

Motor neuron loss is accompanied by reactive gliosis (Leigh and Swash, 1991), intracytoplasmic neurofilament abnormalities, and axonal spheroids (Carpenter, 1968; Gonatas et al. 1992; Hirano et al. 1984; Leigh et al. 1991). In end-stage disease, there is significant loss of large myelinated fibers in the corticospinal tracts and ventral roots as well as evidence of Wallerian degeneration and atrophy of the myelinated fibers (Delisle and Carpenter, 1984). ALS rarely affects cognitive functions. Electromyogram shows signs of diffuse denervation with generally preserved nerve conduction velocities. Although an inflammatory process may be present, new evidence points toward multiple mechanisms that promote neuronal cell death in the CNS as the underlying basis for ALS. It has been observed that more than half of ALS patients die within 2.5 years following the onset of symptoms.

ALS primarily involves anterior horn cells in the spinal cord and cranial motor nerves. Patients may have weakness of bulbar muscles or of single or multiple limb muscle groups. Presentation is not always bilateral or symmetrical. A predominantly bulbar form usually leads to more rapid deterioration and death. Limb weakness is predominantly distal. Weakness and atrophy of the intrinsic hand muscles are prominent. Weakness progresses to involve the forearms, shoulder girdle muscles, and the lower extremities.

Although major recent advances have shed light on its etiology, the key mechanisms in both familial and sporadic ALS remain unknown and no cure is known.

In 1991, researchers identified a link between ALS and chromosome 21. Two years later, a particular gene, copper-zinc superoxide dismutase (SOD1), was identified as being associated with about 20% of the inherited cases in families.

Efforts to find new genes linked to the remainder of familial ALS cases continue. The identification of three separate families with linkage to chromosome 16 (Abalkhail et al. 2003, Ruddy et al. 2003, Sapp et al. 2003) has significantly narrowed the region of interest, and rapid identification of the gene involved is anticipated. In addition, efforts to identify disease genes for chromosomes 18 and 20 are under way. Researchers have just begun to identify other genetic contributors, including one dominant locus on mouse chromosome 13 that can sharply slow initiation of SOD1 mutant mediated disease (Kunst et al. 2000).

Using modern gene mapping methods, researchers have identified a new gene linked to a rare, recessively inherited juvenile or infantile onset form that progresses slowly (Hadano et al. 2001). The gene, localized to chromosome 2, encodes a 184-kD protein (named ALS2 or alsin) with three putative guanine-nucleotide-exchange factor (GEF) domains. Small GTP-binding proteins of the Ras superfamily act as molecular switches in signal transduction, affecting cytoskeletal dynamics, intracellular trafficking, and other important biological processes. GEFs catalyze the dissociation of the tightly bound GDP from the small G protein in response to upstream signals. Although widely expressed, the ALS2 protein is enriched in nervous tissue, where it is peripherally bound to the cytoplasmic face of endosomal membranes, an association that requires the amino-terminal RCC1-like GEF domain

(Yamanaka et al. 2003). The G protein(s) upon which the ALS2 GEFs act has not been identified, although an initial report has shown that ALS2 can act *in vitro* as an exchange factor for Rab5a (Otomo et al. 2003), which functions in endosomal trafficking. All of the disease-causing mutants are highly unstable (Yamanaka et al. 2003), this has led to the conclusion that early-onset motor neuron disease is caused by loss of activity of one or more of the GEF domains of this endosomal GEF.

The striking pathological and clinical similarity between familial and sporadic disease has sparked enthusiasm that the animal models based on mutant SOD1 might provide insight into mechanisms of both sporadic and familial disease. However, to date, there is no direct evidence validating this assumption.

Of the more than 100 mutations of SOD1 in humans, 3 (SOD1G85R, SOD1G37R, and SOD1G93A) have been extensively characterized in transgenic mouse models of ALS (Bruijn and Cleveland, 1996; Gurney, 1994; Ripps et al. 1995). In these mice, the mutant human protein is ubiquitously expressed (under control of the human or mouse SOD1 gene promoter) at levels equal to or several fold higher than the level of endogenous SOD1. Vacuolar pathology that at least partially represents damaged mitochondria is seen in motor neurons of mice or rats expressing high levels of SOD1G93A (Dal Canto and Gurney, 1994; Jaarsma et al. 2001; Howland et al. 2002) and SOD1G37R (Wong et al. 1995), but it is not seen in disease from other mutants (Bruijn et al. 1997; Nagai et al. 2001). This may be due to the higher (by as much as 20-fold) accumulated levels of the catalytically active SOD1G93A and SOD1G37R mutants, as compared with the inactive SOD1G85R, which are required to provoke disease in mice. This difference is even more provocative for the frame shift mutation SOD1G127X truncated in the last 26 residues of SOD1. Despite accumulation to only 1% the level of endogenous wild-type SOD1 in mice (i.e., 1/400 the level required to cause disease from the dismutase-active mutants in mice), this highly toxic mutant provokes disease accompanied by prominent aggregates (Jonsson et al. 2004).

The obvious loss of motor neurons initially focused attention on how mutant SOD1 may act within motor neurons to provoke neuronal degeneration and death. The presence of abnormal protein aggregates or inclusions has been described in many neurodegenerative diseases. For ALS, prominent, intracellular, cytoplasmic inclusions in motor neurons and in some cases within the astrocytes surrounding them are found in each of the prominent mouse models of SOD1-mediated disease and in all reported instances of human ALS (Bruijn et al. 1998). In mouse models, these accumulations are highly immunoreactive for SOD1. At least some misfolded SOD1 aggregates cannot be readily dissociated and are resistant to strong ionic detergents; some also contain covalent adducts to other components that can be detected biochemically in spinal cord extracts of transgenic SOD1 mice long before (Johnston et al. 2000; Wang et al. 2002), or contemporaneous with onset of disease (Bruijn et al. 1997).

An unsolved puzzle is whether these aggregates damage motor neurons (or other cells in the spinal cord), and if so through what mechanism(s). Several possible toxicities of the protein aggregates have been proposed, including aberrant chemistry; loss of protein function through co-aggregation with the aggregates; depletion of protein folding chaperones; dysfunction of the proteasome overwhelmed with undigestable, misfolded protein; and inhibition of specific organelle function, including mitochondria and peroxisomes, through mutant aggregation onto or within such organelles. Consistent with involvement of the ubiquitin-proteasome system, the aggregates are intensely immunoreactive with antibodies to ubiquitin, a feature common to all SOD1 ALS mouse models (Bruijn et al. 1998, Jonsson et al. 2004, Wang et al. 2003) and human SOD1-mediated familial ALS (Bruijn et al. 1998, Kato et al. 2000).

However, as in almost all prominent examples of inherited human neurodegenerative diseases, the mutant gene products are expressed widely. In the case of SOD1, expression is ubiquitous, raising the possibility that the toxic cascade may be achieved wholly or in part by mutant SOD1 action in the adjacent nonneuronal cells (Brujin et al. 2004). In the first set of experiments to address this question, mutant SOD1 was expressed selectively in astrocytes (Gong et al. 2000) or in neurons (Pramatarova et al. 2001; Lino et al. 2002). Although there was astrocytosis in the mice with astrocyte-specific expression, none of the mice in these three sets developed motor neuron disease. Because high mutant SOD1 levels were sustained in the mice-accumulating mutant SOD1 only in the astrocytes, mutant action solely within astroctyes appears insufficient to cause disease. In the case of neuron-specific expression [using either neurofilament (Pramatarova et al. 2001) or neural-specific enolase promoters (Lino et al. 2002)], no clear outcome emerged. However, the levels of mutant SOD1 may have been too low to yield disease. Further efforts, however, made it clear that toxicity to motor neurons from SOD1 mutants is noncell autonomous, that is, it requires mutant damage not just within motor neurons but also to nonneuronal cells. These findings indicate that motor neuron death is not cell autonomous and depends on surrounding cells. Neurons that express mutant SOD1 protein can be rescued by nearby non-neuronal wild-type cells (Clement et al. 2003).

On the other hand, it has been proposed that SOD1-mediated toxicity in ALS is not due to loss of function but instead to a gain of one or more toxic properties that are independent of the levels of SOD1 activity. The main arguments against the importance of loss of dismutase function are that (a) SOD1 null mice do not develop motor neuron disease (Reaume et al. 1996) and (b) removal of the normal SOD1 genes in mice that develop motor neuron disease from expressing a dismutase inactive mutant (SOD1G85R) does not affect onset or survival (Bruijn et al. 1998). In addition, levels of SOD1 activity do not correlate with disease in mice or humans; in fact, some mutant enzymes retain full dismutase activity (Borchelt et al. 1994; Bowling et al. 1995). Finally, chronic increase in the levels of wild-type SOD1 (and dismutase activity) either has no effect on disease (Bruijn et al. 1998) or accelerates it (Jaarsma et al. 2000).

Although the primary toxicities of the familial ALS–linked mutations of SOD1 remain unresolved, the final event in the death cascade has been partially clarified. Activation of caspase-1, one of the early events in the mechanism of toxicity of SOD1 mutants, occurs months prior to neuronal death and phenotypic disease onset (Pasinelli et al. 2000). A central feature in cell death mediated by mutant SOD1 is the activation of caspase-3, one of the major cysteine-aspartate proteases responsible for degradation of many key cellular constituents in apoptotic cell death. Caspase-3 activation occurs in motor neurons and astrocytes (Pasinelli et al. 2000) contemporaneous with the first stages of motor neuron death in all three of the best-studied mouse models. For SOD1G93A, release of cytochrome C from mitochondria is followed by activation of caspase-9 (Guegan et al. 2001), which may be the effector for the subsequent activation of caspases-3 and -7. *In vitro*, this temporal cascade of caspase activation occurs within the same neuronal cell (Pasinelli et al. 2000), although this has not been firmly established in mice.

Thus, it is established that a common step toward toxicity of mutant SOD1 is a sequential activation of at least two caspases, which act more slowly here than they do during the apoptotic death processes of development. Moreover, apparent inhibition of one or more caspases in this cascade is beneficial: Despite a short half-life once in aqueous solution, long-term intrathecal administration of the pan-caspase inhibitor (N-benzylocarbonyl-Val-Ala-Asp-fluoromethylketone or zVAD-fmk) prolongs the life of SOD1G93A mice by approximately 27 days (Li et al. 2000). Further evidence that proteins of the cell death pathways are important for SOD1-mediated neuron death includes the demonstration that increasing expression of the antiapopotic factor Bcl2 slows disease onset and survival of SOD1G93A mice by three to four weeks (Kostic et al. 1997).

Another candidate contributor to motor neuron disease is peripherin, an intermediate protein expressed in spinal motor neurons, peripheral sensory neurons, and autonomic nerves. Corbo and Hays (1992) found peripherin with neurofilament proteins in the majority of axonal inclusions in motor neurons of ALS patients. Increasing expression of the major peripherin isoform (peripherin 58) in motor neurons of transgenic mice leads to late-onset motor neuron disease accompanied by disruption of neurofilament assembly and organization (Beaulieu et al. 1999).

On the other hand, Glutamate-mediated excitotoxicity from repetitive firing and/or elevation of intracellular Ca^{2+} by calcium-permeable glutamate receptors has long been implicated in ALS neuronal death. In motor neurons, the bulk of the glutamate is actively cleared from the synapse by the glial glutamate transporter, EAAT2, presumably helping to prevent excitotoxicity (Tanaka et al. 1997). Evidence for abnormal glutamate handling in ALS first arose from the discovery that the cerebrospinal fluid of ALS patients had increased glutamate levels (Rothstein et al. 1990; Shaw et al. 1995), a finding now reported in 40% of sporadic ALS patients. Direct measurement of functional glutamate transport in ALS revealed a marked diminution in the affected brain regions, which was the result of pronounced loss of the astroglial EAAT2 protein (Rothstein et al. 1995).

Excitotoxicity has provided one of the few mechanistic links between sporadic and SOD1 mutant–mediated ALS. SOD1 mutants in rodent disease models induce a focal loss of EAAT2 selectively from the astrocytes within the portion of the spinal cord containing the motor neuron cell bodies (Howland et al. 2002). Thus, glutamate excitotoxicity is likely to be an important contributor to neuronal death. Indeed, the only FDA-approved therapy in ALS, Riluzole, functions by decreasing glutamate toxicity.

Other important factor related to the neuronal death in ALS is the presence of microglia and inflammation. In ALS tissues, there is strong activation and proliferation of microglia in regions of motor neuron loss (Kawamata et al. 1992; Ince et al. 1996). Recent studies showed that expression of proinflammatory mediators [TNF-α, Interleukin-1B, and cyclooxygenase 2 (COX-2)] is an early event in mouse models of ALS (Alexianu et al. 2001; Elliott, 2001; Hensley et al. 2002).

ALS is a complex disease with multiple causes, making the discovery of effective pharmacologic therapies challenging. Despite the impressive list of therapies attempted in both mouse and humans, as we mentioned earlier, Riluzole is currently the only FDA-approved compound that may slow disease progression and extend survival, although its effect on both is generously described as modest. Nonpharmacologic therapies such as maintaining nutrition and attending to respiratory function have more significant effects than does any currently available medication, but none of these therapeutic efforts significantly slows disease progression. Many neurologists recommend vitamin supplements and antioxidants, and many discuss with patients medications such as minocycline and celecoxib that have shown some efficacy in mouse models. Clinical trials for several of these compounds are ongoing. As in other neurodegenerative diseases, ALS therapies are advancing in three overlapping areas: (a) small molecules including "off-the-shelf" compounds; (b) delivering protein, DNA, or RNA; and (c) novel gene therapy approaches including viral vectors and stem cells (Brujin et al. 2004).

Given the convergence of multiple pathways leading to disease, various therapies targeting different processes may be the most effective. Although the exact combination of therapies is not yet known, new insights into disease mechanisms and anticipated discoveries of new genes responsible for ALS foster hopes that therapies to significantly slow this disease are within reach.

Final Considerations

In this chapter, we reviewed common themes taking place in several neurodegenerative disorders. Slowly progressive neurodegenerative diseases are probably not the result of a single hit-and-run event, but rather a several-step process linking environmental, epigenetic and genetic events. Thus, the next generation of drug treatment most focus on combined therapies selective for several critical events that distinguish these disorders. Lowering the burden of mitochondrial injury, protein aggregation, oxidative and nitrosative stress, inflammatory response and heavy metal accumulation in the brain so as to re-establish neurotransmission and block excitotoxicity may establish beneficial in the treatment of several neurodegenerative diseases.

References

Abalkhail, H., Mitchell, J., Habgood, J., Orrell, R. and de Belleroche, J. (2003) A new familial amyotrophic lateral sclerosis locus on chromosome 16q12.1-16q12.2. *Am J Hum Genet*. 73: 383–89.

Adams, Jr. J.D., Chang, M.L. and Klaidman, L (2001) Parkinson's disease-redox mechanisms. *Curr Med Chem*. 8:809–814.

Alexianu, M.E., Kozovska, M. and Appel, S.H. (2001) Immune reactivity in a mouse model of familial ALS correlates with disease progression. *Neurology*, 57: 1282–89.

Anandatheerthavarada, H.K., Biswas, G., Robin, M.A. and Avadhani, N.G. (2003) Mitochondrial targeting and a novel transmembrane arrest of Alzheimer's amyloid precursor protein impairs mitochondrial function in neuronal cells. *J Cell Biol.* 161: 41–54.

Ankarcrona, M., Dypbukt, J.M., Bonfoco, E., Zhivotovsky, B., Orrenius, S., Lipton, S.A. and Nicotera, P. (1995) Glutamate-induced neuronal death: a succession of necrosis or apoptosis depending on mitochondrial function. *Neuron.* 15:961–973.

Aylward, E.H., Li, Q., Stine, O.C., Ranen, N., Sherr, M., Barta, P.E. Bylsma, F.W., Pearlson, G.D. and Ross, C.A. (1997) Longitudinal change in basal ganglia volume in patients with Huntington's disease. *Neurology*, 48:394-399.

Beaulieu, J.M., Nguyen, M.D. and Julien, J.P. (1999) Late onset death of motor neurons in mice overexpressing wild-type peripherin. *J Cell Biol.* 147: 531–44.

Becher, M.W., Kotzuk, J.A., Sharp, A.H., Davies, S.W., Bates, G.P., Price, D.L. and Ross, C.A. (1998) Intranuclear neuronal inclusions in Huntington disease and dentatorubral and pallidoluysian atrophy: correlation between the density of inclusions and IT15 CAG triplet repeat length. *Neurobiol Dis.* 4:387-397.

Ben, S.D., Zuk, R. and Glinka, Y. (1995) Dopamine neurotoxicity: inhibition of mitochondrial respiration. *J Neurochem.* 64:718–723.

Betarbet, R., Sherer, T.B., MacKenzie, G., Garcia-Osuna, M., Panov, A.V. and Greenamyre, J.T (2000) Chronic systemic pesticide exposure reproduces features of Parkinson disease. *Nat Neurosci.* 3:1301–1306.

Bishop, G.M. and Robinson, S.R. (2004) Physiological roles of amyloid-beta and implications for its removal in Alzheimer disease. *Drugs Aging*, 21:621-30.

Borchelt, D.R., Lee MK, Slunt, H.S., Guarnieri M, Xu ZS, Wong PC, Brown RH, Price D.L., Sisodia, S.S. and Cleveland, D.W. (1994) Superoxide dismutase 1 with mutations linked to familial amyotrophic lateral sclerosis possesses significant activity. *Proc Natl Acad Sci USA*, 91: 8292–96.

Bossy-Wetzel, E., Schwarzenbacher, R. and Lipton, S.A. (2004) Molecular pathways to neurodegeneration. *Nat Med.* 10 Suppl: S2-9.

Bowling, A.C., Barkowski, E.E., McKenna-Yasek, D., Sapp, P., Horvitz, H.R., Beal, M.F. and Brown, R.H. (1995) Superoxide dismutase concentration and activity in familial amyotrophic lateral sclerosis. *J Neurochem.* 64: 2366–69.

Braak, H. and Braak, E. (1997) Staging of Alzheimer-related cortical destruction. *Int Psychogeriatr.* 9 [Suppl 1]:257–261.

Brion, S., Mirol, J. and Psimaras, A. (1973) Recent findings in Pick's disease. In: Progress in Neuropathology, Ed. Zimmerman, H.M., Vol. 2, pp. 421–452. New York: Grune and Stratton.

Brown, R.C., Lockwood, A.H. and Sonawane, R. (2005) Neurodegenerative Diseases: An Overview of Environmental Risk Factors. *Environ Health Persp*, 113: 1250-1256.

Bruijn, L.I. and Cleveland, D.W. (1996) Mechanisms of selective motor neuron death in ALS: insights from transgenic mouse models of motor neuron disease. *Neuropathol Appl Neurobiol.* 22: 373–87.

Bruijn, L.I., Becher, MW, Lee, M.K., Anderson, K.L., Jenkins NA, Copeland N, Sisodia S., Rothstein, J., Borchelt, D. and Price, D. (1997) ALS-linked SOD1 mutant G85R mediates damage to astrocytes and promotes rapidly progressive disease with SOD1-containing inclusions. *Neuron.* 18: 327–38.

Bruijn, L.I., Houseweart, M.K., Kato, S., Anderson, K.L., Anderson, S.D., Ohama, E., Reaume, A.G., Scott, R.W. and Cleveland, D.W. (1998) Aggregation and motor neuron toxicity of an ALS-linked SOD1 mutant independent from wild-type SOD1. *Science*, 281: 1851–54.

Brujin, L.I., Miller, T.M. and Cleveland, D.W. (2004) Unraveling the mechanisms involved in motor neuron degeneration in ALS. *Ann Rev Neurosci*. 27: 723-749.

Buee, L., Bussiere, T., Buee-Scherrer, V., Delacourte, A. and Hof, P.R. (2000) Tau protein isoforms, phosphorylation and role in neurodegenerative disorders. *Brain Res Brain Res Rev*. 33:95-130.

Burn, D.J. and Jaros, E. (2001). Multiple system atrophy: cellular and molecular pathology. *Mol. Pathol.* 54:419–426.

Bush, A.I. (2002) Metal complexing agents as therapies for Alzheimer's disease. *Neurobiol Aging*, 23: 1031−1038.

Bush, A.I. (2003) Copper, zinc, and the metallobiology of Alzheimer disease. *Alzheimer Dis Assoc Disord.* 17: 147−150.

Butterfield, D.A. and Kanski, J. (2001) Brain oxidation in agerelated neurodegenerative disorders that are associated with aggregated proteins. *Mech Ageing Dev.* 122, 945–962.

Butterfield, D.A. and Boyd-Kimball, D. (2004) Amyloid beta-peptide(1-42) contributes to the oxidative stress and neurodegeneration found in Alzheimer disease brain. *Brain Pathol.* 14(4):426-32.

Butterworth, N.J., Williams, L., Bullock, J.Y., Love, D.R., Faull, R.L. and Dragunow, M. (1998) Trinucleotide (CAG) repeat length is positively correlated with the degree of DNA fragmentation in Huntington disease striatum. *Neurosci.* 87: 49.

Carpenter, S. (1968) Proximal axonal enlargement in motor neuron disease. *Neurology*, 18: 841–51.

Casley, C.S., Landb, J.M., Sharpea, M.A., Clarka, J.B., Duchenc, M.R. and Canevari, L. (2002a) beta-Amyloid fragment 25−35 causes mitochondrial dysfunction in primary cortical neurons. *Neurobiol Dis*. 10: 258−267.

Casley, C.S., Canevari, L., Land, J.M., Clark, J.B. and Sharpe, M.A. (2002b) beta-Amyloid inhibits integrated mitochondrial respiration and key enzyme activities. *J Neurochem*. 80: 91−100

Cedazo-Mínguez, A. and Cowburn, R.F. (2001) Apolipoprotein E: a major piece in the Alzheimer's disease puzzle. *J Cell Mol Med.* 5: 254–266.

Cha, J.H.J. (2000) Transcriptional dysregulation in Huntington's disease. *Trends Neurosci.* 23:387–392.

Chapman, P.F., White, G.L., Jones, M.W., Cooper-Blacketer D, Marshall VJ, Irizarry M, Younkin L, Good MA, Bliss TVP, Hyman, B.T., Younkin, S.G. and Hsiao, K.K. (1999) Impaired synaptic plasticity and learning in aged amyloid precursor protein transgenic mice. *Nat Neurosci.* 2: 271−276.

Checkoway, H., Farin, F.M., Costa-Mallen, P., Kirchner, S.C. and Costa, L.G. (1998) Genetic polymorphisms in Parkinson's disease. *Neurotoxicol.* 19:635–643.

Cherny, R.A., Atwood, C.S., Xilinas, M.E., Gray DN, Jones WD, McLean CA, Barnham KJ, Volitakis I, Fraser FW, Kim Y, Huang, X., Goldstein, L.E., Moir, R.D., Lim, J.T., Beyreuther, K., Zheng, H., Tanzi, R.E., Masters, C.L. and Bush, A.I. (2001) Treatment with a copper-zinc chelator markedly and rapidly inhibits beta-amyloid accumulation in Alzheimer's disease transgenic mice. *Neuron.* 30: 665–676.

Clement, A.M., Nguyen MD, Roberts EA, Garcia ML, Boillée S, Rule M, McMahon, A.P., Doucette, W., Siwek, D., Ferrante, R.J., Brown, R.H., Julien, J.P., Goldstein, L.S.B. and Cleveland, D.W. (2003) Wild-type nonneuronal cells extend survival of SOD1 mutant motor neurons in ALS mice. *Science*, 302: 113–117.

Cooper, A.J., Sheu, K.F., Burke, J.R., Strittmatter, W.J., Gentile, V., Peluso, G. and Blass, J.P. (1999) Pathogenesis of inclusion bodies in (CAG)n/Qn-expansion diseases with special reference to the role of tissue transglutaminase and to selective vulnerability. *J Neurochem.* 72: 889-899.

Corbo, M. and Hays, A.P. (1992) Peripherin and neurofilament protein coexist in spinal spheroids of motor neuron disease. *J Neuropathol Exp Neurol.* 51: 531–37.

Cummings, C.J., Reinstein, E., Sun, Y., Antalffy, B., Jiang, Y., Ciechanover A, Orr HT, Beaudet, A.L. and, Zoghbi, H.Y. (1999) Mutation of the E6-AP ubiquitin ligase reduces nuclear inclusion frequency while accelerating polyglutamine-induced pathology in SCA1 mice. *Neuron.* 24: 879–892.

Dal Canto, M.C. and Gurney, M.E. (1994) Development of central nervous system pathology in a murine transgenic model of human amyotrophic lateral sclerosis. *Am J Pathol.* 145: 1271–79.

Delisle, M.B. and Carpenter, S. (1984) Neurofibrillary axonal swellings and amyotrophic lateral sclerosis. *J Neurol Sci.* 63: 241–50.

DiFiglia, M., Sapp, E., Chase, K.O., Davies, S.W., Bates, G.P., Vonsattel, J.P. and Aronin, N. (1997) Aggregation of huntingtin in neuronal intranuclear inclusions and dystrophic neurites in brain. *Science*, 277:1990-1993.

Dong, J., Atwood, C.S., Anderson, V.E., Siedlak, S.L., Smith, M.A., Perry, G. and Carey, P.R. (2003) Metal binding and oxidation of amyloid-beta within isolated senile plaque cores: Raman microscopic evidence. *Biochem*, 42: 2768–2773.

Elbaz A, Bower JH, Maraganore DM, McDonnell SK, Peterson BJ, Ahlskog JE, Schaida DJ, Rocca WA. (2002) Risk tables for parkinsonism and Parkinson's disease. *J Clin Epidemiol*, 55:25–31.

Elliott, J.L. (2001) Cytokine upregulation in a murine model of familial amyotrophic lateral sclerosis. *Brain Res Mol Brain Res.* 95: 172–78.

Ferrante, R.J., Kowall, N.W., Beal, M.F., Richardson, E.P., Bird, E.D. and Martin, J.B. (1985) Selective sparing of a class of striatal neurons in Huntington's disease. *Science*, 230:561-563.

Fusco, F.R., Chen, Q., Lamoreaux WJ, Figueredo-Cardenas G, Jiao Y, Coffman JA, Surmeier, D.J., Honig, M.G., Carlock, L.R. and Reiner, A. (1999) Cellular localization of huntingtin in striatal and cortical neurons in rats: lack of correlation with neuronal vulnerability in Huntington's disease. *J Neurosci.* 19: 1189.

Gabby, M., Tauber, M., Porat, S. and Simantov, R. (1996) Selective role of glutathione in protecting human neuronal cells from dopamine-induced apoptosis. *Neuropharmacology*, 35:571–578.

Gaeta, A. and Hider, R.C. (2005) The crucial role of metal ions in neurodegeneration: the basis for a promising therapeutic strategy. *British J Pharmacology*, 146: 1041–1059.

Gómez-Isla, T., Price, J.L., McKeel, D.W., Morris, J.C., Growdon, J.H. and Hyman, B.T. (1996) Profound loss of layer II entorhinal cortex neurons occurs in very mild Alzheimer's disease. *J Neurosci.* 16:4491– 4500.

Gonatas, N.K., Stieber, A., Mourelatos, Z., Chen, Y., Gonatas, J.O., Appel, A.P., Hays, W.F., Hickey, J. and Hauw, J. (1992) Fragmentation of the Golgi apparatus of motor neurons in amyotrophic lateral sclerosis. *Am J Pathol.* 140: 731–37.

Gong, Y.H., Parsadanian, A.S., Andreeva, A., Snider, W.D. and Elliott, J.L. (2000) Restricted expression of G86R Cu/Zn superoxide dismutase in astrocytes results in astrocytosis but does not cause motoneuron degeneration. *J Neurosci.* 20: 660–65.

Gotz, J., Schild, A., Hoerndli, F. and Pennanen, L. (2004) Amyloid-induced neurofibrillary tangle formation in Alzheimer's disease: insight from transgenic mouse and tissue-culture models. *Int J Dev Neurosci.* 22:453-65.

Graham, D.G. (1978) Oxidative pathways for catecholamines in the genesis of neuromelanin and cytotoxic quinones. *Mol Pharmacol.* 14:633–643.

Grundke-Iqbal, I., Iqbal, K., Quinlan, M., Tung, Y.C., Zaidi, M.S. and Wisniewski, H.M. (1986) Microtubule-associated protein tau. A component of Alzheimer paired helical filaments. *J Biol Chem.* 261: 6084–6089.

Grünewald, T. and Beal, M.F. (1999) Bioenergetics in Huntington's Disease. *Ann NY Acad Sci.* 893:203-213.

Guegan, C., Vila, M., Rosoklija, G., Hays, A.P. and Przedborski, S. (2001) Recruitment of the mitochondrial-dependent apoptotic pathway in amyotrophic lateral sclerosis. *J Neurosci.* 21: 6569–76.

Gurney, M.E. (1994) Transgenic-mouse model of amyotrophic lateral sclerosis. *N Engl J Med.* 331: 1721–22.

Guttman, M., Kish, S.J. and Furukawa, Y. (2003) Current concepts in the diagnosis and management of Parkinson's disease. *CMAJ*, 168: 293-301.

Hadano, S., Hand, C.K., Osuga, H., Yanagisawa, Y., Otomo A, Devon RS, Miyamoto N, Showguchi-Miyata J, Okada Y, Singaraja R, Figlewicz, D.A., Kwiatkowski, T., Hosler, B.A., Sagie, T., Skaug, J., Nasir, J., Brown RH, Scherer SW, Rouleau GA, Hayden, M.R. and Ikeda, J.E. (2001) A gene encoding a putative GTPase regulator is mutated in familial amyotrophic lateral sclerosis 2. *Nat Genet.* 29: 166–73.

Halliwell, B. and Gutteridge, J. (1985) Oxygen radicals and the nervous system. *Trends Neurosci.* 8:22–29.

Hardy, J. (1997) Amyloid, the presenilins and Alzheimer's disease. *Trends Neurosci.* 20: 154−159.

Hardy, J., Cookson, M.R. and Singleton, A. (2003) Genes and parkinsonism. *Lancet Neurol.* 2: 221−228.

Harper, P.S. (1992) The epidemiology of Huntington's Disease. *Hum Genet.* 89: 365.

Harjes, P. and Wanker, E.E (2003) The hunt for huntingtin function: interaction partners tell many different stories. *Trends Biochem Sci.* 28:425–433.

Hastings, T.G., Lewis, D.A. and Zigmond, M.J. (1996) Role of oxidation in the neurotoxic effects of intrastriatal dopamine injections. *Proc Natl Acad Sci USA*, 93:1956–1961.

Hedreen, J.C., Peyser, C.E., Folstein, S.E. and Ross, C.A. (1991) Neuronal loss in layers V and VI of cerebral cortex in Huntington's disease. *Neurosci Lett.* 133:257-261.

Hensley, K., Floyd, R.A., Gordon, B., Mou, S., Pye, Q.N., Stewart, C., West, M. and Williamsonm K. (2002) Temporal patterns of cytokine and apoptosis-related gene expression in spinal cords of the G93A-SOD1 mouse model of amyotrophic lateral sclerosis. *J Neurochem*. 82: 365–74.

Herrup, K., Neve, R., Ackerman, S.L. and Copani, A. (2004) Divide and Die: Cell Cycle Events as Triggers of Nerve Cell Death. *J Neurosci*. 24:9232–9239.

Hirano, A., Nakano, I., Kurland, L.T., Mulder, D.W., Holley, P.W. and Saccomanno, G. (1984) Fine structural study of neurofibrillary changes in a family with amyotrophic lateral sclerosis. *J Neuropathol Exp Neurol.* 43: 471–80.

Howland, D.S., Liu, J., She Y, Goad B, Maragakis NJ, Kim B, Erickson J, Kulik J, DeVito, L., Psaltis G, DeGennaro, L.J., Cleveland, D.W. and Rothstein, J.D. (2002) Focal loss of the glutamate transporter EAAT2 in a transgenic rat model of SOD1 mutant-mediated amyotrophic lateral sclerosis (ALS). *Proc Natl Acad Sci USA*, 99: 1604–9.

Hughes, A.J., Daniel, S.E., Kilford, L. and Lees, A.J. (1992) Accuracy of clinical diagnosis of idiopathic Parkinson's disease: a clinico-pathological study of 100 cases. *J Neurol Neurosurg Psychiatry*, 55:181–184.

Hyman, B.T., Van Hoesen, G.W., Damasio, A.R. and Barnes, C.L. (1984) Alzheimer's disease: cell-specific pathology isolates the hippocampal formation. *Science*, 225:1168 – 1170.

Ince, P.G., Shaw, P.J., Slade, J.Y., Jones, C. and Hudgson, P. (1996) Familial amyotrophic lateral sclerosis with a mutation in exon 4 of the Cu/Zn superoxide dismutase gene: pathological and immunocytochemical changes. *Acta Neuropathol* (Berlin), 92: 395–403.

Iqbal, K., Zaidi, T., Bancher, C. and Grundke-Iqbal, I. (1994) Alzheimer paried helical filaments. Restoration of the biological activity by dephosphorylation. *FEBS Lett.* 349:104-8.

Iqbal, K., Alonso Adel, C., Chen, S., Chohan MO., El-Akkad, E., Gong, C.X., Khatoon, S., Li B., Liu, F., Rahman, A., Tanimukai, H. and Grundke-Iqbal, I. (2005) Tau pathology in Alzheimer disease and other tauopathies. *Biochim Biophys Acta*, 1739:198-210.

Jaarsma, D., Haasdijk, E.D., Grashorn, J.A., Hawkins, R., van Duijn, W., Verspagetb, H.W., London, J. and Holstege, J.C. (2000) Human Cu/Zn superoxide dismutase (SOD1) overexpression in mice causes mitochondrial vacuolization, axonal degeneration, and premature motoneuron death and accelerates motoneuron disease in mice expressing a familial amyotrophic lateral sclerosis mutant SOD1. *Neurobiol Dis*. 7: 623–43.

Jaarsma, D., Rognoni, F., van Duijn, W., Verspaget, H.W., Haasdijk, E.D. and Holstege, J.C. (2001) CuZn superoxide dismutase (SOD1) accumulates in vacuolated mitochondria in transgenic mice expressing amyotrophic lateral sclerosis-linked SOD1 mutations. *Acta Neuropathol* (Berlin), 102: 293–305.

Jack, C.R. and Petersen, R.C. (2000) Structural imaging approaches to Alzheimer's disease. In: Early diagnosis and treatment of Alzheimer's disease (Scinto LFM, Daffner KR, eds), pp 127–148. Totowa, NJ: Human.

Jenner, P. and Olanow, C.W. (1996) Oxidative stress and the pathogenesis of Parkinson's disease. *Neurology*, 47:161–70.

Jenner, P. (2003) Oxidative stress in Parkinson's disease. *Ann Neurol* 53 [Suppl 3]:S26 –S36.

Jonsson, P.A., Ernhill, K., Andersen, P.M., Bergemalm, D., Brannstrom, T., Gredal, O., Nilsson, P. and Marklund, S.L. (2004) Minute quantities of misfolded mutant superoxide dismutase-1 cause amyotrophic lateral sclerosis. *Brain*, 127: 73–88.

Johnston, J.A., Dalton, M.J., Gurney, M.E. and Kopito, R.R. (2000) Formation of high molecular weight complexes of mutant Cu, Zn-superoxide dismutase in a mouse model for familial amyotrophic lateral sclerosis. *Proc Natl Acad Sci USA*, 97: 12571–76.

Kato, S., Takikawa, M, Nakashima, K., Hirano A, Cleveland DW, Kusaka H, Shibata N, Kato, M., Nakano, I. and Ohama, E. (2000) New consensus research on neuropathological aspects of familial amyotrophic lateral sclerosis with superoxide dismutase 1 (SOD1) gene mutations: inclusions containing SOD1 in neurons and astrocytes. *Amyotroph Lateral Scler Other Motor Neuron Disord*. 1: 163–84.

Kaur, D., Yantiri, F., Rajagopalan, S., Kumar, J., Mo, J.Q., Boonplueang, R., Viswanath, V., Jacobs, R., Yang, L., Beal, M.F., DiMonte, D., Volitaskis, I., Ellerby, L., Cherny, R.A., Bush, A.I. ans Andersen, J.K. (2003) Genetic or pharmacological iron chelation prevents MPTP-induced neurotoxicity in vivo: a novel therapy for Parkinson's disease. *Neuron*, 37: 899–909.

Kawamata, T., Akiyama, H., Yamada, T. and McGeer, P.L. (1992) Immunologic reactions in amyotrophic lateral sclerosis brain and spinal cord tissue. *Am J Pathol*. 140: 691–707.

Khachaturian, Z.S. (1985) Diagnosis of Alzheimer's disease. *Arch Neurol*. 42:1097–1105.

Kim, H.J., Chae, S.C., Lee, D.K., Chromy, B., Lee, S.C., Park, Y.C., Klein, W.L., Krafft, G.A. and Hong, S.T. (2003) Selective neuronal degeneration induced by soluble oligomeric amyloid beta protein. *FASEB J*. 17:118-20.

King, M.E. (2005) Can tau filaments be both physiologically beneficial and toxic? *Biochim Biophys Acta*, 1739(2-3): 260-267.

Kirkitadze, M.D., Bitan, G. and Teplow, D.B. (2002) Paradigm shifts in Alzheimer's disease and other neurodegenerative disorders: the emerging role of oligomeric assemblies. *J Neurosci Res*. 69:567-77.

Kish, S.J., Shannak, K. and Hornykiewicz, O. (1988) Uneven pattern of dopamine loss in the striatum of patients with idiopathic Parkinson's disease. *N Engl J Med*. 318: 876-80.

Köpke, E., Tung, Y.C., Shaikh, S., Alonso, A del C., Iqbal, K. and Grundke-Iqbal, I. (1993) Microtubule-associated protein tau. Abnormal phosphorylation of a non- paired helical filament pool in Alzheimer disease. *J Biol Chem*. 268, 24374–24384.

Kostic, V., Jackson-Lewis, V., de Bilbao, F., Dubois-Dauphin, M. and Przedborski, S. (1997) Bcl-2: prolonging life in a transgenic mouse model of familial amyotrophic lateral sclerosis. *Science*, 277: 559–62.

Kunst, C.B., Messer, L., Gordon, J., Haines, J. and Patterson, D. (2000) Genetic mapping of a mouse modifier gene that can prevent ALS onset. *Genomics*, 70: 181–89.

Langston, J.W., Ballard, P.A., Tetrud, J.W. and Irwin, I. (1983) Chronic parkinsonism in humans due to a product of meperidine analog synthesis. *Science*, 219:979–80.

Lanska, D.J. (1995) George Huntington and hereditary chorea. *J Child Neurol*. 10: 46.

Leigh, P.N. and Swash, M. (1991) Cytoskeletal pathology in motor neuron diseases. *Adv Neurol*. 56: 115–24.

Leigh, P.N., Whitwell, H., Garofalo, O., Buller, J., Swash, M., Martin, J.E., Gallo, J.M., Weller, R.O. and Anderton, B.H. (1991) Ubiquitin-immunoreactive intraneuronal inclusions in amyotrophic lateral sclerosis. Morphology, distribution, and specificity. *Brain*, 114: 775–88.

Li, H. and Dryhurst, G. (1997) Irreversible inhibition of mitochondrial complex Iby 7-(2-aminoethyl)-3,4-dihydro-5-hydroxy-2H-1,4-benzothiazine-3-carboxylic acid (DHBT-1): a putative nigral endotoxin of relevance to Parkinson's disease. *J Neurochem.* 69:1530–1541.

Li, M., Ona, V.O., Guegan, C., Chen, M., Jackson-Lewis, V., Andrews, L.J., Olszewski, A.J., Stieg, P.E., Lee, J.P., Przedborski, S. and Friedlander, R.M. (2000) Functional role of caspase-1 and caspase-3 in an ALS transgenic mouse model. *Science*, 288: 335–39.

Lindwall, G. and Cole, R.D. (1984) Phosphorylation affects the ability of tau protein to promote microtubule assembly. *J Biol Chem.* 259: 5301–5305.

Lino, M.M., Schneider, C. and Caroni, P. (2002) Accumulation of SOD1 mutants in postnatal motoneurons does not cause motoneuron pathology or motoneuron disease. *J Neurosci.* 22: 4825–32.

Lipton, S.A. (2004) Concepts: turning down but not off—neuroprotection requires a paradigm shift in drug development. *Nature*, 428: 473.

Liu, Q., Smith, M.A., Avila, J., DeBernardis, J., Kansal M, Takeda A, Zhu X, Nunomura A, Honda K, Moreira PI, Oliveira CR, Santos, M.S., Shimohama, S., Aliev, G., de la Torre, J., Ghanbari, H.A., Siedlak, S.L., Harris, P.L, Sayre, L.M. and Perry, G. (2005) Alzheimer specific epitopes of tau represent lipid peroxidation-induced conformations. *Free Radic Biol Med.* 38:746-54.

Lustbader, J.W., Cirilli, M., Lin, C., Wei Xu H, Takuma K, Wang N, Caspersen C, Chen X, Pollak S, Chaney M, Trinchese, F., Liu, S., Gunn-Moore, F., Lue, L.F., Walker, D.G., Kuppusamy, P., Zewier, Z.L., Arancio, O., Stern, D., ShiDu Yan, S. and Wu, H. (2004) ABAD directly links Abeta to mitochondrial toxicity in Alzheimer's disease. *Science.* 304: 448−452.

Martindale, D., Hackam, A., Wieczorek, A., ellerby L, Wellington C, Mccutcheon K, Singaraja R, Kazemi-esfarjani P, Devon R, Kim, S.U., Bredesen, D.E., Tufaro, F. and Hayden, M.R. (1998) Length of huntingtin and its polyglutamine tract influences localization and frequency of intracellular aggregates. *Nat Genet.* 18: 150.

Lovell, M.A., Robertson, J.D., Teesdale, W.J., Campbell, J.L. and Markesbery, W.R. (1998). Copper, iron and zinc in Alzheimer's disease senile plaques. *J Neurol Sci.* 158: 47−52.

McCullagh, C.D., Craig, D., McIlroy, S.P. and Passmore, A.P. (2001) Risk factors for dementia. *Adv Psychiatric Treatm.* 7:24–31.

McDowell. I. (2001) Alzheimer's Disease: insights from epidemiology. *Aging*, 13: 143–162.

Miller, M.L. and Johnson, G.V.W. (1995) Transglutaminase cross-linking of the t protein. *J Neurochem.* 65:1760-70.

Moolman, D.L., Vitolo, O.V., Vonsattel, J.P.G. and Shelanski, M.L. (2004) Dendrite and dendritic spine alterations in Alzheimer models. *J Neurocytol.* 33: 377–387.

Mulder, D.W., Kurland, L.T., Offord, K.P. and Beard, C.M. (1986) Familial adult motor neuron disease: amyotrophic lateral sclerosis. *Neurology*, 36: 511–17.

Nagai, M., Aoki, M., Miyoshi, I., Kato, M., Pasinelli, P., Kasai, N., Brown, R.H. and Itoyama, Y. (2001) Rats expressing human cytosolic copper-zinc superoxide dismutase transgenes with amyotrophic lateral sclerosis: associated mutations develop motor neuron disease. *J Neurosci.* 21: 9246–54.

National Institute on Aging (1997) Consensus recommendations for the postmortem diagnosis of Alzheimer's disease. The National Institute on Aging, and Reagan Institute Working Group on Diagnostic Criteria for the Neuropathological Assessment of Alzheimer's Disease. *Neurobiol Aging*, 18:S1–S2.

Nemes, Z., Devreese, B., Steinert, P.M., Van Beeumen, J. and Fesus, L. (2004) Cross-linking of ubiquitin, HSP27, parkin, and alpha-synuclein by gamma-glutamyl-epsilon-lysine bonds in Alzheimer's neurofibrillary tangles. *FASEB J.* 18:1135-7.

Nicklas, W.J., Vyas, I. and Heikkila, R.E. (1985) Inhibition of NADH-linked oxidation in brain mitochondria by 1-methyl-4-phenyl-pyridine, a metabolite of the neurotoxin 1-methyl-4-phenyl-1,2,5,6-tetrahydropyridine. *Life Sci.* 36:2503–8.

Nunomura, A., Perry, G., Pappolla, M.A., Wade, R., Hirai, K., Chiba, S. and Smith, M.A. (1999) RNA oxidation is a prominent feature of vulnerable neurons in Alzheimer disease. *J Neuroscience.* 19:1959-64.

Nuydens, R., De Jong, M., Nuyens, R., Cornelissen, F. and Geerts, H. (1995) Neuronal kinase stimulation leads to aberrant tauphosphorylation and neurotoxicity. *Neurobiol Aging.* 16:465-75.

Olanow, C.W. (1990) Oxidation reactions in Parkinson's disease. *Neurology*, 40:32–37.

Olanow, C.W. (1993) A radical hypothesis for neurodegeneration. *Trends Neurosci.* 16: 439–44.

Olanow, C.W. and Tatton, W.G. (1999) Etiology and pathogenesis of Parkinson's disease. *Annu Rev Neurosci.* 22:123–144.

Otomo, A., Hadano, S, Okada T, Mizumura H, Kunita R, Nishijima H, Showguchi-Miyata, J., Yanagisawa, Y., Kohiki, E., Suga, E., Yasuda, M., Osuga, H., Nishimoto, T., Narumiya, S. and Iked, J.E. (2003) ALS2, a novel guanine nucleotide exchange factor for the small GTPase Rab5, is implicated in endosomal dynamics. *Hum. Mol Genet.* 12: 1671–87.

Panov, A., Gutekunst, C.A., Leavitt, B.R., Hayden, M.R., Burke, J.R., Strittmatter, J. and Greenamyre, J.T. (2002) Early mitochondrial calcium defects in Huntington's disease are a direct effect of polyglutamines. *Nat Neurosci.* 5: 731-736.

Pasinelli, P., Houseweart, M.K., Brown, R.H., Jr. and Cleveland, D.W. (2000) Caspase-1 and -3 are sequentially activated in motor neuron death in Cu, Zn superoxide dismutase-mediated familial amyotrophic lateral sclerosis. *Proc Natl Acad Sci USA*. 97: 13901–6.

Perry, G., Raina, A.K., Nunomura, A., Wataya, T., Sayre, L.M. and Smith, M.A. (2000) How important is oxidative damage? Lessons from Alzheimer's disease. *Free Radic Biol Med.* 28:831-4.

Pigino, G., Morfini, G., Pelsman, A., Mattson, M.M., Brady, S.T. and Busciglio, J. (2003) Alzheimer presenilin 1 mutations impair kinesin-based axonal transport. *J Neurosci.* 23: 4499–4508.

Pramatarova, A., Laganiere, J., Roussel, J., Brisbois, K. and Rouleau, G.A. (2001) Neuron-specific expression of mutant superoxide dismutase 1 in transgenic mice does not lead to motor impairment. *J Neurosci.* 21: 3369–74.

Price, J.L., Davis, P.B., Morris, J.C. and White, D.L. (1991) The distribution of tangles, plaques, and related immunohistochemical markers in healthy aging and Alzheimer's disease. *Neurobiol Aging.* 12:295–312.

Price, D.L. and Sisodia, S.S. (1998) Mutant genes in familial Alzheimer's disease and transgenic models. *Ann Rev Neurosci.* 21: 479–505.

Prolla, T.A. and Mattson, M.P (2001) Molecular mechanisms of brain aging and neurodegenerative disorders: lessons from dietary restriction. *Trends Neurosci.* 24:S21–S31.

Prusiner, S.B. (2001) Neurodegenerative diseases and prions. *N Engl J Med.* 344, 1516–1526.

Przedborski, S., Vila, M. and Jackson-Lewis, V. (2003) Neurodegeneration: What is it and where are we? *J Clin Invest.* 111:3–10.

Raber, J., Huang, Y. and Ashford, J.W. (2004) ApoE genotype accounts for the vast majority of AD risk and AD pathology. *Neurobiol Aging.* 25:641–650.

Reaume, A.G., Elliott, J.L., Hoffman, E.K., Kowall, N.W., Ferrante, R.J., Siwek, D.R., Wilcox, H.M., Flood Dorothy, G.M., Beal, F., Brown, R.H., Scott, R.W., and Snider, W.D. (1996) Motor neurons in Cu/Zn superoxide dismutase-deficient mice develop normally but exhibit enhanced cell death after axonal injury. *Nat Genet.* 13: 43–47.

Ripps, M.E., Huntley, G.W., Hof, P.R., Morrison, J.H. and Gordon, J.W. (1995) Transgenic mice expressing an altered murine superoxide dismutase gene provide an animal model of amyotrophic lateral sclerosis. *Proc Natl Acad Sci USA.* 92: 689–93.

Rosas, H.D., Liu, A.K., Hersch, S., Glessner, M., Ferrante, R.J., Salat, D.H., van der Kouwe, A., Jenkins, B.G., Dale, A.M. and Fischl, B. (2002) Regional and progressive thinning of the cortical ribbon in Huntington's disease. *Neurology.* 58:695-701.

Roses, A.D. (1996) Alipoprotein E in neurology. *Curr Opin Neurol.* 4: 265–270.

Rothstein, J.D., Tsai, G., Kuncl, R.W., Clawson, L., Cornblath, D.R., Drachman, D.B., Pestronk, A., Stauch, B.L. and Coyle, J.T. (1990) Abnormal excitatory amino acid metabolism in amyotrophic lateral sclerosis. *Ann Neurol.* 28: 18–25.

Rothstein, J.D., Van Kammen, M., Levey, A.I., Martin, L.J. and Kuncl, R.W. (1995) Selective loss of glial glutamate transporter GLT-1 in amyotrophic lateral sclerosis. *Ann Neurol.* 38: 73–84.

Rubinsztein, D.C. and Easton, D.F. (2000) Apolipoprotein E genetic variation and Alzheimer's disease: a meta-analysis. *Dement Geriatr Cogn Disord.* 10:1999–209.

Rubinsztein, D.C. (2002) Lessons from animal models of Huntington's disease. *Trends Genet.* 18: 202-209.

Ruddy, D.M., Parton, M.J., Al-Chalabi, A., Lewis, C.M., Vance, C., Smith, B.N., Leigh, P.N., Powell, J.F., Siddique, T., Meyjes, E.P., Baas, F., De Jong, V. and Shaw, C.E. (2003) Two families with familial amyotrophic lateral sclerosis are linked to a novel locus on chromosome 16q. *Am J Hum Genet.* 73: 390–96.

Sapp, E., Penney, J., Young, A., Aronin, N., Vonsattel, J.P., and Difiglia, M. (1999) Axonal transport of N-terminal huntingtin suggests early pathology of corticostriatal projections in Huntington disease. *J Neuropathol Exp Neurol.* 58: 165.

Sapp, P.C., Hosler, B.A., McKenna-Yasek, D., Chin, W., Gann, A., Genise, H., Gorenstein, J., Huang, M., Sailer, W., Scheffler, M., Valesky, M., Haines, J.L., Pericak-Vance, M.,

Siddique, T.H., Horvitz, R. and Brown, R.H. (2003) Identification of two novel loci for dominantly inherited familial amyotrophic lateral sclerosis. *Am J Hum Genet.* 73: 397–403.

Sayre, L.M., Moreira, P.I., Smith, M.A. and Perry, G. (2005) Metal ions and oxidative protein modification in neurological disease. *Ann Ist Super Sanità.* 41:143-164.

Saudou, F., Finkbeiner, S., Devys, D. and Greenberg, M.E. (1998) Huntingtin acts in the nucleus to induce apoptosis but death does not correlate with the formation of intranuclear inclusions. *Cell*, 95: 55.

Selkoe, D.J. (1999) Translating cell biology into therapeutic advances in Alzheimer's disease. *Nature*, 399: A23−A31.

Selkoe, D.J. (2002) Alzheimer's disease is a synaptic failure. *Science*, 298, 789−791.

Selkoe, D.J. (2005) Defining molecular targets to prevent Alzheimer disease. *Arch Neurol.* 62:192-5.

Shastry, B.S. (2003) Neurodegenerative disorders of protein aggregation. *Neurochem Internat.* 43, 1–7.

Shaw, P.J., Forrest, V., Ince, P.G., Richardson, J.P. and Wastell, H.J. (1995) CSF and plasma amino acid levels in motor neuron disease: elevation of CSF glutamate in a subset of patients. *Neurodegeneration.* 4: 209–16.

Sherer, T.B., Betarbet, R., Stout, A.K., Lund, S., Baptista, M., Panov, A.V., Cookson, M.R. and Greenamyre, J.T (2002) An in vitro model of Parkinson's disease: linking mitochondrial impairment to altered α-synuclein metabolism and oxidative damage. *J Neurosci.* 22: 7006 –7015.

Sieradzan, K.A. and Mann, D.M. (2001) The selective vulnerability of nerve cells in Huntington's disease. *Neuropathol Appl Neurobiol.* 27: 1–21.

Singer, T.P., Castagnoli, N., Ramsay, R.R. and Trevor, A.J. (1987) Biochemical events in the development of parkinsonism induced by 1-methyl-4-phenyl-1,2,3,6-tetrahydropyridine. *J Neurochem.* 49:1–8.

Sipione, S. and Cattaneo, E. (2001) Modeling Huntington's disease in cells, flies, and mice. *Mol Neurobiol.* 23: 21-51.

Stacy, M. and Jankovic, J. (1992) Differential diagnosis of Parkinson's disease and the parkinsonism plus syndromes. *Neurol Clin.* 10:341–359.

Stewart, W.F., Kawas, C., Corrada, M. and Metter, E.J. (1997) Risk of Alzheimer's disease and duration of NSAID use. *Neurology.* 48: 626−632.

Stephan, A., Laroche, S. and Davis, S. (2001) Generation of aggregated beta-amyloid in the rat hippocampus impairs synaptic transmission and plasticity and causes memory deficits. *J Neurosci.* 21: 5703−5714.

Stoessl, A.J. (1999) Etiology of Parkinson's disease. *Can J Neurol Sci.* 26(Suppl 2):S5-12.

Sulkava, R., Haltia, M., Paetau, A., Wikstrom, J. and Palo, J. (1983) Accuracy of clinical diagnosis in primary degenerative dementia: correlation with neuropathological findings. *J. Neurol. Neurosurg. Psychiatry.* 46:9–13.

Tanaka, K., Watase, K., Manabe, T., Yamada, K., Watanabe, M., Takahashi, K., Iwama, H., Nishikawa, T., Ichihara, N., Kikuchi, T., Okuyama, S., Kawashima, N., Hori, S., Takimoto, M. and Wada, K. (1997) Epilepsy and exacerbation of brain injury in mice lacking the glutamate transporter GLT-1. *Science.* 276: 1699–702.

Tanner, C.M., Ottman, R., Goldman, S.M., Ellenberg, J., Chan, P., Mayeux, R. and Langston, J.W. (1999) Parkinson disease in twins: an etiologic study. *JAMA*. 281:341–346.

Terry, R.D., Masliah, E., Salmon, D.P., Butters, N., Deteresa, R., Hill, R., Hansen, L.A. and Katzman R. (1991) Physical basis of cognitive alterations in Alzheimer's disease: Synapse loss is the major correlate of cognitive impairment. *Annals Neurol.* 30: 572–580.

Tomlinson, B.E. (1977) The pathology of dementia. *Contemp Neurol Ser.* 15:113–153.

Valente, E.M., Abou-Sleiman, M., Caputo, V., Muqit, M.M.K., Harvey, K., Gispert, S., Ali, Z., Del Turco, D., Bentivoglio, A.R., Healy, D.G., Albanese, A., Nussbaum, R., González-Maldonado, R., Deller, T., Salvi, S., Cortelli, P., Gilks, W.P., Latchman, D.S., Harvey, R.J., Dallapiccola, B., Auburger, G. and Wood, N.W. (2004) Hereditary early-onset Parkinson's disease caused by mutations in PINK1. *Science*. 304: 1158−1160.

van Dellen, A., Grote, H.E. and Hannan, A.J. (2005) Gene-environment interactions, neuronal dysfunction and pathological plasticity in Huntington's disease. *Clin Exp Pharmacol Physiol.* 32:1007–1019.

Vonsattel, J.P., Myers, R.H., Stevens, T.J., Ferrante, R.J., Bird, E.D. and Richardson, E.P.J. (1985) Neuropatholigical classification of Huntington's disease. *J Neuropathol Exp Neurol.* 44: 559.

Wang, J., Xu, G., Gonzales, V., Coonfield, M., Fromholt, D., Copeland, N.G., Jenkins, N.A. and Borchelt, D.R. (2002) Fibrillar inclusions and motor neuron degeneration in transgenic mice expressing superoxide dismutase 1 with a disrupted copper-binding site. *Neurobiol Dis.* 10: 128–38.

Wang, J., Slunt, H., Gonzales, V., Fromholt, D., Coonfield, M., Copeland, N.G., Jenkins, N.A. and Borchelt, D.R. (2003) Copper-binding-site-null SOD1 causes ALS in transgenic mice: aggregates of non-native SOD1 delineate a common feature. *Hum Mol Genet.* 12: 2753–64.

Walsh, D.M., Klyubin, I., Fadeeva, J.V., Cullen, W.K., Anwyl, R., Wolfe, M.S., Rowan, M.J. and Selkoe, D.J. (2002) Naturally secreted oligomers of amyloid beta protein potently inhibit hippocampal long-term potentiation in vivo. *Nature*. 416: 535−539.

Weggen, S., Eriksen, J.L., Das, P., Sagi, S.A., Wang, R., Pietrzik, C.U., Findlay, K.A., Smith, T.E., Murphy, M.P., Bulter, T., Kang, D.E., Marquez-Sterling, N., Golde, T.E. and Koo, E.H. (2001) A subset of NSAIDs lower amyloidogenic Abeta42 independently of cyclooxygenase activity. *Nature*. 414: 212−216.

Weingarten, M.D., Lockwood, A.H., Hwo, S.Y. and Kirschner, M.W. (1975) A protein factor essential for microtubule assembly. *Proc Natl Acad Sci USA*. 75: 1858–1862.

Weisgraber, K.H. and Mahley, R.W. (1996) Human apolipoprotein E: the Alzheimer's disease connection. *FASEB J.* 10:1485–1494.

Wellington, C.L., Ellerby, L.M., Hackam, A.S., Margolis, R.L., Trifiro, M.A., Singaraja, R., Mccutcheon, K., Salvesen, G.S., Propp, S.S., Bromm, M., Rowland, K.J., Zhang, T., Rasper, D., Roy, S., Thornberry, N., Pinsky, L., Kakizuka, A., Ross, C.A., Nicholson, D.W., Bredesen, D.E. and Hayden, M.R. (1998) Caspase cleavage of gene products associated with triplet expansion disorders generates truncated fragments containing the polyglutamine tract. *J Biol Chem.* 273: 9158.

Wisniewsky, H.M., Coblentz, J.M. and Terry, R.D. (1972) Pick's disease. A clinical and ultrastructural study. *Arch Neurol.* 26, 97–108.

Wong, P.C., Pardo, C.A., Borchelt, D.R., Lee, M.K., Copeland, N.G., Jenkinse, N.A., Sisodiad, S.S., Cleveland, D.W. and Price, D.L. (1995) An adverse property of a familial ALS-linked SOD1 mutation causes motor neuron disease characterized by vacuolar degeneration of mitochondria. *Neuron*. 14: 1105–16.

Xu, J., Kao, S.Y., Lee, F.J.S., Song, W., Jin, L.W. and Yankne, B.A. (2002) Dopamine-dependent neurotoxicity of α-synuclein: a mechanism for selective neurodegeneration in Parkinson disease. *Nat Med*. 8: 600−606.

Yamanaka, K., Vande, V.C., Bertini, E., Boespflug-Tanguy, O. and Cleveland, D.W. (2003) Unstable mutants in the peripheral endosomal membrane component ALS2 cause early onset motor neuron disease. *Proc Natl Acad Sci.* 100: 16041–46.

In: Progress in Neurodegeneration
Editors: M.R. Avila-Costa and V.A. Martinez
ISBN 978-1-60741-317-2

Chapter 3

Metals, Toxicity and Neurodegeneration

Maria Rosa Avila-Costa[1], Leonardo Reynoso Erazo[2], Ana Luisa Gutierrez-Valdez[1] and Jose Luis Ordoñez Librado[1]
1) Dept. of Neuroscience, Neuromorphology Lab
2) Research Division, Health Education Team. UNAM, Mexico

Until the 1990s, the neuroscience research community paid scant attention to the metabolism of metal ions. Apart from a great deal of work done on calcium (Ca^{2+}), and some on magnesium, the neurobiology of metal ions did not arouse much interest as they were not notably linked to major disease syndromes. This outlook seems set to change dramatically over the coming decade, with a growing number of publications pointing the way to a seminal relationship between Iron (Fe), Lead (Pb), Copper (Cu), Vanadium (V), Manganese (Mn) and Zinc (Zn) in the generation (or defense) of oxygen radicals and protein that mediate the major neurological diseases (Bush, 2000).

It seems that as brain ages, distinct alterations in oxidative and inflammatory events occur that are associated with pathological lesions characteristic of neurodegenerative diseases. Aluminum (Al), Cadmium (Cd), Mn, Fe, and other trace redox-active transition metals may be involved in mediating these processes and thus may be involved in the neuropathology of disorders such as Parkinson disease (PD), Alzheimer disease (AD), amyotrophic lateral sclerosis (ALS) and others.

Recently it has been discovered that transition metals bind to proteins involved in neurodegeneration and this association appears to preserve metal redox activity in a manner that is consistent with a pro-oxidant and free radical generating action (Figure 1). The binding of multivalent metals to colloidal Al may parallel this phenomenon. Alterations in the levels of metals-containing metalloenzymes, involved in processing partially reduced oxygen intermediates, as well as the antioxidant status of cells, may also contribute to altered redox homeostasis in neurodegenerative diseases. Nonetheless, even in familial forms of ALS, linked to mutations in superoxide dismutase (SOD), it is not clear whether an altered enzyme activity or, indirectly, a disturbance in transition metal homeostasis is involved in the disease pathogenesis (Campbell et al. 2001).

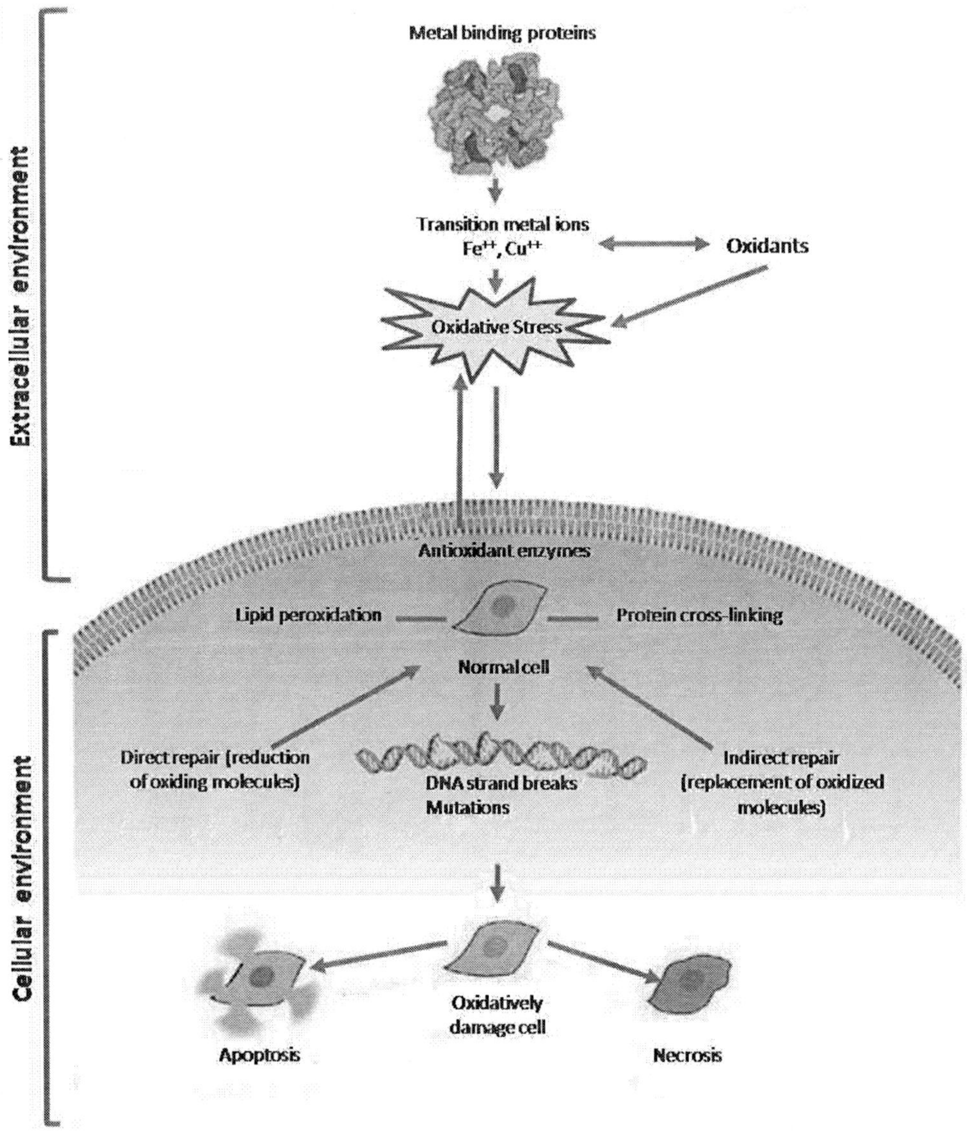

Figure 1. Transition metal/particle surface complexes may initiate a common chronic oxidative stress pathways culminating in neurodegeneration.

Metal ions are vital life elements that participate in numerous metabolic junctions in every living cell (Ross et al. 2006). Metal ion homeostasis is maintained through highly regulated processes of uptake, storage and secretion. A specific set of transporters function in each cellular compartment to provide a delicate balance of transport activities across their membranes. Because metal ions are vital for several life processes, and their action also inflicts damage on DNA and proteins, their proper distribution is vital and a slight alteration in their activity could cause severe disease. For example, abnormal Fe uptake has been implicated in the most common hereditary disease hemochromatosis, as well as in neurological diseases such as PD, AD, Friedreich ataxia and others (Babcock et al. 1997; Andrews and Levy, 1998; Askwith and Kaplan, 1998; Haacke et al. 2005); Cu accumulation in the liver and nervous system is related to Wilson disease (Hershey et al. 1983). Apparently, any aberration in the cellular metal ion concentrations may cause a shortage of a vital metabolic element or inflict damage that could lead to cell death. Therefore, limited activity of a single metal ion transporter may cause growth arrest, and excess activity of the same transporter may be toxic, leading to cell death.

Moreover, cells throughout the body rely on transition metals to regulate a wide range of metabolic and signaling functions, most significantly the modulation and transport of oxygen. They are almost invariably involved in the reduction of oxygen directly (oxidases) or indirectly by way of hydrogen abstraction (dehydrogenases). As critical components of such enzymes, these elements can impose conformational changes and/or serve as catalytic centers for redox reactions with molecular oxygen or endogenous peroxides. However, the very essential nature of these elements allows them to invade and disrupt controls of their cellular uptake, transport, compartmentalization, and binding to cell components. When these metals are liberated from their catalytic sites within enzymes, and are sequestered abnormally to other ligands, the consequences can be deleterious. Under such circumstances, irreversible oxidative changes to both the metal-complexing ligand itself, and neighboring molecules that intercept the reactive intermediates of the original reaction, may initiate a cascade of events of oxidative stress to inflict damage upon critical biological processes. By this means, mitochondrial dysfunction, excitotoxicity, and rises in cytosolic free Ca^{2+} may be promoted. Thus, the toxic consequences of transition metals may be regarded as an aberrant expression of their normal physiological function (HaMai et al. 2001).

Given the high rate of oxidative metabolism of the brain, and correspondingly, its relatively high concentration of transition metals, the organ is particularly prone to such aberrant activity. The study of rare incidences of direct exposures of cerebral tissue to elevated levels of these metals may offer insights into more subtle changes in their intracellular disposition that may be encountered in more common forms of neurodegenerative disease such as AD and PD.

Metals are key constituents of well-characterized metalloproteins such as hemoglobin, ceruloplasmin, and ferritin. They are often bound to protein via sulphydryl groups of amino acids such as cysteine and methionine. Body stores and the concentrations of metals such as Zn and Fe in cells and body fluids are well regulated and essential to protein function, mainly the enzymes. Barriers to metal entry into the nervous system exist both at the level of the blood-brain barrier (BBB) and the barrier between blood and cerebrospinal fluid (CSF; Zheng et al. 2003). CSF may be considered an ultrafiltrate of blood and thus contains extracellular fluid surrounding the brain and spinal cord, which is vital to the function of nervous cells. Some data exist on the metal concentrations in CSF (Basun et al. 1991; Hershey et al. 1983).

Because metal ions are critical for several metabolic processes and are poisonous at moderate or high concentrations, it is not surprising that growing numbers of neurological diseases are connected to metal ion homeostasis. The interplay of specific and general metal ion transporters and the connection between Cu and Fe transport cause multifarious alterations in metal ion homeostasis in each genetic aberration (Radisky and Kaplan, 1999).

As we mentioned in Chapter 2, activated inflammatory processes and the accumulation of protein deposits within brain parenchyma, which lead to neuronal damage and loss, characterize neurodegenerative diseases. There is often progressive brain degeneration long before the development of a cascade of symptoms over 2 to 20 years that leads to increasing disability, ultimately contributing to death. The dominant risk factor of neurodegenerative diseases is age. There is an age-related increase of Al, Cu, Fe, and Zn), but not Mn, in the brain (Massie et al. 1979; Bartzokis et al. 1997; Zecca et al. 2001).

The purpose of this chapter is to evaluate recent findings implicating redox-active metal ions in promoting oxidative events and cell death. The connection between reactive oxygen moieties in promoting mitochondrial alterations, glial activation and the resulting pro-inflammatory cascade and the different types of cell death will be discussed. Those processes are present in the aged and diseased brain and metal-induced exacerbation of these events may contribute to the pathological lesions characteristic of neurodegeneration.

3.1. Entrance of Metals to the Brain

To produce neurodegeneration, it is assumed that the neurotoxic metal enters the central nervous system (CNS). Many metals that have been suggested to contribute to neurodegenerative diseases are considered to be essential for human health in trace amounts (Cu, Fe, Mn, V and Zn). For Pb there is circumstantial evidence for essentiality that is not yet convincing (Uthus and Seaborn, 1996; Goyer and Clarkson, 2001). There is no good evidence for essentially in mammals for Al or Hg. Because metal species, such as the free metal ion and complexes of the metal with an amino acid or protein, such as transferrin, are quite hydrophilic, they would not be expected to be able to distribute across the BBB at a rate that is sufficient to meet the requirements of the brain (Yokel, 2006). This is due to properties of the BBB that greatly limit diffusion of non-lipophilic substances into and out of the brain (Pardridge, 2003). As brain glucose demand greatly exceeds the rate of glucose diffusion across the BBB, Glut-1 mediates brain glucose uptake to meet the brain's needs. Similarly, the leucine system mediates the bidirectional transport of leucine and phenylalanine across the BBB to provide the brain the amino acids it requires and remove excess amino acids and similar substances when they accumulate as metabolic products. Therefore, it is anticipated that metal distribution across the BBB might be transporter mediated (Yokel, 2006).

Metals are most often absorbed from the gastrointestinal tract, across the lungs, or through the skin. They then enter systemic circulation. The metal could then enter the CNS from the blood across the BBB or from the blood by crossing the choroid plexus (CP) into the CSF, from which it can diffuse into the CNS. Alternatively, there is evidence that some metals can be taken up by exposed sensory nerves in the nasal cavity and possibly enter the brain, as discussed below.

The mechanisms of distribution of substances across cell membranes include diffusion and carrier mediated transport (Pardridge, 2003). Flux of substances between cells is primarily by paracellular diffusion. Brain entry through the intact BBB might be achieved by the very limited diffusion through the paracellular pathway. Diffusion through endothelial cell membranes is generally limited to small (usually < 700 dalton) lipophilic substances. There are exceptions. Elemental Hg is readily absorbed from the lung and diffuses across the BBB. Organomercurials such as methyl and dimethyl Hg are absorbed from the lung, gastrointestinal tract and via the percutaneous route and readily distribute across the BBB. Carrier mediated transport includes equilibrative, and energy-dependent transporters that are able to move substrates unidirectionally and against a concentration gradient, and receptor-mediated mechanisms which may operate by facilitated diffusion and are often bidirectional (Yokel, 2006).

Membrane transporters are often quite substrate specific. For example: the divalent metal transporter (DMT1, DCT1, Nramp2) transports divalent metals, but not metals in other valence states. Another example of substrate specificity is the much greater binding affinity of transferrin for Fe^{+3} than Fe^{+2} (Harris, 1983). As a result, transferrin-receptor mediated endocytosis (TfR-ME) for Fe^{+3} is much greater than for Fe^{+2}. Furthermore, the affinity of the transferring receptor for halotransferrin (diferric transferrin) is considerably greater than for monoferric transferrin (Huebers et al. 1983). TfR-ME is also believed to play a role in the transport of other trivalent metals into the brain, such as Al and Mn.

Other way that the metals enter the brain is the Choroid Plexuses (CP). The CP are capillary networks in the two lateral, the third and the fourth ventricles of the brain. They are surrounded by a monolayer of epithelial cells that have tight junctions. They are highly vascular and are the sites of CSF production.

In addition to the routes of distribution from blood to brain through the BBB and CP, it has been known for some time that proteins can distribute from the nasal cavity into the olfactory neuron, the only site where the CNS is exposed to the environment, and then trans-synaptically beyond the olfactory neuron into other brain regions (Oliver and Fazakerley, 1998). Trans-synaptic movement is involved in the distribution of tetanus virus from muscle to spinal cord ganglia, the site of its toxicity. It is known from studies in rats and pike fish that Mn, and perhaps nickel, can enter the brain directly by this route (Brenneman et al. 2000; Dorman et al. 2002; Rao et al. 2003). There is some disagreement on the extent of this mechanism of brain entry. It appears that this is not the major route of brain entry of metals. Other metals, such as Cd, cobalt, Fe, inorganic Hg and Zn enter the olfactory bulb from the nasal cavity but their distribution beyond that into other brain regions in the fish or rat is much less or not detectible (Rao et al. 2003; Persson et al. 2003a; 2003b). However, there are results in fish suggesting some distribution of inorganic Hg and tributyl tin into the brain by routes other than through the BBB, which were suggested to be uptake by water-exposed sensory nerves (Rouleau et al. 2003).

3.2. Metal Transport and Toxicity to the Brain Barriers

As we mentioned above, many metals (Ca, Cu, magnesium, Mn, Fe, Zn, cobalt, and molybdenum) are essential and required for optimal CNS function. They play important roles in brain function as catalysts, second messengers, and gene expression regulators. Being essential cofactors for functional expressions of many proteins, trace elements are needed to activate and stabilize enzymes, such as SOD, metalloproteases, protein kinases, and transcriptional factors containing Zn finger proteins. Clearly, metals must be supplied to the CNS at an optimal level, because both deficiency and excess can result in aberrant CNS function. Thus, transport of metal ions across the BBB is the first step in regulating their CNS levels. A series of active or receptor-mediated transport systems inherent to the BBB vasculature serve to control the transport of these metals into the brain, maintaining their optimal concentrations. Other metals that are nonessential (i.e., have no known functional attributes), such as Hg and Pb, can also readily gain access to the CNS (Zheng et al. 2003).

At least nine metals have been found to accumulate in the BBB (Zheng, 2001). Clinically, poisoning with Pb, Hg, and arsenic have been demonstrated to induce vascular destruction and cerebral hemorrhage. The damage to the endothelial structure of the BBB is a fundamental cause of leakage of blood-borne materials to surrounding brain parenchyma.

Transport of metal ions from the blood to the brain involves the crossing of BBB. Studies of Fe transport in the brain showed that the BBB permeability of Fe–transferrin is similar to that of albumin (Morris et al. 1992). The experiments suggested that uptake of Fe into the brain involves the transport of Fe from Fe-loaded blood transferrin to a brain-derived transferrin for extracellular Fe transport within the brain (Morris et al. 1992). Subsequently, Fe is taken up via transferrin receptors into the brain cells. Perturbation in Fe homeostasis was observed by chronic treatment of rats with chlorpromazine, a medication for schizophrenic patients (Ben-Shachar et al. 1994). This observation suggests that some of the side effects caused by psychotic drugs may be due to perturbation of metal ion homeostasis. Mn is also readily taken up into the CNS, most likely as a free ion (Murphy et al. 1991). However, plasma proteins such as albumin and transferrin that bind Mn^{2+} affect its transport. Except for its function in metalloproteins, there is no indication of specific neuronal modulation by Mn. However, the observation that Mn^{2+} and Fe^{2+} affect the taste behavior of Drosophila suggests that metal ions may function in novel signal transduction pathways in the brain (Orgad et al. 1998).

On the other hand, recent studies demonstrated that Zn can be considered as a neurotransmitter because it is accumulated in presynaptic vesicles of excitatory neurons, is released with synaptic activity, interacts with some ionotropic receptors and is taken back by a specific transporter (Assaf and Chung, 1984; Howell et al. 1984; Peters et al. 1987; Westbrook and Mayer, 1987; Seguela et al. 1996; Sensi et al. 1997; McMahon and Cousins, 1998). Zn can interact strongly with a variety of ligands including sulfur in cysteine, nitrogen in histidine and oxygen in acidic amino acids. Therefore, it is likely to be bound to serum proteins, and it may cross the BBB as a ligand of amino acids or other components that bind Zn. In light of the recent observation of stress-related breaks in BBB (Kaufer et al. 1998), the mechanism of metal ion accumulation in the brain should be re-evaluated.

Research during the past several decades clearly points to a critical role for the brain barrier systems in chemical induced neurotoxicities. However, the linkage between barrier dysfunction and the etiology of various neurological disorders remains unclear, owing to the lack of rigorous research effort in this area.

It is important to stand out that blood–brain interfaces protect brain tissues against organic and inorganic chemicals by means of different complementary mechanisms. Nevertheless, toxic metals could bypass these mechanisms and inflict damage to the brain parenchyma. The effectiveness of the barrier systems may also be compromised either in pathological situations or following toxic insults by compounds that target the blood–brain interfaces. Cumulative evidence has revealed that the brain barriers, the gatekeepers of the cerebral compartment, are subject to toxic insults from heavy metal exposure. Because of the special roles of brain barriers in overall brain development and function, it is reasonable to postulate that injury to the barriers may contribute to metal-induced neurotoxicities. Our knowledge about this aspect remains developing. It is, therefore, imperative that future investigations address how brain barriers sequester toxic metals, whether toxic metals alter barrier functions, and what neurological consequences may ensue.

Some transition metals Fe and Cu as free ions provide oxidation/reduction chemistry that can produce free radicals and reactive oxygen species. This has been implicated in the oxidative injury seen in some neurodegenerative diseases. Elevated Fe, Cu and Zn have been seen in the senile plaques of AD. This might contribute to neurodegenerative diseases through increased oxidative injury, Aβ aggregation and/or other mechanisms (Qian and Wang, 1998; Christen, 2000).

The biochemical pathways to cell death in chronic and acute forms of neurodegeneration are poorly understood, limiting the ability to develop effective therapeutic approaches. As details of the apoptotic and necrotic pathways have been revealed, an appreciation for the decisive role that mitochondria play in life-death decisions for the cell has grown. As a result, the need has arisen to reevaluate the significance to cell viability of oxidative stress, reactive oxidative species generation, mitochondrial Ca^{2+} sequestration, and the membrane permeability transition.

3.3. Oxidative Stress

All aerobic organisms produce at least minimal levels of reactive oxygen species (ROS), mostly arising from the side-production of superoxide anion (O_2^-) during the reduction of molecular oxygen by mitochondria. That oxidative stress has been frequently implicated in neurodegeneration (reviewed in Smith et al. 1988) reflects the selective vulnerability of the central nervous system arising from increased dioxygen (O_2) utilization. Additionally, Hydrogen Peroxide (H_2O_2), produced by oxidases such as monoamine oxidase (MAO), can result in greater oxidative stress susceptibility in tissues enriched in these enzymes.

The generation of ROS requires the activation of molecular oxygen. At the quantum level, O_2 is in a triplet spin-state and consequently interactions with most organic molecules are spin-forbidden (that is, they are energetically unfavorable). As utilization of O_2 is a prerequisite for most life forms, this intrinsic feature of O_2 must be overcome. Organisms have evolved a range of metalloenzymes to take advantage of the interactions between O_2 and metal ions to activate molecular oxygen as ROS; the subsequent free radicals are an intrinsic part of normal metabolism. As ROS are also toxic, cells have developed highly elaborate means of regulating both metal ion interactions and the generation of ROS. The same properties that cells harness for beneficial means become destructive when the regulatory processes breakdown (Barnham et al. 2004).

ROS are produced by a number of different pathways, including direct interactions between redox-active metals and oxygen species via reactions such as the Fenton and Haber–Weiss reactions, or via indirect pathways involving the Ca^{2+} activation of metallo-enzymes such as phospholipases, nitric oxide synthase and xanthine oxidase (Lewen et al. 2000). Ca^{2+} is crucial to signal transduction and as such is both sensitive to a number of different stimuli and able to elicit a variety of different cellular responses. There is a large body of evidence documenting disruption of Ca^{2+} homeostasis in neurodegenerative diseases, leading to a breakdown in a large number of cellular processes. However, most of these different signaling pathways rely on feedback signaling and it is consequently difficult to differentiate cause and effect (Barnham et al. 2004). An example of this is the interplay between Ca^{2+} signaling and ROS generation: although increases in intracellular calcium have been reported to induce the production of ROS (Lewen et al. 2000), at subtoxic levels ROS have been shown to be important cell signalers and will induce an increase in cytosolic Ca^{2+} (Suzuki et al. 1997; Neill et al. 2002). This has lead to much debate as to whether the observed disruption of Ca^{2+} homeostasis is a cause or consequence of ROS generation.

In recent years, the upsurge in research on nitric oxide (NO) as a common second messenger in neurotransmission and signaling has resulted in recognition of its enzymatic release from activated microglia (macrophages in the CNS), along with $O2^-$. Accumulating levels of diffusible NO and $O2^-$ give rise to peroxynitrite ($ONOO^-$). $ONOO^-$ and related reactive nitrogen species (RNS) are capable of both oxidation chemistry and nitration of the aromatic side-chains of tyrosine and tryptophan (Alvarez and Radi, 2003), resulting in a condition known as “nitrosative stress”. There seems to be growing acceptance that ROS- and RNS-mediated damage seen in the central nervous system may reflect underlying neuroinflammatory processes (Floyd, 1999).

As we mentioned earlier, in some neurodegenerative diseases there is compelling evidence for a role of a metabolic imbalance and resulting oxidative stress that is superimposed on hereditary factors, and likely play a larger role in the spontaneously occurring forms of these diseases (Sayre et al. 2001; Andersen, 2004). The link between oxidative stress and neuronal death is complex, but there is support for a contributory role of oxidative stress induced neurotoxicity in AD and PD (Facheris et al. 2004).

The condition of oxidative stress in each of these disease states is accompanied to a varying degree by a dyshomeostasis of metal ions, including the redox-active transition metals Fe and Cu as well as redox inactive metal ions such as Zn. The CNS is particularly vulnerable to oxidative stress on account of the high rate of O_2 utilization, the relatively poor concentrations of antioxidants and related enzymes, and the high content of polyunsaturated lipids, the most vulnerable biomacromolecule to oxidation. There is also an accumulation of Fe in the brain as a function of age, which can be a potent catalyst for oxidative species formation (Sayre et al. 2005).

Oxidative stress is defined as the imbalance between biochemical processes leading to production of ROS and those responsible for the removal of ROS, the so-called cellular antioxidant cascade. Tissues that become subject to oxidative stress witness steady state levels of ROS-mediated damage to all biomacromolecules (polynucleotides, proteins, lipids, and sugars) that can lead to a critical failure of biological functions and ultimately cell death. Several neurodegenerative disorders are associated with oxidative stress that is manifested by lipid peroxidation, protein oxidation and other markers (Sayre et al. 2005), since brain relies much on aerobic respiration and it consumes large quantity of oxygen, it is particularly susceptible to oxidative stress.

Mitochondria are essential organelles for neuronal function because the high-energy requirement makes them highly dependent on aerobic oxidative phosphorylation. However, this event is the major source of ROS. All aerobic organisms produce at least minimal levels of ROS, mostly arising from the side-production of O_2^- by reaction of molecular oxygen with sites in the electron-transport chain where reducing equivalents accumulate. O_2^-, which is also generated from the respiratory burst of neutrophils, can be subsequently transformed to the other classical ROS species H_2O_2 and hydroxyl radical ($OH^{\bullet}$). ROS can also arise from mutationally altered or damaged metalloenzymes involved in oxidative metabolism. Reactive species generated by mitochondria have several cellular targets including mitochondrial components themselves (lipids, proteins and DNA). The lack of histones in mitochondrial DNA (mtDNA) and diminished capacity for DNA repair render mitochondria an easy target to oxidative stress events. Mitochondrial dysfunction and free radical-induced oxidative damage have been implicated in the pathogenesis of PD and AD, as well as other neurodegenerative disorders.

Usually considered as the chief instigator of oxidative stress damage, the $OH^{\bullet}$ reacts non-discriminately with all biomacromolecules at diffusion-controlled rates, i.e., within nm distances from its site of generation. $OH^{\bullet}$ can be produced by gamma radiation, but is most commonly generated physiologically by the Fenton reaction between reduced transition metals (usually Fe^{2+} or Cu^+ and H_2O_2). Re-reduction of the resulting oxidized transition metal ions (Fe^{3+} or Cu^{2+}) can be effected by cellular reductants such as vitamin C or thiols. In contrast to $OH^{\bullet}$, superoxide radical ($O_2^{\bullet -}$) is chemically unreactive, except at lower pH, where it exists as the hydroperoxy radical (H_2O–HO_2). However, O_2^- can serve as the reductant of oxidized metal ions for the production of $OH^{\bullet}$ from H_2O_2, the so-called Haber-Weiss reaction. Under normal conditions, damage by ROS is kept in check by an efficient antioxidant cascade, including both enzymatic and non-enzymatic entities. Important in the former regard are cytosolic Cu-Zn superoxide dismutase (CuZnSOD) and mitochondrial Mn superoxide dismutase (MnSOD), which convert superoxide to O_2 and H_2O_2. The latter, also the normal by-product of oxygen reduction by oxidases such as MAO, is removed by catalase and peroxidases, which have ubiquitous tissue distribution, but can result in greater oxidative stress susceptibility in tissues lacking sufficient activity of these enzymes (Sayre et al. 2005).

As a general principle, the chemical origin of the majority of ROS is the reaction of molecular oxygen with the redox-active metals (Halliwell and Gutteridge, 1999). The ability of these metal ions to occupy multiple valence states and undertake facile redox cycling, thereby activating molecular oxygen has been utilized by a variety of enzymes. However, unregulated redox-active metals will inappropriately react with O_2 to generate ROS. Barnham et al. (2004) have proposed that the proteins implicated in several age-dependent neurodegenerative disorders (Aβ in AD, α-synuclein in PD, SOD1 in ALS, frataxin in Friedreich ataxia), might abnormally present Cu^{2+} or Fe^{3+} ligands for inappropriate reaction with O_2. These proteins might have some aspect of their function subserved by these metal ions, which normally occupy higher-affinity, embedded, redox-shielded binding sites. As metal concentration rises in the brain with age, the probability increases that a redox-competent, low-affinity metal-binding site will recruit a metal ion from the normally redox-silent cellular pool. In this manner, proteins such as Aβ can harness endogenous biometals to foster the release of inappropriate redox activity and ROS generation.

3.3.1. Metal Ions and Oxidative Stress

Cu, Mn, Fe, and other trace redox-active transition metals are essential in most biological reactions, e.g., in the synthesis of DNA, RNA, and proteins, and as cofactors of numerous enzymes, particularly those involved in respiration. Thus their deficiency can lead to disturbances in central nervous system and other organ function. However, accumulation of redox-active transition metals in tissues in excess of the capacity of the cellular complement of metalloproteins (catalytic, transport, storage) can be cytotoxic as a consequence of their participation in an array of cellular disturbances characterized by oxidative stress and increased free radical production (Sayre et al. 1999; 2005). Only trace levels of such circulating excess of redox-active transition metal ions are required for ROS generation, generally reflecting reaction of the reduced metal oxidation states with dioxygen or hydrogen peroxide (Fenton reaction). Although the metal ions are thereby converted to their oxidized forms, these can be re-reduced by superoxide ion or cellular reducing agents such as ascorbate.

Common to both AD and PD is a dyshomeostasis of both redox-active and redox-inactive metal ions. In general, the loss of homeostasis of Fe and Cu in the brain is accompanied by severe neurological consequences. A role of oxidative stress in AD and PD is consistent with the finding that the areas of the brain affected by these diseases contain abnormally high levels of redox-active metals, particularly Fe. However, it is not known whether the metal excesses are a cause of oxidative stress and neurodegeneration or a by-product of the neuronal cell loss. Alterations in the levels of "anti-oxidant" metalloenzymes likely also contribute to altered redox homeostasis in neurodegenerative diseases (Corson et al. 1999) though it is unclear whether this reflects altered enzyme activities or, indirectly, a disturbance in transition metal homeostasis.

In addition to oxidative damage to proteins, oxidative stress conditions and the occurrence of Fe or Cu-mediated Fenton chemistry results also in oxidative damage to nucleic acids, in particular RNA. 8-Hydroxyguanosine (8-OHG), a marker of nucleic acid oxidation, is commonly observed in the cytoplasm of the neurons that are particularly vulnerable to degeneration in AD (Nunomura et al. 1999; 2001). 8-OHG is likely to form at the site of hydroxyl radical production, most likely by the reaction of H_2O_2 with reduced Cu or Fe bound to nucleic acid bases (the oxidized metal ion thereby generated is re-reduced by cellular reductants such as ascorbate or O_2^-). RNA oxidation is also seen in vulnerable neurons in PD as well as in dementia with Lewy bodies (Nunomura et al. 2002), suggesting that it might represent one of the fundamental abnormalities in metal-associated neurodegenerative diseases.

3.3.2. Oxidative Stress in AD

Genetic evidence from cases of familial AD indicates that Aβ metabolism is linked to the disease (Hardy, 1997; Price et al. 1998). AD is characterized by the deposition of amyloid plaques, the major constituent being the amyloid-β peptide (Aβ) that is cleaved from the membrane-bound amyloid precursor protein (APP) (Glenner and Wong, 1984). Although the function of APP is unknown, recent evidence suggests it functions in maintaining Cu homeostasis (Barnham et al. 2004).

Recent results have highlighted the importance of Zn in amyloid plaque formation; for example, age- and female-sex-related plaque formation in Tg2576 transgenic mice was reduced by genetic ablation of the Zn transporter protein, which is required for Zn transport into synaptic vesicles (Lee et al. 2002). The plaques could be described as metallic sinks because remarkably high concentrations of Cu (400 μM), Zn (1 mM) and Fe (1 mM) have been found within the amyloid deposits in AD-affected brains (Smith et al. 1997; Lovell et al. 1998).

In vitro studies have shown that low μM levels of Zn will induce protease-resistant aggregation and precipitation of Aβ (Huang et al. 1997). Cu and Fe also induce peptide aggregation that is exaggerated at acidic pH (Huang et al. 1997). These metals are normally found at high concentrations in the region of the brain most susceptible to AD neurodegeneration. During neurotransmission, high concentrations of Zn (300 μM) and Cu (30 μM) are released, which might explain why Aβ precipitation into amyloid commences in the synapse (Lee et al. 2002).

The cause of the neuronal cell loss in AD might be related to oxidative stress from excessive free-radical generation (Smith et al. 1997; Perry et al. 2002; Bush, 2000; 2003). It has been proposed that the major source of oxidative stress and free-radical production in the brain in AD is the transition metals Cu and Fe (Bush, 2000) when bound to Aβ. Synthetic Aβ is toxic to cells in the presence of Cu, but this toxicity is inhibited by extracellular catalase, which implicates H_2O_2 in the toxic pathway (Behl et al. 1994). When Cu^{2+} or Fe^{3+} coordinate Aβ, extensive redox chemical reactions take place that reduce the oxidation state of both metals and produce H_2O_2 from O_2 in a catalytic manner (Cuajungco et al. 2000). The formal reduction potential of Cu^{2+} to Cu^{+} by Aβ is highly positive and characteristic of strongly reducing cupro-proteins (Huang et al. 1999). The generation of H_2O_2 in the presence of the reduced form of the metal creates conditions in which Fenton chemistry occurs with the generation of highly toxic $OH^{\bullet}$ (Huang, 1999) (Figure 2).

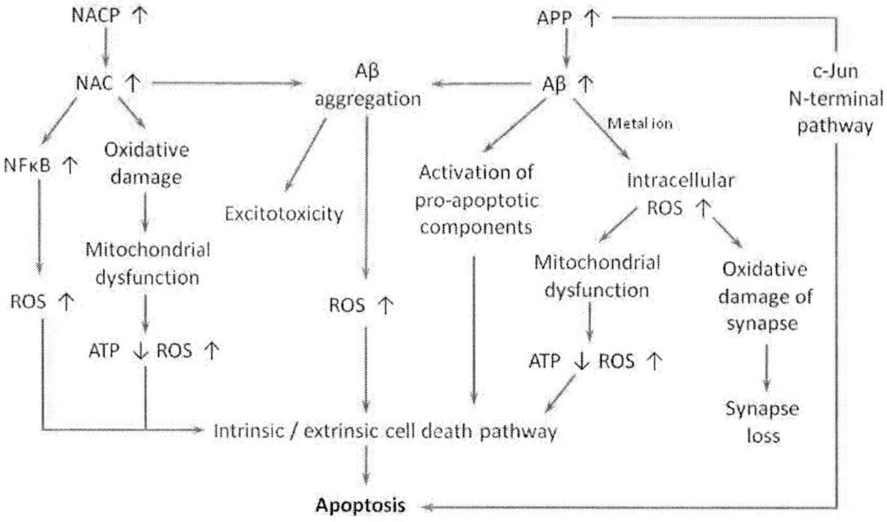

Modified from Loh et al. 2006.

Figure 2. Oxidative stress in Alzheimer disease.

3.3.3. Oxidative Stress in PD

PD is characterized by the loss of dopaminergic neurons of the Substantia Nigra *pars compacta* (SNc) and the deposition of intracellular inclusion bodies. The principal protein component of these deposits is α-synuclein (Spillantini et al. 1997), which is ubiquitously expressed in the brain; mutations of α-synuclein (A30P and A53T) contribute to familial forms of the disease (Gasser, 2001).

A characteristic feature of the neurons within the SNc is the age-dependent accumulation of neuromelanin (Marsden, 1983). In PD, these neuromelanin-containing cells are most likely to be lost (Hirsch et al. 1988). Neuromelanin is a dark brown pigment that accumulates metal ions, particularly Fe. Although the composition of neuromelanin has not been rigorously characterized, it is known that it consists primarily of the products of dopamine redox chemistry (Wakamatsu et al. 2003; Zecca et al. 2003).

Dopamine is an essential neurotransmitter, but as it is a catechol it is also a good metal chelator, and a potential electron donor (that is, a metal reductant). Dopamine coordinates metals such as Cu^{2+} and Fe^{3+} (Gerard et al. 1994), reduces the oxidation state of the metal, and subsequently engenders the production of H_2O_2, setting up conditions for Fenton chemistry. The purpose (if any) of neuromelanin is unknown, but it has been postulated that it protects against dopamine-induced redox-associated toxicity (Smythies, 1996; Sulzer et al. 2000). At low Fe concentrations, neuromelanin is known to have antioxidant properties, but at higher metal loads is prooxidant (Ben-Shachar et al. 1991). Another postulated role for neuromelanin is as an iron-storage molecule. Double et al. (2003) have shown that neuromelanin isolated from human SNc has both high- and low-affinity Fe^{3+}-binding sites, and that the Fe bound to neuromelanin is redox active (Faucheux et al 2003). The oxidative stress associated with PD could be the result of a breakdown in the regulation of dopamine (neuromelanin)/Fe biochemistry (Barnham et al. 2004).

A diverse array of evidence is emerging that α-synuclein has a role in modulating the activity of dopamine. The A53T mutation associated with familial PD impairs vesicular storage of dopamine (Lotharius and Brundin, 2002), which leads to the accumulation of dopamine in the cytoplasm and subsequent generation of ROS through its interaction with Fe, a process that increases with age. The mutations in α-synuclein have been shown to alter the expression of dihydropteridine reductase, which indirectly regulates the synthesis of dopamine (Baptista et al. 2003). Co-immunoprecipitation experiments have shown that α-synuclein forms stable complexes with the human dopamine transporter, thereby inhibiting uptake of dopamine by its transporter (Baptista et al. 2003) and that α-synuclein can regulate dopamine synthesis by inhibiting tyrosine hydroxylase (Perez et al. 2002). The link between α-synuclein and redox chemistry associated with iron-bound dopamine/neuromelanin has been given further credence by a study showing that initiation of Lewy body formation coincides with α-synuclein deposition exclusively within lipofuscin and neuromelanin deposits (Braak et al. 2001); in addition, α-synuclein crosslinked to neuromelanin has been reported (Fasano et al. 2003). The breakdown in α-synuclein-modulated dopamine homeostasis is consistent with the recent observation that the pathogenicity of mutant α-synuclein is dopamine dependent (Xu et al. 2002).

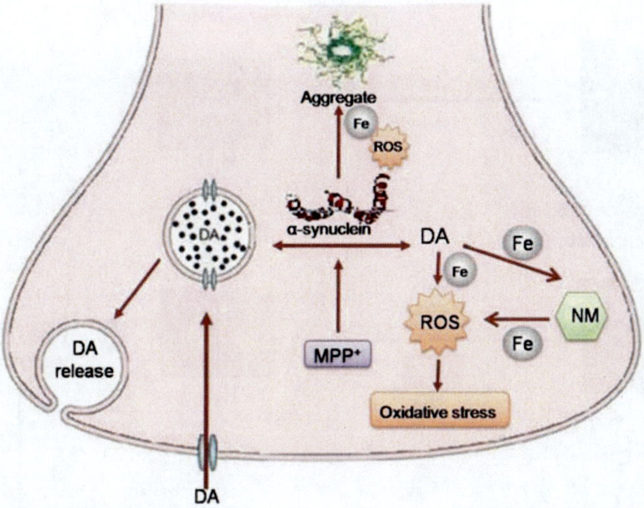

Modified from Barnham et al. 2004.

Figure 3. Oxidative stress in Parkinson disease. Dopamine (DA) required for neuronal signalling is vesicle bound and redox inert; when DA is released from the vesicle into the cytoplasm it is able to coordinate Fe and undergo redox reactions that result in the formation of neuromelanin (NM) and reactive oxygen species (ROS). Neuromelanin will also coordinate Fe and produce ROS. The equilibrium between vesicle bound dopamine and cytoplasmic dopamine is regulated by α-synuclein; mutations in this protein shift the dopamine equilibrium in favour of the cytoplasm. In the presence of Fe and under conditions of oxidative stress α-synuclein will aggregate and form deposits. MPP, 1-methyl-4-phenyl pyridine.

In addition to the regulation of dopamine by α-synuclein, studies have shown a direct interaction of α-synuclein with metal ions, leading to protein aggregation (Uversky et al. 2001; Yamin et al. 2003). Methionine oxidation inhibits α-synuclein aggregation; however, in the presence of certain metal ions aggregation and fibrillization of α-synuclein still occurs (Yamin et al. 2003), which highlights a potential role of metals and oxidative stress in the deposition of α-synuclein (Figure 3).

3.3.4. Oxidative Stress in ALS

ALS is distinguished by the loss of the lower motor neurons of the spinal cord and upper motor neurons in the cerebral cortex; as in AD and PD there are both sporadic and familial forms of the disease. Like other neurodegenerative diseases, ALS is characterized by the deposition of a misfolded protein in neural tissue, in this instance Cu/Zn SOD (Brujin et al. 1998). There are more than 100 mutations of SOD associated with the familial forms of the disease. Through transgenic mouse studies it has been shown that these mutations lead to a toxic gain of function by SOD (Gaudette et al. 2000). The nature of this gain of function is widely debated, and there are two main theories: one suggests that the toxicity is due to misfolded aggregated forms of SOD, whereas the other proposes that SOD becomes a pro-oxidant protein generating ROS (Barnham et al. 2004).

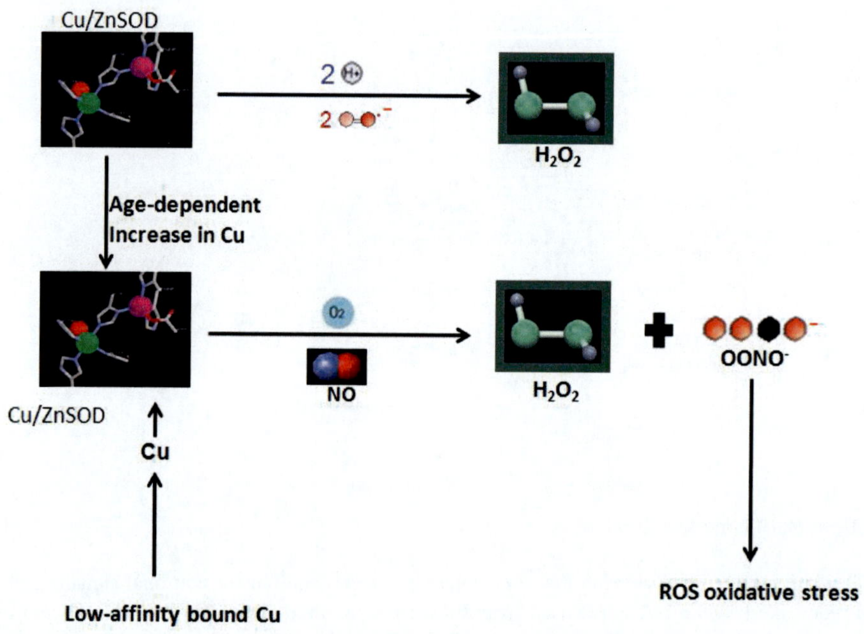

Figure 4. Oxidative stress in amyotrophic lateral sclerosis. The normal function of superoxide dismutase (SOD) is to convert toxic superoxide radicals into H_2O_2 that are subsequently inactivated by catalase. In amyotrophic lateral sclerosis this antioxidant protein is converted into a pro-oxidant protein. With age-dependent increases in copper levels (Cu), low-affinity copper sites on SOD such as Cys111 are occupied; these sites are redox active and give rise to aberrant redox chemistry and subsequent reactive oxygen species (ROS) generation.

Mutations of SOD, a cupro-enzyme that detoxifies the ROS superoxide, can convert the protein from an anti-oxidant to a pro-oxidant capable of causing oxidative insults. Evidence that inappropriate metal-mediated redox chemistry is central to the progression of ALS includes the observation that Cu chelators inhibit the course of the disease in both cell culture and mouse models (Cherny et al. 2001). Moreover, Cu can be incorrectly incorporated into the Zn site (Goto et al. 2000), and *in vitro* studies (Liu et al. 2000) with an H46R mutant SOD linked to familial ALS, and which has no SOD activity, have shown that a surface-exposed cysteine residue in SOD is also capable of coordinating Cu and is redox active. As the concentration of Cu rises with age (Lovell et al. 1998), the possibility that these lower-affinity metal-binding sites are occupied is increased and the hypothesis that these sites might be responsible for aberrant redox chemistry, and the generation of ROS and subsequent toxicity, remains untested (Bush, 2002) (Figure 4).

3.3.5. Oxidative Stress in HD

Unlike the neurodegenerative disorders discussed above, Huntington disease is entirely genetically determined. The condition is autosomal dominant and results in deterioration of neurons in the caudate and putamen with a late onset of action and progressive development of behavioral abnormalities, cognitive impairment, and involuntary jerky movements that gave rise to its historical name, Huntington chorea. All these clinical features are physical manifestations of a mutation in a gene on chromosome 4 encoding the huntingtin protein. Despite identification of the genetic defect, the role of the mutant huntingtin protein in neuronal degeneration remains unclear, as is the factor underlying the selective vulnerability of striatal neurons. Most recent experimental evidence links the pathogenesis of cell death in HD to a "gain of function" of the mutant huntingtin protein that appears to involve energetic defects, oxidative damage, and excitotoxicity (Estrada et al. 2008). Evidence for oxidative stress is provided by immunocytochemical studies showing increases in the levels of a number of markers of oxidative damage in HD brain, including nitrotyrosine, MDA adducts, and 8-hydroxydeoxyguanosine (8-OHdG), as well as markers of cellular response to oxidative stress such as $OH^{\bullet}$ (Browne and Beal, 2006; Ryu et al. 2006). In addition, transgenic animal models for the disease have been developed that exhibit much of the classical human phenotype, and are characterized by free radical production. Although there is now abundant evidence for oxidative damage, whether oxidative stress is a causative element of neurodegeneration or a secondary factor of the cell death cascade has so far been difficult to determine.

Oxidative damage does not occur in isolation but participates in the complex interplay between excitotoxicity, apoptosis and inflammation. Free radicals alter the structural and functional integrity of cells by a variety of mechanisms, including lipid peroxidation, sulfhydryl oxidation, proteolysis and shearing of the nuclear material, thereby causing the severe cell injury, and subsequently lead to cell death by neither necrosis nor apoptosis. Furthermore, free radicals produced can damage the mitochondrial respiratory components, which further increase the free radical production, which can potentiate the whole damaging effect. Therefore, modulation of oxidative stress could be a potential therapeutic approach for various disorders.

3.4. Mitochondrial Alterations

One of the cellular mechanisms for resistance to metal ion toxicity includes their sequestration within the organelles. Large amounts of various metal ions may be accumulated by mitochondria. Some metal ions interact with important functional groups of a variety of enzymes in the matrix or in the inner mitochondrial membrane.

Mitochondria are key regulators of cell survival and death and a dysfunction of mitochondrial energy metabolism leads to reduced ATP production, impaired Ca^{2+} buffering, and increased generation of ROS (Figure 5) (Beal, 2005). Increased production of ROS damages cell membranes through lipid peroxidation and further accelerates the high mutation rate of mitochondrial DNA (mtDNA) (Petrozzi et al. 2007). Accumulation of mtDNA mutations enhances normal aging, leads to oxidative damage, causes energy failure and increased production of ROS, in a vicious cycle. As we mentioned above, the brain is especially vulnerable to oxidative damage because of its high content of easily peroxidizable unsaturated fatty acids, high oxygen consumption rate, and relative paucity of antioxidant enzymes compared with other organs (Nunomura et al. 2006). Further, conditions of excessive oxidative stress and mitochondrial Ca^{2+} overload can favor mitochondrial permeability transition pore (MPT) state in which the proton-motive force is disrupted (Crompton, 2004). The ultimate result of the activation of the MPT pore is the release of cytochrome c and the induction of caspase-mediated apoptosis (Stavrovskaya and Kristal, 2005).

The mtDNA is particularly susceptible to oxidative damage. Because of its vicinity with ROS source (electron transport chain) and because it is not protected by histones and is inefficiently repaired, mtDNA shows a high mutation rate. In the last decades, it was hypothesized that somatic mtDNA mutations acquired during ageing contribute to the physiological decline that occurs with ageing and ageing-related neurodegeneration (Lin and Beal, 2006).

As we mentioned above, dysfunction of the mitochondria leads to a deceased ATP production, impaired intracellular Ca^{2+} buffering and the generation of ROS. Elevation of the concentration of free cytosolic Ca^{2+} is critical for many types of neuronal cell death (Figure 5). Under Ca^{2+} overload due to activation of Ca^{2+} permeant excitatory amino acid (EAA) receptors play a central role in initiating the progression to neuronal death via excitotoxicity. Ca^{2+} overload can lead to the opening of the mitochondrial MPT and to the deposition of Ca^{2+} phosphate precipitates. High Ca^{2+} concentration may also increase the rate of mitochondrial ROS generation. PD is an example of neurodegenerative disease, which manifests increased susceptibility of neurons to excitotoxic cell death. The pathogenesis of PD remains unclear but there is increasing evidence that impairment of mitochondrial functions, oxidative damage and inflammation are contributing factors. Drugs that target the mitochondria may therefore represent the best hope for disease modifying therapies in PD. Substantial evidence suggests complex I deficiency in PD. Complex I defect may result in oxidative stress. BJ-1 mutation, PINK-1 gene mutation and MPT opening are also implicated in PD. Familial multisystem degeneration with parkinsonism is associated with the 11778 mtDNA mutation. The role of mitochondria is further confirmed from the fact that therapies targeting inflammation and mitochondrial dysfunction are efficacious in PD (Trushina and McMurray, 2007).

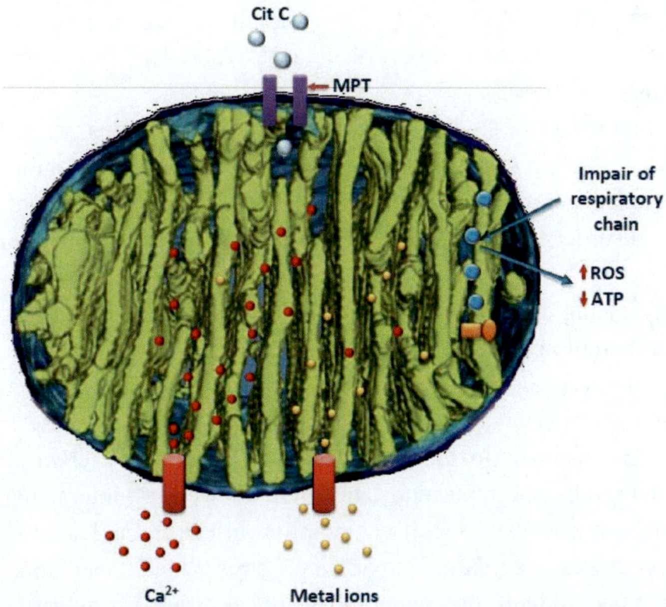

Figure 5. Potential targets for ROS-generated oxidative damage in mitochondria. Miochondria-mediated neuronal dysfunction could be caused by disruption of the mitochondrial membrane potential, sensitized MPT, excitotoxin-induced Ca^{2+} influx, and diminished ATP production. In experimental models, mitochondrial toxins (such as metals) impair respiratory enzyme activities and ATP generation and increase ROS. This is consistent with impaired electron transport chain activities in some neurodegenerative diseases. In addition, the release of cytochrome c from damaged/dysfunctional MPT triggers the activation of apoptotic cascade and release of initiator and executioner caspases (caspase-9 and -3), resulting in neuronal cell death. Calcium channels and ATP synthesis are also affected.

The defective ATP production and increased $O_2^{\cdot-}$ radicals may induce mitochondria dependent cell death because damaged mitochondria are unable to meet the energy demands of the cell. A deficiency in the terminal complex of the mitochondrial electron transport chain, cytochrome oxidase, in platelet mitochondria has also been reported in AD. AD brain mitochondria demonstrated a generalized depression in activity of all electron transport chain complexes and have increased sensitivity towards oxygen free radicals (Manash et al. 2006; Trushina and McMurray, 2007).

3.5. Common Mechanisms of Neurotoxicity

Most of the toxic metals exist as cations, and as such, can react with most ligands present in living cells. These include such common ligands as sulphydril, phosphate, amino and carboxyl. Thus they have the potential to inhibit enzymes, disrupt cell membranes, damage structural proteins, affect the genetic code in nucleic acids and promote cell death (Clarkson, 1987). There are two types of cell death, necrosis and apoptosis, the presence of either kind of cell death depends on the type and duration of the aggression and is preceded by multiple mechanisms such as excitotoxicity, glial activation and inflammation. As we mentioned above, mitochondria is the most susceptible organelle, and it seems that relatively mild mitochondrial injury, where ATP levels are maintained near normal, results in mainly apoptotic cell death. More extensive injury that causes ATP depletion shifts the form of cell death toward necrosis.

Recent data obtained from studies of human postmortem brains, as well as animal and cell culture models, suggest that apoptosis and other forms of programmed cell death contribute to neurodegeneration (Heidenreich, 2003). Necrosis may also play a role in neurodegenerative disorders but is probably more important in acute and extreme insults to the brain.

There is an increasing amount of experimental evidence that oxidative stress is a causal factor in the neuropathology of several adult neurodegenerative disorders, as well as in stroke, trauma, and seizures. At the same time, excessive or persistent activation of glutamate-gated ion channels may cause neuronal degeneration in these same conditions. Thus, two broad mechanisms--oxidative stress and excessive activation of glutamate receptors--are converging and represent sequential as well as interacting processes that provide a final common pathway for cell vulnerability in the brain. The broad distribution in brain of the processes regulating oxidative stress and mediating glutamatergic neurotransmission may explain the wide range of disorders in which both have been implicated. Yet differential expression of components of the processes in particular neuronal systems may account for selective neurodegeneration in certain disorders (Figure 6).

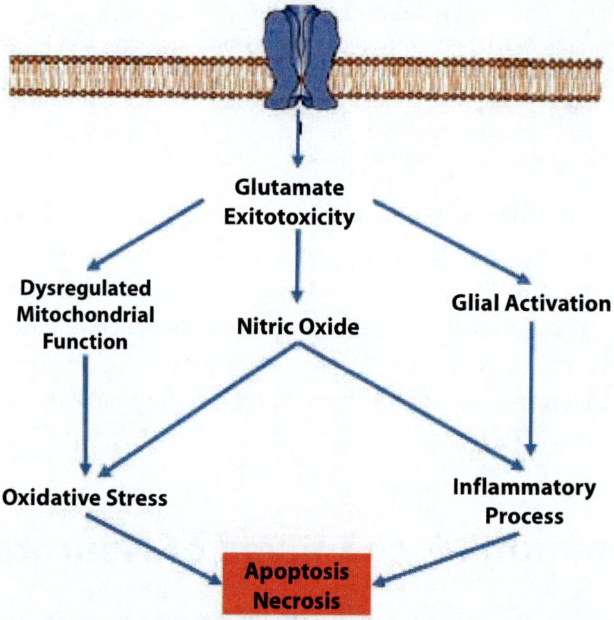

Figure 6. Common pathways of neurotoxicity.

3.5.1. Excitotoxicity

The broad spectrum of acute and chronic neurologic disease appears to preclude the existence of a common mechanism of pathogenesis. In the last two decades, however, a substantial body of work has suggested that the mechanism of neuronal death in many acute and chronic neurologic diseases may have important common elements. In particular, speculation now focuses on the role of excitotoxins, endogenous compounds that act via excitatory amino acid neurotransmitter receptors, as instruments of neuronal death in both acute and chronic neurologic diseases.

After its release from synaptic terminals, glutamate activates three different receptor subtypes in postsynaptic neurons: N-methyl-D-aspartate (NMDA) receptors; non-NMDA receptors, sensitive to α-amino-3-hydroxy-5-methyl-4-isoxazolepropionic acid (AMPA) and kainic acid; and metabotropic receptors. Activation of AMPA receptors induces the influx of sodium ions and the subsequent depolarization of the plasma membrane, promoting the extrusion of the magnesium ion normally blocking the NMDA receptor channel. Metabotropic receptors are coupled to G-proteins and induce the activation of second messenger systems such as inositol-3-phosphate, triggering the release of Ca^{2+} from the endoplasmic reticulum (Estrada et al. 2008).

Homeostasis of glutamatergic neurotransmission depends on the coordinated activity of its different components, and failure in any one of them can lead to excitotoxic neuronal damage. Impairment of glutamate removal after its synaptic release leads to the accumulation of the amino acid in the synaptic cleft. The subsequent sustained activation of glutamate receptors may trigger excitotoxic neuronal death, in particular during ATP limiting conditions (Massieu et al. 2003).

This type of cell death is dependent on the increase in the intracellular concentration of Ca^{2+} after its influx through the NMDA receptor channel. Calcium-activated enzymes such as proteases, endonucleases and phospholipases contribute to the degradation of different cell components and neuronal death (Figure 7). Intracellular concentration of Ca^{2+} is regulated by energy-dependent systems (cytoplasmic and endoplasmic reticulum Ca^{2+}-ATPases). Massive influx of Ca^{2+} ions during excitotoxicity leads to intramitochondrial Ca^{2+} overload, altering its activity and disrupting ATP production (Dugan and Kim-Han, 2006). As a consequence, ATP-dependent Ca^{2+} loading and extrusion mechanisms are impaired leading to a sustained increase in intracellular Ca^{2+}. Another key event involved in excitotoxic death is the generation of free radicals, as a result of mitochondrial dysfunction and the activation of Ca^{2+}-dependent enzymes such as xanthine oxidase and nitric oxide synthase (Lafon-Cazal et al. 1993).

In the last few decades, excitotoxic cell death has been intensively studied, using both intact and isolated neuronal systems. In cortical cell cultures, experimental evidence showed that this process occurred in two main patterns: (i) rapidly triggered excitotoxicity induced by brief, intense stimulation of large numbers of NMDA receptors, and (ii) slowly triggered excitotoxicity induced by prolonged stimulation of AMPA/kainic acid receptors (Choi, 1992). Indeed, based on time of observation, the injury by brief and intense exposure to excitatory amino acids can be distinguished into two components such as acute and delayed mechanism of excitotoxicity (Choi, 1991).

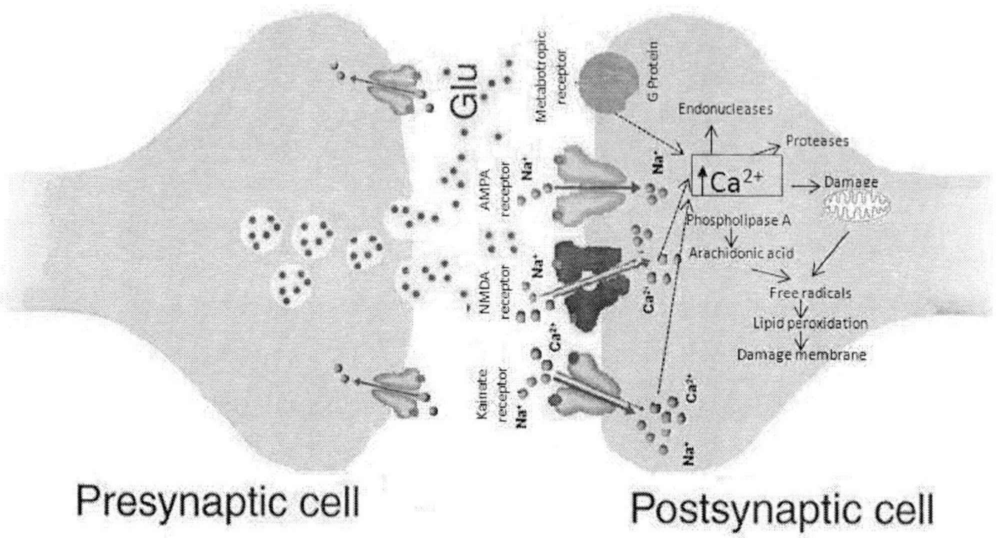

Figure 7. Mechanisms involved in excitotoxic cell death.

In all instances, both morphological and biochemical findings from different experimental models suggest that the exposure to glutamate agonists produces a loss of cellular volume homeostasis, leading to delayed neurodegenerative damage associated with the degeneration of organelles and pyknosis. Nevertheless, an accumulation of evidence demonstrated the occurrence of different mechanisms, such as the apoptotic processes under excitotoxic insults (Tenneti and Lipton, 2000).

Thus, depending on experimental conditions and cell types, excitotoxic neuronal death may either be apoptotic or necrotic (Leist and Nicotera, 1998). Furthermore, it has been suggested that in cortical cell cultures the exposure to relatively short duration or low concentrations of NMDA induces cell death characterized by apoptotic features; in contrast, the intense exposure to high concentrations of glutamate agonist produces cell necrosis (Bonfoco et al. 1995).

3.5.2. Apoptosis

Apoptotic cell death should be considered as an ongoing, *normal* event in the control of cell populations. However, apoptosis can also be induced by a variety of toxicants, including many of the inorganic compounds, resulting in the loss of affected cell populations. Apoptosis essentially occurs when cellular damage, including damage to genetic material, has exceeded the capacity for repair. What has often received less attention is the concept that environmental agents, including metals, can impair apoptosis, and that suppression of the apoptotic response could facilitate aberrant cell accumulation, which may be a critical step in the pathogenesis of malignancy or autoimmunity. Thus, for many toxic metals, disorders of cell accumulation may be a crucial aspect of their toxicity, but these disorders have just recently begun to be recognized.

Apoptosis is characterized by nuclear chromatin condensation/fragmentation, cell shrinkage, disruption of mitochondria, and formation of apoptotic bodies (Kerr et al. 1995). At present, in addition to morphologic features, apoptosis is defined by complex biochemical processes involving mitochondria, activation of a family of cysteine-aspartate proteases (caspases) that cleave to various proteins, and endonuclease-induced nuclear fragmentation (Figure 8) (Nagata, 2000).

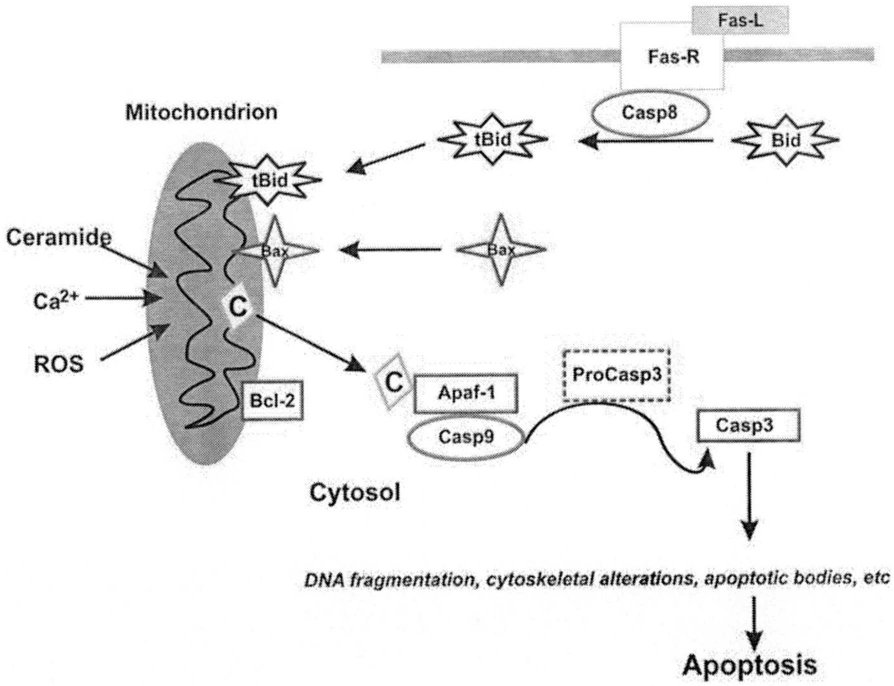

Modified from Fiskum et al. 2003.

Figure 8. Mitochondrial participation in apoptosis. Agents, such as Ca^{2+}, ROS and ceramide, as well as proapoptotic proteins, such as Bax, stimulate the release of other proapoptotic proteins—for example, cytochrome c (C in figure)—from the mitochondrial intermembrane space into cytosol. Cytochrome c and procaspase 9 together with apoptosis activating factor 1 (Apaf-1) from multiprotein complex (apoptosome) that activates caspase 9, which then cleaves procaspase 3, forming active caspase 3. This caspase, together with other caspases activates, proteolytically degrades a variety of proteins, causing the molecular and morphological alterations characteristic of apoptosis. Cell surface "death receptors" such as Fas, can activate caspase 3 directly through activation of caspase 8 (not shown). The antiapoptotic protein Bcl-2 is capable of inhibiting the release of cytochrome c mediated by either Ca^{2+} or Bax.

There are two principal pathways leading to apoptotic cell death. These include the "extrinsic" or death receptor-initiated pathway and the "intrinsic" or mitochondrial pathway (Figure 9). The extrinsic pathway originates with binding of death-promoting ligands (such as Fas ligand, FasL) to their cognate death receptors (such as Fas). There is a large family of death ligands and death receptors, and the FasL/Fas system is outlined as a prototype. Ligand binding induces oligomerization of death receptors and promotes their association with adapter molecules such as Fas-associated death domain protein (FADD) (Chinnaiyan et al. 1996). The receptor-FADD interaction occurs via protein-protein binding pattern known as the death domain (Feinstein et al. 1995). The initiator caspase, procaspase 8, is then recruited to the death-inducing signaling complex via binding to the death effector domain of FADD. The resulting proximity of multiple procaspase-8 molecules facilitates their autocatalytic cleavage to the active protease caspase-8 (Muzio et al. 1998; Earnshaw et al. 1999).

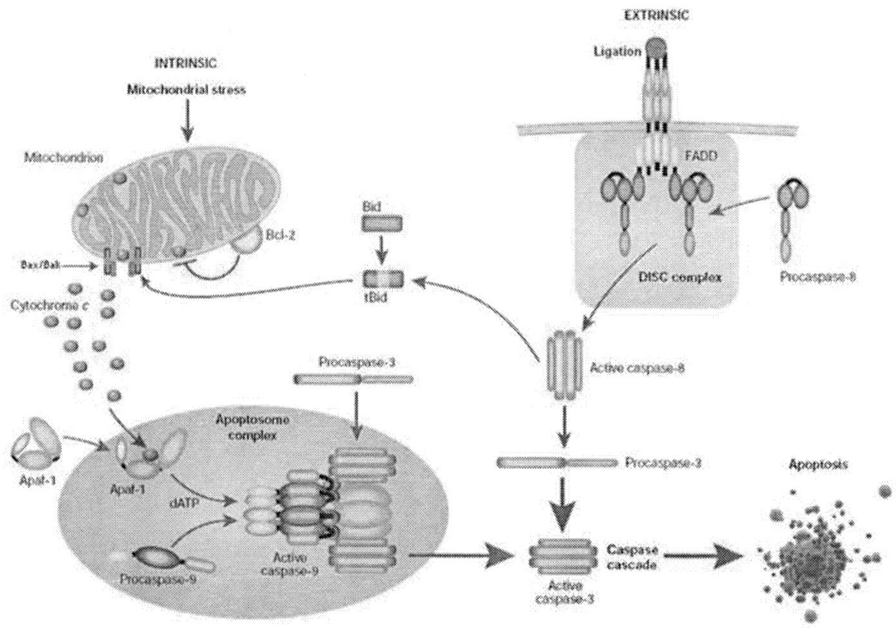

Modified from Heindenreich, 2003.

Figure 9. Apoptotic extrinsic and intrinsic pathways.

The intrinsic pathway is initiated by the release of cytochrome c from mitochondria and its subsequent association with apoptosis activating factor-1 (Apaf-1) and procaspase-9 (Cai et al. 1998; Wang et al. 2001). This large protein complex (the apoptosome) promotes the activation of caspase-9 (Zou et al. 1999). Each of the above initiator caspases, 8 and 9, cleave downstream executioner caspases, such as caspase-3, from the pro-form to the active protease. Activation of the executioner caspases then results in the cleavage of critical cellular proteins and apoptosis (Heidenreich, 2003). Among activated proteases, caspase-3 has been shown to be required for DNA fragmentation and morphologic changes of apoptosis (Jänicke et al. 1998).

The intrinsic death pathway is regulated by both pro- and antiapoptotic members of the Bcl-2 family (Tsujimoto, 1998). Bax and Bak are pro-apoptotic members of the Bcl-2 family that appear to serve a redundant function in making the mitochondrial membrane permeable to cytochrome c (Wei et al. 2001). Cytochrome c release from mitochondria occurs by formation of Bax- or Bak containing "pore" in the outer mitochondrial membrane that permits passage of small proteins (Polster et al. 2001). The proapoptotic function of Bax is attenuated by antiapoptotic members of the Bcl-2 family (Bcl-2, Bcl-X_L) that heterodimerize with Bax and sequester it away from mitochondria (Otter et al. 1998). Conversely, BH3 domain-only Bcl-2 family members, including Bim, Bid, Dp5/Hrk, and Bad, promote the proapoptotic effects of Bax by binding to Bcl-2, thus freeing Bax to incorporate to mitochondrial membrane (Zong et al. 2001).

Besides the sequestration of Bcl-2 away from Bax by BH3-only proteins, another critical step required for initiation of cytochrome c release is the active translocation of Bax from the cytoplasm to mitochondria through the opening of the mitochondrial permeability pore. Some apoptotic stimuli are capable of opening the mitochondrial pore resulting in disruption of the mitochondrial membrane potential, a decline in ATP production, and entry of solutes and water into the mitochondrial matrix. Ultimately, mitochondrial swelling and rupture of the outer membrane can occur (Heidenreich, 2003).

Aberrant apoptotic mechanisms are thought to contribute significantly to many neurodegenerative disorders. Recent findings indicate that components of both the extrinsic (death receptors ligands) and intrinsic (Bcl-2 family members) death pathways are regulated at the level of expression during neurodegeneration or neuronal injury *in vivo* (Kitamura et al. 1998). Moreover, transgenic animal models or spontaneously occurring mutants of specific death receptor signaling molecules or Bcl-2 family members provide further evidence that these pathways are involved in neuronal injury (Heidenreich, 2003).

Concerning metals-inducing apoptosis, it has been demonstrated that caspase-3-mediates Zn^{2+}-induced apoptosis (Schrantz et al. 2001) and Jimenez Del Rıo and Velez-Pardo (2004) provide evidence that redox-active Fe^{2+}, Mn^{2+}, Cu^{2+}, and Zn^{2+} ion-induced apoptosis in lymphocytes by $H_2O_2/OH^{\cdot}$ generation, resulting in mitochondria depolarization, caspase-3 activation, and nuclear fragmentation. These authors suggest that metals might trigger alternative pathways of death in a specific-cell fashion. In this way, it has been reported that excessive tissue accumulation of redox-active metals can be cytotoxic, in particular because perturbations in metal homeostasis result in an array of cellular disturbances characterized by oxidative stress and cell death. The biological importance of this disarray can be seen in recessive autosomal neurodegeneration with brain Fe accumulation 1 disease (NBIA 1, formerly Hallenvorden–Spatz syndrome, OMIM: 234200) (Galvin et al. 2000), and in dominant autosomal neuroferritinopathy disorder (NFP) (Curtis et al. 2001), characterized both by excessive accumulation of Fe in basal ganglia and with progressive movement disorder. Interestingly, these two neuropathologic features are also present in PD patients (Dexter et al. 1989; Takanashi et al. 2001). These data clearly suggest that Fe may be a primary or secondary event in the process that leads to neuronal death. Similarly, Cu (Hoogenraad and Van den Hamer, 1996) and Mn (Yamada et al. 1986) accumulation might be also associated with the neurodegenerative process (Waggoner et al. 1999). Because of these findings, Jimenez Del Rıo and Carlos Velez-Pardo (2004) suggest that metal ions such as Fe^{2+} and Mn^{2+} in presence of O_2, as we mentioned above, may directly react with Fe^{2+} or Mn^{2+} to produce OH· by Fenton reaction. Overproduction of $OH^{\cdot}$ may then be able to alter mitochondria transmembrane potential to induce liberation of different apoptogenic factors and subsequent activation of caspase-3 resulting in disassembly and fragmentation of nuclear chromatin, typical of apoptosis. Concerning Cu chemistry, one should bear in mind that Cu^{2+} ions must first be reduced (e.g., by available intracellular reductants such as vitamin C and/or reductases) into Cu^{+}, which in turn might reduce molecular O_2 into $O2^{\cdot-}$ and finally react with H_2O_2 via Fenton reaction. Interaction of Zn with oxygen is less clear, although it is able to generate H_2O_2 and $OH^{\cdot}$ (Lee et al. 2002). In fact, they demonstrated that $OH^{\cdot}$ is the central mediator in metal-induced apoptosis and that this latter event is independent on NF-κB and p53 transcription. The significance of these results is the ability to directly monitor metals and their association with $H_2O_2/OH^{\cdot}$ generation provides evidence for understanding death mechanisms in metabolic as well as in neurologic disorders, wherein deregulation of metal ions serves either as catalyzer or direct effector of oxidative stress.

3.5.3. Necrosis

One of the major reasons that the importance of apoptosis in neurodegeneration is questioned is the overlap between apoptosis and necrosis in neuronal biology. Necrotic cells have swollen nuclei, swollen mitochondria, and loss of plasma membrane integrity. Necrosis is a passive form of cell death that typically results from injury or from excess calcium influx during excitotoxicity. Excitotoxicity occurs in multiple pathological situations including ischemia, seizures, and head trauma or by the exposure to excitotoxins such as metals. The dividing line that separates necrosis from apoptosis has been emphasized for years owing to the clear distinct features that classify both events. However, death in neurons can be biphasic, beginning with necrosis and then showing delayed apoptosis. Moreover, if apoptosis is blocked, for instance by overexpressing the neuroprotective transcription factor NF-κB, the mode of cell death often simply switches from apoptosis to necrosis (Furukawa and Mattson, 1998). The ability of neurons to switch from apoptotic death to necrotic death raises the possibility that antiapoptotic treatments, such as caspase inhibitors, will block apoptosis but not prevent cell death. Moreover, in neurodegenerative disorders such as AD, Aβ is able to kill via multifaceted pathways, which raises the possibility that both types of cell death contribute to the neurodegenerative events (Wolozin and Behl, 2000).

It is important to note that, contrary to apoptosis, necrosis does not involve the mobilization of molecular mechanisms that evolved to specifically facilitate cell death. Rather, death is produced by cellular mechanisms that operate within the cell under normal conditions; under exceptional conditions or when extensive damage is inflicted, these mechanisms turn "rogue" and demolish the cell. In addition to understanding the physiological functions of these biochemical pathways, it is necessary to elucidate the means by which these are deregulated.

Acute and chronic injury to the nervous system triggers a large network of morphologic and metabolic changes that play a role in two crucial physiological processes: protection against infectious agents and repair of the damaged tissue (Raivich et al. 1996). The injured neurons assume a state of emergency, rapidly change their gene expression and stimulate nearby microglia and astrocytes for support (Raivich et al. 1995). This activation of microglia and astrocytes is a graded, stereotypic response, which is commonly observed in stroke and ischemia, in neurodegenerative diseases, after direct or indirect axonal injury or during inflammation due to infectious or autoimmune disease (Kreutzberg, 1996). They are accompanied, also in a graded manner, by production of proinflammatory cytokines, functional changes in brain vascular endothelia and a recruitment of cells of the immune system into the damaged tissue.

3.5.4. Glial Activation

Neuroglia represented by astrocytes, oligodendrocytes and microglial cells provide for numerous vital functions (see chapter 1). Glial cells shape the micro-architecture of the brain matter; they are involved in information transfer by virtue of numerous plasmalemmal receptors and channels; they receive synaptic inputs; they are able to release 'glio'transmitters and produce long-range information exchange; finally they act as pluripotent neural precursors and some of them can even act as stem cells, which provide for adult neurogenesis. Recent advances in gliology emphasised the role of glia in the progression and handling of the insults to the nervous system. The brain pathology, is, to a very great extent, a pathology of glia, which, when falling to function properly, determines the degree of neuronal death, the outcome and the scale of neurological deficit. Glial cells are central in providing for brain homeostasis. As a result, glia appears as a brain warden, and as such it is intrinsically endowed with two opposite features: it protects the nervous tissue as long as it can, but it also can rapidly assume the guise of a natural killer, trying to eliminate and seal the damaged area, to save the whole at the expense of the part (Figure 10) (Giaum et al. 2007).

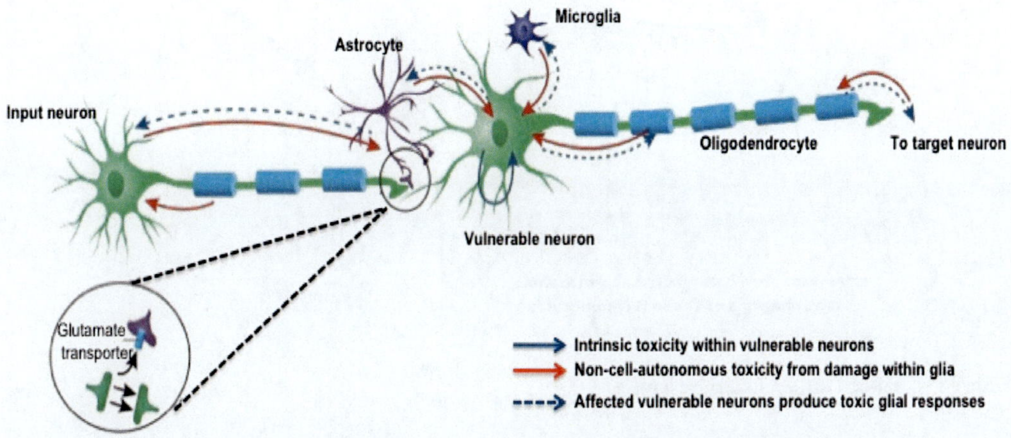

Modified from Lobsiger and Cleveland 2007.

Figure 10. Insights from animal models of diverse human neurodegenerative diseases indicate that disease mechanisms are non-cell-autonomous, requiring the convergence of damage within the vulnerable neurons and their neighboring glial cells. Glia-derived toxicity can strongly influence disease progression (for example, in ALS) or can even contribute to disease initiation. Glutamate-mediated excitotoxicity is a prime example of neuron-glia toxicity that has been proposed to be a significant component in ALS HD, PD, and MS.

Common pathways of neuronal cell death in response to diverse insults, include early disruption of ion homeostasis, increased release and impaired uptake of neurotransmitters (such as glutamate), excessive neuronal activation, cellular swelling, intracellular entry of divalent cations, and release of nitric oxide and free radicals. These changes in cell physiology lead to both apoptotic and necrotic cell death, and set in motion the development of a gliotic scar (Ankarcrona et al. 1995; Back and Schuler, 2004).

The first cell type to respond to injury is the microglia. On activation, these CNS macrophages phagocytose apoptotic cells and necrotic debris; release proinflammatory cytokines, chemokines, and reactive nitrogen species; and up-regulate surface expression of specific receptors, such as major histocompatibility complexes I and II (Figure 11). Microglial release of cytokines, such as tumor necrosis factor, interferon, chemokines and interleukin-8, employ peripheral white blood cells to the site of damage (Mount et al. 2007).

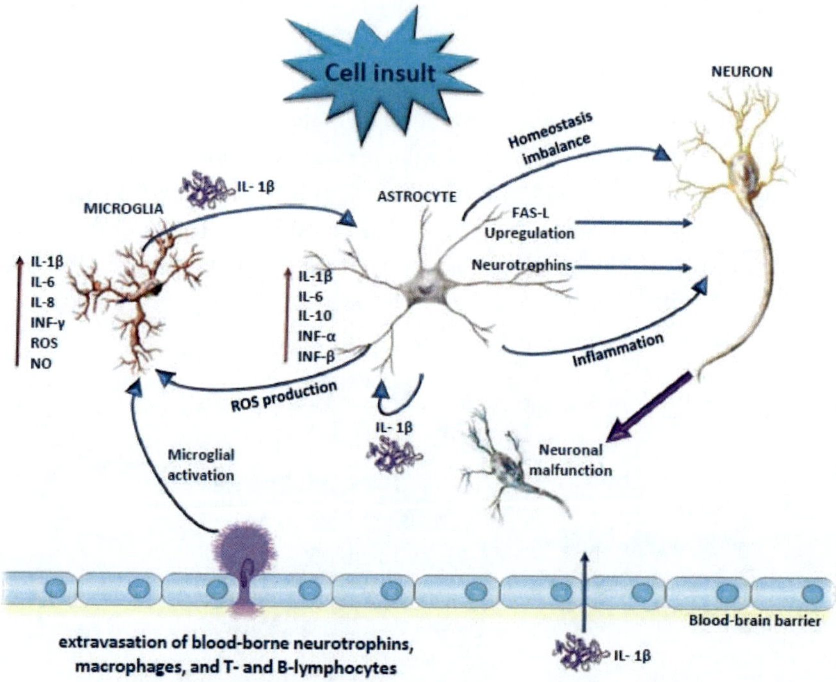

Figure 11. Glial-cell inflammatory response.

Breakdown of the BBB, which is composed of endothelial cells and astrocytes, occurs concurrently with microglial activation. This breakdown appears in response to the release of various cytokines, ROS, glutamate, ATP, bradykinins, histamine, and nitric oxide from neurons, activated microglia, and the endothelial cells themselves. Breakdown of this barrier facilitates the translocation of plasma-derived molecules into the brain (Ballabh et al. 2004). Several studies suggest that this influx of blood-derived molecules is a critical step in the formation of a glial scar. Consistent with this notion, areas of greatest glial scarring are often found near regions of the largest BBB breakdown (Silver and Miller, 2004). Moreover, breakdown of the BBB contributes to the posttraumatic inflammatory response by increasing extravasation of blood-borne neurotrophins, macrophages, and T- and B-lymphocytes, which may trigger further brain damage (Ballabh et al. 2004).

Oligodendrocyte precursor cells (OPC) are recruited by inflammation to the site of injury within 2 days of the injury, and their numbers increase for the following 2 weeks (Levine, 1994). Although these cells are activated by neuronal damage, proliferation requires at least some demyelination of neurons (Di Bello et al. 1999). Because of the close proximity of OPCs to synapses and nodes of Ranvier, as well as the involvement of OPCs in excitatory transmission, it is not clear whether OPCs are responding to a growth factor released by myelin sheath breakdown or whether OPCs are sensitive to changes in neuronal conduction (Bergles et al. 2000).

Astrocytes outnumber neurons by five- to tenfold in the adult brain (Bignami, 1991) and establish numerous small contacts with neurons, neighboring astrocytes, and all the other brain cell types, including the endothelial cells of blood vessels (Rohlmann and Wolf, 1996). These physical interactions allow them to function as metabolic and passive supportive cells of the brain. Astrocytes can also respond to neuronal activity by clearance of glutamate (Glu) from the extracellular space. Specific astrocytic Glu transporters that are predominantly coupled to Na^+-dependent systems mediate astrocytic Glu uptake (for review see Anderson and Swanson, 2000). Glu is transformed into glutamine (Gln) by a Mn-dependent enzyme called glutamine synthetase. Gln in turn can either be converted into GABA by the enzyme glutamic acid dehydrogenase, or be turned back to Glu by glutaminase (Quintanar, 2008).

In response to injury, astrocytes undergo many cellular changes, leading them to adopt a reactive phenotype. As with microglia, the astrocytic response to injury proceeds through several stages and depends on the extent of trauma. Soon after injury, there is a rapid increase in the synthesis of GFAP that can extend far from the actual site of damage (Ridet et al. 1997). This is followed by the appearance of small and slender GFAP-positive processes, which in several days become fully stellarized fibrillary astrocytes. Long-standing hypotheses suggest that reactive astrocytes create a physical barrier between damaged and healthy cells and re-establish an intact BBB (Faulkner et al. 2004). However, given the wide array of signaling systems involved in the astrocyte response to injury, it seems likely that additional roles will emerge. Astrocytes are capable of producing a variety of cytokines, including interleukins (IL-1, IL-6, IL-10), and interferons (IFN-α, IFN-β), tumor necrosis factors (TNF-α, TNF-β), and a variety of growth factors (fibroblast growth factor, platelet-derived growth factor, nerve growth factor, and EGF) (see figure 11) (Dong and Benveniste, 2001; Lau and Yu, 2001).

Astrocytes, in conjunction with microglia, respond to neuronal injury and undergo a series of metabolic and morphological changes that are known as reactive gliosis or astrogliosis (Schipper, 1996). Increased number of activated microglia and enlarged and phagocytic cells that express the cytokines are prominent in reactive gliosis (Mrak et al. 1997). Concomitantly with the proliferation of microglial cells, there is hypertrophy of astrocytes and a marked variation in the expression of cytoplasmic antigens [glial fibrillary acidic protein (GFAP) and vimentin], surface proteins (PSA-NCAM), and growth factors (CNTF; Ridet et al. 1997).

There is considerable and growing evidence that chronic glial activation plays a major role in numerous neurological conditions including AD, PD, ALS, strokes, and inflammatory brain diseases. The release of toxic elements from activated glia, such as cytokines and excitotoxins, is known to produce neurodegeneration.

3.5.5. Inflammation

Inflammatory events are likely to contribute to the pathogenesis of many disorders such as AD, PD, and ALS. Environmental exposures appear to exacerbate the endogenous heightened CNS inflammation that is a result of normal aging and by doing so may accelerate neuronal cell loss (Figure 9).

Environmental factors that can trigger inflammatory events in the CNS are lipopolysaccharide, metals, and particulate matter present in air pollution. These factors may enhance existing age-related inflammation in the CNS and thus accelerate neuronal toxicity (Campbell, 2004).

3.5.6. Cytokines

Cytokines have been implicated as mediators and inhibitors of diverse forms of neurodegeneration. They are induced in response to brain injury and have diverse actions that can cause, exacerbate, mediate and/or inhibit cellular injury and repair.

Cytokines, a diverse group of polypeptides that are generally associated with inflammation, immune activation, and cell differentiation or death, include Interleukins (IL), Interferons (IFN), tumor necrosis factors (TNF), Chemokines and growth factors. They have diverse actions, and most have little or no known function in healthy tissues, but are rapidly induced in response to tissue injury, infection or inflammation. Their involvement in CNS disease is a rapidly growing area of biological and clinical research (Barone and Feuerstein, 1999).

Cytokines such as IL-1, IL-6, and IL-8 are primarily synthesized by activated microglia and macrophages in response to pathogens and trauma (Figure 11). Chronic production of these factors can result in cytotoxicity because they recruit and activate macrophages that produce high concentrations of ROS (Dunn, 1991). IL-1 and IL-6 are both elevated in the brain in some neurodegenerative diseases and postischemia (Cadman et al. 1994). In an animal model of chronic inflammation, induced by infusion of lipopolysaccharide, there was astrogliosis as well as an increase in the levels amyloid-β protein precursor, IL-1, and TNFα mRNA levels. This was subsequently followed by hippocampal cell loss and impairment of spatial memory, all of which mirror changes seen in the AD brain (Wegrzyniak et al. 1998).

Cytokine Actions on Neurons

Direct actions of cytokines on neuronal functions (for example, transmitter release and ion channel activity) can contribute to neuronal injury.

In vivo IL-1 and TNFα acutely enhance neuronal injury, yet TNFα induces the expression of the BCL proteins Bcl2 and Bclx in hippocampal neurons *in vitro* through NFκB activation, which protects against hypoxic injury (Tamatani et al. 1999).

Cytokine Actions on Glia

Neuronal survival is critically dependent on glial function, which can exert both neuroprotective and neurotoxic influences (Giulian et al. 1993). Glial cells are a primary target of cytokines and are activated in response to many cytokines, including TNFα and IL-1. This activation can trigger further release of more cytokines that might enhance or suppress local inflammatory responses and neuronal survival.

TNFα directly affects astrocytes, inducing a slow increase in intracellular Ca^{2+} and marked depolarization, and reducing glutamate-evoked rises in Ca^{2+} (Koller et al. 2001), which would affect synaptic transmission indirectly. The contribution of IL-1 to neurodegeneration might depend on the balance between neuroprotective and neurotoxic factors released from glia. IL-1 induces NGF expression and release in astrocytes (Carman-Krzan et al. 1991), which exerts neuroprotective actions *in vitro* (Strijbos and Rothwel, 1995). Giulian et al. (1993) suggest that IL-1 releases neurotoxins from glia, and recent data indicate that conditioned medium from IL-1-treated glia is neurotoxic to cortical neurons (Allan and Rothwell, 2001). The specific nature of these neurotoxins has not been fully elucidated, although there are several candidates that include nitric oxide (Boje and Arora, 1992), quinolinic acid (Heyes and Nowak, 1990), and acute-phase proteins (for example, β-amyloid precursor protein and α1-antichymotrypsin) (Kordula et al. 2000).

Some metals such as Al enhance neuroinflammatory events in the brain. Intracerebroventricular injection of Al salts leads to activation of glial cells in the rat brain (Platt et al. 2001). Al caused an increase in TNFα production and NF-κB activation in a human glioblastoma cell line. In mice treated for ten weeks with Al lactate in the drinking water, there was a dose-dependent increase in proinflammatory cytokines (Campbell et al. 2002). In these animals, the striatum had the highest number of astrocytes and activated microglia. The use of transgenic animal models of neurodegeneration will help to determine whether aluminum-induced inflammation plays a role in potentiating pathological lesions characteristic of neurological impairments (Campbell, 2004).

3.6. Metals and the Activation of Signalling Pathways

The most significant effect of metals on signaling pathways has been observed in the mitogen-activated protein (MAP) kinase/AP-1 and NF-kB pathways (Donaldson et al. 2003). The nuclear transcription factor NF-kB, is involved in inflammatory responses and AP-1 is important for cell growth and differentiation. P53 is a gene whose disruption is associated with more than half of all human cancers. The p53 protein guards a cell-cycle checkpoint, and inactivation of p53 allows uncontrolled cell division (Meplan et al. 2000).

The activation of AP-1 and NF-kB family of transcription factors is involved in both cell proliferation and apoptosis. The concentration of radicals generated inside cells appears to influence the selective activation of these transcription factors and may therefore help explain the observation that either cell death or cell proliferation may be related to the exposure to metals (Valko et al. 2005).

The stressors, such as metals or H_2O_2 invoke a signal cascade that begins with the activation of MAP kinases family of serine/threonine kinases, regulating processes important in cell damage including proliferation, differentiation, and apoptosis. Three major subfamilies have been identified: extracellular signal-regulated kinases (ERK), c-Jun N-terminal kinases

(JNK), and the p38 kinases (Dong et al. 2002). The induction of AP-1 by H_2O_2, metals, cytokines, and other stressors is mediated mainly by JNK and p38 MAP kinase cascades (Dong et al. 2002). It is known that stressors can activate MAP kinases and thereby AP-1 in several manners.

The other mechanism involves oxidant-mediated inhibition of MAP kinase phosphatases, which leads to increased MAP kinase activation (Valko et al. 2005). Whichever mechanism prevails, activation of MAP kinases leads directly to increased AP-1 activity. The role of cellular oxidants and AP-1 activation in the cell damage is now well documented by a number of experiments (Donaldson et al. 2003). An effect of AP-1 activation is to increase cell proliferation. It has been demonstrated that c-fos and c-Jun are positive regulators of cell proliferation (Rusovici and LaVoie, 2003). Expression of c-fos and c-jun can be induced by a variety of compounds, involving reactive radicals and nongenotoxic and tumor promoting compounds (various metals, carbon tetrachloride, phenobarbital, alcohol, ionizing radiation, asbestos). In addition to affecting cell proliferation, AP-1 proteins also function as either positive or negative regulators of apoptosis (Varfolomeev and Ashkenazi, 2004). Whether AP-1 induces or inhibits apoptosis is dependent upon the balance between the pro- and anti-apoptotic target genes, the stimulus used to activate AP-1 and also on the duration of the stimulus.

A number of reports during recent years indicate that some metals are able to affect the activation or activity of NF-kB transcription factors. NF-kB is an inducible and ubiquitously expressed transcription factor for genes involved in cell survival, differentiation, inflammation, and growth (Baud and Karin, 2001).

Activation of transcription factors is clearly stimulated by signal transduction pathways that are activated by metals, H_2O_2 and other cellular oxidants. Through the ability to stimulate cell proliferation and either positive or negative regulation of apoptosis, transcription factors can mediate many of the documented effects of both physiological and pathological exposure to metals or chemicals that induce ROS and/or other conditions that favor increased cellular oxidants. Through regulation of gene transcription factors, and disruption of signal transduction pathways, ROS are intimately involved in the maintenance of concerted networks of gene expression that may interrelate with neuronal damage development (Valko et al. 2005).

3.6.1. Protein Aggregation in Neurodegeneration

A variety of neurological degenerative disorders are characterized by the presence of neuronal inclusions. These can be found as neurofibrillary tangles in AD, cytoplasmic Lewy bodies occurring in PD, diffuse Lewy bodies disorder, and neurodegeneration with brain Fe accumulation type 1. A different kind of cytoplasmic and nuclear inclusions are described in inherited neurological disorders characterized by abnormal polyglutamine repeats such as HD, inherited spinocerebellar ataxias, or spinobulbar muscular atrophy, while other inclusions occur in familial and sporadic forms of ALS. Cell inclusions in the CNS may not be limited to neurons, involving also glial cells as occurs in multiple system atrophy. The presence of neuronal inclusions is also described in neurological diseases, which are not included among degenerative disorders as transmissible prion diseases or selective types of epilepsy (e.g. Lafora bodies) (Fornai et al. 2005).

Despite many common features, these neuronal (and glial) inclusions possess a distinct shape, structure, and subcellular localization, which somehow rely on the phenotype of the affected neuron and the kind of abnormal proteins which are accumulated. In fact, neuronal inclusions found in PD (Lewy bodies) mainly involve catecholamine containing neurons, they are constantly featured by the presence of α-synuclein and segregate in the cytoplasm, while neuronal inclusions affecting GABA cells (i.e. those occurring in HD) are featured by proteins bearing abnormal amino acid chains and do not segregate in specific cell compartments, although they are present more frequently in the nucleus than the cytoplasm (Fornai et al. 2005).

There is extensive evidence for the association of protein aggregation and metals in many neurodegenerative disorders. Interestingly, metals such as Fe and Cu appear to play an important role in protein aggregation and therefore are likely to provide a link between the two pathological processes of protein aggregation and oxidative damage (Gaeta and Hider, 2005).

The cause of abnormal protein folding (misfolding) and consequent protein accumulation in the brain is still unclear. Nevertheless, genetic and environmental factors as well as age are all involved. For instance, pathogenic mutations in the genes encoding aggregating proteins, such as Aβ-amyloid, α-synuclein and PrP, are responsible for inherited forms of AD, PD and prion disease, respectively. The mutant proteins, in these examples, show an increased tendency to form the so-called amyloid-like fibrils (amyloidogenic activity), the formation of which is pathogenic. Metal ions play a crucial role, acting as mediators of neurotoxicity either by favoring plaque formation or redox cycling. Thus, they provide a suitable pharmacological target for the treatment of neurodegenerative diseases. In particular, bidentate chelators such as hydroxypyridinones and hydroxyquinolines would appear to possess the greatest potential for this purpose (Gaeta and Hider, 2005).

Despite each different type of inclusion shows typical elements, there are many immunocytochemical, biochemical and structural common features. At light microscopy most of them appear as eosinophilic inclusions (apart from Lafora bodies) in all their localizations, while at ultrastructural level they show a fibrillary structure and are not limited by membranes. All types of inclusions are formed by protein material and ubiquitin is frequently present in these aggregates, representing one major common element (Fornai et al. 2005).

Final Considerations

Neurodegenerative disorders are increasing worldwide. The etiology of this kind of disorders is unknown regardless of considerable scientific effort. Genetic variations in metal metabolism and kinetics as well as environmental and occupational exposure to metals need more attention.

Here we reviewed common pathways taking place in neurodegenerative disorders. Slowly progressive neurodegenerative diseases are probably not the result of a single event, but rather several processes involving environmental, epigenic and genetic events. Thus, the next generation of drug treatment needs to focus on combined therapies selective for several decisive events that characterize these disorders. Lowering the dilemma of protein aggregation, oxidative and nitrosative stress, mitochondrial injury, inflammatory response and heavy metal accumulation in the brain so as to re-establish neurotransmission and block excitotoxicity may prove beneficial in the treatment of several neurodegenerative diseases.

Epidemiologic evidence for an association between environmental agents and neurodegenerative disease is uncertain. The amounts of xenobiotics released into the environment are huge by any measure, and the lack of information about their effects on various physiologic systems, including neurodevelopmental processes, represents a major gap in knowledge. To close this gap, the following broad areas of research topics need attention:

a) Better health tracking and monitoring data for chronic diseases.
b) More complete and longitudinal biomonitoring of environmental agents that can be related with specific molecular/biochemical markers of exposure and subsequent health outcome data.
c) More epidemiologic research and testing of environmental agents to better describe their effects on the adult and developing brain, as well as other critical organ systems. Until such time that ethically and scientifically well-designed epidemiologic studies can provide a reasonable confidence that specific environmental agents, either alone or in combination with other agents, cause a given neurodegenerative disorder, research on the environmental contribution to neurodegenerative disease needs to continue.

References

Allan, S.M. and Rothwell, N.J. (2001) Cytokines and acute neurodegeneration. *Nat Rev Neurosci*. 2:734-44.

Alvarez, B. and Radi, R. (2003) Peroxynitrite reactivity with amino acids and proteins. *Amino Acids*. 25:295-311.

Andersen, J.K. (2004) Oxidative stress in neurodegeneration: cause or consequence? *Nat Med.* 10 Suppl: S18-25.

Anderson, C.M. and Swanson, R.A. (2000) Astrocyte glutamate transport: review of properties, regulation and physiological functions. *Glia*. 32: 1–14.

Andrews, N.C. and Levy, J.E. (1998) Iron is hot: an update on the pathophysiology of hemochromatosis. *Blood*. 92: 1845–1851.

Ankarcrona, M., Dypbukt, J.M., Bonfoco, E., Zhivotovsky, B., Orrenius, S., Lipton, S.A. and Nicotera, P. (1995) Glutamate-induced neuronal death: A succession of necrosis or apoptosis depending on mitochondrial function. *Neuron*. 15: 961–973.

Assaf, S.Y. and Chung, S.H. (1984) Release of endogenous Zn^{2+} from brain tissue during activity. *Nature*. 308: 734–736.

Askwith, C. and Kaplan, J. (1998) Iron and copper transport in yeast and its relevance to human disease. *Trends Biochem Sci*. 23: 135–138.

Babcock, M., de Silva, D., Oaks, R., Davis-Kaplan S, Jiralerspong S, Montermini L, Pandolfo, M. and Kaplan, J. (1997) Regulation of mitochondrial iron accumulation by Yfh1p, a putative homolog of frataxin. *Science*. 276: 1709–1712.

Back, T. and Schuler, O.G. (2004) The natural course of lesion development in brain ischemia. *Acta Neurochir Suppl.* 89: 55–61.

Ballabh, P., Braun, A. and Nedergaard, M. (2004) The blood-brain barrier: An overview: Structure, regulation, and clinical implications. *Neurobiol Dis*. 16:1–13.

Baptista, M.J., O'Farrell, C., Daya, S., Ahmad, R., Miller, D.W., Hardy, J., Farrer, M.J. and Cookson, M.R. (2003) Co-ordinate transcriptional regulation of dopamine synthesis genes by α-synuclein in human neuroblastoma cell lines. *J Neurochem*. 85: 957–968.

Barnham, K.J., Masters, C.L. and Bush, A.I. (2004) Neurodegenerative diseases and oxidative stress. *Nat Rev.* 3: 205-214.

Barone, F.C. and Feuerstein, G.Z. (1999) Inflammatory mediators and stroke: new opportunities for novel therapeutics. *J Cereb Blood Flow Metab*. 19: 819–834.

Bartzokis, G., Beckson, M., Hance, D.B., Marx, P., Foster, J.A. and Marder, S.R. (1997) MR evaluation of age-related increase of brain iron in young adult and older normal males. *Magn Reson Imaging*. 15: 29–35.

Basun, H., Forssell, L.G., Wetterberg, L. and Winblad, B. (1991) Metals and trace elements in plasma and cerebrospinal fluid in normal aging and Alzheimer's disease. *J Neural Transm Park Dis Dement Sect*. 3:231–258.

Baud, V. and Karin, M. (2001) Signal transduction by tumor necrosis factor and its relatives. *Trends Cell Biol.* 11: 372-377.

Beal, M.F. (2005) Mitochondria take center stage in aging and neurodegeneration. *Ann Neurol.* 58:495–505.

Ben-Shachar, D., Livne, E., Spanier, I., Leenders, K.L. and Youdim, M.B. (1994) Typical and atypical neuroleptics induce alteration in blood–brain barrier and brain 59FeCl3 uptake. *J Neurochem*. 62: 1112–1118.

Ben-Shachar, D., Riederer, P. and Youdim, M.B. (1991) Iron-melanin interaction and lipid peroxidation: implications for Parkinson's disease. *J Neurochem*. 57: 1609–1614.

Behl, C., Davis, J.B., Lesley, R., and Schubert, D. (1994) Hydrogen peroxide mediates amyloid β protein toxicity. *Cell.* 77: 817–827.

Bergles, D.E., Roberts, J.D., Somogyi, P. and Jahr, C.E. (2000) Glutamatergic synapses on oligodendrocyte precursor cells in the hippocampus. *Nature*. 405:187–191.

Bignami, A. (1991) Glial cells in the central nervous system. Discussions in neuroscience, vol 8. Amsterdam: Elsevier. p 1–45.

Boje, K.M. and Arora, P.K. (1992) Microglial-produced nitric oxide and reactive nitrogen oxides mediate neuronal cell death. *Brain Res*. 587: 250–256.

Bonfoco, E., Krainc, D., Ankarcrona, M., Nicotera, P., and Lipton, S.A. (1995) Apoptosis and necrosis: Two distinct events induced, respectively, by mild and intense insults with N-methyl-daspartate or nitric oxide/superoxide in cortical cell cultures. *Proc Natl Acad Sci USA*. 92: 7162-7166.

Braak, E., Sandmann-Keil, D., Rub, U., Gai, W.P., de Vos, R.A., Steur, E.N., Arai, K. and Braak, H. (2001) α-synuclein immunopositive Parkinson's disease-related inclusion bodies in lower brain stem nuclei. *Acta Neuropathol* (Berl), 101: 195–201.

Brenneman, K.A., Wong, B.A., Buccellato, M.A., Costa, E.R., Gross, E.A. and Dorman, D.C. (2000) Direct olfactory transport of inhaled manganese (54MnCl2) to the rat brain: toxicokinetic investigations in a unilateral nasal occlusion model. *Toxicol Appl Pharmacol.* 169: 238–248.

Browne, S.E. and Beal, M.F. (2006) Oxidative Damage in Huntington's disease Pathogenesis. *Antiox Redox Signaling*, 8: 2061-2073.

Bruijn, L.I., Houseweart, M.K., Kato, S., Anderson, K.L., Anderson, S.D., Ohama, E., Reaume, A.G., Scott, R.W. and Cleveland, D.W. (1998) Aggregation and motor neuron toxicity of an ALS-linked SOD1 mutant independent from wild-type SOD1. *Science*, 281: 1851–54.

Bush, A.I. (2000) Metals and neuroscience. *Curr Opinion Chem Biol.* 4:184–191.

Bush, A.I. (2002) Is ALS caused by an altered oxidative activity of mutant superoxide dismutase? *Nature Neurosci.* 5: 919–920.

Bush, A.I. (2003) The metallobiology of Alzheimer's disease. *Trends Neurosci.* 26:207–214.

Cadman, E.D., Witte, D.G. and Lee, C.M. (1994) Regulation of the release of interleukin-6 from human astrocytoma cells. *J. Neurochem.* 63:980–987.

Cai, J., Yang, J. and Jones, D.P. (1998) Mitochondrial control of apoptosis: The role of cytochrome c. *Biochim et Biophys Acta - Bioenergetics*, 1366: 139-149.

Campbell, A., Smith, M.A., Sayre, L.M., Bondy, S.C. and Perry, G. (2001) Mechanisms by which metals promote events connected to neurodegenerative diseases. *Brain Res Bull.* 55: 125–132.

Campbell, A., Yang, E.Y., Tsai-Turton, M., and Bondy, S.C. (2002) Pro-inflammatory effects of aluminum in human glioblastoma cells. *Brain Res.* 933: 60−65.

Campbell, A. (2004) Inflammation, Neurodegenerative Diseases, and Environmental Exposures. *Ann NY Acad Sci.* 1035: 117–132.

Carman-Krzan, M., Vigé, X. and Wise, B.C. (1991) Regulation by interleukin-1 of nerve growth factor secretion and nerve growth factor mRNA expression in rat primary astroglia cultures. *J Neurochem.* 56: 636–643.

Cherny, R.A., Atwood, C.S., Xilinas ME, Gray DN, Jones WD, McLean CA, Barnham, K.J., Volitakis I, Fraser FW, Kim Y, Huang X, Goldstein, L.E., Moir, R.D., Lim, J.T., Beyreuther K, Zheng H, Tanzi RE, Masters, C.L. and Bush, A.I. (2001) Treatment with a copper-zinc chelator markedly and rapidly inhibits beta-amyloid accumulation in Alzheimer's disease transgenic mice. *Neuron*, 30: 665−676.

Chinnaiyan, A.M., Tepper, C.G., Seldin, M.F., O'Rourke, K., Kischkel, F.C., Hellbardt S, Krammer, P.H., Peter, M.E. and Dixit, V.M. (1996) FADD/MORT1 is a common mediator of CD95 (Fas/APO-1) and tumor necrosis factor receptor-induced apoptosis. *J Biol Chem.* 271:4961-5.

Choi, D. W. (1991) Excitotoxicity, in Excitatory Amino Acid Antagonists. (Meldrum B., eds), pp. 216±236. Blackwell Scientific Publications, London.

Choi, D. W. (1992) Excitotoxic cell death. *J Neurobiol.* 23: 1267±1276.

Christen, Y. (2000) Oxidative stress and Alzheimer disease. *Am J Clin Nutr.* 71: 621S–629S.

Clarkson, T.W. (1987) Metal Toxicity in the Central Nervous System. *Environ Health Persp.* 75: 59-64.

Corson, L.B., Folmer, J., Strain, J.J., Culotta, V.C. and Cleveland, D.W. (1999) Oxidative stress and iron are implicated in fragmenting vacuoles of Saccharomyces cerevisiae lacking Cu,Znsuperoxide dismutase. *J Biol Chem.* 274:27590-6.

Crompton, M. (2004) Mitochondria and aging: a role for the permeability transition? *Aging Cell*, 3:3–6.

Cuajungco, M.P., Goldstein, L.E., Nunomura, A., Smith MA, Lim JT, Atwood CS, Huang, X., Farrag, Y.W., Perry, G. and Ashley, I.B. (2000) Evidence that the β-amyloid plaques of Alzheimer's disease represent the redox-silencing and entombment of Aβ by zinc. *J Biol Chem*. 275: 19439–19442.

Curtis, A.R.J., Fey, C., Morris, C.M., Bindoff, L.A., Ince PG, Chinnery PF, Coulthard A, Jackson MJ, Jackson AP, McHale DP, Hay, D., Barker, W.A., Markham, A.F., Bates, D., Curtis, A. and Burn, J. (2001) Mutation in the gene encoding ferritin light polypeptide causes dominant adult-onset basal ganglia disease. *Nat Genet.* 28:350–354.

Dexter, D.T., Wells, F.R., Lees, A.J., Agid, F., Agid, Y., Jenner, P. and Marsden, C.D. (1989) Increased nigral iron content and alterations in other metal ions occurring in brain in Parkinson's disease. *J Neurochem.* 52:1830–1836.

Di Bello, I.C., Dawson, M.R., Levine, J.M. and Reynolds, R. (1999) Generation of oligodendroglial progenitors in acute inflammatory demyelinating lesions of the rat brain stem is associated with demyelination rather than inflammation. *J Neurocytol.* 28:365–381.

Donaldson, K., Stone, V., Borm, P.J.A., Jimenez, L.A., Gilmour PS, Schins RPF, Knaapen, A.M., Rahman I, Faux, S.P., Brown, D.M. and MacNee, W. (2003) Oxidative stress and calcium signaling in the adverse effects of environmental particles (PM10). *Free Rad Biol Med.* 34: 1369 - 1382.

Dong, Y. and Benveniste, E.N. (2001) Immune function of astrocytes. *Glia*, 36:180–190.

Dong, C., Davis, R.J. and Flavell, R.A. (2002) MAP kinases in the immune response. *Ann Rev Immunol.* 20: 55-72.

Dorman, D.C., Brenneman, K.A., McElveen, A.M., Lynch, S.E., Roberts, K.C. and Wong, B.A. (2002) Olfactory transport: a direct route of delivery of inhaled manganese phosphate to the rat brain. *J Toxicol Environ Health A*, 65: 1493–1511.

Double, K.L., Gerlach, M., Schunemann V, Trautwein, A.X., Zecca, L., Gallorini, M., Youdim, M.B., Riederer, P. and Ben-Shachar, D. (2003) Iron-binding characteristics of neuromelanin of the human substantia nigra. *Biochem Pharmacol.* 66: 489–494.

Dugan, L.L. and Kim-Han, J.S. (2006) Hypoxic-ischemic brain injury and oxidative stress. In: Siegel GJ, Albers RW, Brady ST, Price DL, eds. Basic Neurochemistry: Molecular, Cellular and Medical Aspects. New York: Elsevier Academic Press; pp. 563-569.

Dunn, C.J. (1991) Cytokines as mediators of chronic inflammatory disease. In: Kimball, E. S., ed. Cytokines and inflammation. Boca Raton: CRC Press, Pp. 1–35.

Earnshaw, W.C., Martins, L.M. and Kaufmann, S.H. (1999) Mammalian caspases: structure, activation, substrates, and functions during apoptosis. *Annu Rev Biochem.* 68:383–424.

Estrada, S.A.M., Mejía-Toiber, J. and Massieu, L. (2008) Excitotoxic Neuronal Death and the Pathogenesis of Huntington's disease. *Arch Med Res.* 39: 265-276.

Facheris, M., Beretta, S. and Ferrarese, C. (2004) Peripheral markers of oxidative stress and excitotoxicity in neurodegenerative disorders: tools for diagnosis and therapy? *J Alzheimer's Dis.* 6:177-84.

Fasano, M., Giraudo, S., Coha, S., Bergamasco, B. and Lopiano, L. (2003) Residual substantia nigra neuromelanin in Parkinson's disease is cross-linked to α-synuclein. *Neurochem Int.* 42: 603–606.

Faucheux, B.A., Martin, M.E., Beaumont, C., Hauw. J.J., Agid, Y. and Hirsch, E.C. (2003) Neuromelanin associated redox-active iron is increased in the substantia nigra of patients with Parkinson's disease. *J Neurochem.* 86: 1142–1148.

Faulkner, J.R., Herrmann, J.E., Woo, M.J., Tansey, K.E., Doan, N.B. and Sofroniew, M.V. (2004) Reactive astrocytes protect tissue and preserve function after spinal cord injury. *J Neurosci.* 24:2143–2155.

Feinstein, E., Kimchi, A., Wallach, D., Boldin, M. and Varfolomeev, E. (1995) The death domain: a module shared by proteins with diverse cellular functions. *Trends Biochem Sci.* 20: 342-344.

Fiskum, G., Starkov, A., Polster, B.M. and Chinopoulos, C. (2003) Mitochondrial mechanisms of neural cell death and neuroprotective interventions in Parkinson's disease. *Ann N Y Acad Sci.* 991: 111-119.

Floyd, R.A. (1999) Neuroinflammatory processes are important in neurodegenerative diseases: An hypothesis to explain the increased formation of reactive oxygen and nitrogen species as major factors involved in neurodegenerative disease development. *Free Rad Biol Med.* 26: 1346–1355.

Fornai, F., Soldani, P., Lazzeri, G., di Poggio, A.B., Biagioni, F., Fulceri, F., Batini S, Ruggieri, S. and Paparelli, A. (2005) Neuronal inclusions in degenerative disorders Do they represent static features or a key to understand the dynamics of the disease? *Brain Res Bull.* 65: 275–290.

Furukawa, K. and Mattson, M. (1998) The transcription factor NF-κB mediates increases in calcium currents and decreases in NMDA- and AMPA/kainate-induced currents induced by tumor necrosis factor α in hippocampal neurons. *J Neurochem.* 70:1876-1878.

Gaeta, A. and Hider, R.C. (2005) The crucial role of metal ions in neurodegeneration: the basis for a promising therapeutic strategy. *British J Pharmacology*, 146: 1041–1059.

Gasser, T. (2001) Genetics of Parkinson's disease. *J Neurol.* 248: 833–840.

Galvin, J.E., Giasson, B., Hurtig, H.I., Lee, V.M.Y. and Trojanowski, J.Q. (2000) Neurodegeneration with brain accumulation, type 1 is characterised by α-,β-, and γ-synuclein neuropathology. *Am J Pathol.* 157:361–368.

Gaudette, M., Hirano, M. and Siddique, T. (2000) Current status of SOD1 mutations in familial amyotrophic lateral sclerosis. *Amyotroph Lateral Scler Other Motor Neuron Disord.* 1: 83–89.

Gerard, C., Chehhal, H. and Hugel, R.P. (1994) Complexes of iron(III) with ligands of biological interest: dopamine and 8-hydroxyquinoline-5-sulfonic acid. *Polyhedron.* 13: 591–597.

Giaume, C., Kirchhoff, F., Matute, C., Reichenbach, A. and Verkhratsky, A. (2007) Glia: the fulcrum of brain diseases. *Cell Death and Differentiation*, 14: 1324–1335.

Glenner, G.G. and Wong, C.W. (1984) Alzheimer's disease: initial report of the purification and characterization of a novel cerebrovascular amyloid protein. *Biochem Biophys Res Commun.* 120: 885–890.

Goto, J.J., Zhu, H., Sanchez, R.J., Nersissian, A., Gralla, E.B. and Valentine, J.S. (2000) Loss of in vitro metal ion binding specificity in mutant copper-zinc superoxide dismutases associated with familial amyotrophic lateral sclerosis. *J Biol Chem.* 275: 1007–1014.

Goyer, R.A. and Clarkson, T.W. (2001) Toxic Effects of Metals, in: Casarett and Doull's Toxicology The Basic Science of Poisons, C.D. Klaassen, ed., McGraw-Hill, New York, pp. 811–867.

Giulian, D., Vaca, K. and Corpuz, M. (1993) Brain glia release factors with opposing actions upon neuronal survival. *J Neurosci.* 13: 29–37.

Haacke, E.M., Cheng, N.Y., House, M.J., Liu Q, Neelavalli, J., Ogg RJ, Khan, A., Ayaz, M., Kirsch, W. and Obenaus, A. (2005) Imaging iron stores in the brain using magnetic resonance imaging. *Magn Reson Imaging*, 23:1–25.

Halliwell, B. Gutteridge, J. (1999) Free Radicals in Biology and Medicine (Oxford Univ. Press, Oxford).

HaMai, D., Bondy, S.C., Becaria, A. and Campbell, A. (2001) The Chemistry of Transition Metals in Relation to Their Potential Role in Neurodegenerative Processes. *Curr Topics Med Chem.* 1: 541-551.

Hardy, J. (1997) Amyloid, the presenilins and Alzheimer's disease. *Trends Neurosci.* 20: 154–159.

Harris, W.R. (1983) Thermodynamic binding constants of the zinchuman serum transferrin complex. *Biochem.* 22: 3920–3926.

Heidenreich, K.A. (2003). Molecular Mechanisms of Neuronal cell death. *Ann NY Acad Sci.* 991: 237-250.

Hershey, C.O., Hershey, L.A., Varnes, A., Vibhakar, S.D., Lavin, P. and Strain, W.H. (1983) Cerebrospinal fluid trace element content in dementia: clinical, radiologic, and pathologic correlations. *Neurology*, 33:1350–1353.

Heyes, M.P. and Nowak, T.S. (1990) Delayed increases in regional brain quinolinic acid follow transient ischemia in the gerbil. *J Cereb Blood Flow Metab.* 10: 660–667.

Hirsch, E., Graybiel, A.M. and Agid, Y.A. (1988) Melanized dopaminergic neurons are differentially susceptible to degeneration in Parkinson's disease. *Nature*, 334: 345–348.

Hoogenraad, T.U. and Van den Hamer, C.J.A. (1996) Copper metabolism, copper toxicity and pathogenesis of Wilson's. In: Hoogenraad TU, editor.Wilson's disease. London: W.B. Saunders Co., Ltd. pp. 25–44.

Howell, G.A., Welch, M.G. and Frederickson, C.J. (1984) Stimulation-induced uptake and release of zinc in hippocampal slices. *Nature*, 308: 736–738.

Huang, X., Atwood, C.S., Moir, R.D., Hartshorn, M.A., Vonsattel, J.P., Tanzi, R.E. and Bush, A.I. (1997) Zinc-induced Alzheimer's Aβ 1–40 aggregation is mediated by conformational factors. *J Biol Chem.* 272: 26464–26470.

Huang, X. (1999) The Aβ peptide of Alzheimer's disease directly produces hydrogen peroxide through metal ion reduction. *Biochemistry*, 38, 7609–7616.

Huang, X., Cuajungco, M.P., Atwood, C.S., Hartshorn, M.A., Tyndall, J.D.A., Hanson, G.R., Stokes, K.C., Leopold, M., Multhaup, G., Goldstein, L.E., Scarpa RC, Saunders AJ, Lim,

J., Moirg, R.D., Glabe, C., Bowden EF, Masters, C.L., Fairlie, D.P., Tanzi, R.E. and Bush, A.I. (1999) Cu(II) potentiation of Alzheimer Aβ neurotoxicity. Correlation with cell-free hydrogen peroxide production and metal reduction. *J Biol Chem.* 274: 37111–37116.

Huebers, H.A., Csiba, E., Huebers, E. and Finch, C.A. (1983) Competitive advantage of diferric transferrin in delivering iron to reticulocytes. *Proc Natl Acad Sci USA*, 80: 300–304.

Jänicke, R.U., Sprengart, M.L., Wati, M.R. and Porter, A.G. (1998) Caspase-3 is required for DNA fragmentation and morphological changes associated with apoptosis. *J Biol Chem.* 273: 9357-9360.

Jimenez Del Rıo, M. and Velez-Pardo, C. (2004) Transition Metal-Induced Apoptosis in Lymphocytes Via Hydroxyl Radical Generation, Mitochondria Dysfunction, and Caspase-3 Activation: An In Vitro Model for Neurodegeneration. *Arc Med Res.* 35: 185–193.

Kaufer, D., Friedman, A., Seidman, S. and Soreq, H. (1998) Acute stress facilitates long-lasting changes in cholinergic gene expression. *Nature*, 393: 373–377.

Kerr, J.F.R., Gobe, G.C., Winterford, C.M. and Harmon, B.V. (1995) Anatomical methods in cell death. In: Schwartz LM, Osborne BA, editors. Methods in cell biology: cell death. New York: Academic Press. pp. 1–27.

Kitamura, Y., Shimohama, S., Kamoshima, W., Ota, T., Matsuoka, Y., Nomura, Y., Smith, M.A. and Taniguchi, T. (1998) lteration of proteins regulating apoptosis, Bcl-2, Bcl-x, Bax, Bak, Bad, ICH-1 and CPP32, in Alzheimer's disease. *Brain Res.* 780: 260-269.

Koller, H., Trimborn, M., Von Giesen, H., Schroeter, M. and Arendt, G. (2001) TNFα reduces glutamate induced intracellular Ca^{2+} increase in cultured cortical astrocytes. *Brain Res.* 93: 237–243.

Kordula, T., Bugno, M., Rydel, R.E. and Travis, J. (2000) Mechanism of interleukin-1- and tumor necrosis factor α-dependent regulation of the α1-antichymotrypsin gene in human astrocytes. *J Neurosci.* 20: 7510–7516.

Kreutzberg, G.W. (1996) Microglia — a sensor for pathological events in the CNS. *Trends Neurosci.* 19: 312–318.

Lafon-Cazal, M., Pietri, S., Culcasi, M. and Bockaert, J. (1993) NMDA-dependent superoxide production and neurotoxicity. *Nature*, 364:535-537.

Lau, L.T. and Yu, A.C. (2001) Astrocytes produce and release interleukin-1, interleukin-6, tumor necrosis factor alpha and interferon-gamma following traumatic and metabolic injury. *J Neurotrauma*, 18: 351-359.

Lee, J.Y., Cole, T.B., Palmiter, R.D., Suh, S.W. and Koh, J.Y. (2002) Contribution by synaptic zinc to the gender-disparate plaque formation in human Swedish mutant APP transgenic mice. *Proc Natl Acad Sci USA*, 99: 7705–7710.

Leist, M. and Nicotera, P. (1998) Apoptosis, excitotoxicity, and neuropathology. *Exp Cell Res.* 239: 182-201.

Levine, J.M. (1994) Increased expression of the NG2 chondroitin-sulfate proteoglycan after brain injury. *J Neurosci.* 14:4716–4730.

Lewen, A., Matz, P. and Chan, P.H. (2000) Free radical pathways in CNS injury. *J Neurotrauma*, 17: 871–890.

Lin, M.T. and Beal, M.F. (2006) Mitochondrial dysfunction and oxidative stress in neurodegenerative diseases. *Nature*, 443:787–795.

Liu, H., Zhu, H., Eggers, D.K., Nersissian, A.M., Kym Faull, F., Goto, J.J., Ai, J., Sanders-Loehr, J., Gralla, E.B. and Valentine JS. (2000) Copper2+ binding to the surface residue cysteine 111 of His46Arg human copper-zinc superoxide dismutase, a familial amyotrophic lateral sclerosis mutant. *Biochem*. 39: 8125–8132.

Lobsiger, C.S. and Cleveland, D.W. (2007) Glial cells as intrinsic components of non-cell-autonomous neurodegenerative disease. *Nat Neurosci*. 10: 1355-1360.

Loh, K.P., Huang, S.H., De Silva, R., Tan, B.K. and Zhu, Y.Z. (2006) Oxidative stress: apoptosis in neuronal injury. *Curr Alzheimer Res*. 3: 327-337.

Lotharius, J. and Brundin, P. (2002) Impaired dopamine storage resulting from α-synuclein mutations may contribute to the pathogenesis of Parkinson's disease. *Hum Mol Genet.* 11: 2395–2407.

Lovell, M.A., Robertson, J.D., Teesdale, W.J., Campbell, J.L. and Markesbery, W.R. (1998) Copper, iron and zinc in Alzheimer's disease senile plaques. *J Neurol Sci*. 158: 47–52.

Manash, K.P., Rajan, K.T., Rajinder, and Anup, K.M. (2006) Mitochondria - Role in different diseases: potential for drug development. *CRIPS*, 7: 42-46.

Marsden, C.D. (1983) Neuromelanin and Parkinson's disease. *J Neural Transm Suppl.* 19: 121–141.

Massie, H.R., Aiello, V.R. and Iodice, A.A. (1979) Changes with age in copper and superoxide dismutase levels in brains of C57BL/6J mice. *Mech Ageing Dev*. 10: 93–99.

Massieu, L., Montiel, T., Del Rıo, P., Hernandez, K., Haces, M.L., Garcıa, O., Camacho, A. and Mejía, J. (2003) Role of energy metabolism in neuronal death associated with cerebral ischemia and neurodegenerative diseases, and its prevention by energy substrates. *Recent Res Dev Neurochem.* 81-104.

McMahon, R.J. and Cousins, R.J. (1998) Mammalian zinc transporters. *J Nutr.* 128: 667–670.

Morris, C.M., Keith, A.B., Edwardson, J.A. and Pullen, R.G. (1992) Uptake and distribution of iron and transferrin in the adult rat brain. *J Neurochem.* 59: 300–306.

Meplan, C., Richard, M.J. and Hainaut, P. (2000) Redox signalling and transition metals in the control of the p53 pathway. *Biochem Pharmacol.* 59:25-33.

Mount, M.P., Lira, A., Grimes, D., Smith, P.D., Faucher, S., Slack, R., Anisman, H., Hayley, S. and Park, D.S. (2007) Involvement of interferon-gamma in microglial-mediated loss of dopaminergic neurons. *J Neurosci.* 27: 3328-3337.

Mrak, R.E., Griffin, S.T. and Graham, D.I. (1997) Aging-associated changes in human brain. *J Neuropathol Exp Neurol*. 56: 1269–1275.

Murphy, V.A., Wadhwani, K.C., Smith, Q.R. and Rapoport, S.I. (1991) Saturable transport of manganese (II) across the rat blood–brain barrier. *J Neurochem.* 57: 948–954.

Muzio, M., Stockwell, B.R., Stennicke, H.R., Salvesen, G.S. and Dixit, V.M. (1998) An induced proximity model for caspase-8 activation. *J Biol Chem*. 273:2926-30.

Nagata, S. (2000) Apoptotic DNA fragmentation. *Exp Cell Res.* 256: 12–18.

Neill, S., Desikan, R. and Hancock, J. (2002) Hydrogen peroxide signalling. *Curr Opin Plant Biol.* 5: 388–395.

Nunomura, A., Perry, G., Pappolla, M.A., Wade, R., Hirai, K., Chiba, S. and Smith, M.A. (1999) RNA oxidation is a prominent feature of vulnerable neurons in Alzheimer disease. *J Neurosci.* 19:1959-64.

Nunomura, A., Perry, G., Aliev, G., Hirai, K., Takeda, A., Balraj, E.K., Jones, P.K., Ghanbari, H., Wataya, T., Shimohama, S., Chiba, S., Atwood, C.S., Petersen, R.B. and Smith, M.A. (2001) Oxidative damage is the earliest event in Alzheimer disease. *J Neuropathol Exp Neurol.* 60:759-67.

Nunomura, A., Chiba, S., Kosaka, K., Takeda, A., Castellani, R.J., Smith, M.A. and Perry, G. (2002) Neuronal RNA oxidation is a prominent feature of dementia with Lewy bodies. *Neuroreport*, 13: 2035-9.

Nunomura, A., Honda, K., Takeda, A., Hirai, K., Zhu, X., Smith, M.A. and Perry, G. (2006) Oxidative Damage to RNA in Neurodegenerative Diseases. *J Biomed Biotechnol.* 2006: 82323.

Oliver, K.R. and Fazakerley, J.K. (1998) Transneuronal spread of Semliki Forest virus in the developing mouse olfactory system is determined by neuronal maturity. *Neurosci.* 82: 867–877.

Orgad, S., Nelson, H., Segal, D. and Nelson, N. (1998) Metal ions suppress the abnormal taste behavior of the Drosophila mutant malvolio. *J Exp Biol.* 201, 115–120.

Otter, I., Conus, S., Ravn, U., Rager, M., Olivier, R., Monney, L., Fabbro, D. and Borner, C. (1998) The binding properties and biological activities of Bcl-2 and Bax in cells exposed to apoptotic stimuli. *J Biol Chem.* 273: 6110-20.

Pardridge, W. (2003) Blood-brain barrier drug targeting: The future of brain drug development. *Mol Interventions*, 3: 90–105.

Perez, R.G., Waymire, J.C., Lin, E., Liu, J.J., Guo, F. and Zigmond, M.J. (2002) A role for α-synuclein in the regulation of dopamine biosynthesis. *J Neurosci.* 22: 3090–3099.

Persson, E., Henriksson, J. and Tjälve, H. (2003a) Uptake of cobalt from the nasal mucosa into the brain via olfactory pathways in rats. *Toxicol Lett.* 145: 19–27.

Persson, E., Henriksson, J., Tallkvist, J., Rouleau, C. and Tjälve, H. (2003b) Transport and subcellular distribution of intranasally administered zinc in the olfactory system of rats and pikes. *Toxicol.* 191: 97–108.

Perry, G., Nunomura, A., Hirai, K., Zhu X, Perez, M., Avila, J., Castellani, R.J., Atwood, C.S., Aliev, G., Sayre, L.M., Takeda, A. and Smith, M.A. (2002) Is oxidative damage the fundamental pathogenic mechanism of Alzheimer and other neurodegenerative diseases? *Free Radic Biol Med.* 33:1475–1479.

Peters, S., Koh, J. and Choi, D.W. (1987) Zinc selectively blocks the action of N-methyl-D-aspartate on cortical neurons. *Science*, 236: 589–593.

Petrozzi, L., Ricci, G., Giglioli, N.J., Siciliano, G. and Mancuso M. (2007) Mitochondria and Neurodegeneration. *Bioscience Reports*, 27: 87-104.

Platt, B., Fiddler, G., Riedel, G. and Henderson, Z. (2001) Aluminum toxicity in the rat brain: histochemical and immunocytochemical evidence. *Brain Res Bull.* 55: 257–267.

Polster, B.M., Kinnally, K.W. and Fiskum, G. (2001) BH3 death domain peptide induces cell type-selective mitochondrial outer membrane permeability. *J Biol Chem.* 276: 37887-94.

Price, D.L., Tanzi, R.E., Borchelt, D.R. and Sisodia, S.S. (1998) Alzheimer's disease: genetic studies and transgenic models. *Annu Rev Genet.* 32: 461–493.

Qian, Z.M. and Wang, Q. (1998) Expression of iron transport proteins and excessive iron accumulation in the brain in neurodegenerative disorders. *Brain Res Rev.* 27: 257–267.

Quintanar, L. (2008) Manganese neurotoxicity: A bioinorganic chemist's perspective *Inorganica Chimica Acta*, 361: 875–884.

Radisky, D.C. and Kaplan, J. (1999) Regulation of transition metal transport across the yeast plasma membrane. *J Biol Chem.* 274: 4481–4484.

Raivich, G., Reddington, M., Haas, C.A. and Kreutzberg, G.W. (1995) Peptides in motoneurons, *Prog Brain Res.* 104:3–20.

Raivich, G., Bluethmann, H. and Kreutzberg, G.W. (1996) Signaling molecules and neuroglial activation in the injured central nervous system. *Keio J Med.* 45: 239–247.

Rao, D.B., Wong, B.A., McManus, B.E., McElveen, A.M., James, A.R. and Dorman, D.C. (2003) Inhaled iron, unlike manganese, is not transported to the rat brain via the olfactory pathway. *Toxicol Appl Pharmacol.* 193: 116–126.

Ridet, J.L., Malhotra, S.K., Privat, A. and Gage, F.H. (1997) Reactive astrocytes: cellular and molecular cues to biological function. *Trends Neurosci.* 20:570–577.

Rohlmann, A. and Wolff, J.R. (1996) Subcellular topography and plasticity of gap junction distribution on astrocytes. In: Spray DC, Dermietzel R, editors. Gap junctions in the nervous system. Austin, TX: R.G. Landes Co. p 175–192.

Ross, P.M., Vesterberg, O. and Nordberg, M. (2006) Metals in Motor Neuron Diseases. *Expl Biol Med.* 231:1481-1487.

Rouleau, C., Xiong, Z.H., Pacepavicius, G. and Huang, G.L. (2003) Uptake of waterborne tributyltin in the brain of fish: axonal transport as a proposed mechanism. *Environ Sci Technol.* 37: 3298–3302.

Rusovici, R. and LaVoie, H.A. (2003) Expression and distribution of AP-1 transcription factors in the porcine ovary. *Biol Reprod.* 69: 64-74.

Ryu, J.K., Choi, H.B. and McLarnon, J.G. (2006) Combined minocycline plus pyruvate treatment enhances effects of each agent to inhibit inflammation, oxidative damage, and neuronal loss in an excitotoxic animal model of Huntington's disease. *Neuroscience*, 141:1835-1848.

Sayre, L.M., Perry, G. and Smith, M.A. (1999) Redox metals and neurodegenerative disease. *Curr Opin Chem Biol.* 3:220-5.

Sayre, L.M., Smith, M.A. and Perry, G. (2001) Chemistry and biochemistry of oxidative stress in neurodegenerative disease. *Curr Med Chem.* 8:721-38.

Sayre, L.M., Moreira, P.I., Smith, M.A., and Perry, G. (2005) Metal ions and oxidative protein modification in neurological disease. *Ann Ist Super Sanità*, 41:143-164.

Schipper, H.M. (1996) Astrocytes, brain aging and neurodegeneration. *Neurobiol Aging*, 17: 467–480.

Schrantz, N., Auffredou, M.T., Bourgeade, M.F., Besnault, L., Leca, G. and Vazquez, A. (2001) Zinc-mediated regulation of caspases activity: dose-dependent inhibition or activation of caspase-3 in the human Burkitt lymphoma B cells (Ramos). *Cell Death Differentation*, 8:152–161.

Seguela, P., Haghighi, A., Soghomonian, J.J. and Cooper, E. (1996) A novel neuronal P2x ATP receptor ion channel with widespread distribution in the brain. *J Neurosci.* 16: 448–455.

Sensi, S.L., Canzoniero, L.M., Yu, S.P., Ying, H.S., Koh, J.Y., Kerchner, G.A. and Choi, D.W. (1997) Measurement of intracellular free zinc in living cortical neurons: routes of entry. *J Neurosci.* 17, 9554–9564.

Silver, J. and Miller, J.H. (2004) Regeneration beyond the glial scar. *Nat Rev Neurosci.* 5:146–156.

Smith, M.A., Sayre, L.M. and Perry, G. (1988) Primary involvement of oxidative damage and redox imbalance in Alzheimer and other neurodegenerative diseases. In Redox Regulation of Cell Signaling. Edited by Packer L, Yodoi J. New York: Marcel Dekker, Inc; 115-126.

Smith, M.A., Harris, P.L., Sayre, L.M. and Perry, G. (1997) Iron accumulation in Alzheimer disease is a source of redoxgenerated free radicals. *Proc Natl Acad Sci USA*, 94: 9866–9868.

Smythies, J. (1996) On the function of neuromelanin. *Proc R Soc Lond B Biol Sci.* 263: 487–489.

Spillantini, M.G., Schmidt, M.L., Lee, V.M.Y., Trojanowski, J.Q., Jakes, R. and Goedert, M. (1997) α-synuclein in Lewy bodies. *Nature*, 388, 839–840.

Stavrovskaya, I.G. and Kristal, B.S. (2005) The powerhouse takes control of the cell: is the mitochondrial permeability transition a viable therapeutic target against neuronal dysfunction and death? *Free Radic Biol Med.* 38:687–697.

Strijbos, P.J.L.M. and Rothwell, N.J. (1995) Interleukin-1β attenuates excitatory amino acid-induced neurodegeneration in vitro: involvement of nerve growth factor. *J Neurosci.* 15: 3468–3474.

Sulzer, D., Bogulavsky, J., Larsen, K.E., Behr G., Karatekin, E., Kleinman, M.H., Turro, N., Krantz, D., Edwards, R.H., Greene, L.A. and Zecca, L. (2000) Neuromelanin biosynthesis is driven by excess cytosolic catecholamines not accumulated by synaptic vesicles. *Proc Natl Acad Sci USA*, 97: 11869–11874.

Suzuki, Y.J., Forman, H.J. and Sevanian, A. (1997) Oxidants as stimulators of signal transduction. *Free Radic Biol Med.* 22: 269–285.

Takanashi, M., Mochizuchi, H., Yokomizo, K., Hattori, N., Mori, H., Yamamura, Y. and Mizuno, Y. (2001) Iron accumulation in the substantia nigra of autosomal recessive juvenile parkinsonism (ARJP). *Parkinson Relat Disord.* 7:311–314.

Tamatani, M., Che, Y.H., Matsuzaki, H., Ogawa, S., Okado, H., Miyake, S., Mizuno, T. and Tohyama, M. (1999) Tumor necrosis factor induces Bcl-2 and Bcl-x expression through NFκB activation in primary hippocampal neurons. *J Biol Chem.* 274: 8531–8538.

Tenneti, L. and Lipton, S.A. (2000) Involvement of activated caspase-3-like proteases in N-methyl-d-aspartate-induced apoptosis in cerebrocortical neurons. *J Neurochem.* 74: 134-142.

Trushina, E. and McMurray, C.T. (2007) Oxidative stress and mitochondrial dysfunction in neurodegenerative diseases. *Neuroscience*, 145: 1233-1248.

Tsujimoto, Y. (1998) Role of Bcl-2 family proteins in apoptosis: apoptosomes or mitochondria? *Genes to Cells*, 3: 697–707.

Uthus, E.O. and Seaborn, C.D. (1996) Deliberations and evaluations of the approaches, endpoints and paradigms for dietary recommendations of the other trace elements. *J Nutr.* 126: 2452S–2459S.

Uversky, V.N., Li, J. and Fink, A.L. (2001) Metal-triggered structural transformations, aggregation, and fibrillation of human α-synuclein. A possible molecular NK between Parkinson's disease and heavy metal exposure. *J Biol Chem*. 276: 44284–44296.

Valko, M., Morris, H. and Cronin, M.T.D. (2005) Metals, Toxicity and Oxidative Stress. *Curr Med Chem*. 12: 1161-1208.

Varfolomeev, E.E. and Ashkenazi, A. (2004) Tumor necrosis factor: an apoptosis JuNKie? *Cell*, 116: 491-497.

Waggoner, D.J., Bartnikas, T.B. and Gitlin, J.D. (1999) The role of copper in neurodegenerative disease. *Neurobiol Dis*. 6:221–230.

Wakamatsu, K., Fujikawa, K., Zucca, F.A., Zecca, L. and Ito, S. (2003) The structure of neuromelanin as studied by chemical degradative methods. *J Neurochem*. 86: 1015–1023.

Wang, X. (2001) The expanding role of mitochondria in apoptosis. *Genes Dev*. 15:2922–2933.

Wei, M.C., Zong, W.X., Cheng, E.H., Lindsten, T., Panoutsakopoulou, V., Ross, A.J., Roth, K.A., MacGregor, G.R., Thompson, C.B. and Korsmeyer, S.J. (2001) Proapoptotic BAX and BAK: a requisite gateway to mitochondrial dysfunction and death. *Science*, 27:727-30.

Wegrzyniak, B.H., Dobrzanski, P., Stoehr, J.D. and Wenk, G.L. (1998) Chronic neuroinflammation in rats reproduces components of the neurobiology of Alzheimer's disease. *Brain Res*. 780:294 –303.

Westbrook, G.L. and Mayer, M.L (1987) Micromolar concentrations of Zn^{2+} antagonize NMDA and GABA responses of hippocampal neurons. *Nature*, 328: 640–643.

Wolozin, B. and Behl, C. (2000) Mechanisms of Neurodegenerative Disorders: Part 2: Control of Cell Death. *Arch Neurol*. 57:801-804.

Xu, J., Xu, J., Kao, S., Lee, F., Song, W., Jin, L. and Yankner, B. (2002) Dopamine-dependent neurotoxicity of α-synuclein: a mechanism for selective neurodegeneration in Parkinson disease. *Nature Med*. 8: 600–606.

Yamada, M., Ohno, S., Okayasu, I., Okeda, R., Hatakeyama, S., Watanabe, H., Ushio, K. and Tsukagoshi, H. (1986) Chronic manganese poisoning: a neuropathological study with determination of manganese distribution in the brain. *Acta Neuropathol* (Berl), 70:273–278.

Yamin, G., Glaser, C.B., Uversky, V.N. and Fink, A.L. (2003) Certain metals trigger fibrillation of methionine-oxidized α-synuclein. *J Biol Chem*. 278: 27630–27635.

Yokel, R.A. (2006) Blood-brain barrier flux of aluminum, manganese, iron and other metals suspected to contribute to metal-induced neurodegeneration. *J Alz Dis*. 10: 223–253.

Zecca, L., Gallorini, M., Schunemann, V., Trautwein, A.X., Gerlach, M., Riederer, P., Vezzoni, P. and Tampellini, D. (2001) Iron, neuromelanin and ferritin content in the substantia nigra of normal subjects at different ages: consequences for iron storage and neurodegenerative processes. *J Neurochem*. 76: 1766–1773.

Zecca, L., Zucca, F.A., Costi, P., Tampellini, D., Gatti, A., Gerlach, M., Riederer, P., Fariello, R.G., Ito, S., Gallorini, M. and Sulzer, D. (2003) The neuromelanin of human substantia nigra: structure, synthesis and molecular behaviour. *J Neural Transm Suppl*. 65: 145–155.

Zheng, W. (2001) Toxicology of choroid plexus: a special reference to metal-induced neurotoxicities. *Microsc Res Tech.* 52: 89–103.

Zheng, W., Aschner, M. and Ghersi-Egea, J.F. (2003) Brain barrier systems: a new frontier in metal neurotoxicological research. *Toxicol Appl Pharmacol.* 192:1–11.

Zong, W.X., Lindsten, T., Ross, A.J., MacGregor, G.R. and Thompson, C.B. (2001) BH3-only proteins that bind pro-survival Bcl-2 family members fail to induce apoptosis in the absence of Bax and Bak. *Genes Dev.* 15:1481-6.

Zou, H., Li, Y., Liu, X. and Wang, X. (1999) An APAF-1.cytochrome c multimeric complex is a functional apoptosome that activates procaspase-9. *J Biol Chem.* 274 :11549-56.

In: Progress in Neurodegeneration
Editors: M.R. Avila-Costa and V.A. Martinez
ISBN 978-1-60741-317-2

Chapter 4

Aluminum

Maria Rosa Avila-Costa[1], Laura Colín-Barenque[1], Leonardo Reynoso-Erazo[2], Ana Luisa Gutierrez-Valdez[1] and Jose Luis Ordoñez-Librado[1]

1) Dept. of Neuroscience, Neuromorphology Lab
2) Research Division, Health Education Team. UNAM, Mexico

Aluminum (Al) is an environmentally abundant element to which we are all exposed. The neurotoxicity of this metal has been known for more than a century. More recently, it has been implicated as an etiological factor in some pathologies (including encephalopathy, bone disease, anemia) related to dialysis treatment. In addition, it has been hypothesized to be a cofactor in the etiopathogenesis of some neurodegenerative diseases, including Alzheimer disease (AD), although, despite many studies in several laboratories in different countries, direct evidence is still, so far controversial. Thus, examples of Al neurotoxicity are well recognized in experimental animals and in individuals with renal failure (consequent upon aging, intoxication or renal disease) and there are grounds to link neurodegenerative disorders to Al exposure. Furthermore, an increased concentration of Al in infant formulas and in solutions for home parenteral nutrition has been associated with neurological consequences and metabolic bone disease, characterized by low-bone formation rate, respectively.

In spite of its abundance and ubiquity in the Earth's crust, Al has been attributed no role by nature in living processes. Its poor availability (Martin, 1994) and more probably the unfavorable aspects of its chemistry (Williams, 1999) may have been at the origin of this exclusion. Although formerly characterized as neurotoxic to experimental animals and in a more recent case to man (Mclaughlin et al. 1962), Al was long considered innocuous under usual environmental conditions (Sorenson et al. 1974). It was not until the discovery of its role in the so-called dialysis encephalopathy in the mid-seventies (Alfrey et al. 1976) that attention was called to its toxic properties. In particular, regardless of the large gap between iatrogenic and environmental contamination levels, the Al found in the brains of patients deceased from AD and Parkinson disease (PD)—particularly in senile plaques and neurofibrillary tangles for AD (Crapper et al. 1973; Perl and Brody, 1980; Zatta, 1993) and substantia nigra of PD patients (Yasui et al. 1992)—was suggested to be implicated in the genesis of these neurodegenerative disorders.

Nowadays, Al is extensively used and its alloys and compounds are crucial in many industrial fields. Among them, Al oxide and sulfate are the compounds of greatest importance in technological terms. Curiously, Al phosphide (used as a rodenticide, insecticide and cereal grain fumigant) (Garry et al. 1993), Al fumes and dust, fibrous forms of Al oxide and Al sulfate are substances that appear on lists of toxic chemicals published by agencies devoted to define the relative toxicity risk of materials. There are only a few existing regulations and international guidelines for Al, including the "Drinking water quality guidelines for aluminum, WHO 2004" and the "Carcinogenicity classification for aluminum production, IARC 1987."

4.1. Sources of Aluminum

There is conclusive evidence that Al compounds can reach systemic circulation via different routes, given that ingestion (Walton et al. 1995; Yokel et al. 2001), dermal absorption (Flarend et al. 2001) and intramuscular injection (Flarend et al. 1997) can raise blood Al concentrations. Of particular interest is the inhalation-related intranasal absorption via the olfactory system, because it is the only portion of the central nervous system that has direct exposure to the external environment. Intranasal exposure results in Al distribution along olfactory pathways and uptake into the brain via axonal transport (Perl and Good, 1987; Zatta et al. 1993). Excess accumulation of Al has been documented in the olfactory bulb, cerebral cortex, hippocampus, entorhinal area, and white matter (Zatta et al. 1993). One study found high levels of Al (ranging from 15 to 250 μg g−1) in the rat olfactory bulb (Domingo et al. 1996).

4.1.1. Environmental Exposure

Natural Al occurs in the soil and makes up about 8% of the surface of the earth. Higher concentrations may exist in soil surrounding waste sites associated with certain industries such as coal combustion and Al mining and smelting (U. S. Public Health Service, 1992). Flux of dust from ores and rock materials are the largest source of particle-borne Al (Sorenson et al. 1974). Both natural processes (weathering of aluminosilicate crustal material) and human activities (mining and agriculture) continually add dust particles to the environment (Eisenrich, 1980; Filipek et al. 1987). In the atmosphere, Al is mainly found as aluminosilicates associated with particulate matter and the background levels of Al in the atmosphere generally range from 0.005 to 0.18 mg/m^3 (Sorenson et al. 1974). Al concentrations in natural water normally are small but are found to be higher in the urban areas (Constantini and Giordano, 1991). The increasing amount of Al is being leached continuously by acid rain (Harris et al. 1996) and contributes significantly to environmental inputs.

4.1.2. Dietary Exposure

Many types of foods contain Al because they are grown in soil that contains it. When the soil pH is lower than 4.5-5.0, Al is solubilized in the soil water and absorbed by plant roots (Matsumoto, 2000). Apart from this, food additives also contribute substantial amounts of Al in the diet. Al is present in many manufactured foods and is added to drinking water for purification purposes (Levesque et al. 2000). The common foods with aluminum-containing food additives include some processed cheese, baking powders, cake mixes, frozen dough, pancake mixes, etc. Another significant dietary source of Al is soy-based milk products, which contribute 2.1 mg/day based on the typical intake of an infant (Yokel and McNamara, 2001). Leaching of Al from beverage cans and cookware can also occur (Lin et al. 1997). The average concentration of Al in cola drinks was found to be 0.1 mg/g. Dry tea leaves have 555-1009 μg/g Al and a typical infusion has 4.5-6.0 μg/ml Al. Infusion of coffee has an Al concentration 0.04-0.30 μg/ml (Koch et al. 1988). It has been estimated that about 20% of daily intake of Al comes from cooking utensils made of Al (Lin et al. 1997). Lione (1983) reported that an average human daily consumes 3-100 mg Al through foods and drinks.

4.1.3. Iatrogenic Exposure

Introduction of Al directly into the bloodstream via high-aluminum dialysate or consumption of large oral doses of aluminum-containing phosphate binders or antacids is the major cause of Al overload through medication. Al is a well-known contaminant of intravenous solutions (Wilhelm et al. 2001). Baydar et al. (1997) found that Al concentration in the enteral nutrition formulas and the parenteral solutions ranges groups, e.g., workers of Al refining and metal industries, people employed in printing and publishing and in automotive dealerships and service stations, and individuals involved in fabricated metal products (U. S. Public Health Service, 1992).

4.1.4. Occupational Exposure

During the past two decades a considerable number of studies reported cognitive changes and possible impairment and other occupational hazards in relation to exposure to Al dusts and fumes (Bast-Pattersen et al. 1994; McLachlan, 1995). Arbour (1991) had reported an unusual case of molten Al inhalation. Schlesinger et al. (2000) also recently supported the concept of gradual accumulation and long-term retention of Al within the respiratory tract of individuals repeatedly exposed in occupational settings. Both objective neurophysiological and neuropsycological measures and subjective symptomatology indicated mild but unequivocal dose-dependent increase in Al burden in the body in Al welders and current mild steel welders (Riihimaki et al. 2000).

4.2. Aluminum Entry

Aluminum absorption seems to be very low, but many factors can enhance it in animals and humans (Deng et al. 1998). One of the significant routes of Al absorption is through the gut (Ittel, 1993). Intestinal absorption of Al per se is very poor (Flaten et al. 2001) and as low as 0.1%, but many organic dietary components are potential chelators of Al and may enhance its absorption (Venturini-Soriano and Berthon, 2001). Moore et al. (2000) have shown diminished ability of the gastrointestinal tract to exclude Al in Alzheimer disease, which increases the possibility of systemic exposure.

Exposure to Al through dusts and aerosols is of considerable importance in miners, smelters, and other metal workers. It has been suggested that the inhaled Al accumulates in the brain through absorption via the olfactory system (Perl and Good, 1987; Exley et al. 1996; Zatta et al. 1993) or systemized (a) through the lung epithelia (Gitelman et al. 1995) and (b) via the gastrointestinal tract when particulates are swallowed (Rollin et al. 1993). It has been estimated that about 3% of Al is absorbed into the blood from the lung (Jones and Benett, 1986). In addition to olfactory nerve uptake, transsynaptic Al distribution is also suggested through animal experimentation (Divine et al. 1999). According to Yokel and McNamara (2001), pulmonary Al absorption appears to be more efficient than gastrointestinal absorption.

Dermal applications of Al compounds in cosmetic, antiperspirant, and health care products generally do not induce harmful effects on skin or other organs (Sorenson et al. 1974). Specific conditions showed intradermal penetration (as in elephantiasis) (Blundell et al. 1989) and histiocytic accumulation (as in hyperhydrosis) (Barr et al. 1993) of Al compounds. Underarm application of ^{26}Al chlorohydrate was associated with up to 0.012% absorption of the Al through the skin (Flarend et al. 2001). Thus the skin may be a minor route of Al entry into the body.

4.2.1. Distribution

The total body burden of Al in healthy human subjects is approximately 30-50 mg (Ganrot, 1986). This total body Al is in a flux between different systemic compartments (Exley et al. 1996). Unequal distribution of Al to various tissues has been reported in normal, aluminum-exposed humans and in aluminum-treated experimental animals (Yokel and McNamara, 2001). In the general population, of the total amount of Al of the body, about one half is in the skeleton and about one-fourth is in the lungs (Ganrot, 1986). Whatever the route of exposure, the brain is an important target of Al in the body. The gray matter contains about twice the concentration found in the white matter (Arieff et al. 1979). Al is also found in human skin (Alfrey, 1980), lower gastrointestinal tract (Tipton and Cook, 1963), lymph nodes (Hamilton et al. 1973), adrenals (Tipton and Cook, 1963), and parathyroid glands (Cann et al. 1979).

Distribution of the metal in different target organs varies with route, dose, and duration of exposure (Nayak, 2002). Little is known about the temporal pattern of pulmonary clearance of Al compounds from the lungs with repeated exposure or the potential subsequent translocation to other organs (Schlesinger et al. 2000). Lungs, hilar lymph nodes, liver, and spleen are the main targets of Al accumulation with inhalation exposure (Teraoka, 1981). Following oral exposure retention of Al is reported in brain (preferentially hippocampus), bone, kidneys, muscle, and heart (Chan et al. 1988).

With increasing age, the Al concentrations of lungs, liver, kidneys, and brain increase (Nayak, 2002). The distribution of Al is influenced by parathyroid hormone and 1,25 dihydroxy vitamin D. Parathyroid hormone was shown to increase the Al concentration in liver; 1,25 dihydroxy vitamin D enhances Al uptake in heart and muscles. But when applied simultaneously, the Al content of bone, liver, and brain decrease (Hirschberg et al. 1985).

Al accumulates in the lysosome (Galle, 1987), cell nucleus (Lukiw et al. 1992), and chromatin (De Boni et al. 1974). An association between intranuclear Al and formation of neurofibrillary tangles (an indication of Al neurotoxicity) has been suggested (Uemura, 1984). It has also been reported that Al present in the lysosome may be associated with dementia (van Rensberg et al. 1997).

Al toxicity is well documented but the mechanism of action is poorly understood (Han et al. 2000). Toxic effects of Al on brain, liver, skeletal muscles, heart, and bone marrow are well established (Galle, 1987). Similarly, there is lack of information on its cellular sites of action (Levesque et al. 2000). The intralysosomal accumulating mechanism of Al in kidney enables elimination of the metal. But in other tissues it has some specific toxic effects. In extrarenal tissues, slowly accumulating Al can cause large deposits and impair proper cell functioning or even cell survival.

The normal level of Al in the human brain is below 2 μg/g (w/w) (Andrási et al. 2005), and Al concentrations may increase up to 23 μg/g (w/w) in the case of dialysis encephalopathy (Alfrey et al. 1976). Notably, in addition to oral Al administration induced subacute encephalopathy in non-dialysed uraemic children (Nathan and Pedersen, 1980), overt signs of Al neurotoxicity have been seen in recent times. Reusche et al. (2001) reported a subacute, iatrogenic aluminum-induced encephalopathy after reconstructive otoneurosurgery, and the onset of a severe cerebral congophilic angiopathy was linked to previous exposure to extremely high concentrations of Al in drinking water (100–600 mg L−1) (Exley and Esiri, 2006). In both cases, post-mortem determinations of Al content in brain samples showed Al concentration ranging from 0.75 μg/g (frontal white matter) to 49 μg/g (choroid plexus). It is known that Al can cross the blood–brain barrier (Yokel, 2002) and accumulates in nerve (Shi and Haug, 1990; De Stasio et al. 1994; Nagasawa et al. 2006) and glial cells (De Stasio et al. 1994; Aremu and Meshitsuka, 2005; Nagasawa et al. 2006), reaching up to micromolar concentrations within the cells (Gonçalves and Silva, 2007).

There is sufficient evidence that Al is present in the brain and should be considered as a bridging step to neurotoxicity, since when at the target site; the neurotoxic potency of Al is reasonably high (Simonsen et al. 1994).

4.3. Mechanisms by Which Aluminum Causes Neuronal Damage

As typical hard metal ions, Al displays a strong affinity for oxygen donors, especially those carrying negative charges. Inorganic phosphate as well as the numerous phosphate-containing biomolecules is therefore privileged ligands for this metal (Zatta et al. 2002).

There are many reports indicating that Al^{3+} interacts with cell membranes (see reviews in Meiri et al. 1993), which constitute the primary targets of this trivalent cation to induce membrane structural and functional perturbations (Takano and Shimmen, 1999; Jones and Benett, 1986). Thus, the interaction of Al^{3+} with cell membranes for instance resulted in the production of osmotic fragility (Zatta et al. 1989), anisocytosis, poikilocytosis and shape changes (Zatta et al. 1997; Suwalsky et al. 2001), alterations in membrane dynamics (Weis and Haug, 1989), and decrease in membrane fluidity (Zatta et al. 1997) (Figure 1). Other cellular models include neurons and synapses of invertebrate and vertebrate (Zsiros et al. 1998), brain endothelial cells (Vorbrodt et al. 1994), dorsal root ganglion neurons (Platt and Busselberg, 1994), and the blood-brain barrier (Banks and Kastin, 1989). Also, it has been reported functional perturbation include changes in resting membrane potential and input resistance (Zsiros et al. 1998; Csoti and Erdelyi, 2000), voltage-activated ionic channels (Meiri et al. 1993), transmitter secretion (Meiri et al. 1993), Ca^{2+}-ATPase activity (Vorbrodt et al. 1994), alkaline band formation (Takano and Shimmen, 1999), transmembrane potential difference and short-circuit current (Suwalsky et al. 2001). In addition, Al displaces membrane-associated ions of physiological significance (Deleers, 1985), altering the physiological functionality of the plasma membrane.

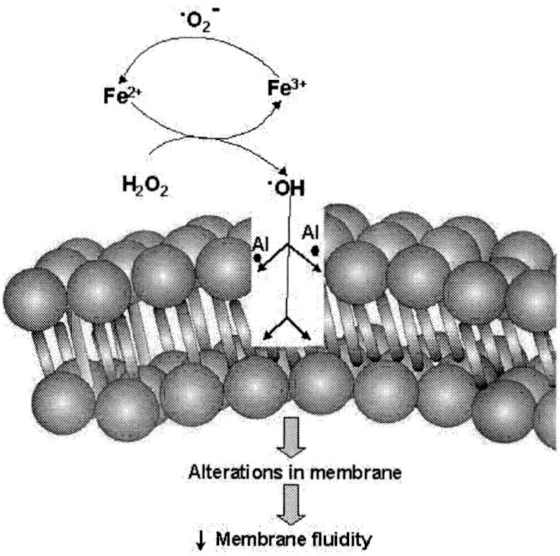

Modified from Zatta et al. 2002.

Figure 1. The peroxidation of membrane lipids: effects of aluminum and iron.

These and other alterations of cell membrane activities might be: (i) direct interaction of Al with proteins forming ion channels, receptors and enzymes; (ii) induction of structural alterations in the lipid matrix; or (iii) action on the lipid protein interfaces. To elucidate among these alternatives, and given the structural complexity of native cell membranes, lipid bilayers have been widely used as cell membrane models (Zatta et al. 2002).

Al^{3+} ion also affects vesicle fusions (Deleers et al. 1986) and alters membrane permeability (Zambenedetti et al. 1994). According to Jones and Kochian (1997), plasma membrane—not enzymatic binding domains—is the most likely site of Al interaction and, therefore, the site of toxic effects.

4.3.1. Aluminum Oxidation

Since Al exists only as a trivalent cation, it is incapable of promoting redox reactions (Campbell, 2001). However Al is capable of enhancing iron-based oxidation *in vitro* and *in vivo*. While Al alone did not promote the formation of ROS in cerebral tissues Al and iron promoted the generation of ROS over concentrations ranging from 50 μM to 1mM (Bondy and Kirstein, 1996).

The first mention of the capacity of Al^{3+} ions to enhance membrane oxidative damage dates back to 1985 (Gutteridge et al. 1985). Al salts were shown to accelerate peroxidation of membrane lipids induced by Fe^{2+} salts at acidic pH values. It was suggested that Al^{3+} ions produced a subtle rearrangement in the membrane structure that facilitated the oxidative action of iron; a relative lack of effect being observed on already disorganized micelles (see Figure 1) (Gutteridge et al. 1985). The later report of the increased peroxidizability of brain homogenates from Al intoxicated mice (Fraga et al. 1990) was interpreted as a possible consequence of a higher rate of production of oxidative reactions *in vivo*, this being probably due to a direct interaction of Al with cell components (Zatta et al. 2002). A higher binding of Al to the membrane was predicted to cause a greater rearrangement of the membrane phospholipids, rendering these more accessible to the attack of free radicals. Interestingly, the direct addition of Al to brain homogenates from control animals resulted in the dual effect: antioxidant-like in the absence of iron, or pro-oxidant in the presence of iron (Fraga et al. 1990). Further investigations on mouse brain membranes by the same group provided evidence that Al^{3+} was the species involved in the promotion of Fe^{2+}-induced lipid peroxidation (Oteiza et al. 1993) and that membrane integrity was necessary for the manifestation of the Al stimulatory effect.

The effect of Al on iron-induced lipid peroxidation has been reported in mice brain homogenate in a concentration and time-dependent manner, with protein oxidative modifications being enhanced at high, but suppressed at low concentrations of Al^{3+} ions (Toda and Yase, 1998), which is reminiscent of the dual effect reported above (Fraga et al. 1990). A significantly higher content of brain myelin galactolipids was observed in Al intoxicated mice following pre-natal and early post-natal exposure with respect to controls, a significant correlation being found between concentration of myelin galactolipids and lipid peroxidation. This synergistic effect of Al with galactolipids was confirmed on liposomes, where it induced higher phase separation and membrane rigidification (Verstraeten et al. 1998). Al was also shown to enhance lipid peroxidation of microsomes from rat liver under acidic conditions and to attenuate the antioxidant action of flavonoids under neutral conditions (Yoshino et al. 1999). With reference to humans, Al has been shown to induce lipid peroxidation and aggregation of blood platelets, there being a correlation between the two phenomena through stimulation of the lipogenase pathway (Neiva et al. 1997).

4.3.2. Impairment of Glia Function

Recent findings have implicated astrocytes as the principal target for Al toxic action (Campbell et al. 2001; Struys-Ponsar et al. 2000). For example, Al salt was more toxic to the glioblastoma cells than neuroblastoma cells (Campbell et al. 2001) and Al treatment *in vivo* or *in vitro* severely impairs astrocyte function (Struys-Ponsar et al. 2000).

There is paucity of data on direct evidence of the consequence of aluminum-impaired astrocytes on neuronal vulnerability. However, Sass et al. (1993) clearly showed that prior treatment of astrocytes with Al impairs the ability of astrocytes to promote neuronal survival, and it was suggested that Al may cause astrocytes to: (i) secrete a factor that makes neurons more susceptible to glutamate-induced toxicity; (ii) secrete a neurotoxic factor in the presence of glutamate; or (iii) reduce secretion of a factor that protects neurons from glutamate excitotoxicity.

The reason why prior treatment of astrocytes with Al impairs their ability to promote neuronal survival is possibly because Al may interfere with any of the ways by which astrocytes protect neurons from toxic injuries and/or modulate the efficacy of neural transmission (Aremu and Meshitsuka, 2005). Theiss and Meller (2002) had reported that Al significantly impaired the gap junctional intercellular communication in cultured astrocytes. Effect of Al on cytoskeletal elements was said to be responsible for the observation. This may be a possible way that Al may impair the capability of astrocytes to buffer ions and transmitters in the extracellular environment of neurons, and, as a consequence of the limited functionality of the astrocytes, Al could affect the physiological activity of neurons (Theiss and Meller, 2002; Ullian et al. 2004).

Impairment of astrocytes metabolism was also shown to worsen Al accumulation by astrocyte, and it was suggested that Al could compromise astrocytes via apoptosis (Aremu and Meshitsuka 2006).

Recently, Aremu and Meshitsuka (2005) summarized the main effects of Al on astrocyte function; these are particularly relevant in the context of the tripartite synapse, a junction of presynaptic terminal and postsynaptic membrane enveloped by an astrocytic process (Halassa et al. 2007). Neurotransmission is among the most specialized functions of nervous system, and neurotoxicity may be the most obvious form of Al toxicity.

4.3.3. Inflammatory Response

Several studies have described the inflammatory action of Al in a variety of tissues. Low doses of Al administered in parenteral nutrition formula in rats caused portal inflammation. This inflammation correlated with the duration of exposure and the amount of Al accumulated in liver (Demircan et al. 1998).

There is limited direct evidence of aluminum-induced inflammatory responses in the CNS. Extended treatment of rabbits with Al lactate increased the concentrations of glial fibrillary acidic protein (GFAP) in the frontal cortex (Yokel and O'Callaghan, 1998). The level of tumor necrosis factor-α (TNF-α), a cytokine implicated in neuronal damage, was significantly increased in the brain of Al-exposed mice in drinking water compared with controls (Tsunoda and Sharma, 1999).

4.3.4. Apoptosis

Among the various effects of Al on the nervous system, neuronal apoptosis may underlie the mechanism of the selective neuronal loss, which is a characteristic symptom of neurodegenerative diseases. Numerous studies suggest that Al causes the death of neurons *in vitro* as well as *in vivo*. Ghribi et al. (2001) observed the apoptotic neuronal loss after the intracisternal administration of Al complexes to the rabbit brain. They also revealed that glial cell -derived neurotrophic factor markedly prevents Al-induced apoptosis (Gauthier et al. 2000).

4.3.5. Conformational Changes of Proteins

Another crucial effect of Al is Al-induced conformational changes in biologically important proteins (Gillette-Guyonnet et al. 2005). Al has a property of firmly binding to various residues of proteins, such as Tyr, His, or phosphorylated amino acids, and has been used as a cross-linker and in tanning agents for leather. Thus, Al binds to various proteins, causes their aggregation or conformational changes, and inhibits the protein degradation by proteases (Kawahara, 2005). The strong binding of Al^{3+} to phosphorylated amino acids promotes the self-aggregation of highly phosphorylated cytoskeletal proteins such as neurofilament or microtubule-associated proteins (MAPs).

Recent lines of evidence suggest that diverse human disorders are considered to arise from the misfolding and aggregation of the underlying protein including several neurodegenerative diseases, such as AD, PD, prion diseases (including bovine spongiform encephalopathy (BSE) and Creutzfeld-Jacob Disease (CJD)) and triplet repeat diseases (including Huntington disease), as well as cystic fibrosis and systemic amyloidosis.

This concept of "conformational disease" may explain the common mechanism underlie these various disorders (Carrell and Lomas, 1997). The underlying disease-related proteins (Aβ in AD, prion protein in prion disease, α-synuclein in PD, polyglutamine in Huntington disease) exhibit similarities in terms of the formation of amyloid fibrils with β-pleated sheet structures and the induction of apoptotic degeneration. Exley et al. (1993) first demonstrated by circular dichroism spectroscopy that Al causes conformational changes of Aβ. These results suggest that Al-aggregated Aβs have a strong affinity to membrane surfaces and are scarcely degraded by proteases.

The aggregation and fibrillation of α-synuclein have been implicated in the formation of abnormal inclusions termed Lewy bodies and considered as a key step in the pathogenesis of

PD and several other neurodegenerative disorders (dementia with Lewy bodies). Uversky et al. (2001) found that Al and other metals markedly promote the aggregation of α-synuclein. Furthermore, Al has also been reported to bind and cause the conformational changes of other disease-related proteins such as Amyloid beta precursor protein (AβPP), the PHF-tau protein, and the ABri (British amyloid peptide) (Kawahara, 2005).

4.3.6. Neurotransmission Impairment

Learning difficulties and memory deficit were repeatedly observed in Al treated animals.

Impaired acquisition and retention of active and passive avoidance were reported in cats (Crapper and Dalton, 1973), rats (Connor et al. 1988) and rabbits (Rabe et al. 1982). Conditioned eye blink/nictitating membrane response was acquired in Al treated rabbits only after prolonged training sessions (Pendlebury et al. 1987).

Some investigators related these behavioral deficiencies to formation of neurofibrillary tangles within nerve cells and tried to correlate them with cell death in different brain regions. Others related the behavioral difficulty to synaptic malfunction (Meiri et al. 1993).

Several investigators argue that Al impairs intracellular signal transduction pathways, due to binding to G proteins and alteration of the GTPase activity (G-protein-mediated second messenger systems), modification of the adenylate cyclase and phosphodiesterase activities (cyclic AMP second messenger system) and competitive inhibition of phosphatidylinositol 4,5-bisphosphate (PIP2) hydrolysis by PIP2-specific phospholipase C (phosphoinositide second messenger-producing system) (Shafer and Mundy, 1995).

Another important outcome of the presence of Al at the synapse could be the reduction of effective concentrations of neurotransmitters with biological activity, due to binding of Al to neurotransmitter molecules. It has been shown that Al^{3+} binds at two metal-binding sites of enkephalin, involving the Tyr1CO and Leu5COO$^-$ groups of the neuropeptide in dimethyl sulphoxide solution (Mazarguil et al. 1982). Moreover, it has been demonstrated that Al interferes with the metabolism of acetyl-CoA leading to a decrease in mitochondrial acetyl-CoA content (Gonçalves and Silva 2007).

It has also been observed that Al can induce a strong and long-lasting impairment of neuronal signal transduction pathways associated with glutamate (NMDA) receptors, including the decline of glutamate-induced, activation of nitric-oxide synthase and nitric oxide-induced activation of guanylate cyclase, formation of cyclic GMP, proteolysis of the microtubule-associated protein-2, disaggregation of microtubules and neuronal death (Gonçalves and Silva 2007).

The so called cholinergic hypothesis (Szutowicz, 2001) is mainly based on the assumption that cholinergic neurons are particularly vulnerable to Al insult because they utilize acetyl-CoA not only for energy production but also for acetylcholine synthesis. Thus, intense cholinergic activity should cause an energy deficit and an acetyl-CoA deficit should depress cholinergic activity more than other neurotransmitter systems.

The disturbance of monoamine neurotransmitter metabolism in the brain, as a result of inhibition of dihydropteridine reductase by Al, has been also suggested to be a putative source of clinical signs of intoxication by this metal (Altindag et al. 2003).

In recent studies, aluminum-induced degradation in the ability to learn and memorize was linked to the configurational changes in synapse (Colomina et al. 2002; Jing et al. 2004). The observed abnormal synaptic ultrastructure was characterized by a reduction in the thickness of postsynaptic density, enlargement of synaptic cleft and attenuation of synaptic curvature. Moreover, a clear shift from convex to concave synapses, a diminution of perforated synapses and an increase of flat type synapses were also observed in the hippocampus and frontal cortex. The alterations in synaptic interfacial structural parameters were more prominent in the hippocampus than in the frontal cortex, where a tiny increase in Al concentration produces a much larger modification in the relative amount of each type of synaptic interfacial structure. The potential functional implications of morphological changes of synapses are not completely understood. However, it is plausible to assume that the depicted aluminum-induced abnormal synaptic ultrastructure will cause profound alterations in the synchronization and probability of neurotransmitter release and attenuates capacity for plasticity (Gonçalves and Silva, 2007).

Finally, Al hydroxide gel/alumina cream is used to produce animal models of epilepsy, which corresponds to an abnormal synchronization of electrical neuronal activity that can be manifested as an alteration in mental state, tonic or clonic movements, convulsions, and various other congnitive symptoms. These temporary abnormal electrophysiological phenomena are also signs of intoxication by Al and reflect an imbalance between excitatory and inhibitory brain circuits (Feria-Velasco et al. 1980). Using this model, it has been shown that Al provokes a widespread, severe specific reduction of GABA (Banks and Kastin, 1989), the major inhibitory neurotransmitter in adult mammalian brain, and loss of GABAergic neurons in the absence of significant cytoskeleton alterations (Franceschetti et al. 1990).

It is not clear whether abnormal synaptic ultrastructure and membrane electrical properties reflect neurotransmission impairment by exposure to Al or by conditioning synaptic transmission. In both cases, the efficiency of communication between neurons, which is crucial to the normal functioning of the central and peripheral nervous systems, is compromised by Al (Gonçalves and Silva, 2007).

4.4. Aluminum and Neurodegeneration

Although the link between Al and the pathogenesis of neurodegenerative diseases was still controversial, the new lines of evidence make it difficult to contradict the link. It is widely accepted that the adverse effects of Al on the nervous system cause cognitive deficiency and dementia when it enters the brain. Owing to several specific chemical characteristics of Al, it is not an essential element and causes more than 200 biologically important reactions (Martin, 1997).

By chronological order, the following pioneering studies on the human neurological effects of Al exposure have appeared: (1) the incidence of neurological symptoms and subclinical neurotoxic effects among miners treated with the prophylactic McIntyre powder and welding, potroom, smeltery and foundry workers chronically exposed to Al since 1962 (Mclaughlin et al. 1962); (2) the Al connection to dialysis encephalopathy since 1972 (Alfrey et al. 1972; 1976); (3) the possible role of Al in the etiology of neurodegenerative diseases, like Alzheimer dementia since 1973 (Crapper et al. 1973); (4) the onset of neurological symptoms following accidental ingestion of Al compounds since 1986 (Chopra et al. 1986); and (5) the occurrence of Al related endemic neurodegenerative diseases, namely the amyotrophic lateral sclerosis (ALS) and parkinsonian syndromes of Guam (Perl et al. 1982), Avyu and Japoi people of West New Guinea (Gadjusek and Salazar, 1982) and Kii Peninsula of Honshu Island in Japan (Yasui et al. 1991).

Several neurological manifestations have been attributed to Al intoxication in humans. These include memory loss, tremor, jerking movements, impaired coordination, sluggish motor movement, ataxia, myoclonic jerks, and generalized convulsions with status epilepticus (Zatta et al. 1991; Crapper McLachlan and DeBoni, 1980). The neuropathological conditions that have been associated with elevated Al in brain include Alzheimer-type senile and presenile dementia, Down syndrome with manifested AD, Guam and Kii peninsula, ALS, Parkinsonian dementia with neurofibrillary degeneration, neurofibrillary degeneration adjacent to harmatoma, dialysis encephalopathy, striatonigral syndrome, alcohol dementia with patchy demyelination, senile plaques of AD, and aged brain (Crapper et al. 1973; Zatta et al. 1993).

The first case of 'aluminosis' was reported by Lapresle et al. (1975) in an alcoholic subject with mental disturbances and progressive neurological deteriorations. Later, several articles reported an abnormal accumulation of metal ions, especially iron and Al, in human subjects affected by certain neurological diseases. Brain Al concentrations were found to be significantly higher in patients dying with dialysis encephalopathy (Lapresle et al. 1975; Nathan and Pedersen, 1980; McClure and Smith, 1984).

On the other hand, it is well documented that one of the most prominent signs of AD is the accumulation of neurofibrillary tangles. Therefore, a number of studies to probe whether exposure to Al is associated with neurofibrillar abnormalities have appeared. Garruto et al. (1989) showed neurofibrillary changes in *Cynomolgus* monkeys fed on a high Al, low calcium diet, simulating conditions associated with endemic neurodegenerative diseases. Neurofibrillary degeneration in rabbit brain was shown to occur after subcutaneous administration of soluble Al salts (Boni et al. 1976). In many other studies, for instance the location of Al deposits in tangle-bearing neurons by laser microprobe mass analysis (Good et al. 1992a), the simultaneous occurrence of neurofibrillary tangle formation and accumulation of Al in the brain have been reported. Recently, new results on the effect of Al on Aβ42 permeation at the striatum and thalamus and sequestration by brain endothelial cells, as well as on the formation of β-sheets of Aβ42 have questioned whether Al participated in the neurotoxic action of amyloid beta-peptide (Drago et al. 2007).

As mentioned in chapters 2 and 3, a common feature of the AD and PD brain is the occurrence of increased levels of iron and Al, two metals involved in membrane lipid peroxidation. At least part of Al neurotoxicity is likely to be due to Al stimulation of iron-induced oxidative damage to neurons (Shinobu and Beal, 1977). The properties of the two metals make them likely to act synergistically in the peroxidation process: whereas the binding of Al to the neuronal membrane is expected to facilitate its attack by iron-induced free radicals (Gutteridge et al. 1985), the resulting oxidation of the membrane will in turn increase its binding to Al, thus aggravating oxidation (Amador et al. 1999). Thus, the pro-oxidant effect of Al on the lipid peroxidation of brain homogenates of Al intoxicated mice as observed by some authors was interpreted in terms of direct interaction of the Al with the cell membrane (Fraga et al. 1990).

Considering Al as a potential risk factor of some neurodegenerative diseases, this possibility requires that at least some of the brain dysregulations associated with neurodegeneration may be due to its damaging influence. Regardless of its concentration, Al may *a priori* interfere with a number of metabolic processes. For example, the Al^{3+} ion can directly compete with, and even substitute for, several essential metal ions *in vivo* (Zatta et al. 2002). Ca^{2+} is its first target in this respect as judged from aluminium-induced osteomalacia (McClure and Smith, 1984) and other instances such as gastrointestinal absorption (van der Voet, 1992) or cell gate regulation (Mundy et al. 1997). Another essential metal known to be subject to Al^{3+} competition is iron, not only as Fe^{3+} for its substitution in transferrin and ferritin, but even as Fe^{2+} in iron gastrointestinal absorption (Berthon, 1996). Strong competition is also expected between Al^{3+} and Mg^{2+} ions given their close chemical resemblance (Martin, 1986; 1994).

4.4.1. Dialysis Encephalopathy

A disease entity in which dementia figures prominently and where AI may well acts as a neurotoxin is Dialysis Encephalopathy (DE). This syndrome tends to develop in uremic patients treated by dialysis who are exposed to an excessive Al load in dialysate water and to Al-containing phosphate binders. High serum and brain levels of Al have been found in these patients and decreasing Al loads as well as chelating Al (with desferoxamine) tend to improve the clinical symptoms of those already afflicted and to lessen the risk for development of the syndrome (Lapresle et al. 1975; Alfrey et al. 1976; Nathan and Pedersen, 1980; McClure and Smith, 1984 and others).

4.4.2. Encephalopathy with Epilepsy

Only rare case reports point directly to Al involvement in human neurological disease. In two reported cases of Al encephalopathy in which high concentrations of brain AI were found postmortem, convulsive activity was prominent and manifested as focal and generalized epileptic seizures which accompanied progressive mental and motor deterioration (McLaughlin et al. 1962; Lapresle et al. 1975). As it has been mentioned, the epileptogenic properties of AI compounds in experimental animals are well established. In terms of the mechanism involved, much emphasis has been focus on the formation of a "Gliotic Scar" at the site of AI application that may act as an epileptogenic focus (Hoeppner and Morrell, 1986). A report of Franceschetti et al. (1990) suggests a more specific action of Al at the cellular and synaptic level.

4.4.3. Alzheimer Disease

The first connection between aluminium-induced neuronal changes and AD was drawn by Klatzo et al. (1965) in who found that the intracisternal administration of Al phosphate to New Zealand white rabbits led to the production of intraneuronal bodies which after silver staining appeared remarkably similar to the neurofibrillary tangles of AD. This finding was serendipitous, since their investigations of the immune response of the central nervous system had at one point required the intracerebral introduction of an antigen prepared in Holt's adjuvant (Savory et al. 2001), the latter containing Al phosphate. As a result, within two days the rabbits developed neurological symptoms so severe as to require their sacrifice. Examination of the brain tissue revealed characteristic aluminum-induced neurofibrillary degeneration, consisting of argyrophilic fibrillary inclusions, located predominantly in the neuronal cell bodies and proximal neurites (dendrites and axon hillock) (Klatzo et al. 1965). Light microscopic examination indicated a close resemblance between the aluminum-induced neurofibrillary aggregates and excessive numbers of neurofibrillary tangles that are a major histopathologic characteristic of AD. However, other hallmarks of AD, such as neuritic plaques, do not appear in the experimental Al-induced encephalopathy.

The amyloid peptide (Aβ) is considered to play a key role in AD. Its sequence of up to 40–43 residues is constitutively produced by normal cells (Vyas and Duffy, 1995) from a transmembrane glycoprotein known as the amyloid protein precursor (APP). Neurotrophic in its original soluble state, Aβ becomes neurotoxic following aggregation and precipitation (Yankner et al. 1990). *In vitro* aggregation of human Aβ has been obtained through interaction with metal ions such as Al (Kawahara et al. 1994; Chong and Suh, 1995), as well as via metal-catalyzed oxidation, this process mainly affecting His and Tyr residues (Dyrks et al. 1992). On the other hand, Aβ reversible aggregation properties would contribute to maintain regional structural integrity (Atwood et al. 1998) at local injury sites where Aβ deposits rapidly appear following injury (Roberts et al. 1991). The inverse correlation found in AD patients between Cerebro-Spinal Fluid Aβ levels and dementia severity (Hock et al. 1998) would also suggest some Aβ protective function.

As already mentioned, oxidative stress and inflammation are a determining component of AD pathogenesis, and β-amyloid and oxidative damage are inextricably linked *in vivo* (Smith et al. 1998). In addition to metal-mediated Aβ aggregation and amyloidogenicity through oxidation (Dyrks et al. 1992), various oxidative processes are involved in the expression of Aβ neurotoxicity (Berthon, 2000).

Considering numerous studies about Al neurotoxicity, Kawahara (2005) proposes a hypothesis as to the involvement of Al and other trace metals in the etiology of AD (Figure 2). Al binds to Iron Regulatory Protein (IRP) and influences the expression of AβPP as well as that of ferritin. Abnormal expression of AβPP will lead to the increased amount of Aβ. Normally secreted Aβ is degraded by various proteases. However, Aβ, which is aggregated in the presence of trace metals, including Al, Zn, Fe, and Cu, could be resistant to proteases and accumulates in the brain. The aggregated Aβ could be easily incorporated into membranes resulting in the formation of ion channels. The abnormal calcium influx through amyloid channels cause the phosphorylation of tau, the depletion of neurotrophic factors such as BDNF, and the formation of free radicals, and finally induces neuronal death. Al can influence these neurodegenerative pathways. Meanwhile, abnormal iron metabolism induced by Al leads to an altered concentration of free iron ions, cause oxidative damage and membrane lipid peroxidation, and also finally leads to neuronal death. This working hypothesis may be useful for the profound understanding about the link between Al and AD.

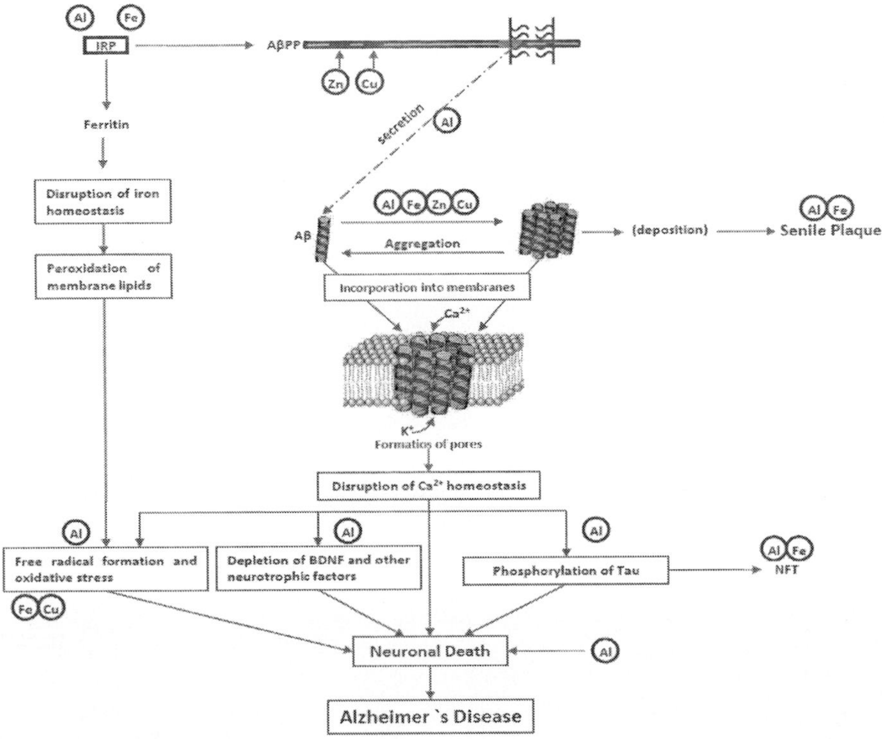

Modified from Kawahara, 2005.

Figure 2. Hypothesis regarding the implications of Al and other trace metals in the pathogenesis of Alzheimer disease. Al binds to IRP and influences the expression of AβPP as well as ferritin. Abnormal expression of AβPP will lead to the increased amount of Aβ. Normally secreted Aβ is degraded by various proteases. However, Aβ, which is aggregated in the presence of trace metals, including Al, Zn, Fe, and Cu, could be resistant to proteases and accumulates in the brain. The aggregated Aβ could be easily incorporated into membranes resulting in the formation of ion channels. The abnormal calcium influx through amyloid channels cause the phosphorylation of tau, the depletion of neurotrophic factors such as BDNF, and the formation of free radicals, and finally induces neuronal death. Al implicates these neurodegenerative pathways by inhibiting of BDNF-induced increase in intracellular calcium levels, by accelerating the phosphorylation of tau, and by stimulating iron-induced lipid peroxidation. Meanwhile, abnormal expression of ferritin caused an altered concentration of free iron ions, and thus, will cause oxidative damage and membrane lipid peroxidation. These events also finally lead to neuronal death. Various genetic and environmental factors may contribute to these pathways. It is possible that Al and other metals are implicated in various stages of these degenerative processes and finally link to the pathogenesis of Alzheimer disease.

4.4.4. Parkinson Disease

Al is a well known neurotoxic agent that is overaccumulated in the substantia nigra (SN) of patients affected by PD (Good et al. 1992b). Although the role of Al in neurodegenerative diseases is yet to be clearly understood, the metal ion is known to substantially alter the activity of several key enzymes in the central nervous system such as dopamine-beta-hydroxylase from bovine adrenal gland with a mixed type mechanism following the Michaelis-Menten equation (Milanese et al. 2001).

Good et al. (1992b) found important concentrations of iron and Al in neuromelanin-containing neurons of the substantia nigra of PD patients concluding that neuromelanin within SNc neurons selectively binds iron and Al promoting the reduction of Fe^{2+} to Fe^{3+} which, in the presence of Al and peroxides generated from the metabolism of dopamine, leads to the formation of cytotoxic radicals and cell degeneration. The finding of excess iron and Al within neuromelanin granules of patients with PD supports the notion that such a scenario might contribute to the pathogenesis of PD and raises the possibility that chelators which bind iron and Al and prevent them from participating in these reactions might be a useful therapy.

A combination of PD and dementia is frequent among the Chamorro people of Guam. Guam Parkinson-Dementia (GPD) is a combination of tauopathy and synucleinopathy. Neurofibrillary tangles are seen in the cortex and subcortical nuclei, including the SN, and α-synuclein inclusions are present in the entorrhinal cortex, amygdala, and SN. The appearance of this disease in successive generations implicates genetic factors. However, in recent years, the incidence of GPD has declined dramatically. Epidemiological work and animal experiments support the hypothesis that GPD is caused by a toxic amino acid in the seed of a cycad plant that is used to make flour and is a staple in the local diet. Another hypothesis implicates other environmental factors leading to deficiencies of calcium and magnesium, and high concentrations of Al and other minerals in drinking water, with Al deposition in neurons (Crapper et al. 1973; Zatta et al. 1993). Al can disrupt the neuronal cytoskeleton and cause neurofibrillary pathology. The decline of GPD in recent years has been attributed to changes in diet and improved nutrition. GPD underlines the important role of genetics and environmental neurotoxins in the pathogenesis of neurodegenerative diseases (Deloncle and Guillard 2005).

4.4.5. Multiple Sclerosis

Multiple sclerosis (MS) is a chronic, immune-mediated demyelinating disease of the central nervous system of as yet unknown etiology. A consensus of opinion has suggested that the disorder is the result of the interplay between environmental factors and susceptibility genes. Exley et al. (2006) have used a battery of analytical techniques to determine if the urinary excretion of i) markers of oxidative damage; ii) iron and iii) Al are altered. Urinary concentrations of oxidative biomarkers, MDA and TBARS, were not found to be useful indicators of inflammatory disease in MS. However, urinary concentrations of another potential marker for inflammation and oxidative stress, iron, were significantly increased. Urinary concentrations of Al were also significantly increased such that the levels of Al excretion in the former were similar to those observed in individuals undergoing metal chelating therapy. Increased excretion of iron in urine supported a role for iron dysmetabolism in MS. Levels of urinary Al excretion similar to those seen in Al intoxication suggested that this metal may be a hitherto unrecognized environmental factor associated with the etiology of MS.

Figure 3 summarizes aluminum-inducing neurodegeneration.

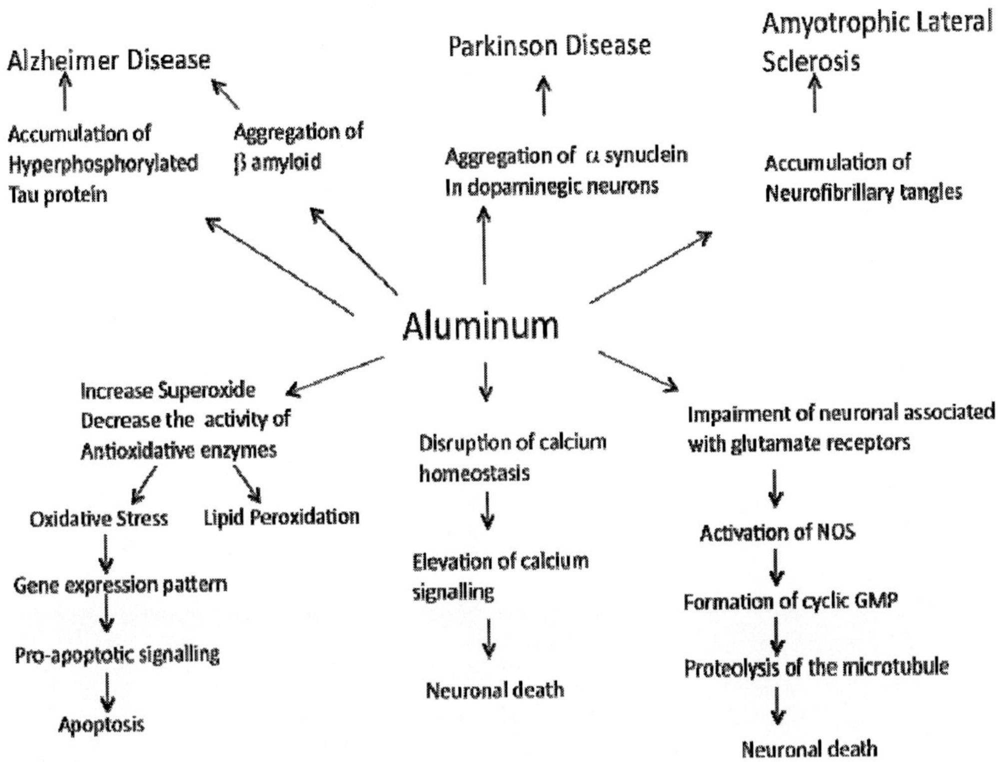

Figure 3. Aluminum and its relation with neurodegeration.

Final Considerations

Although the link between Al and the pathogenesis of neurodegenerative diseases remains controversial, the recent evidences make it difficult to deny the connection, and numerous studies described here support the "aluminum hypothesis".

The knowledge of the several deleterious effects of Al on the central nervous system is on the one hand vast and on the other scarce to explain how disturbances in learning and memory are caused and how altered motor control, increased muscle tone, myoclonic jerks, seizures, convulsions and death. It is incontrovertible that neurotransmission impairment occurs during intoxication by Al, but whether this metal primarily interferes with the metabolism of neurotransmitter precursors, the transduction cascades and modulatory mechanisms of neurotransmission or by targeting specific molecular components of the neurotransmission machinery is unknown. The precise sequences of events at the cellular and subcellular levels that follow exposure to a toxic dose remain to be revealed. The human nervous system is one of the most complex organs in terms of both structure and function. It is composed of more than 10^{11} neurons of different types; a typical neuron may have 10^3–10^4 synapses to other cells, more than 10^2 different substances are used as neurotransmitters and the synapses are plastic, changing their function and structure in response to preceding experiences. In fact, a small number of synapses have been studied under Al exposition until now. Thus considerably more research is needed to identify the mechanisms of Al neurotoxicity.

Furthermore, it is widely accepted that Al is a neurotoxin, and could cause cognitive deficiency and dementia when it enters the brain. In particular, Al can affect infants, elderly people, and patients with impaired renal functions. Therefore, it should be emphasized that unnecessary exposure to Al should be avoided. Further detailed research on the neurotoxic characteristics of Al, including cellular effects, metabolisms, its bioavailability, and in particular, metal-metal interactions are required.

References

Alfrey, A.C., Mishell, J.M., Birks, J., Contiguglia, S.R., Rudolph, H., Lewin, E. and Holmes, J.H. (1972) Syndrome of dyspraxia and multifocal seizures associated with chronic hemodialysis. *Trans Am Soc Artif Intern Org*. 18:257–261.

Alfrey, A.C., LeGendre, G.R. and Kaehny, D. (1976) The dialysis encephalopathy syndrome. Possible aluminium intoxication. *N Engl J Med*. 294: 184–188.

Alfrey, A.C. (1980) Aluminum metabolism in uremia. *Neurotoxicology*, 1: 43-53.

Altindag, Z.Z., Baydar, T., Engin, A.B. and Sahin, G. (2003) Effects of the metals on dihydropteridine reductase activity. *Toxicol In Vitro*, 17: 533–537.

Amador, F.C., Santos, M.S. and Oliveira, C.R. (1999) Lipid peroxidation facilitates aluminum accumulation in rat brain synaptosomes. *J Toxicol Environ Health* A, 58: 427-435.

Andrási, E., Páli, N., Molnár, Z. and Kösel, S. (2005) Brain aluminum, magnesium and phosphorus contents of control and Alzheimer-diseased patients. *J Alzheimer's Dis*. 7: 273–284.

Arbour, P. (1991) Molten aluminum inhalation in the nose and ethamoid sinus. Report of an unusual case. *Rhinology*, 29: 239-241.

Aremu, D.A. and Meshitsuka, S. (2005) Accumulation of aluminum by primary cultured astrocytes from aluminum amino acid complex and its apoptotic effect. *Brain Res*. 1031: 284–296.

Aremu, A.D. and Meshitsuka, S. (2006) Some aspects of astroglial functions and aluminum implications for neurodegeneration. *Brain Res. Rev*. 52(1):193-200.

Arieff, A.I., Cooper, J.D., Armstrong, D. and Larowitz, V.C. (1979) Dementia, renal failure and brain aluminum. *Ann Intern Med*. 90: 741-747.

Atwood, C.S., Moir, R.D., Huang, X., Scarpa, R.C., Bacarra, N.M.E., Romano, D.M., Hartshorn, M.A., Tanzi, R.E. and Bush, A.I. (1998) Dramatic aggregation of Alzheimer by Cu(II) is induced by conditions representing physiological acidosis. *J Biol Chem.* 273: 12817 -12826.

Banks, W.A. and Kastin, A.J. (1989) Aluminum-induced neurotoxicity: Alterations in membrane function at the blood-brain barrier. *Neurosci Biobehav Rev.* 13: 47-53.

Barr, R.J., Alpern, K.S. and Jay, S. (1993) Histiocytic reaction associated with topical aluminum chloride (Drysol reaction). *J Dermatol Surg Oncol.* 19: 1017-1021.

Bast-Pattersen, R., Drablos, P.A., Goffeng, L.O., Thomassen, Y. and Torres, C.G. (1994) Neuropsychological deficit among elderly workers in aluminum production. *Am J Ind Med.* 25:649-662.

Baydar, T., Aydin, A., Duru, S., Isimer, A. and Sahin, G. (1997) Aluminum in enteral nutrition formulas and parenteral solutions. *J Toxicol Clin Toxicol.* 35: 277-281.

Berthon, G. (1996) Chemical speciation studies in relation to aluminium metabolism and toxicity. *Coord Chem Rev.* 149: 241-280.

Berthon, G. (2000) Does human βA4 exert a protective function against oxidative stress in Alzheimer's disease? *Med Hypoth.* 54: 672-677.

Blundell, G., Henderson, W.J. and Price, E.W. (1989) Soil particles in the tissues of the foot in endemic elephantiasis of the lower legs. *Ann Trop Med Parasitol.* 83:381-385.

Bondy, S.C. and Kirstein, S. (1996) The promotion of iron-induced generation of reactive oxygen species in nerve tissue by aluminum. *Mol Chem Neuropathol.* 27:185-94.

Boni, U.D., Otvos, A., Scott, J.W. and Crapper, DR. (1976) Neurofibrillary degeneration induced by systemic aluminum. *Acta Neuropathol* (Berl), 35: 285–294.

Campbell, A. (2001) The potential role of aluminium in Alzheimer's disease. *Nephrol Dial Transplant*, 17: 17-20.

Campbell, A., Hamai, D. and Bondy, S.C. (2001) Differential toxicity of aluminum salts in Human cell lines of neural origin: implications for neurodegeneration. *NeuroToxicology*, 22: 63–71.

Cann, C.E., Prussin, S.G. and Gordan, G.S. (1979) Aluminum uptake by the parathyroid gland. *J Clin Endocrinol Metab.* 49: 543-545.

Carrell, R.W. and Lomas, D.A. (1997) Conformational disease. *Lancet*, 350: 134–138.

Chan, J.C.M., Jacob, M., Brown, S., Savory, J. and Wills, M.R. (1988) Aluminum metabolism in rats; Effects of vitamin D dihydrocortisol, 1,25-dihydroxyvitamin D and phosphate binders. *Nephron.* 48: 61-64.

Chong, Y.H. and Suh, Y.H. (1995) Aggregation of amyloid precursor proteins by aluminum in vitro. *Brain Res.* 670:137 -141.

Chopra, J.S., Kalra, O.P., Malik, V.S., Sharma, R. and Chandna, A. (1986) Aluminium phosphide poisoning: a prospective study of 16 cases in one year. *Postgrad Med J.* 62: 1113–1116.

Colomina, M.T., Roig, J.L., Sanchez, D.J. and Domingo, J.L. (2002) Influence of Age on Aluminum-Induced Neurobehavioral Effects and Morphological Changes in Rat Brain. *Neurotoxicology*, 23: 775–781.

Connor, D.J., Jope, R.S. and Harrell, L.E. (1988) Chronic oral aluminium administration to rats: cognition and cholinergic parameters. *Pharmac Biochem Behav.* 31:467-474.

Constantini, S. and Giordano, R. (1991) In ''Aluminum in Chemistry, Biology and Medicine'' (M. Nicolini, P. F. Zatta, and B. Corain, Eds.), pp. 21-29. Cortina International, Verona.

Crapper, D.R., Krishnan, S.S. and Dalton, A.J. (1973) Brain aluminum distribution in Alzheimer's disease and experimental neurofibrillary degeneration. *Trans Am Neurol Assoc.* 98: 17–20.

Crapper, D.R. and Dalton, A.J. (1973) Aluminium induced neurofibrillary degeneration, brain electrical activity and alterations in acquisition and retention. *Physiol Behav.* 10: 935-945.

Crapper McLachlan, D.R. and DeBoni, W. (1980) Aluminum in human brain disease—An overview. *Neurotoxicology*, 1: 3–16.

Csoti, T. and Erdelyi, L. (2000) Time-dependent actions of aluminates on membrane and action potentials of snail neurons. *Acta Biol Hung*,. 51: 317-324.

De Boni, U., Scott, J.W. and Crapper, D.R. (1974) Intracellular aluminum binding; a histochemical study. *Histochemistry*, 40: 31-37.

Deleers, M. (1985) Cationic atmosphere and cation competition binding at negatively charged membranes: Pathological implications of aluminum. *Res Commun Chem Pathol Pharmacol.* 49: 277-294.

Deleers, M., Servais, J.P. and Wulfert, E. (1986) Neurotoxic cations induce membrane rigidification and membrane fusion at micromolar concentrations. *Wulfert Biochim Biophys Acta*, 855: 271-276.

Deloncle, R. and Guillard, O. (2005) Mechanism of Alzheimer's disease: Arguments for a neurotransmitter-aluminium complex implication. *Neurochem Res.* 15: 1239-1245.

Demircan, M., Ergun, O., Coker, C., Yilmaz, F., Avanoglu, S., and Ozok, G. (1998) Aluminum in total parenteral nutrition solutions produces portal inflammation in rats. *J Pediatr Gastroenterol Nutr.* 26: 274-278.

Deng, Z., Coudray, C., Gouzoux, L., Mazur, A., Rayssiguier, Y. and Pepin, D. (1998) Effect of oral aluminum and aluminum citrate on blood level and short-term tissue distribution of aluminum in the rat. *Biol Trace Elem Res.* 63: 139-147.

De Stasio, G., Mercanti, D., Ciotti, M.T., Dunham, D., Droubay, T.C., Tonner, B.P., Perfetti, P. and Margaritondo, G. (1994) Aluminium in rat cerebellar primary cultures: Glial cells and GABAergic neurones. *Neuroreport*, 5: 1973–1976.

Divine, K.K., Lewis, J.L., Grant, P.G. and Bench, G. (1999) Quantitative particle-induced X-ray emission imaging of rat olfatory epithelium applied to the permeability of rat epithelium to inhaled aluminum. *Chem Res Toxicol.* 12: 575-581.

Domingo, J.L., Llorens, J., Sanchez, D.J., Gomez, M., Llobet, J.M. and Corbella, J. (1996) Age-related effects of aluminum ingestion on brain aluminum accumulation and behavior in rats. *Life Sci.* 58:1387–1395.

Drago, D., Folin, M., Baiguera, S., Tognon, G., Ricchelli, F. and Zatta, P. (2007) Comparative effects of Aβ(1-42)-Al complex from rat and human amyloid on rat endothelial cell cultures. *J. Alzheimer's Dis.* 11:33–44.

Dyrks, T., Dyrks, E., Hartmann, T., Masters, C. and Beyreuther, K. (1992) Amyloidogenicity of βA4 and βA4-bearing amyloid protein precursor fragments by metal-catalyzed oxidation. *J Biol Chem.* 267: 18210-18217.

Eisenrich, S.J. (1980). Atmospheric input of trace metals to Lake Michigan (USA). *Water Air Soil Pollut.* 13: 287-302.

Exley, C., Burgess, E., Day, J.P., Jeffery, E.H., Melethil, S. and Yokel, R.A. (1996) Aluminum toxicokinetics. *J Toxicol Environ Health*, 48: 569-584.

Exley, C. and Esiri, M.M. (2006) Severe cerebral congophilic angiopathy coincident with increased brain aluminium in a resident of Camelford, Cornwall, UK. *J Neurol Neurosurg Psychiatr.* 77: 877–879.

Exley, C., Mamutse, G., Korchazhkina, O., Pye, E., Strekopytov, S., Polwart, A. and Hawkins, C. (2006) Elevated urinary excretion of aluminium and iron in multiple sclerosis. *Multiple Sclerosis*, 12:533-540.

Exley, C., Price, N.C., Kelly, S.M. and Birchall, J.D. (1993) An interaction of β-amyloid with aluminium in vitro. *FEBS letters,* 324: 293–295.

Feria-Velasco, A., Olivares, N., Rivas, F., Velasco, M. and Velasco, F. (1980) Alumina cream-induced focal motor epilepsy in cats. IV. Thickness and cellularity of layers in the perilesional motor cortex. *Arch Neurol.* 37: 287–290.

Filipek, L.H., Nordstorm, D.K. and Ficklin, W.H. (1987) Interaction of acid mine drainage with waters and sediments of West Squaw Creek in the West Shasta mining district. *California Environ Sci Technol.* 21: 388-396.

Flarend, R., Bin, T., Elmore, D. and Hem, S.F. (2001) A preliminary study of the dermal absorption of aluminium from antiperspirants using aluminium-26. *Food Chem Toxicol.* 39:163–168.

Flarend, R.E., Hem, S.L., White, J.L., Elmore, D., Suckow, M.A., Rudy, A.C. and Dandashli, E.A. (1997) In vivo absorption of aluminium-containing vaccine adjuvants using 26Al. *Vaccine*, 15:1314-1318.

Flaten, T.P. (2001) Aluminium as a risk factor in Alzheimer's disease, with emphasis on drinking water. *Brain Res Bull*, 55: 187-196.

Fraga, C.G., Oteiza, P.I., Golub, M.S., Gershwin, M.E. and Keen, C.L. (1990) Effects of aluminum on brain lipid peroxidation. *Keen Toxicol Lett.* 51: 213-219.

Franceschetti, S., Bugiani, O., Panzica, F., Tagliavini, F. and Ayanzini, G. (1990) Changes in excitability of CA1 pyramidal neurons in slices prepared from $AICl_3$ treated rabbits. *Epilepsy Res.* 6: 39-48.

Gadjusek, C.D. and Salazar, A.M. (1982). Amyothrophic Lateral Sclerosis and Parkinsonian Syndromes in high incidence among Avyu Japoi people of West New Guinea. *Neurology*, 32: 107-115.

Galle, P. (1987) The toxicity of aluminum. *World Sci.* 13: 26-35.

Ganrot, P.O. (1986) Metabolism and possible health effects of aluminum. *Environ Health Perspect.* 65: 363-441.

Garry, V.F., Good, P.F., Manivel, J.C. and Perl, D.P. (1993) Investigation of a fatality from nonoccupational aluminum phosphide exposure: Measurement of aluminum in tissue and body fluids as a marker of exposure. *J Lab Clin Med.* 122: 739–747.

Garruto, R.M., Shankar, S.K., Yanagihara, R., Salazar, A.M., Amyx, H.L. and Gajdusekm D.C. (1989) Low-calcium, high-aluminum diet-induced motor neuron pathology in cynomolgus monkeys. *Acta Neuropathol.* 78: 210–219.

Gauthier, E., Fortier, I., Courchesne, F., Pepin, P., Mortimer, J. and Gauvreau, D. (2000) Aluminum forms in drinkingwater and risk of Alzheimer's disease. *Environ Res.* 84: 234–246.

Gitelman, H.J., Aldernan, F.R., Jurs-Lasky, M. and Rockette, H.E. (1995) Serum and urinary aluminum levels of workers in the aluminum industry. *Ann Occup Hyg*. 39: 181-191.

Gonçalves, P.P. and Silva, V.S. (2007) Does neurotransmission impairment accompany aluminium neurotoxicity? *J Inorg Biochem.* 101: 1291–1338.

Good, P.F., Perl, D.P., Bierer, L.M. and Schmeidler, J. (1992a) Selective accumulation of aluminum and iron in the neurofibrillary tangles of Alzheimer's disease: A laser microprobe (LAMMA) study. *Ann Neurol.* 31: 286–292.

Good, P.F., Olanow, C.W. and Perl, D.P. (1992b). Neuromelanin-containing neurons of the substantia nigra accumulate iron and aluminum in Parkinson's disease: A LAMMA study. *Brain Res.* 593(2): 343-346.

Gillette-Guyonnet, S., Andrieu, S., Nourhashemi, F., de La Gueronniere, V., Grandjean, H. and Vellas, B. (2005) Cognitive impairment and composition of drinking water in women: findings of the EPIDOS Study. *Am J Clin Nutr.* 81: 897–902.

Ghribi, O., DeWitt, D.A., Forbes, M.S., Arad, A., Herman, M.M. and Savory, J. (2001) Cyclosporin A inhibits Al-induced cytochrome c release from mitochondria in aged rabbits. *J Alzheimer's Dis.* 3: 387-391.

Gutteridge, J.M.C., Quinlan, G.J., Clark, I. and Halliwell, B. (1985) Aluminium salts accelerate peroxidation of membrane lipids stimulated by iron salts. *Biochim Biophys Acta*, 835: 441-447.

Halassa, M.M., Fellin, T. and Haydon, P.G. (2007) The tripartite synapse: roles for gliotransmission in health and disease. *Trends Mol Med.* 13: 54–63.

Hamilton, E.I., Miniski, M.J. and Cearly, J.J. (1973) The concentration and distribution of some stable elements in healthy human tissues from the United Kingdom: An environmental study. *Sci Total Environ.* 1: 341-374.

Han, J., Han, J. and Dunn, M.A. (2000) Effect of dietary aluminum on tissue nonheme iron and ferritin levels in the chick. *Toxicology*, 142: 97-109.

Harris, W.R., Berthon, G., Day, J.P., Exley, C., Flaten, T.P., Forbes, W.F., Kiss, T., Orvig, C. and Zatta, P.F. (1996) Speciation of aluminum in biological system. *J Toxicol Environ Health*, 48: 543-568.

Hirschberg, R., von Herrath, D., Voss, R., Bossaller, W., Mauelshagen, U., Pauls, A. and Schaefer, K. (1985) Organ distribution of aluminum in uremic rats: Influence of parathyroid hormone and 1,25-dihydroxy vitamin D3. Miner. *Electrolyte Metab.* 11: 106-110.

Hock, C., Golombowski, S., Müller-Spahn, F., Naser W, Beyreuther K, Mönning U, Schenk D, Vigo-Pelfrey C, Bush AM, Moir R, Tanzi, R.E., Growdon, J.H. and Nitsch, R.M. (1998) Cerebrospinal fluid levels of amyloid precursor protein and amyloid β-peptide in Alzheimer's disease and major depression - Inverse correlation with dementia severity. *Nitsch Eur Neurol.* 39: 111-118.

Hoeppner, T.J. and Morrell, F. (1986) Control of scar formation in experimentally induced epilepsy. *Expl Neurol.* 94: 519-536.

Ittel, T.H. (1993) Determinants of gastrointestinal absorption and distribution of aluminum in health and uraemia. *Nephrol Dial Transplant*, 8(Suppl. 1): 17-24.

Jing, Y., Wang, Z. and Song, Y. (2004) Quantitative study of aluminum-induced changes in synaptic ultrastructure in rats *Synapse*, 52: 292–298.

Jones, K.C. and Benett, B.G. (1986) Exposure of man to environmental aluminum- An exposure commitment assessment. *Sci Total Environ*. 52: 65-82.

Jones, D.L. and Kochian, L.V. (1997) Aluminum interaction with plasma membrane lipids and enzyme metal binding sites and its potential role in Al cytotoxicity. *FEBS Lett*, 400: 51 -57.

Kawahara, M. (2005) Effects of aluminum on the nervous system and its possible link with neurodegenerative diseases. *J Alz Dis*. 8: 171–182.

Kawahara, M., Muramoto, K., Bobayashi, K., Mori, H. and Kuroda, Y. (1994) Aluminum promotes the aggregation of Alzheimer's amyloid β-protein in vitro. *Biochem Biophys Res Commun*. 198: 531 -535.

Klatzo, I., Wisniewski, H.M. and Streicher, E.J. (1965) Experimental production of neurofibrillary degeneration: 1. light microscopic observations. *Neuropathol. Exp Neurol*. 24: 187–199.

Koch, K.R., Pougnet, M.A.B., Villiers, S.D. and Montegudo, F. (1988) Increased urinary excretion of aluminum after drinking tea. *Nature*, 333: 122.

Lapresle, J., Duckett, S., Galle, P. and Cartier, L. (1975) Clinical, anatomical and biophysical data on a case of encephalopathy with aluminum deposits. *Comp. Rend. Soc. Biol*. 169: 282-285.

Levesque, L., Mizzen, C.A., McLachlan, D.R. and Fraser, P.E. (2000) Ligand specific effects on aluminum incorporation and toxicity in neurons and astrocytes. *Brain Res*. 877: 191-202.

Lin, J.L., Yang, Y.J., Yang, S.S. and Leu, M.L. (1997) Aluminum utensils contribute to aluminum accumulation in patients with renal disease. *Am J Kidney Dis*. 30: 653-658.

Lione, A. (1983) The prophylactic reduction of aluminum intake. *Food Chem Toxicol*. 21: 103-109.

Lukiw, W.J., Krishnan, B., Wong, L., Kruck, T.P., Bergeron, C. and Crapper McLachlan, D.R. (1992) Nuclear compartmentalization of aluminum in Alzheimer's disease (AD). *Neurobiol Aging*, 13: 115-121.

Martin, R.B. (1986) The chemistry of aluminum as related to biology and medicine. *Clin Chem*. 32: 1797-1806.

Martin, R.B. (1994) Aluminum: A Neurotoxic Product of Acid Rain. *Accounts Chem Res*. 27: 204-210.

Martin, R.B. (1997) Chemistry of aluminum in the central nervous system, in: Mineral and Metal Neurotoxicology, M. Yasui et al. Eds, CRC Press, NY, pp. 75–81.

Mazarguil, H., Haran, R. and Laussac, J.P. (1982) The binding of aluminium to [Leu5]-enkephalin. An investigation using 1H, 13C and 27Al NMR spectroscopy. *Biochim Biophys Acta*, 717:465–472.

Matsumoto, H. (2000) Cell biology of aluminum toxicity and tolerance in higher plants. *Int Rev Cytol*. 200: 1-46.

McClure, J. and Smith, P.S.J. (1984) The localization of aluminium and other elements in bone tissue of a case of renal osteodystrophy with an associated dialysis encephalopathy syndrome. *Pathol.* 142: 293-299.

McLachlan, D.R.C. (1995) Aluminum and the risk of Alzheimer's disease. *Environmetrics*, 6: 233-275.

Mclaughlin, A.I., Kazantzis, G., King, E., Teared Porter, R.J. and Owen, R. (1962) Pulmonary fibrosis and encephalopathy associated with the inhalation of aluminium dust. *Br J Ind Med.* 19:253-63.

Milanese, M., Lkhayat, M.I. and Zatta, P. (2001) Inhibitory effect of aluminum on dopamine beta-hydroxylase from bovine adrenal gland. *J Trace Elements in Med Biol.* 15: 139.

Meiri, H., Banin, E., Roll, M. and Rousseau, A. (1993) Toxic effects of aluminium on nerve cells and synaptic transmission. *Prog Neurobiol.* 40: 89-121.

Moore, P.B., Day, J.P., Taylor, G.A., Ferrier, I.N., Fifield, L.K. and Edwardson, J.A. (2000) Absorption of aluminium-26 in Alzheimer's disease, measured using accelerator mass spectrometry. *Dement Geriatr Cogn Disord.* 11: 66-69.

Mundy, W.R., Freudenrich, T.M. and Kodavanti, P.R.S. (1997) Aluminum potentiates glutamate-induced calcium accumulation and iron-induced oxygen free radical formation in primary neuronal cultures. *Kodavanti Mol Chem Neuropathol.* 32: 41-57.

Nagasawa, K., Akagi, J., Koma, M., Kakuda, T., Nagai, K., Shimohama, S. and Fujimoto, S. (2006) Transport and toxic mechanism for aluminum citrate in human neuroblastoma SH-SY5Y cells. *Life Sci.* 79: 89–97.

Nathan, E. and Pedersen, S.E. (1980) Dialysis encephalopathy in a non-dialysed uraemic boy treated with aluminium hydroxide orally. *Acta Paediatr Scand.* 69: 793–796.

Nayak, P. (2002) Aluminum: Impacts and Disease. *Environ Res Section A*, 89: 101-115.

Neiva, T.J.C., Fries, D.M., Monteiro, H.P., D'Amico, E.A. and Chamone, D.A.F. (1997) Aluminum induces lipid peroxidation and aggregation of human blood platelets. *Chamone Braz J Med Biol Res.* 30: 599 -604.

Oteiza, P.I., Fraga, C.G. and Keen, C.L. (1993) Aluminum has both oxidant and antioxidant effects in mouse brain membranes. *Arch Biochem Biophys.* 300: 517-521.

Pendlebury, W.W., Beal, M.F., Kowall, N.W. and Solomon, P.R. (1987) Results of immunocytochemical neurochemical and behavioral studies in aluminium-induced neurofilamentous degeneration. *J neural Trans.* 24: 213-217.

Perl, D.P., Gajdusek, D.C. and Garruto, R.M. (1982) Intraneuronal aluminium accumulation in amyotrophic lateral sclerosis and Parkinsonism-dementia of Guam. *Science*, 217: 1053–1055.

Perl, D.P. and Brody, A.R. (1980) Alzheimer's disease: X-ray spectrometric evidence of aluminum accumulation in neurofibrillary tangle-bearing neurons. *Science,* 208: 297-299.

Perl, D.P. and Good, P.F. (1987) Uptake of aluminium into central nervous system along nasal-olfactory pathways. *Lancet*, 1: 1028.

Platt, B. and Busselberg, D. (1994) Actions of aluminum on voltage-activated calcium channel currents. *Cell Mol Neurobiol.* 14: 819 -829.

Rabe, A., Lee, M.H., Shek, J., and Wisniewski, H.M. (1982) Learning deficit in immature rabbits with aluminium induced neurofibrillary changes. *Expl Neurol.* 76: 441-446.

Reusche, E., Pilz, P., Oberascher, G., Lindner, B., Egensperger, R., Gloeckner, K., Trinka, E. and Iglseder, B. (2001) Subacute fatal aluminum encephalopathy after reconstructive otoneurosurgery: A case report. *Hum Pathol.* 32: 1136–1140.

Riihimaki, V., Hanninen, H., Akila, R., Kovala, T., Kuosma, E., Paakkulainen, H., Valkonen, S. and Engstrom, B. (2000) Body burden of aluminum in relation to central nervous sysem function among metal inert-gas welders. *Scand J Work Environ Health*, 26: 118-130.

Roberts, G.W., Gentleman, S.M., Lynch, A. and Graham, D.I. (1991) βA4 amyloid protein deposition in brain after head trauma. *Lancet*, 338: 1422-1423.

Rollin, H.B., Theodorou, P. and Kilroe-Smith, T.A. (1993) Changes in the concentration of copper and zinc in body fluids and tissues of rabbits following the inhalation of low concentrations of Al_2O_3 dust. *S African J Sci.* 89: 246-249.

Sass, J.B., Ang, L.C. and Juurlink, B.H.J. (1993) Aluminum pretreatment impairs the ability of astrocytes to protect neurons from glutamate mediated toxicity. *Brain Res,* 621: 207–214.

Savory, J., Ghribi, O., Forbes, M.S. and Herman, M.M. (2001) Aluminium and neuronal cell injury: inter-relationships between neurofilamentous arrays and apoptosis. *J Inorg Biochem.* 87: 15–19.

Shafer, T.J. and Mundy, W.R. (1995) Effects of aluminum on neuronal signal transduction: mechanisms underlying disruption of phosphoinositide hydrolysis. *Gen Pharmacol.* 26:889- 895.

Shinobu, L.A. and Beal, M.F. (1977) The role of oxidative processes and metal ions in aging and Alzheimer disease. In: J.R. Connor, Editor, Metals and Oxidative Damage in Neurological Disorders, Plenum Press, New York pp. 237–275.

Schlesinger, R.B., Snyder, C.A., Chen, L.C., Gorczynski, J.E. and Menache, M. (2000). Clearance and translocation of aluminum oxide (alumina) from the lungs. *Inhal Toxicol.* 12: 927-939.

Shi, B. and Haug, A. (1990) Aluminum uptake by neuroblastoma cells. *J Neurochem.* 55: 551–558.

Simonsen, L., Johnsen, H., Lund, S.P., Matikainen, E., Midtgard, U. and Wennberg, A. (1994) Methodological approach to the evaluation of neurotoxicity data and the classification of neurotoxic chemicals. *Scand J Work Environ Health*, 20: 1–12.

Smith, M.A., Hirai, K., Hsiao, K., Pappolla, M.A., Harris, P.L.R., Siedlak, S.L., Tabaton, M. and Perry, G. (1998) Amyloid-β deposition in Alzheimer transgenic mice is associated with oxidative stress. *J Neurochem.* 70: 2212-2215.

Sorenson, J.R.J., Campbell, I.R., Tepper, L.B. and Lingg, R.D. (1974) Aluminum in the environment and human health. *Environ Health Perspect.* 8: 3-95.

Struys-Ponsar, C., Guillard, O. and van den Bosch de Aguilar, P. (2000) Effects of aluminum exposure on glutamate metabolism: a possible explanation for its toxicity. *Exp Neurol.* 163: 57–164.

Suwalsky, M., Ungerer, B., Villena, F., Norris, B., Cárdenas, H. and Zatta, P. (2001) Effects of AlCl3 on toad skin, human erythrocytes, and model cell membranes. *Brain Res Bull*, 55: 205-210.

Szutowicz, A. (2001) Aluminum, NO, and nerve growth factor neurotoxicity in cholinergic neurons. *J Neurosci Res*. 66:1009-1018.

Takano, M. and Shimmen, T. (1999) Effects of aluminum on plasma membrane as revealed by analysis of alkaline band formation in internodal cells of Chara corallina. *Cell Struct Funct.* 24: 131-137.

Theiss, C. and Meller, K. (2002) Aluminum impairs gap junctional intercellular communication between astroglial cells in vitro. *Cell Tissue Res.* 310:143–154.

Teraoka, H. (1981) of 24 elements in the internal organs of normal males and metallic workers in Japan. *Arch Environ Health*, 36: 155-165.

Tipton, I.H. and Cook, M.J. (1963) Trace elements in human tissue. Part II. Adult subjects from the United States. *Health Phys*. 9: 103-145.

Toda, S. and Yase, Y. (1998) ffect of aluminum on iron-induced lipid peroxidation and protein oxidative modification of mouse brain homogenate. *Biol Trace Elem Res.* 61: 207-217.

Tsunoda, M. and Sharma, R.P. (1999) Modulation of tumor necrosis factor alpha expression in mouse brain after exposure to aluminum in drinking water. *Arch Toxicol.* 73: 419-26.

Uemura, E. (1984) Intranuclear aluminum accumulation in chronic animals with experimental neurofibriullary changes. *Exp Neurol.* 85: 10-18.

Ullian, E.M., Harris, B.T., Wu, A., Chan, J.R. and Barres, B.A. (2004) Schwann cells and astrocytes induce synapse formation by spinal motor neurons in culture. *Mol Cell Neurosci.* 25: 241–251.

U. S. Public Health Service. (1992). ''Toxicological Profile for Aluminum and Compounds.'' (Contact No. 205-88-0608). Prepared by Clement International Corp.

Uversky, V.N., Li, J. and Fink, A.L. (2001) Metal-triggered structural transformations, aggregation, and fibrillation of human alpha-synuclein. A possible molecular link between Parkinson's disease and heavy metal exposure. *J Biol Chem*. 276: 44284–44296.

van der Voet, G.B. (1992) Intestinal absorption of aluminium. In: Chadwick D, Whelan J (eds) Aluminum in Biology and Medicine. Ciba Foundation Symposium 169. Wiley, Chichester, pp. 109–122.

van Rensberg, S.J., Daniels, W.M., Potocnic, F.C., van Zyl, J.M., Taljaard, J.J. and Emsley, R.A. (1997) A new model for pathophysiology of Alzheimer's disease: Aluminum toxicity is exacerbated by hydrogen peroxide and attenuated by an amyloid protein fragment and melatonin. *S Afr Med J.* 87: 1111-1115.

Verstraeten, S.V., Keen, C., Golub, M.S. and Oteiza, P.I. (1998) Membrane composition can influence the rate of Al^{3+}-mediated lipid oxidation: Effect of galactolipids. *Oteiza Biochem J.* 333: 833-838.

Venturini-Soriano, M. and Berthon, G. (2001) Aluminum speciation studies in biological fluids. Part 7. A quantitative investigation of aluminum(III)-malate complex equilibria and their potential implications for aluminum metabolism and toxicity. *J Inorg Biochem.* 85: 143-154.

Vorbrodt, A.W., Trowbridge, R.S. and Dobrogowska, D.H. (1994) Cytochemical study of the effect of aluminium on cultured brain microvascular endothelial cells. *Histochem J.* 26: 119 -126.

Vyas, S.B. and Duffy, L.K. (1995) Interaction of synthetic Alzheimer β-protein-derived analogs with aqueous aluminum: A low-field ^{27}Al NMR investigation. *J Protein Chem.* 14: 633-644.

Walton, J., Tuniz, C., Fink, D., Jacobsen, G. and Wilcox, D. (1995) Uptake of trace amounts of aluminum into the brain from drinking water. *Neurotoxicology*, 16: 187–190.

Weis, C. and Haug, A. (1989) Aluminum-altered membrane dynamics in human red blood cell white ghosts. *Haug Thrombosis Res*. 54: 141 -149.

Wilhelm, M., Hoelzer, J., Luebbers, K., Stoehr, G. and Ohmann, C. (2001) Aluminum balance in intensive care patients. *J Trace Elem Med Biol.* 14: 223:227.

Williams, R.J.P. (1999) What is wrong with aluminium? The J.D. Birchall memorial lecture. *J Inorg Biochem*. 76: 81-88.

Yankner, B.A., Duffy, L.K. and Kirschner, D.A. (1990) Neurotrophic and neurotoxic effects of amyloid β protein: Reversal by tachykinin neuropeptides. *Science*, 250: 279 -282.

Yasui, M., Kihira, T. and Ota, K. (1992) Calcium, magnesium and aluminum concentrations in Parkinson's disease. *Neurotoxicology*, 13:593-600.

Yasui, M., Yase, Y., Ota, K., and Garruto, R.M. (1991) Aluminum deposition in the central nervous system of patients with amyotrophic lateral sclerosis from the Kii Peninsula of Japan. *Neurotoxicology*, 12: 615–620.

Yokel, R.A. and O'Callaghan, J.P. (1998) An aluminum-induced increase in GFAP is attenuated by some chelators. *Neurotoxicol Teratol.* 20: 55-60.

Yokel, R.A. (2002) Brain uptake, retention, and efflux of aluminum and manganese. *Environ Health Perspect*, 110: 699–704.□

Yokel, R.A. and McNamara, P.J. (2001) Aluminum toxicokinetics: An updated minireview. *Pharmacol Toxicol*. 88: 159-167.

Yokel, R.A., Rhineheimer, S.S., Brauer, R.D., Sharma, P., Elmore, D. and McNamara, P.J. (2001) Aluminum bioavailability from drinking water is very low and is not appreciably influenced by stomach contents or water hardness. *Toxicology*, 161: 93–101.

Yoshino, M., Ito, M., Haneda, M., Tsubouchi, R. and Murakami, K. (1999) Prooxidant action of aluminum Ion - Stimulation of iron-mediated lipid peroxidation by aluminum. *Biometals*, 12 :237-240.

Zambenedetti, P., Tisato, F., Corain, B. and Zatta, P.F. (1994) Reactivity of Al(III) with membrane phospholipids: A NMR approach. *Zatta Biometals*, 7: 244 -252.

Zatta, P.F. (1993) Controversial aspects of aluminum(III) accumulation and subcompartmentation in Alzheimer's disease. *Trace Elem Med.* 10: 120-128.□

Zatta, P., Favarato, M. and Nicolini, M. (1993) Deposition of aluminium in brain tissues of rats exposed to inhalational of aluminium acetylacetonate. *NeuroReport*, 4: 1119–1122.

Zatta, P., Kiss, T., Suwalsky, M. and Berthon, G. (2002) Aluminium(III) as a promoter of cellular oxidation. *Coord Chem Rev*. 228: 271-284.

Zatta, P.F., Nicolini, M. and Corain, B. (1991) Aluminum (III) toxicity and blood–brain barrier permeability. In: M. Nicolini, P. F. Zatta and B. Corain, Editors, Aluminum in Chemistry, Biology and Medicine, Cortina International, Verona, pp. 97–112.

Zatta, P., Perazzolo, M. and Corain, B. (1989) Tris acetylacetonate aluminium(III) induces osmotic fragility and acanthocyte formation in suspended erythrocytes. *Corain J Inorg Biochem.* 65: 109-114.

Zatta, P., Zambenedetti, P., Toffoletti, A., Corvaja, C. and Corain, B. (1997) Aluminum (III) induces alterations on the physical state of the erythrocytic membrane: An ESR evaluation. *Corain J Inorg Biochem.* 65: 109-114.

Zsiros, V., Rojik, I., Kovacs, T., Csoti, T. and Erdelyi, L. (1998) Comparative electrophysiological aspects of aluminum actions on central neurons and neuronal synapses of invertebrate and vertebrate animals. *Erdelyi Neurotoxicology*, 19: 553 -556.

In: Progress in Neurodegeneration
Editors: M.R. Avila-Costa and V.A. Martinez
ISBN 978-1-60741-317-2

Chapter 5

Manganese

Jose Luis Ordoñez-Librado
Department of Neuroscience, Neuromorphology Lab
UNAM, Mexico

Manganese (**Mn**) is the 12th most common element on earth's crust (Sistrunk *et al.*, 2007) and the fourth most widely used metal in the world (Olanow, 2004; Uchino *et al.*, 2007). In the environment exists in a number of physical and chemical forms, and are known eleven oxidation states from -3 to +7 (Hazell *et al.*, 2002; Takeda, 2003).

Mn is a trace element with a remarkable physiological and physiopathological importance (Huang, 1989), it has been found as Mn^{2+}, Mn^{3+} and Mn^{4+} in both, animals and humans (Archibald and Tyree, 1987) being Mn^{2+} the predominant form in biological systems (Aschner *et al.*, 2005a). It is an integral part of metalloproteinases such as hydrolases, kinases, decarboxylases and transferases (Villalobos *et al.*, 2001; Hazell, 2002; Struve *et al.*, 2007); activates a Mn-dependent ATPase (Doherty *et al.*, 1983) and it is essential for cAMP synthesis (Cooper and Caldwell, 1990). Furthermore it is involved in the formation of bone and in amino acids, lipid and carbohydrate metabolism (Wedler, 1993; Jae-Hoon *et al.*, 2006).

Mn ion is a normal dietary component of food and water (Jankovic, 2005), is present in several dietary sources including nuts (18–46 ppm), grains (0.4–40 ppm), legumes (2.2–6ppm), fruits (0.2–10 ppm), and vegetables (0.4–6.6ppm). Poultry, meat, refined foods, and dairy products contain small amounts (0.2–0.5 ppm) (Barceloux, 1999). In humans, the average daily consumption of Mn is 2.3 and 1.8 mg/day for men and women, respectively and 1–3.5% of ingested Mn is absorbed into the blood (Sistrunk *et al.*, 2007).

Environmental exposures to Mn occur most commonly through drinking contaminated water, from exposure to organo-manganese agricultural chemicals, and more recently, from environmental deposition of methylcyclopentadienyl manganese tricarbonyl (MMT), which is used as an anti-knock additive to gasoline (Vézer *et al.*, 2005). Occupational exposure is through industrial emissions associated with ferroalloy production, iron and steel foundries, coke ovens and power plant combustion emissions (Schneider *et al.*, 2006). The background levels of Mn in rural and urban areas without point sources of Mn range from about 0.01–0.07 mg Mn/m^3, while average ambient air levels near industrial sources range from0.22–0.30 mg Mn/m^3. The anthropogenic sources of Mn in ambient air include the combustion of fossil fuels (20%) and the emission from industrial sources (80%). Erosion of soil is the most important natural source in the ambient air, but little data are available to estimate the contribution of Mn from this source (Barceloux, 1999).

Mn deficiency can lead to multiple problems such as stunted growth, skeletal defects, abnormal glucose tolerance (Erikson *et al.*, 2005a) and seizure activity (Critchfield *et al.*, 1993). Clinically significant Mn deficiency occurs rarely in humans (Erikson *et al.*, 2005a). In contrast, exposure to excessive amounts of Mn is more prevalent and is associated with a variety of psychiatric and motor disturbances (Schneider *et al.*, 2006). Impairments of central nervous system (CNS) activity in Mn poisoning were first described at the beginning of the 20th century. Some studies demonstrated that victims of poisoning showed neuropsychological disorders at early stages, with psychomotor excitation, speech disorders, and compulsive behavior, which could include disordered movements, laughing, singing, etc (Shukakidze *et al.*, 2003). Chronic poisoning of the same patients demonstrated mask-like facies, tremor, gait difficulty, muscular hypertonia and extrapyramidal parkinsonian syndrome (Hill and Switzer, 1984).

5.1. Manganese Absorption

It is generally assumed that uptake of Mn across the intact skin is very limited. On the other hand, the gut closely regulates ingested Mn; only about 1-5% is absorbed into the body by gastrointestinal tract under normal conditions. Studies in animals indicate that some Mn that is retained for shorter periods is localized in the liver and intestinal tract and eliminated through biliary excretion (Dobson *et al.*, 2004). The concentration of Mn in the diet is known to influence the amount that is absorbed from the gastrointestinal tract and the amount that is eliminated in the bile. When dietary Mn levels are high, adaptive changes are presented, this mechanism include reduced gastrointestinal absorption, enhanced liver metabolism, and increased biliary and pancreatic excretion of this metal (Aschner *et al.*, 2005b).

In the case of inhalation exposure, transport of Mn is more efficient than others routes because the inhaled Mn bypasses the physiological system that regulates its absorption producing higher target doses than equal Mn uptake by other routes (Roels *et al.*, 1997). The systems that regulate Mn absorption include the normal function of the gastrointestinal tract, hepatobilliary system, transferrin (Tf) and other blood transport proteins (Aschner *et al.*, 2005b). For these reasons, inhalation is the most relevant route of exposure, since dermal absorption is extremely limited and oral absorption of Mn is under tight physiological regulation (Aschner *et al.*, 2005a; Struve *et al.*, 2007).

There are a number of factors that influence pulmonary deposition of inhaled Mn particles. Particle size heavily influences the site at which Mn particles are deposited within the respiratory tract. Lung deposition is favored following the inhalation of particles less than 0.1 μm as well as those ranging in size from 0.8 to 3 μm. The soluble fraction of Mn particles deposited in the lower airways may be absorbed from the lung, while insoluble components may be cleared or translocated to other sites within the respiratory system. The rate at which Mn is absorbed from the lung may influence Mn organ delivery (Aschner *et al.*, 2005b).

It is clear that the route of exposure can influence Mn distribution, metabolism, and potential toxicity (Roels *et al.*, 1997). In the case of the brain, inhalation exposure is more efficient transporting Mn. Pharmacokinetic factors that may contribute to the increased efficiency of brain Mn delivery following inhalation include greater Mn absorption from the lungs and slower clearance of absorbed Mn from the circulation. Moreover, inhalation exposure to soluble forms of Mn results in higher brain Mn concentration compared with insoluble form of Mn. One study has shown that after intratracheal instillation, a surrogate for inhalation exposure, Mn concentrations were higher in brain following the administration of the soluble salt $MnCl_2$, than following the administration of the insoluble oxide MnO_2. Striatal Mn concentrations increased by 205% and 48% following $MnCl_2$ and MnO_2 administration, respectively (Salehi *et al.*, 2003).

5.2. Manganese Transport

The affinity of divalent Mn toward endogenous ligands is relatively low. It does not avidly complex with sulfhydryl (—SH) groups or amines, and it shows little variation in its stability constants for endogenous complexing ligands such as glycine, cysteine, riboflavin and guanosine. Absorbed Mn is transported to other organs via the iron-binding proteins; within the plasma approximately 80% of Mn is bound to ß-globulin and albumin, and a small fraction of trivalent Mn is bound to Tf (Aschner *et al.*, 2005b; Aschner *et al.*, 2007).

Mn readily crosses the brain barriers and is largely distributed to the CNS (Aschner *et al.*, 1999). These barrier systems are the blood–brain and the blood–cerebrospinal fluid barriers (Takeda, 2003). In normal plasma concentrations, Mn enters the brain mainly across the cerebral capillaries. However, when Mn plasma concentration increases, entry of Mn across the choroids plexus becomes more important (Murphy *et al.*, 1991).

Both divalent and trivalent Mn may be transported into the brain, and binding affinity for ligands plasma is important for understanding its mechanism of transport (Scheuhammer and Cherian, 1985). Divalent Mn can be transported into brain capillary endothelial cells and choroidal epithelial cells via undefined divalent metal transporter DMT- 1, DCT-1 or nramp-2 which have an unusually broad range of substrates, e.g. Fe^{2+}, Zn^{2+}, Mn^{2+}, Cu^{2+} and Co^{2+} (Salehi *et al.* 2006). It is known that the striatum is a brain region that is rich in DMT-1 (Burdo *et al.*, 1999). Others transport systems, i.e. calcium channels, Na/Ca exchanger, active calcium uniporter, and Na/Mg antiporter have been reported for divalent Mn uptake (Takeda, 2003).

Trivalent Mn binds to Tf to be transported across the brain barriers via the receptor-mediated endocytosis (Roth and Garrick, 2003). Mn is then released from the complex in the endothelial cell interior by endosomal acidification, and the apo-Tf Tf complex is returned to the luminal surface (Morris *et al.*, 1992). Mn released within the endothelial cells is subsequently transferred to the adluminal cell surface for release into the extracellular fluid (Moos, 1996). There is the possibility that this secreted Mn into the brain extracellular fluid binds again to Tf secreted from oligodendrocytes. Thus, Mn exists as non-Tf-bound and Tf-bound forms in the extracellular fluid in the brain, and because neurons express Tf receptors on the surface, Tf-bound Mn appears to be taken up by neurons via the receptor-mediated endocytosis (Takeda, 2003). The distribution of Tf receptors in relationship to CNS Mn accumulation is striking. The pallidum, thalamic nuclei and substantia nigra contain the highest concentrations of Mn (Aschner *et al.*, 2007). Other brain cells that also take up Mn for usage and storage are the oligodendrocytes and astrocytes (Moos, 1996).

Even when Mn is transported bounded to proteins, it has been reported that when it is intravenously injected into experimental animals, is rapidly removed from the blood and a portion of Mn is quickly transported into the brain (Sotogaku, *et al.*, 2000). This may be due to the presence of non-protein-mediated Mn uptake mechanisms in the brain barrier systems. Although the proportion of divalent and trivalent Mn in the plasma is unclear, a large (approximately 60%) of divalent Mn is reported to be non-protein-bound, probably the free ion in the plasma. Non-protein-bound Mn appears to be transported into the brain more rapidly than Tf-bound Mn (Takeda, 2003; Zheng *et al.*, 2003).

Finally, other route from which Mn has access to the brain is by intranasal administration; by means of this via Mn may circumvent the brain barriers and then pass directly into the CNS via olfactory pathways (Tjälve *et al.*, 1996) increasing its bioavailability (Frumkin and Solomon, 1997). Moreover, it has been reported that Mn can enter the brain through retrograde transport via the olfactory neuronal pathway and the trigeminal system (Dorman *et al.*, 2004).

5.3. Manganese and the Central Nervous System

As it has been mentioned the CNS is an important target of chronic Mn exposure (Jae-Hoon *et al.*, 2006) and the turnover there is much slower than in other parts of the body (Vezér *et al.*, 2005). In this system Mn has an essential role in the normal functioning of several enzymes including glutamine synthetase, mitochondrial superoxide dismutase (SOD), and phosphoenolpyruvate carboxykinase (Jae-Hoonet *et al.*, 2006). Mn is eliminated from brain slowly; its half-live is 223 to 267 days. Furthermore, ^{54}Mn was detected in the lungs for 500 days after exposure, suggesting that they served as an uptake reservoir for the brain; these data indicate the strong possibility that long residence times in the lung provide a continuing source of brain exposure (Weiss, 2002).

It has been reported that following exposure to this metal, the main areas of the mammalian brain targeted are the basal ganglia. The basal ganglia are a group of subcortical nuclei that consists of the striatum (caudate-putamen), the globus pallidus (GP) divided in the external segment (GPe) and internal segment (GPi), the subthalamic nucleus (STN) and the substantia nigra (SN), which is divided into dorsal *pars compacta* (SNc) and the ventral *pars reticulata* (SNr; Sistrunk *et al.*, 2007). These nuclei are responsible for integrating and coordinating information from various brain regions associated with motor movement (Marsden and Obeso 1994; Chesselet and Delfs 1996). The complexity of those nuclei is exemplified by the interaction of several neurotransmitters that there converge (Figure 1); in particular, the interaction between glutamate (Glu), dopamine (DA) and γ-aminobutyric acid

(GABA), which has been suggested to be an important neuronal substrate for functions such as motor activity, emotion and cognition (Mora *et al.* 2008). It is known that the nerve terminals from the cerebral cortex signal the striatum using Glu as an excitatory neurotransmitter, while the terminals from the SNc signal it using DA. At the same time, the striatum sends a feedback signal to the SNc using GABA as an inhibitory neurotransmitter, and it forwards information to the GP, also using GABA. Thus, it becomes clear that the levels of these three neurotransmitters, DA, Glu and GABA, are closely interrelated; consequently, an alteration on the production or release of one of these neurotransmitters inevitably will affect the others (Quintanar, 2008). On the other hand, diverse studies have demonstrated that Mn induces metabolic changes in DA, GABA and Glu regulating their release in the CNS (Villalobos et al., 2001; Gonzalez-Reyes et al. 2007; Quintanar, 2008). For that reason, it is likely that the alterations observed in human and animal exposed to Mn are due to the interactions of these basal ganglia neurotransmitters.

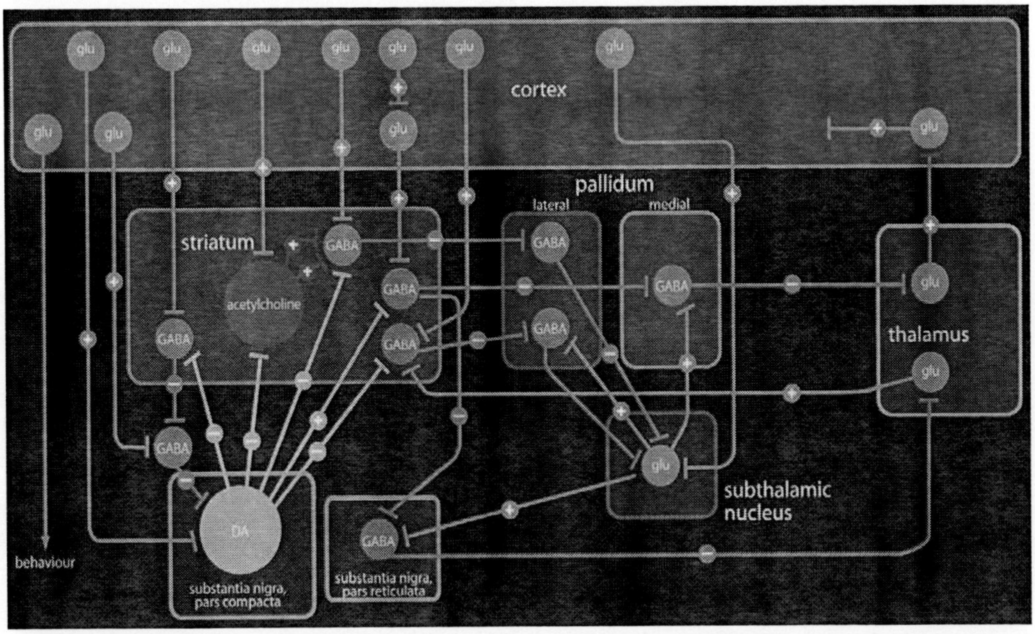

Modified from Carlsson, 1995.

Figure 1.-Stimulatory glutamate (Glu) projections from the cortex signal the striatum and activate dopaminergic pathways (DA) in the SNc; the DA pathway sends projections mainly to the striatum. In turn, the striatum sends a feedback signal to the SNc using GABA as an inhibitory neurotransmitter, and it forwards information to the GPe and GPi/SNr.

5.3.1. γ-Aminobutyric Acid (GABA)

Although it has been known for many years that the early site of Mn accumulation is the GP, most research related to Mn neurotoxicity has focused on the dopaminergic system. However, it has been reported that Mn have effects on GABA concentrations and motor activity (Gwiazda *et al.*, 2002), for example, it has been observed that Mn inhibited the Ca^{2+}-dependent release of GABA and GABA transport is inhibited by $MnCl_2$ in rat forebrain synaptosomes, providing a link between Mn exposure and perturbations in GABA regulation (Fitsanakis *et al.*, 2006).

Some studies have shown enhanced GABA levels after Mn treatment. Reaney *et al.* (2006) found that rats exposed to Mn^{3+} showed a significant increase in GABA levels of ~15-30% over controls in the GP and tended to increase in the striatum (15-45%) though significant differences were not observed in this nucleus. Another report showed that rats exposed to 20 mg Mn/Kg per day (~30 times normal intake) had significantly increased brain GABA concentrations (Lipe *et al.*, 1999). Also Gwiazda *et al.* (2002) inform that low levels of Mn (three injections of 4.8 mg Mn/kg for 5 weeks) increase striatal GABA concentration.

In contrast to these reports, another authors show GABA decrease after Mn exposure. Struve *et al.* (2007) found that monkeys exposed to $MnSO_4$ 1.5 mg Mn/m^3 showed decrease in pallidal GABA concentration. In the same way Lai *et al.* (1984) showed that rats exposed to 6mg Mn/Kg per day (~10 times normal intake), led to a significant decrease in GABA concentrations. Furthermore, eighteen-month-old rats that received intraperitoneal injections of $MnCl_2$ showed a significant decrease in GAD (the biosynthetic enzyme for GABA) mRNA levels in GP (68.2% of control); in contrast, no differences were found in striatum in terms of GAD mRNA levels (Tomás-Camardiel *et al.*, 2002).

Conflicting data exists regarding whether Mn accumulation leads to decreases or increases in regional GABA levels. Nevertheless, it is clear that the GABAergic systems in the basal ganglia are affected (Erikson and Aschner, 2003). However, there is still a debate in the literature; whether the observed GABA alterations are due to direct perturbation of the GABA-producing systems, or if they are secondary to the alteration of other neurotransmitter levels. It is conceivable that Mn would alter any neurotransmitter release event, it is also capable of concentrating in synaptosomes, interfere with Ca^{2+} homeostasis, and block voltage-dependent Ca^{2+} channels, which in turn would affect Ca^{2+} dependent neurotransmitter release (Quintanar, 2008).

5.3.2. Glutamate (Glu)

The basal ganglia receive a massive and converging glutamatergic innervations from all cerebral cortices except from primary visual and auditory areas, which represents under physiological conditions, the main driving force for striatal neuron activity (Stern et al., 1998) and that, during pathological events can trigger both necrotic and apoptotic neuronal death (Greene and Greenamyre 1996; Calabresi et al., 2000).

There is consensus that Mn exposure results in increased extracellular Glu levels, and it has been shown that Glu-mediated mechanisms play a crucial role in the development of Mn-induced striatal neurotoxicity *in vivo*. The injurious effects of Mn injection into the rat striatum have been found to be blocked by prior removal of the corticostriatal glutamatergic inputs or by the treatment with Glu receptor antagonists. How Mn intoxication can lead to altered glutamate mediated transmission in the striatum is unknown, however it has been proposed that chronic treatment with Mn might trigger excitotoxic events by altering corticostriatal synaptic transmission at pre- or postsynaptic levels (Brouillet *et al.*, 1993; Verity, 1999).

On the other hand, the most accredited cellular effect of Mn, in fact, is the impairment of energy metabolism resulting from mitochondrial dysfunction and free radical production. A defective energy metabolism, in turn, is known to alter excitatory transmission in a number of ways and in particular by causing abnormal release of Glu, by impairing Glu reuptake processes, and by increasing postsynaptic responses to glutamate receptor activation (Centonze *et al.*, 2001).

In summary, Mn exposure causes the increase of Glu in the basal ganglia which contributes to neuronal degeneration (this process will be detailed later on).

5.3.3. Dopamine (DA)

Research on the neurobiological consequences of Mn toxicity conducted in the last two decades has been predominantly directed at the effects of Mn on DA metabolism and its associated behavioral alterations. DA is a catecholaminergic neurotransmitter found predominately in the CNS; it is synthesized from the amino acid tyrosine, which is converted to L-dihydroxyphenylalanine (L-DOPA) by the enzyme tyrosine hydroxylase (TH), in turn, L-DOPA is converted to DA by the enzyme DOPA decarboxylase (or aromatic amino acid decarboxylase; AADC) (Bahena-Trujillo et al., 2000). This neurotransmitter mediates a variety of brain functions including motor behavior, motivation and working memory (Vezér et al., 2007; Cools et al., 2008)

Some studies have reported seemingly conflicting results on the dopaminergic effects of Mn, including decreases, increases or no change in basal ganglia DA concentrations in Mn treated animals (Gwiazda *et al.*, 2007). Schneider *et al.* (2006) reported that in *Cynomolgus macaque* monkeys exposed to $MnSO_4$ (10–15mg/Kg/week) for a period lasting 272 ± 17days, there was not decrease in the content of DA in the striatum; for it, these authors thought that the nigrostriatal pathway is not impaired. On the other hand, some other authors have found that acute exposure to Mn is associated with an increase in DA, which is manifested as hyperactivity (Walowitz and Roth, 1999; Salehi *et al.*, 2003).

In contrast, Ponzoni *et al.*, (2000) established that the unilateral microinjection of $MnCl_2$ into the SNc of rats reduces DA concentrations and the number of TH positive cells. Moreover, the same treatment causes a significant reduction in dihydrophenil acetic acid (DOPAC). These data suggest alterations in DA synthesis and degeneration of dopaminergic neurons. Some other works reported that Mn inhibited tyrosine hydroxylation in rat striatal slices, proposing that TH, the key enzyme of dopamine biosynthesis, is inhibited in dopaminergic neurons soon after exposure to certain amounts of Mn (Hirata *et al.*, 2001). Furthermore, Erikson *et al.* (2007) found that TH protein levels were significantly decreased in the GP, caudate and putamen in monkeys exposed to $MnSO_4$.

The possible explanation of the DA reduction is related at least in part, in the direct oxidation of this neurotransmitter by Mn (Sistrunk *et al.*, 2007). It has been demonstrated that Mn enhances oxidative stress in presence of L-DOPA, indicating that Mn is a potent dopaminergic neurotoxicant (Masashi *et al.*, 2005). Also, it has been observed that Mn^{2+} stimulates DA autoxidation in the dopaminergic neurons, a processes accompanied by an increase in the formation of quinones and protein bound cysteinyl-DA and DOPAC (Desole *et al.*, 1997; Ramesh *et al.*, 2002).

As it has been mentioned, direct actions of Mn on dopaminergic neurons have been described. However, it is important to stand out that the modification in the content of other neurotransmitters can result in dopaminergic alteration. For example, it has been postulated that the neurotoxic effects of Mn on striatal dopamine may be indirectly mediated via abnormal striatal Glu and/or GABA metabolism, and that temporally changes in areas that are known to avidly accumulate Mn precede the well described effects of Mn on dopaminergic function. Specifically, it is hypothesized that Mn accumulation in the GP causes decreased GABAergic efferent firing from the GP into the subthalamic nuclei. Consequently, glutamatergic projections into the substantia nigra that originate from the subthalamic nuclei, will fire in an uncontrolled manner causing dysregulation of dopaminergic output into the striatum from the substantia nigra (Erikson and Aschner, 2003)

The data summarized here suggests that Mn exposure alters the content of the main neurotransmitters inside the basal ganglia. Modifications in the function of these nuclei cause movement disorders such as Parkinson disease (PD). In the same way, clinically Mn neurotoxicity is characterized by gait abnormalities, postural instability, micrographia, dystonia, rigidity, and bradykinesia (Aschner *et al.*, 2005a), this condition is known as manganism, which bears many similarities to PD.

5.4. Manganism

Manganism or Mn poisoning, was first described by Couper in 1837 in 5 patients who worked in a Mn ore crushing plant in France. Many studies were reported subsequently and provided analyses of neurological disturbances arising as a result of Mn exposure. Three clinical phases distinguish the development of chronic Mn neurotoxicity in humans. The early stage of Mn neurotoxicity starts gradually with nonspecific symptoms including fatigue, somnolence, mental excitability, emotional instability, anorexia, and muscular pain, all symptoms of psychiatric nature ("*locura manganica*" or "*manganese madness*"). This phase is followed by the intermediate phase in which objective symptoms such as speech disorders, clumsiness in movements, and masked facies appear. Finally, the late (or established) phase is characterized by extrapyramidal changes that include muscular rigidity, hypokinesia, and tremor of the upper limbs (Walowitz and Roth, 1999; Normandin *et al.*, 2002; Shukakidze *et al.*, 2003).

Although the motor signs exhibited by Mn-exposed humans correspond in part to those seen in PD, great discrepancy exists about Mn-inducing PD, including the specificity of Mn-damaging the GP or SN. Enough differences are visible to question the widely held proposition that Parkinson and manganism are virtually identical. Barbeau *et al*, (1976) suggested that the syndrome more closely resembled a dystonia than classical PD.

Neuropathology associated with manganism generally involves the GP and SNr, with less extensive involvement of the caudate nucleus, putamen, STN, and SNc. These observations suggest that the dopaminergic system is an important target for Mn (Struve *et al.*, 2007). It is conceivably not surprising that idiopathic PD and manganism share similarities in their clinical presentations. Both diseases affect the basal ganglia, and both include perturbations of the dopaminergic system. However, their pathological etiologies are quite different. For example, idiopathic PD results from the loss of the pigmented dopaminergic neurons in the SNc, accompanied by Lewy bodies, which consist in abnormally aggregated protein α-synuclein found largely in dopaminergic neurons (Weiss, 2002). On the other hand, Mn initially accumulates in the GP, a nucleus with GABA projections. Additionally, it has been hypothesized that at least some of the cell death observed in animals given high doses of Mn may be due to excitotoxic lesions from high extracellular levels of Glu (Erikson and Aschner, 2003).

As noted earlier, the GP, on the basis of both chemical analyses and MRI, appears to accumulate Mn in greater quantities than other basal ganglia structures and it is the site of lesions produced by Mn. However, studies show that DA, the principal neurotransmitter that is severely reduced in the striatum of PD patients, is also reduced by Mn both *in vivo* (Parenti *et al.*, 1986) and *in vitro* (Vescovi *et al.*, 1991) exposure paradigms. It has been reported that 15 parkinsonian patients who were professional welders (Racette *et al.*, 2001) and workers occupationally exposed to Mn exhibit decreases in DA in the basal ganglia, and some of these observations have been reproduced in non-human primate models (Reaney *et al.*, 2006). Tomás-Camardiel *et al.* (2002) found that primates exposed to Mn for 18 months, presented a marked degeneration of the SNc. Of interest, PD-like symptoms causes by chronic Mn intoxication respond to L-DOPA in some cases (Huang *et al.*, 1989), this fact indicates a possible presynaptic lesion of the nigrostriatal dopaminergic pathway (Chin-Chang *et al.*, 2003). For it, clinical signs of manganism are often difficult to distinguish of those of PD (Normandin *et al.*, 2002) since the features can be overlapped pathophysiologicaly (Racette *et al.*, 2005).

Although it has been reported that humans exposed to high levels of Mn manifest behavioral and motor alterations, few experiments have been done with animals. In this way, it has been reported that chronic Mn exposure through drinking water, intrastriatal and intrathecal exposures induce motor and other behavioral effects in rats (Normandin *et al.*, 2002).

In our laboratory we found an important loss of TH-positive neurons in mice exposed to $MnCl_2/MnOAc_3$ mixture, the loss of these neurons was approximately 67.58% (Ordoñez-Librado *et al.*, 2008). It is well known that the clinical signs of PD appear when the patients lose more than 60% of the dopaminergic neurons of the SNc and 80% of the striatal DA (Au *et al.*, 2005). However in some other studies, exposure of experimental animals to Mn does not appear to affect the structural integrity of dopaminergic nigrostriatal neurons (Normandin and Hazell, 2002).

These contradictory results may be attributable to diverse factors such as dosage, route of Mn administration and age of the experimental animals. However, among the most important conclusions that have been obtained, there is the fact that the alterations in the DA concentration depend on the time of exposition to Mn (Ali *et al.* 1995). Acute exposure to Mn is associated with an increase in DA neurotransmission, which is also manifested as hyperactivity, while long-term exposure results in a loss of DA in the brain, and the concomitant neuronal cell damage could be expressed as decrease in motor activity (Ponzoni *et al.*, 2000; Salehi *et al.*, 2003). Clinical studies have revealed that psychiatric symptoms might emerge in the early phase of manganism in the absence of motor effects. However, the latter disorders are more commonly reported in the recognized stage. Motor effects have been attributed to basal ganglia dysfunction (Salehi *et al.*, 2003).

These facts suggest that Mn reduces DA only at elevated concentrations after chronic exposures. This raises the intriguing possibility that Mn exposure may both induce atypical parkinsonism (based on pallidal effects) and contribute to more typical parkinsonism (based on dopaminergic effects). While a distinction has been made between the motor effects induced by Mn exposure and those observed in PD (Calne *et al.*, 1994), some studies propose a higher incidence of PD in workers occupationally exposed to Mn (Racette *et al.*, 2005). Thus, it is probable that Mn exposure could be a risk factor of PD, since Mn is capable of accelerate the depletion of striatal DA after sustained occupational exposure and precipitate the appearance of PD like-symptoms (Gwiazda *et al.*, 2007).

On the other hand, as it has been mentioned, Mn is able to alter the function of the nigrostriatal pathway. To date, it remains elusive how and why Mn targets dopaminergic neurons (Anderson *et al.*, 2007). Neurotoxins, such as 6-hydroxydopamine (6-OHDA), 1-methyl-4-phenylpyridium (MPP+), paraquat (pesticide) and maneb are selectively lethal to dopaminergic neurons because they are transported by dopamine transporter (DAT) (Petzinger *et al.*, 2006) and it has been hypothesized that Mn is also transported into dopaminergic neurons via this transporter (Ingersoll *et al.*, 1999).

A study carried out by Erikson *et al.* (2005b) in which they examined the potential role of DAT in Mn accumulation in the brain using a knockout mouse model, found a significant decrease (40%) in Mn accumulation in the striatum of the DAT-KO mice receiving Mn injection compared to the Wild Type mice receiving Mn injection. Additionally the administration of cocaine, a DAT inhibitor, and reserpine, which decreases extracellular DA concentrations by inhibiting vesicular reuptake in the synapse to rats injected with Mn intrathecally, caused a significant decrease in Mn accumulation in the GPe (Ingersoll *et al.*, 1999). In another study with isolated synaptosomes from the striatum which were incubated with GBR12909, a specific inhibitor of DAT, desipramine used to inhibit norepinephrine transporter or fluoxetine to inhibit serotonin transporter, it has been observed that only those synaptosomes treated with the specific DAT inhibitor, significantly blocked Mn accumulation *in vitro*. No effect was seen in those synaptosomes treated with either desipramine or fluoxetine. These data suggest that during Mn toxicity, DAT is involved in the facilitation of the specific accumulation of Mn into the basal ganglia (Anderson *et al.*, 2007). Furthermore, Chin-Chang *et al.* (2003) evaluated the function of DAT in the nigrostriatal dopaminergic pathway in previously documented chronic manganism patients. These authors found a significantly higher activity of DAT in symptomatic manganism patients compared with PD patients.

Nonetheless, some authors have proposed that Mn may not be transported directly by DAT, but rather, its transport may be affected by interactions of DAT protein with other putative Mn transporters, such as DMT-1 (Burdo *et al.*, 1999). DMT-1 is emerging as an important protein for cellular transport of iron and Mn (Roth and Garrick, 2003). It is known that the striatum is a brain region that is rich in DMT-1 (Burdo *et al.*, 1999). Recent studies have shown that DMT-1 levels increase due to low brain iron and that this increase coincides with elevated Mn concentrations (Erikson *et al.*, 2004). In their study, Erikson *et al.* (2005b) showed that neither DMT-1 nor iron concentrations were abnormal in the DAT-KO mice when compared to WT, this demonstrates that the significant attenuation of Mn accumulation observed in the striatum of DAT-KO mice is most likely due to the lack of a functioning DAT. In this way it could be that DAT blockade in the basal ganglia causes an alteration in DMT-1 functioning thereby decreasing regional Mn transport, but this remains to be elucidated.

The data mentioned previously prove that Mn accumulates especially in the basal ganglia and induces movement disorders, however, once Mn has reached and enters the neurons of these nuclei, how does it exert its toxicity? And how does it cause neuronal death? In order to explain the pathogenesis of Mn toxicity, several putative mechanisms have been proposed, which include oxidative stress process (Villalobos *et al.*, 2001; Zhang *et al*, 2004; Zhang *et al.*, 2007), activation or liberation of proapoptotic molecules (Masashi *et al.*, 2005), excitotoxicity (Fitsanakis *et al.*, 2006) and neuroinflammation (Filipov *et al.*, 2005).

5.5. Oxidative Stress

It has been demonstrated that higher concentrations of Mn may significantly accelerate the autoxidation of DA and other catecholamines, which concurrently amplify the formation of reactive oxygen species (ROS) and toxic quinines (Sloot *et al.*, 1996; Sava *et al.*, 2004). Mn oxidation state is important in determining the functionality and toxicity in biological systems (Thompson and Orvig, 2003). Mn normally exists in both Mn^{2+} and Mn^{3+} species within the blood (Harris and Chen, 1994); several studies showed that Mn^{2+} and Mn^{3+} utilize different transport mechanisms for cellular uptake and transport across the blood brain barrier; DMT-1 is known to transport Mn^{2+} species and Tf receptor-mediated endocytosis transports Mn^{3+}-Tf species (Zheng *et al.*, 2003). It is well known that Mn^{2+} induces cellular toxicity (Sun *et al.*, 1993), for example Linert *et al.* (1996) have shown that Mn^{2+} significantly enhances the rate of DA auto-oxidation *in vitro*. Also, it has been reported that Mn^{2+} stimulates DA autoxidation in the dopaminergic neurons, a processes accompanied by an increase in the formation of quinones and protein bound cysteinyl-DA and cysteinyldihydroxyphenylacetic acid (DOPAC; Ramesh *et al.*, 2002). However, some reports indicate that Mn^{2+} is a potent inhibitor of cellular membrane lipid peroxidation and an effective scavenger of superoxide anion ($O_2^{\bullet-}$) produced during spontaneous dopamine auto-oxidation (Villalobos *et al.*, 2001), furthermore Mn is a constituent of superoxide dismutase (MnSOD), which converts $O_2^{\bullet-}$ to hydrogen peroxide (H_2O_2) inside the mitochondria (Fridovich, 1995). In this way, these studies propose an antioxidant effect of Mn^{2+} contrary to the well-documented pro-oxidant properties (Sziráki *et al.*, 1998). On the other hand, Archibald and Tyree (1987) suggest that Mn^{3+} efficiently oxidized DA compared to Mn^{2+}. Oxidation of DA by Mn^{3+} neither produced nor required oxygen, and Mn^{3+} is far more than efficient Mn^{2+}, Mn^{4+}, O_2 or H_2O_2 in oxidizing catecholamines. In this way, Ali *et al.*, (1995) demonstrated that *in vitro* exposure to either $MnCl_2$ (Mn^{2+}) or $MnOAc_3$ (Mn^{3+}) produced a dose-dependent increase of ROS in striatum; however, $MnOAc_3$ produced a significant increase of ROS at much lower concentrations.

It has been proposed that Mn^{2+} might be converted to Mn^{3+}, which may in turn attack catecholamine neurotransmitters. More than one theory exists about the formation of Mn^{3+} and the oxidation of catecholamines. Donaldson *et al.*, (1982) suggested that regardless of the valence state of Mn that enters the brain, spontaneous oxidation and dismutation, peroxidative activity, or $O_2^{\bullet-}$ mediated oxidation, would give rise to Mn^{3+}. Furthermore, the presence of two hydroxyl groups at adjacent 3- and 4-positions on DA and DOPAC will allow Mn^{2+} and Mn^{3+} to redox cycle and generate semiquinones and orthoquinone by the sequential oxidation of catecholamines via several one electron transfer reactions (Donaldson, 1987). For that reason, many researchers have speculated that Mn^{2+}-oxidation to Mn^{3+} by oxygen free radicals is the initiating event in Mn toxicity.

A second theory is that Mn^{2+} is not oxidized to Mn^{3+} as initial event. HaMai *et al.* (2001) demonstrated that trace amounts of Mn^{3+} are necessary to cause formation of ROS. According to a mechanism proposed by these authors, the pro-oxidant activity of Mn^{2+} is dependent on trace amounts of Mn^{3+}, which may facilitate a small portion of Mn^{2+} to oxidize to Mn^{3+}. This synergistic relationship between Mn^{2+} and Mn^{3+} results in continuous redox cycling. The significance of Mn^{3+} in catalytic promotion of oxidative events is supported by the observation that Mn^{3+}, but not Mn^{2+}, accelerates the oxidation of ferrous iron. Low micromolar concentrations of Mn^{3+} were sufficient to trigger an immediate oxidation of ferrous iron, whereas Mn^{2+} at concentrations of 100-fold higher did not promote the conversion of ferrous to ferric ion (HaMai and Bondi, 2004a). In a second work, HaMai and Bondy (2004b) explained that the increased toxicity of Mn^{3+} could be by the four unpaired d orbital electrons which results in a highly unstable atom with an elevated redox potential compared to the more thermodynamically stable Mn^{2+}. Furthermore, it has been demonstrated that after Mn overexposure, this metal is accumulated inside the mitochondria, inhibiting complex I, leading to altered oxidative phosphorylation and generating ROS, mainly $O_2^{\bullet-}$. It has established that Mn^{3+} has a stronger affinity for complex I than Mn^{2+}, but Mn^{2+} is the predominant ion *in vivo* (Aschner *et al.*, 2005b). Therefore, it is possible that Mn^{3+} damages the mitochondrial electron transport system, and the $O_2^{\bullet-}$ generated can react with the abundant Mn^{2+} and oxidized to Mn^{3+} (Archibald and Tyree, 1987; Sistrunk et al., 2007).

5.6. Apoptotic Neuronal Death

A commonly evoked mechanism for Mn-induced neuronal death is apoptosis, a programmed cell death that involves DNA cleavage, mitochondrial dysfunction and activation of signalling processes that eventually lead to proteases activation (like caspases) and cell death (Quintanar, 2008)

Some studies have revealed that 60%–70% of the accumulated Mn is sequestered in mitochondria whereas the rest is localized in the cytosol (Wedler *et al.*, 1989), and 97–99% of intramitochondrial Mn^{2+} is bound, probably to the inner mitochondrial membrane (Gunter *et al.*, 2004). Mn accumulates within mitochondria (~100 nm) preferentially via the uniporter mechanism; however, Mn is slowly transported out of mitochondria via the sodium independent (and energy-dependent) efflux mechanisms. Mn in the form of Mn^{2+} appears to inhibit both Na^+-dependent and Na^+-independent Ca^{2+} efflux, whereas Ca^{2+} does not appear to inhibit Mn^{2+} efflux from brain mitochondria. Mn^{2+} inhibition of Ca^{2+} efflux is thought to increase the probability of the mitochondria undergoing mitochondrial permeability transition (in which the mitochondria swell and relatively large pores in the membrane open to allow release of substances) as well as decrease oxidative phosphorylation; hence, cellular energy is depleted since decreased oxidative phosphorylation causes decreased ATP synthesis (McKinney et al., 2004).

During the opening of the transition pore, cytochrome c is released into the cytoplasm (Anantharam *et al.*, 2002) and the liberation of this molecule from the mitochondria is an early event that occurs during programmed cell death (Müller-Höcker, 1992; Crompton, 1999; Sava *et al.*, 2004). ROS also, has been shown to induce cytochrome c release from mitochondria in both neuronal and non-neuronal systems (Lee and Wei, 2000).

It is well known that cytochrome c release from mitochondria leds to its interaction with cytosolic factors, apoptosis protease-activating factors (Apaf-1) and caspases-9. Binding of Apaf-1 to cytochrome c in the cytosol led to recruitment of other Apaf-1 molecule and procaspase-9 to generate a large complex called apoptosome. Once assembled, procaspase-9 is activated to caspases-9 via autocatalysis, which dissociates from the complex to proteolytically activate downstream caspases such as caspases-3. In this process cytochrome c release is critical because Apaf-1 is unable to bind procaspase-9 in its absence. Activated caspases-3 could activate the caspase-6 and 7. Caspase-7 releases caspase activated DNase (CAD), leading to DNA degradation (Singh and Dikshit, 2007).

It has been demonstrated that Mn may trigger apoptotic-like neuronal cell death secondary to mitochondrial dysfunction in dopaminergic neurons (Panov *et al.*, 2004). Masashi *et al.* (2005) demonstrated that Mn activates the mitochondrial- dependent caspase cascade mainly by means of the activation of caspase-3. Biochemical consequences of caspase-3 activation are proteolytic cleavage of cellular targets associated with apoptosis. Poly (ADP-ribose) polymerase, a DNA cleaving enzyme, has been established as one of the important apoptotic substrates of caspase-3 (Earnshaw *et al.*, 1999; Schultz and Andreasen, 1999). In the same way, Latchoumycandane *et al.*, (2005) examined the effect of Mn on the activity of caspase-3 in N27 mesencephalic clonal cells. These authors report that the treatment of N27 cells with 300 μM Mn resulted in a significant increase in cytosolic cytochrome c, and that exposure to Mn induced a time-and dose-dependent increase in caspase-3 activity compared with controls.

Caspase-12, the most recently characterized member of the ICE (interleukin-1b converting enzyme) subfamily of caspases, mediates endoplasmic reticulum (ER) specific apoptosis. Caspase-12 is localized in ER as a latent zymogen and activated only by the ER stress inducers. It has been demonstrated that Mn exposure resulted in the immediate activation of caspase-12 during the apoptotic death of SN4741 cells. Some other studies using SN4741 cells showed the activation of caspase-1 and caspase-3 during the dopaminergic cell death. Also, Caspase-8/Caspase-9 and Caspase-1-like activities were significantly increased at 8h and 12h after Mn treatment respectively, and the downstream effector caspase-3-like activity was most significantly activated after 24h. Bcl-2 was the first identified antiapoptotic gene in the Bcl-2 family. Bcl-2 protein is located primarily in the ER, nuclear membrane and outer mitochondrial membrane. As expected, Mn-induced dopaminergic cell death was dramatically reduced in all Bcl-2 overexpressed dopaminergic cell lines (Chun *et al.*, 2001).

5.7. Excitotoxicity

Glu is the most abundant excitatory neurotransmitter in the brain. It has been proposed that it plays several major roles in brain function such as cognition, learning and memory as well as development of the CNS, including synapse induction and elimination, cell migration, differentiation and death (Cooper *et al.*, 1990). After its release from synaptic terminals, Glu activates different receptor subtypes in postsynaptic neurons: N-methyl-D-aspartate (NMDA), α-amino-3-hydroxy-5-methyl-4-isoxazolepropionic acid (AMPA)/kainic acid, and metabotropic receptors. The extracellular concentration of Glu is highly regulated through specific Na^+-dependent high affinity transporters located both, in neurons and glial cells, however, impairment of Glu removal after its synaptic release leads to the accumulation of the amino acid in the synaptic cleft. The subsequent sustained activation of Glu receptors may trigger excitotoxic neuronal death. This type of cell death is dependent on the increase in the intracellular concentration of Ca^{2+} after its influx through the NMDA receptor channel. Ca^{2+}-activated enzymes such as proteases, endonucleases and phospholipases contribute to the degradation of different cell components and neuronal degeneration (Estrada-Sánchez *et al.* 2008).

It has been shown that Glu play a crucial role in the development of Mn-induced striatal toxicity *in vivo* (Brouillet *et al.*, 1993; Verity, 1999; Centonze *et al.*, 2001). Previous studies have shown that accumulation of this metal in the brain produces a type of neuropathology consistent with an excitotoxic mechanism, suggesting that a glutamatergic mechanism may be involved in the process of neurotoxicity of Mn (Hazell, 2002; Fitsanakis *et al.*, 2006). In this way it has been shown that Glu levels are elevated in the basal ganglia of Mn-exposed rats (Erikson and Aschner, 2003). Furthermore, chronic treatment with Mn might trigger excitotoxic events by altering corticostriatal synaptic transmission at pre-or postsynaptic levels, and these two events might coexist and cooperate to induce striatal damage (Centonze *et al.*, 2001). Mn also causes the activation of glutamate-gated cation channels, e.g. NMDA receptor, which contributes to neuronal degeneration. Some studies indicate that Mn^{2+} can either substitute Ca^{2+} in the exocytotic process, or induce the release of Ca^{2+} from intracellular stores, possibly the ER (Takeda *et al.*, 2003).

On the other hand, re-uptake of Glu from the synaptic cleft is an important function of astrocytes and is essential for maintaining normal levels of this neurotransmitter in the extracellular space (Cholet *et al.*, 2002). GLAST (glutamate/aspartate transporter) and glutamate transporter-1 (GLT-1) are the most prominent astrocytic Glu transporters (Erikson *et al.* 2007). These two transporters bind one Glu molecule and three sodium ions externally and translocated them into the cytoplasm, while one potassium ion is extruded to the extracellular space (Tzingounis and Wadiche, 2007). Glu taken up by glial cells is metabolized to glutamine, which in turn is released to the extracellular space and taken up by neurons, where it is again converted to Glu to replenish the neurotransmitter pool (Estrada-Sánchez *et al.*, 2008).

It has been demonstrated that the astrocytes treatment with Mn leads to a decrease in Glu transport which does not appear to be due to a generalized toxic effect on these cells (Hazell and Norenberg, 1997; Hazell, 2002), neither appear to be due to an energy-related functional disturbance in the transport mechanism since ATP levels were unaffected (Hazell and Norenberg, 1997). The mechanism by which Mn causes a decrease in Glu uptake in astroglial cells is not known, but it has been proposed that is caused by the alteration in Glu transporters. Erikson *et al.* (2002) have reported that altered Glu uptake by astrocytes is linked to Mn dose-dependent decreased GLAST expression. GLT-1 has also been shown to be affected to Mn exposure, indicating possible mechanisms for Mn-associated decrease in Glu uptake (Mutkus *et al.*, 2005). More recently, Erikson *et al.* (2007) informed that GLT-1 mRNA was significantly decreased in the GP and striatum of monkeys exposed to Mn.

In summary, Mn exposure causes the increase of Glu in the basal ganglia, likewise appears to affect the gene and protein expression of both Glu transporters in the striatum and GP, in addition Mn causes the activation of NMDA receptor. Therefore, we can assume from these findings that Mn may alter Glu homeostasis in the basal ganglia contributing to neuronal degeneration by excitotoxicity.

5.8. Neuroinflammation

Studies performed at the end of the 20th century have demonstrated the important role of neuroglia in the development of neurological derangements (Shukakidze *et al.*, 2002). It has been established that an uncontrolled or chronic inflammatory response, while it is an essential defense against infection, may cause irreversible tissue damage (Filipov *et al.*, 2005).

Reactive microglia and astrocytes are known to secrete different proinflammatory cytokines and cytotoxic molecules, such as interleukin-1 beta (IL-1β), tumor necrosis factor-alfa (TNF-α) nitric oxide (NO) (Tomás-Camardiel *et al.*, 2002), ROS and other substances inducing not only neuron degeneration but also secondary activation of dedifferentiated astrocytes (Shukakidze *et al.*, 2002; Zhang *et al.*, 2007). Although a few of these factors are thought to contribute to tissue repair, some of them are believed to work via mechanisms not yet fully understood to induce neurodegeneration (Liu *et al.*, 2002).

Recent evidence suggests that Mn neurotoxicity involves the activation of microglia and/or astrocytes (Chang and Liu, 1999). Tomás-Camardiel *et al.* (2002) reported that typical morphological features of reactive microglial cells were observed in response to Mn administration in the SNc (mainly) and striatum. These Mn-induced changes were regionally specific, since no significant changes were observed in other brain areas as the ventral tegmental area (VTA), nucleus accumbens and GP. Mn induced-activation of microglial cells in the nigrostriatal system could be a consequence of tissue specific damage, a specific activation of glial cells or both. Interestingly, some studies have shown that the distribution of microglia in the brain is not uniform and the midbrain region that encompasses the basal ganglia is particularly enriched in microglial cells (Kim *et al.*, 2000).

Mn activation of microglia and/or astrocytes is manifested with an increase in NO production, via increased expression of inducible nitric oxide synthase (iNOS). Jae-Hoon *et al.* (2006) have demonstrated that Mn exposure induced concentration-dependent increase of iNOS protein expression in BV2 cells (murine microglial cells). Several *in vitro* investigations showed a potentiating effect of Mn on NO production. These reports has shown that primary murine astrocytes or C6 glioma cells exposed to Mn combined with IL-1β and IFN-γ (Spranger *et al.*, 1998), or with lipopolysaccharide (LPS) (Barhoumi *et al.*, 2004) induced elevated NO production. Others showed the potentiation of NO production by higher concentration of Mn in LPS-stimulated N9 murine microglia cell line (Chang and Liu, 1999; Filipov *et al.*, 2005).

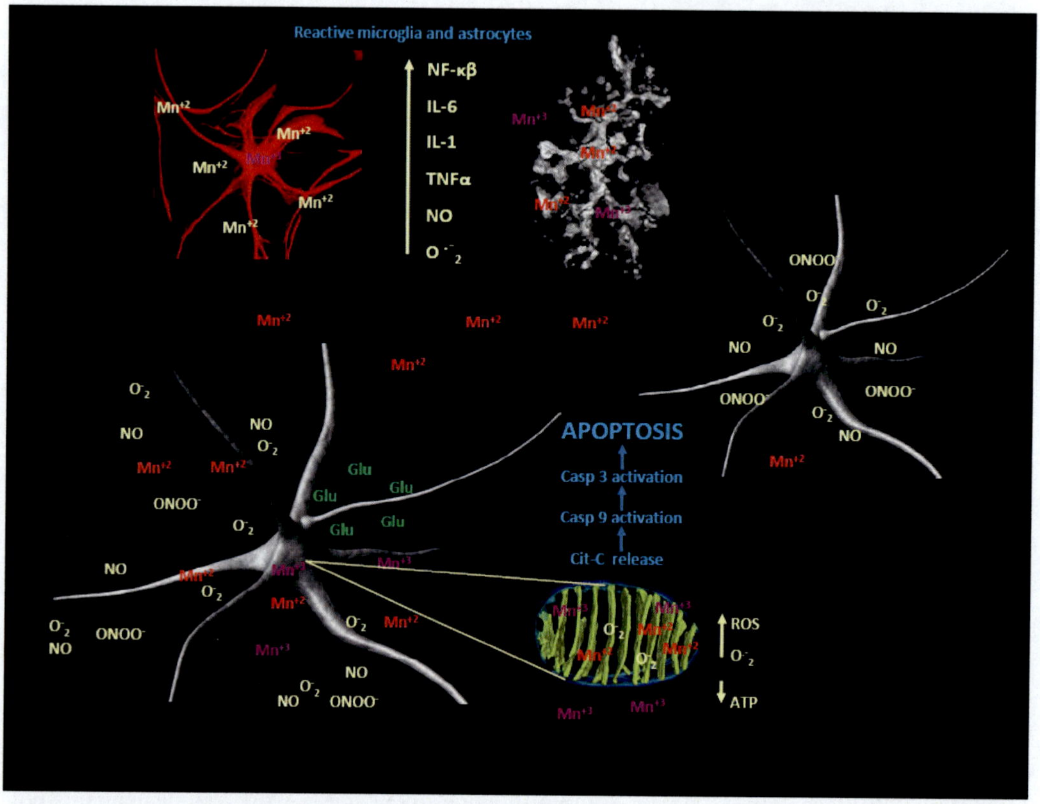

Figure 2. Possible role of the Mn in neuronal death. After overexposure Mn is accumulated inside the mitochondria increasing ROS production, decrease in ATP synthesis and cytochrome c release. Also, some studies have shown that Glu levels are elevated in the basal ganglia of Mn-exposed rats; increased extracellular Glu concentration could play a key role in Mn-induced neuronal cell death. Finally, microglia and astrocytes are known to be activated in presence of Mn, and secrete different proinflammatory cytokines and cytotoxic molecules, such as IL-1β, IL-6, TNF-

In this way, as it has been mentioned, Mn exposure leads to the generation of ROS such as $O_2^{\bullet -}$, $OH^{\cdot}$ and H_2O_2. It has been established that $O_2^{\bullet -}$ can react with NO to generate the highly toxic radical peroxynitrite ($ONOO^-$), which can be converted to an equally toxic hydroxyl radical and nitrate radical (Barzilai *et al.*, 2001). These data suggest that reactive nitrogen intermediates contribute to neuronal cell death, possibly by inhibiting components of the mitochondrial respiratory chain and, as a result, compromising the cells energy; also it has been proven that $ONOO^-$ can nitrate and hydroxylate aromatic amino acids as well as react with lipids, proteins and DNA (Chang and Liu, 1999). Even more interesting is the fact that neurons are susceptible to NO and $ONOO^-$ exposure, whereas the principal producers of NO/$ONOO^-$ in the brain, the microglia and the astrocytes, are relatively resistant (Heales *et al.*, 1999; Tomás-Camardiel *et al.*, 2002).

In this chapter we show some advances to understand the selective vulnerability of the basal ganglia to Mn. Although the exact mechanisms of Mn neurotoxicity are still being unraveled, the neurotoxic effects of Mn have been associated with direct action of this metal on neuronal cells, mitochondrial dysfunction which in turn, can induce an oxidative stress state or the release of proapoptotic molecules, induction of neurotransmitter imbalance (excitotoxicity) and neuroimmflamation; these processes are summarized in the Figure 2.

References

Ali, S.E., Duhart, H.M., Newport, G.D., Lipe, G.W. and Slikker, W. Jr. (1995) Manganese-Induced Reactive Oxygen Species: Comparison between Mn^{+2} and Mn^{+3}. *Neurodegeneration*, 4:329-334.

Anderson, J.G., Cooney, P.T. and Erikson, K.M. (2007) Inhibition of DAT function attenuates manganese accumulation in the globus pallidus. *Environ Toxicol Pharmacol.* 23:179-184.

Anantharam, V., Kitazawa, M., Wagner, J., Kaul, S. and Kanthasamy, A.G. (2002) Caspase-3-dependent proteolytic cleavage of protein kinase C delta is essential for oxidative stress-mediated dopaminergic cell death after exposure to methylcyclopentadienyl manganese tricarbonyl. *J Neurosci.* 22:1738-51.

Archibald, F.S. and Tyree, C. (1987) Manganese poisoning and the attack of trivalent manganese upon catecholamines. *Arch Biochem Biophys.* 256:638-650.

Aschner, M., Vrana, K. and Zheng, W. (1999) Manganese uptake and distribution in the central nervous system. *Neurotoxicology*, 20:173-180.

Aschner, M.A., Erikson, K.M. and Dorman, D.C. (2005a) A review of manganese toxicokinetics. *Crit Rev Toxicol.* 35:1-32.

Aschner, M., Erikson, K.M. and Dorman, D.C. (2005b) Manganese Dosimetry: Species Differences and Implications for Neurotoxicity. *Crit Rev Toxicol.* 35:1-32.

Aschner, M., Guilarte, T.R., Schneider, J.S. and Zheng, W. (2007) Manganese: Recent advances in understanding its transport and neurotoxicity *Toxicol App Pharmacol.* 221:131-147.

Au, W.L., Adams, J.R., Troiano, A.R. and Stoessl, A.J. (2005) Parkinson's disease: in vivo assessment of disease progression using positron emission tomography. *Mol Brain Res.* 134:24-33.

Bahena-Trujillo, R., Flores, G. and Arias-Montaño, J.A. (2000) Dopamina: síntesis, liberación y receptores en el Sistema Nervioso Central. *Rev Biomed.* 11:39-60.

Barbeau, A., Inoué, N. and Cloutier, T. (1976) Role of manganese in dystonia. *Adv Neurol.* 14:339-352.

Barceloux, D.G. (1999) Manganese. *Clinical Toxicol.* 37:293-307.

Barhoumi, R., Faske, J., Liu, X. and Tjalkens, R.B. (2004) Manganese potentiates lipopolysaccharide-induced expression of NOS2 in C6 glioma cells through mitochondrial-dependent activation of nuclear factor kappa B. *Mol Brain Res.* 122:167-179.

Barzilai, A., Melamed, E. and Shirvan, A. (2001) Is there a rationale for neuroprotection against Dopamine toxicity in Parkinson´s disease? *Cell Mol Neurobiol.* 21: 215-230.

Brouillet, E.P., Shinobu, L., McGarvey, U., Hochberg, F. and Beal, M.F. (1993) Manganese injection into rat striatum produces excitotoxic lesions by impairing energy metabolism. *Exp Neurol.* 120:89–94.

Burdo, J.R., Martin, J., Menzies, S.L., Dolan, K.G., Romano, M.A., Fletcher, R.J., Garrick, M.D., Garrick, L.M. and Connor, J.R. (1999) Cellular distribution of iron in the brain of the Belgrade rat. *Neuroscience*, 93:1189-1196.

Calabresi, P., Centonze, D. and Bernardi, G. (2000) Cellular factors controlling neuronal vulnerability in the striatum. A lesson from the striatum. *Neurology*, 55: 1249–1255.

Calne, D.B., Chu, N.S., Huang, C.C., Lu, C.S. and Olanow, C.W. (1994) Manganism and idiopathic parkinsonism: similarities and differences. *Neurology*, 44:1583-1586.

Carlsson, A. (1995) Neurocircuitries and neurotransmitter interactions in schizophrenia. *Int Clin Psychopharmacol.* 3:21-28.

Centonze, D., Gubellini, P., Bernardi, G. and Calabresi, P. (2001) Impaired Excitatory Transmission in the Striatum of Rats Chronically Intoxicated with manganese *Exp Neurol.* 172:469-476.

Chang, Y.J. and Liu, L.Z. (1999) Manganese potentiates nitric oxide production by microglia. *Mol Brain Res.* 68:22-28.

Chesselet, M.F. and Delfs, J.M. 1996 Basal ganglia and movement disorders: an update. *Trends Neurosci.* 19:417-22.

Chin-Chang, H., Yi-Hsin, W., Chin-Song, L., Nai-Shin, C. and Tzu-Chen, Y. (2003) Dopamine transporter binding in chronic manganese intoxication. *J Neurol.* 250:1335-1339.

Cholet, N., Pellerin, L., Magistretti, P.J. and Hamel, E. (2002) Similar perisynaptic glial localization for the Na+, K+ -ATPase α2 subunit and the glutamate transporters GLATS and GLT-1 in the rat somatosensory cortex. *Cerebral Cortex*, 12: 515-525.

Chun, H.S., Lee, H. and Son, J.H. (2001) Manganese induces endoplasmic reticulum (ER) stress and activates multiple caspases in nigral dopaminergic neuronal cells, SN4741. *Neurosci Lett.* 316:5-8.

Cools, R., Gibbs, S.E., Miyakawa, A., Jagust, W. and D'Esposito. (2008) Working Memory Capacity Predicts Dopamine Synthesis Capacity in the Human Striatum. *J Neurosci.* 28:1208 –1212.

Cooper, D.M.F. and Caldwell, K.K. (1990) Signal transduction mechanisms for adenosine. In: Adenosine and adenosine receptors. Williams M. (Eds), The Humana Press, Clifton, New Jersey 105-141.

Couper, J. (1837) On the effects of black oxide of manganese which inhaled into the lung. *Br Ann Med Pharm.* 1:41-42.

Critchfield, J.W., Carl, G.F. and Keen, C.L. (1993) The influence of manganese supplementation on seizure onset and severity and brain monoamines in the genetically epilepsy prone rat. *Epilepsy Res.* 14:3-10.

Crompton, M., Virji, S., Doyle, V., Johnson, N. and Ward, J.M. (1999) The mitochondrial permeability transition pore. *Biochem Soc Symp*. 66:167-79.

Desole, M.S., Esposito, G., Migheli, R., Sircana, S., Delogu, M.R., Fresu, L., Miele, M., De Natale, G. and Miele, E. (1997) Glutathione deficiency potentiates manganese toxicity in rat striatum and brainstem and in PC12 cells. *Pharmacol Res.* 36: 285-292.

Dobson, A.W., Erikson, K.M. and Aschner, M. (2004) Manganese neurotoxicity. *Ann NY Acad Sci.* 1012: 115-128.

Doherty, J.D., Lourte, C.J. and Trams, E.G. (1983) Mn^{2+} stimulated ATPase in rat brain. *Neurochem Res*. 8: 493-500.

Donaldson, J., McGregor, D. and LaBella, F. (1982) Manganese neurotoxicity: a model for free radical mediated neurodegeneration? *Can J Physiol Pharmacol.* 60:1398.

Donaldson, J. (1987) The physiopathologic significance of manganese in brain: its relation to schizophrenia and neurodegenerative disorders. *Neurotoxicology*, 8:451-462.

Dorman, D.C., McManus, B.E., Parkinson, C.U., Manuel, C.A., McElveen, A.M. and Everitt, J.L. (2004) Nasal toxicity of manganese sulfate and manganese phosphate in young male rats following subchronic (13-week) inhalation exposure. *Inhal Toxicol.* 16:481-488.

Earnshaw, W.C., Martins, L.M. and Kaufmann, S.H. (1999) Mammalian caspases: structure, activation, substrates, and functions during apoptosis. *Annu Rev Biochem*. 68:383-424.

Eriksonm, K.M., Suberm, R.L. and Aschner, M. (2002) Glutamate/aspartate transporter (GLAST), taurine transporter and metallothionein mRNA levels are differentially altered in astrocytes exposed to manganese chloride, manganese phosphate or manganese sulfate. *Neurotoxicology*, 23:281-288.

Erikson, K.M. and Aschner, M. (2003) Manganese neurotoxicity and glutamate-GABA interaction. *Neurochem Int.* 43:475-480.

Erikson, K.M., Syversen, T., Stennes, E. and Aschner, M. (2004) Globus pallidus: a target brain region for divalent metal accumulation associated with dietary iron deficiency. *J Nutr Biochem.* 15:335-341.

Erikson, K.M., Syversen, T., Aschner, J.L. and Aschner, M. (2005a) Interactions between excessive manganese exposures and dietary iron-deficiency in neurodegeneration. *Environ Toxicol Pharmacol.* 19:415-421.

Erikson, K.M., John, C.E., Jones, S.R. and Aschner, M. (2005b) Manganese accumulation in striatum of mice exposed to toxic doses is dependent upon a functional dopamine transporter. *Environ Toxicol Pharmacol.* 20:390-394.

Erikson, K.M., Dorman, D.C., Lash, L.H. and Aschner, M. (2007) Manganese Inhalation by Rhesus Monkeys is Associated with Brain Regional Changes in Biomarkers of Neurotoxicity. *Toxicol Sci.* 97:459-466.

Estrada-Sánchez, A.M., Mejía-Toiber, J. and Massieu, L. (2008) Excitotoxic neuronal death and the pathogenesis of Huntington's disease. *Arch Med Res*. 39:265-276.

Filipov, N.M., Seegal, R.F. and Lawrence, D.A. (2005) Manganese Potentiates *InVitro* Production of Proinflammatory Cytokines and Nitric Oxide by Microglia Through a Nuclear Factor kappa B-Dependent Mechanism. *Toxicol Sci.* 84:139-148.

Fitsanakis, V.A., Au, C., Erikson, K.M. and Aschner, M. (2006) The effects of manganese on glutamate, dopamine and gamma-aminobutyric acid regulation. *Neurochem Int.* 48:426-433.

Fridovich, I. (1995) Superoxide radical and superoxide dismutase. *Rev Biochem.* 64:97-112.

Frumkin, H. and Solomon, G. (1997) Manganese in U.S. gasoline supply. *Am J Ind Med.* 31:107-115.

Gonzalez-Reyes, R.E., Gutierrez-Alvarez, A.M. and Moreno, C.B. 2007 Manganese and epilepsy: a systematic review of the literature. *Brain Res Rev.* 53:332-336.

Greene, J.G. and Greenamyre, J.T. (1996) Bioenergetics and glutamate excitotoxicity. *Prog Neurobiol.* 48: 613–634.

Gunter, T.E., Miller, L.M., Gavin, C.E., Eliseev, R., Salter, J., Buntinas, L., Alexandrov, A., Hammond, S. and Gunter, K.K. (2004) Determination of the oxidation states of manganese in brain, liver, and heart mitochondria. *J Neurochem.* 88:266-280.

Gwiazda, R., Lee, D., Sheridan, J. and Smith, D.R. (2002) Low cumulative manganese exposure affects striatal GABA but not dopamine. *Neurotoxicology*, 23:69-76.

Gwiazda, R., Lucchini, R. and Smith, D. (2007) Adequacy and consistency of animal studies to evaluate the neurotoxicity of chronic low-level manganese exposure in humans. *J Toxicol Environ Health Part* A, 70: 594-605.

HaMai, D., Campbell, A. and Bondy, S.C. (2001) Modulation of oxidative events by multivalent manganese complexes in brain tissue. *Free Rad Biol Med.* 31:763-768.

HaMai, D. and Bondy, S.C. (2004a) Pro-or anti-oxidant manganese: a suggested mechanism for reconciliation. *Neurochem Int.* 44:223-229.

HaMai, D. and Bondy, S.C. (2004b) Oxidative basis of Manganese Neurotoxicity. *Ann NY Acad Sci.* 1012:129-141.

HaMai, D. and Bondy, S.C. (2004b) Oxidative basis of Manganese Neurotoxicity. *Ann NY Acad Sci.* 1012:129-141.

Harris, W.R. and Chen, Y. (1994) Electron paramagnetic resonance and difference ultraviolet studies of Mn2+ binding to serum transferrin. *J Inorg Biochem.* 54:1-19.

Hazell, A.S. and Norenberg, M.D. (1997) Manganese Decreases Glutamate Uptake in Cultured Astrocytes. *Neurochem Res.* 22: 1443-1447.

Hazell, A.S. (2002) Astrocytes and manganese neurotoxicity. *Neurochem Int.* 41:271-277.

Heales, S.J., Bolanos, J.P., Stewart, V.C., Brookes, P.S., Land, J.M. and Clark, J.B. (1999) Nitric oxide, mitochondria and neurological disease. *Biochim Biophys Acta*, 1410:215-228.

Hill, J.M. and Switzer, R.C. (1984) The regional distribution and cellular localization of iron in the rat brain. *Neuroscience*, 11:595-603.

Hirata, Y., Kiuchi, K. and Nagatsu, T. (2001) Manganese mimics the action of 1-methyl-4-phenylpyridinium ion, a dopaminergic neurotoxin, in rat striatal tissue slices. *Neurosci Lett.* 311:53-56.

Huang, C.C., Chu, N.S., Lu, C.S., Wang, J.D., Tsai, J.L., Tzeng, J.L., Walters, E.C. and Calne, D.B. (1989) Chronic manganese intoxication. *Arch Neurol.* 46:1104 -1106.

Ingersoll, R.T., Montgomery, Jr. E.B. and Aposhian, H.V. (1999) Central nervous system toxicity of manganese. II. Cocaine or reserpine inhibit manganese concentration in the rat brain. *Neurotoxicology*, 20:467-476.

Jae-Hoon, B., Byeong-Churl, J., Seong-Il, S., Eunyoung, H., Hyung, H.B., Sung-Soo, K., Mi-young, L. and Dong-Hoon, S. (2006) Manganese induces inducible nitric oxide synthase (iNOS) expression via activation of both MAP kinase and PI3K/Akt pathways in BV2 microglial cells. *Neurosci Lett.* 398:151-154.

Jankovic, J. (2005) Searching for a relationship between manganese and welding and Parkinson's disease. *Neurology*, 64:2021-2028.

Kim, W.G., Mohney, R.P.,Wilson, B., Jeohn, G.H., Liu, B. and Hong, J.S. (2000) Regional difference in susceptibility to lipopolysaccharide induced neurotoxicity in the rat brain: role of microglia. *J Neurosci.* 20:6309-6316.

Lai, J.C., Leung, T.K. and Lim, L. (1984) Differences in the neurotoxic effects of manganese during development and aging: some observations on brain regional neurotransmitter and non-neurotransmitter metabolism in a developmental rat model of chronic manganese encephalopathy. *Neurotoxicology*, 5: 37-47.

Latchoumycandane, C., Anantharam, V., Kitazawa, M., Yang, Y., Kanthasamy, A. and Kanthasamym, A.G. (2005) Protein kinase Cδ is a key downstream mediator of manganese-induced apoptosis in dopaminergic neuronal cells. *J Pharmacol Exp Ther.* 313: 46-55.

Lee, H.C. and Wei, Y.H. (2000) Mitochondrial role in life and death of the cell. *J Biomed Sci.* 7:2-15.

Linert, W., Herlinger, E., Jameson, R.F., Kienzl, E., Jellinger, K. and Youdim, M.B. (1996) Dopamine, 6-hydroxydopamine, iron, and dioxygen-the irmutual interactions and possible implication in the development of Parkinson's disease. *Biochim Biophys Acta*, 1316:160-168.

Lipe, G.W., Duhart, H., Newport, G.D., Slikker, W. Jr. and Ali, S.F. (1999) Effect of manganese on the concentration of amino acids in different regions of the rat brain. *J Environ Sci Health* B, 34: 119-132.

Liu, B., Gao, H.M., Wang, J.Y., Jeohn, G.H., Cooper, C.L. and Hong, J.S. (2002) Role of nitric oxide in inflammation-mediated neurodegeneration. *Ann NY Acad Sci.* 962:700-715.

Marsden, C.D. and Obeso, J.A. (1994) The functions of the basal ganglia and the paradox of stereotaxic surgery in Parkinson's disease. *Brain*, 117:877-97.

Masashi, K., Anantharam, V., Yang, Y., Hirata, Y., Kanthasamy, A. and Kanthasamy, A.G. (2005) Activation of protein kinase C?by proteolytic cleavage contributes to manganese-induced apoptosis in dopaminergic cells: protective role of Bcl-2 *Biochem Pharmacol.* 69:133-146.

McKinney, A.M., Filice, R.W., Teksam, M., Casey, S., Truwit, C., Clark, H.B., Woon, C. and Liu, H.Y (2004) Diffusion abnormalities of the globi pallidi in manganese neurotoxicity. *Neuroradiology*, 46: 291–295.

Moos, T. (1996) Immunohistochemical localization of intraneuronal transferrin receptor immunoreactivity in the adult mouse central nervous system. *J Comp Neurol.* 375:675-692.

Mora, F., Segovia, G. and del Arco, A. (2008) Glutamate–dopamine–GABA interactions in the aging basal ganglia. *Brain Res Rev.* 58:340-53.

Morris, C.M., Candy, J.M., Keith, A.B., Oakley, A., Taylor, G., Pullen, R.G.L., Bloxham, C.A., Gocht, A. and Edwardson, J.A. (1992) Brain iron homeostasis. *J Inorg Biochem.* 47: 257-265.

Müller-Höcker, J., Schneiderbanger, K., Stefani, F.H. and Kadenbach, B. (1992) Progressive loss of cytochrome c oxidase in the human extraocular muscles in ageing--a cytochemical-immunohistochemical study. *Mutat Res.* Sep;275(3-6):115-24.

Murphy, V.A., Wadhwani, K.C., Smith, Q.R. and Rapoport, S.I. (1991) Saturable transport of manganese (II) across the rat blood-brain barrier. *J Neurochem.* 57: 948-954.

Mutkus, L., Aschner, J.L., Fitsanakis, V. and Aschner, M. (2005). The in vitro uptake of glutamate in GLAST and GLT-1 transfected mutant CHO-K1 cells is inhibited by manganese. *Biol Trace Elem Res.* 107: 221–230.

Normandin, L. and Hazell Nandedkar, A.S. (2002) Manganese Neurotoxicity: An Update of Pathophysiologic Mechanisms. *Metabol Brain Dis.* 17: 375-387.

Normandin, L., Panisset, M. and Zayed, J. (2002) Manganese neurotoxicity: Behavioral, pathological, and biochemical effects following various routes of exposure. *Rev Environ Health*, 17:189-217.

Olanow, C.W. (2004) Manganese-induced parkinsonism and Parkinson's disease. *Ann N Y Acad Sci.* 1012:209-223.

Ordoñez-Librado, J.L., Gutiérrez-Valdez, A.L., Colín-Barenque, L., Anaya-Martínez, V., Díaz-Bech, P. and Avila-Costa, M.R. (2008) Inhalation of divalent and trivalent manganese mixture induces a Parkinson's disease model: Immunocytochemical and behavioral evidences. *Neuroscience*, 155: 7–16.

Panov, A.V., Andreeva, L. and Greenamyre, J.T. (2004) Quantitative evaluation of the effects of mitochondrial permeability transition pore modifiers on accumulation of calcium phosphate: comparison of rat liver and brain mitochondria. *Arch Biochem Biophys.* 424: 44-52.

Parenti, M., Flauto, C., Parati, E., Vescovi, A. and Groppetti, A. 1986 Manganese neurotoxicity: effects of L-DOPA and pargyline treatments. *Brain Res.* 367: 8-13.

Petzinger, G.M., Fisher, B., Hogg, E., Abernathy, A., Arevalo, P., Nixon, K. and Jakowec, M.W. (2006) Behavioral motor recovery in the 1-methyl-4-phenyl-1,2,3,6-tetrahydropyridine-lesioned squirrel monkey (Saimiri sciureus): changes in striatal dopamine and expression of tyrosine hydroxylase and dopamine transporter proteins. *J Neurosci Res.* 83:332-347.

Ponzoni, S., Guimaraes, F.S., Del Bel, E.A. and García –Caraisco, N. (2000) Behavioral effects of intrnigral microinjections of manganese chloride: Interantion with nitric oxide. *Prog Neuro-psichophmmacot Biol Psychiat.* 24:307-325.

Quintanar, L. (2008) Manganese neurotoxicity: a bioinorganic chemist's perspective. *Inorganica Chimica Acta*, 361:875-884.

Racette, B.A., McGee-Minnich, L., Moerlein, S.M., Mink, J.W., Videen, T.O. and Perlmutter, J.S. (2001) Welding-related parkinsonism: clinical features, treatment, and pathophysiology. *Neurology*, 56:8-13.

Racette, B., Antenor, J., McGee-Minnich, L., Moerlein, S., Videen, T., Kotagal, V. and Perlmutter, J. (2005). [18F]FDOPA PET and clinical features in parkinsonism due to manganism. *Mov Disord.* 20: 492-496.

Ramesh, G.T., Ghosh, D. and Gunasekar PG. (2002). Activation of early signaling transcription factor, NF-kB following low-level manganese exposure. *Toxicol Lett.* 136:151-158.

Reaney, S.H., Bench, G. and Smith, D.R. (2006) Brain Accumulation and Toxicity of Mn(II) and Mn(III) Exposures. *Toxicol Sci.* 93:114-124.

Roels, H., Meiers, G., Delos, M., Ortega, I., Lauwerys, R., Buchet, J.P. and Lison, D. (1997) Influence of the route of administration and the chemical form (MnCl2, MnO2) on the absorption and cerebral distribution of manganese in rats. *Arch Toxicol.* 71:223-230.

Roth, J.A. and Garrick, M.D. (2003) Iron interactions and other biological reactions mediating the physiological and toxic actions of manganese. *Biochem Physiol.* 66:1-13.

Salehi, F., Krewski, D., Mergler, D., Normandin, L., Kennedy, G., Philippe, S. and Zayed, J. (2003) Bioaccumulation and locomotor effects of manganese phosphate/sulfate mixture in Sprague-Dawley rats following subchronic (90 days) inhalation exposure. *Toxicol App Pharmacol.* 191:264-271.

Salehi, F., Normandin, L., Krewski, D., Kennedy, G., Philippe, S. and Zayed, J. (2006) Neuropathology, tremor and electromyogram in rats exposed to manganese phosphate/sulfate mixture. *J Appl Toxicol.* 26:419-26.

Sava, V., Mosquera, D., Song, S., Cardozo-Pelaez, F. and Sánchez-Ramos, J.R. (2004) Effects of melanin and manganese on DNA damage and repair in PC12-derived neurons. *Free Rad Biol Med.* 36:1144-1154.

Scheuhamme, A.M, and Cherian, M.G. (1985) Binding of manganese in human and rat plasma. *Biochim Biophys Acta*, 840:163-169.

Schneider, J.S., Decamp, E., Koser, A.J., Fritz, S., Gonczi, H., Syversen, T. and Guilarte, T.R. (2006) Effects of chronic manganese exposure on cognitive and motor functioning in non-human primates. *Brain Res.* 1118:222-231.

Schultz, S.K. and Andreasen, N.C. (1999) Schizophrenia. *Lancet*, 353:1425-1430.

Shukakidze, A., Lazriev, I.L., Khetsuriani, R.G. and Bikashvili, T.Z. (2002) Changes in Neuroglial Ultrastructure in Various Parts of the Rat Brain during Manganese Chloride Poisoning. *Neurosci Behav Physiol.* 32: 561-566.

Shukakidze, A., Lazriev, I. and Mitagvariya, N. (2003) Behavioral Impairments in Acute and Chronic Manganese Poisoning in White Rats. *Neurosci Behav Physiol.* 33:263-267.

Singh, S. and Dikshit, M. (2007) Apoptotic neuronal death in Parkinson's disease: Involvement of nitric oxide. *Brain Res Rev.* 54:233-250.

Sistrunk, S.C., Ross, M.K. and Filipov, N.M. (2007) Direct effects of manganese compounds on dopamine and its metabolite Dopac: An in vitro study. *Environ Toxicol Pharmacol.* 23:286-296.

Sloot, W.N., Korf, J., Koster, J.F., DeWit, L.E. and Gramsbergen, J.B. (1996) Manganese-induced hydroxyl radical formation in rat striatum is not attenuated by dopamine depletion or iron chelation in vivo. *Exp Neurol.* 138:236-245.

Sotogaku, N., Oku, N. and Takeda, A. (2000) Manganese concentration in mouse brain after intravenous injection. *J Neurosci Res.* 61:350-356.

Spranger, M., Schwab, S., Desiderato, S., Bonmann, E., Krieger, D. and Fandrey, J. (1998) Manganese augments nitric oxide synthesis in murine astrocytes: A new pathogenetic mechanism in manganism? *Exp Neurol.* 149:277-283.

Stern, E.A., Jaeger, D. and Wilson, C.J. (1998) Membrane potential synchrony of simultaneously recorded striatal spiny neurons in vivo. *Nature*, 394: 475–478.

Struve, M.F., McManus, B.E., Wong, B.A., and Dorman, D.C. (2007) Basal Ganglia Neurotransmitter Concentrations in Rhesus Monkeys Following Subchronic Manganese Sulfate Inhalation. *Am J Ind Med.* 50:772-778.

Sun, A.Y., Yang, W.L. and Kim, H.D. (1993) Free radical and lipid peroxidation in manganese-induced neuronal cell injury. *Ann NY Acad Sci.* 679:358-363.

Sziráki, I., Mohanakumar, K.P., Rauhala, P., Kim, H.G., Yeh, K.J. and Chiueh, C.C. (1998) Manganese: a transition metal protects nigrostriatal neurons from oxidative stress in the iron-induced animal model of Parkinsonism. *Neuroscience*, 85:1101-1111.

Takeda, A. (2003) Manganese action in brain function. *Brain Res Rev.* 41:79-87.

Takeda, A., Sotogaku, N. and Oku, N. (2003) Influence of manganese on the release of neurotransmitters in rat striatum. *Brain Res.* 965:279-282.

Thompson, K.H. and Orvig, C. (2003) Boon and bane of metal ions in medicine. *Science*, 300:936-9.

Tjälve, H., Henriksson, J., Tallkvist, J., Larsson, B.S. and Lindquist, N.G. (1996) Uptake of manganese and cadmium from the nasal mucosa into the central nervous system via olfactory pathways in rats. *Pharmacol Toxicol.* 79:347-356.

Tomás-Camardiel, M., Herrera, A.J., Venero, J.L., Sánchez-Hidalgo, M.C., Cano, J. and Machado, A. (2002) Differential regulation of glutamic acid decarboxylase mRNA and tyrosine hydroxylase mRNA expression in the aged manganese-treated rats. *Mol Brain Res.* 103:116-129.

Tzingounis, A.V. and Wadiche, J.I. (2007) Glutamate transporters: confining runaway excitation by shaping synaptic transmission. *Nat Rev Neurosci.* 8:935-947.

Uchino, A., Noguchi, T., Nomiyama, K., Takase, Y., Nakazono, T., Nojiri, J., and Kudo S. (2007) Manganese accumulation in the brain: MR imaging. *Neuroradiology*, 49:715-720.

Verity, M.A. (1999) Manganese neurotoxicity: A mechanistic hypothesis. *Neurotoxicology*, 20:489-497.

Vescovi, A., Facheris, L., Zaffaroni, A., Malanca, G., and Parati, E.A. (1991) Dopamine metabolism alterations in a manganese-treated pheochromocytoma cell line (PC12). *Toxicology*, 67:129-42.

Vezér, T., Papp, A., Hoyk, Z., Varga, C., Naray, M. and Nagymajtenyi, L. (2005) Behavioral and neurotoxicological effects of subchronic manganese exposure in rats. *Environ Toxicol Pharmacol.* 19:797-810.

Vezér, T., Kurunczi, A., Náray, M., Papp, A. and Nagymajtényi, L. (2007) Behavioral Effects of Subchronic Inorganic Manganese Exposure in Rats. *Am J Ind Med.* 50:841–852.

Villalobos, V., Suárez, J., Estévez, J., Novo, E. and Bonilla, E. (2001) Effect of Chronic Manganese Treatment on Adenosine Tissue Levels and Adenosine A2a Receptor Binding in Diverse Regions of Mouse Brain. *Neurochem Res.* 26:1157-1161.

Walowitz, J.L. and Roth, J.A. (1999) Activation of ERK1 and ERK2 Is Required for Manganese-Induced Neurite Outgrowth in Rat Pheochromocytoma (PC12) Cells. *J Neurosci Res.* 57:847-854.

Wedler, F.C., Ley, B.W. and Grippo, A.A. (1989) Manganese (II) dynamics and distribution in glial cells cultured from chick cerebral cortex. *Neurochem Res.* 14:1129-1135.

Wedler, F.C. (1993) Biological significance of manganese in mammalian systems. *Prog Med Chem.* 30:89-133.

Weiss, B. (2002) Calculating the Economic Benefits of Reductions in Manganese Air Concentrations. En: Workshop on the Benefits of Reductions in Exposure to Hazardous Air Pollutants: Developing Best Estimates of Dose-Response Functions. EPA-SAB-EC-WKSHP-02-001, United State Environmental Protection Agency, January.

Zhang, S., Fu, J. and Zhou, Z. (2004) In vitro effect of manganese chloride exposure on reactive oxygen species generation and respiratory chain complexes activities of mitochondria isolated from rat brain. *Toxicol in Vitro*, 18:71-77.

Zhang, P., Hatter, A. and Liu, B. (2007) Manganese chloride stimulates rat microglia to release hydrogen peroxide. *Toxicol Lett.* 173:88-100.

Zheng, W., Aschner, M. and Ghersi-Egeac, J.F. (2003) Brain barrier systems: a new frontier in metal neurotoxicological research. *Toxicol App Pharmacol.* 192:1-11.

In: Progress in Neurodegeneration
Editors: M.R. Avila-Costa and V.A. Martinez
ISBN 978-1-60741-317-2

Chapter 6

Cadmium

Jose Luis Ordoñez-Librado and Maria Rosa Avila-Costa
Department of Neuroscience, Neuromorphology Lab. UNAM, Mexico

6.1. General Description

Cadmium (Cd) was discovered as an element in 1817 and its industrial use was quite minor until 50 years ago. The first effect on human health related to Cd exposure, was the damage to the lungs of Cd-exposed workers, published in 1938 (Nordberg, 2004). Is a soft, ductile, silver-white metal that belongs together with zinc (Zn) and mercury to group IIb in the Periodic Table. It has relatively low melting (320.9 °C) and boiling (765 °C) points and a relatively high vapor pressure.

In the air Cd is rapidly oxidized into Cd oxide. However, when reactive gases or vapor such as carbon dioxide, water vapor, sulfur dioxide, sulfur trioxide or hydrogen chloride are present, Cd vapor reacts to produce Cd carbonate, hydroxide, sulfite, sulfate or chloride, respectively. These compounds may be formed in chimneystacks and emitted to the environment. Several inorganic Cd compounds are quite soluble in water e.g. acetate, chloride and sulfate, whereas Cd oxide, carbonate and sulfide are almost insoluble (WHO, 1992).

Cd is one of the heavy metals with a number of toxicological implications. In modern technology, there are numerous applications of Cd and its derivatives resulting in risk of occupational and/or environmental exposure (Elinder, 1992). Occupational exposure is found in industries smelting and processing the metal, in battery manufacturing, etc. (ATSDR, 1999).

Cadmium and Cd compounds have a wide variety of industrial applications such as electroplating, pigments, plastics and Cd-Ni batteries. Cd is known to produce a variety of health hazards in humans and experimental animals due to its ability to induce severe alterations in various organs and tissue including the nervous system, following either acute or chronic exposure (Manca et al. 1991). It has been shown to cause severe damage to a variety of organs, including the lung, (Manca et al. 1991), liver, kidney, (Manca et al. 1991; Casalino et al. 2002) testis (Manca et al. 1991; Morselt, 1991) and brain (Méndez-Armenta et al. 2001; 2003).

Cd produces neurotoxicity with a complex pathology that includes neurological dysfunction in adults and children (Shukla and Singhal, 1984; Hart et al. 1993), changes in neurochemistry (Gupta et al. 1993; Antonio et al. 1999) and behavioral alterations in young rats, (Holloway and Thor, 1988). Experimental studies have reported histopathological damage with extensive hemorrhagic lesions, destruction of fibers, edema and pyknotic cells in the brain of developing rats (Méndez-Armenta et al. 2001) and young rabbits compared with adults after Cd exposure (Gabbiani et al. 1967). It has been proposed that the age of the animal plays an important role on the neurotoxic effects of Cd (Gupta et al. 1993). Moreover, it seems that Cd exposure produces neurological symptoms include headache and vertigo, parkinsonian-like symptoms, slowing of visuomotor functioning, peripheral neuropathy, decreased equilibrium, and decreased ability to concentrate (Shukla and Singhal, 1984; Okuda et al. 1997; Viaene et al. 2000).

6.1. Sources

Cd is a relatively rare element (0.2 mg/kg in the earth crust) and is not found in the pure state in the nature. It occurs mainly in association with the sulfide ores of Zn, lead and cupper (Cu). Cd has only been produced commercially in the twentieth century. It is a byproduct of the Zn industry; its production is thus determined essentially by the production of Zn. Before the First World War Cd was not usually recovered from Zn plants or other nonferrous metals plants, which resulted in an uncontrolled contamination of the environment for decades. The average annual production of Cd throughout the world increased from only 20 tons in the 1920s to about 12 000 tons in the period 1960–1969, 17 000 tons in 1970–1984; since 1987 it has fluctuated around 20 000 tons (Friberg et al. 1985; Alonso et al. 2001; Leroyer et al. 2001).

The pattern of Cd uses has changed in recent years. In the past Cd was mainly used in the electroplating of metals and in pigments or stabilizers for plastics. In 1960, the engineering coatings and plating sector accounted for over half the Cd consumed worldwide, but in 1990 this had declined to less than 8%. Nowadays, cadmium-nickel battery manufacture consumes 55% of the Cd output and it is expected that this application will expand with the increasing use of rechargeable batteries and their potential use for electric vehicles. For instance, the demand for Cd in nickel-cadmium batteries moved from 3000 tons in 1980 to 9000 tons in 1990. This rapid growth has more than offset declining trends for pigments (20%), plating (8%) and stabilizers (10%). In many respects Cd has become a vital component of modern technology, with countless applications in the electronics, communications, power generation and aerospace industries (Yates, 1992).

Most of the Cd that occurs in air is associated with particulate matter in the breathable range (diameter 0.1–1 μm). Cd is emitted to the atmosphere predominantly as elemental Cd and Cd oxide and from some sources as Cd sulfide (coal combustion and nonferrous metal production) or Cd chloride (refuse incineration). The residence time of Cd in air is relatively short (days to weeks) but sufficient to allow long-range transport in the atmosphere (WHO, 2000). Information on the concentrations and deposition rates of atmospheric Cd is available mostly from northern European countries, in particular Norway, Denmark, Sweden, Netherlands, Belgium and Germany. A review of measurement data available for the period 1980–1988 in these countries gives mean annual levels of around 0.1 ng/m^3 in remote areas (e.g. northern Norway), 0.1–0.5 ng/m^3 in rural areas, 1–10 ng/m^3 in urban areas and 1–20 ng/m^3 in industrial areas; levels of up to 100 ng/m^3 can be found in the proximity of emission sources. Mean annual wet deposition rates of Cd are 0.02–0.08 mg/m^2 in remote areas, 0.04–0.4 mg/m^2 in rural areas, 0.2–3.3 mg/m^2 in urban areas, and 0.8–3.3 mg/m^2 in industrial areas. Mean annual dry depositions vary similarly from less than 1 mg/m^2 in remote areas up to more than 500 mg/m2 around emission sources (Jensen and Bro-Rasmussen, 1991).

This metal is not eliminated from water ecosystems; it accumulates in sediments and is released in water during heavy rainfall, snowmelt, and run-off episodes (Olsvik et al. 2000).

6.2. Routes of Exposure

6.2.1. Air

Assuming a daily inhalation of 20m^3 of air and indoor concentrations similar to those outdoors, the average amount of Cd inhaled daily by humans in rural, urban and industrialized areas should not exceed 0.01, 0.2 and 0.4 μg, respectively. Deposition of inhaled Cd in the lungs varies between 10% and 50% depending on the size of airborne particles. Absorption of Cd in the lung depends on the chemical nature of the particles deposited. It is around 50% for Cd oxide but considerably less for insoluble salts such as Cd sulfide (WHO, 2000).

Cigarette smoking may represent an additional source of Cd, which may equal or exceed that from food. Depending on the brand cigarettes produced in Europe or the United States of America contain Cd at a concentration of 0.5–2 μg/g (dry weight) of tobacco, of which 10% can be absorbed (Friberg et al. 1985).

6.2.2. Drinking-water

Drinking water contains very low concentrations of Cd, usually in the range 0.01–1 μg/L. Levels of up to 5 μg/L have been reported occasionally and, on rare occasions, of up to 10 μg/L (WHO, 1992). In polluted areas, well water may contain very high concentrations of Cd (exceeding 25 μg/liter) (Lauwerys et al. 1990). Such unusual situations except the intake of Cd via drinking water, based on a water consumption of two liters per day are thus very low.

6.2.3. Food

For nonsmokers, food constitutes the principal environmental source of Cd. The lowest concentrations are found in milk (around 1 μg/kg). The concentration of Cd is in the range 1-50 μg/kg in meat, fish and fruit and 10-300 μg/Kg in staple foods such as wheat, rice and potatoes. The highest Cd levels (100-1000 μg/Kg) are found in the internal organs (kidney and liver) of mammals and in certain species of mussels, scallops and oysters. When grown on a cadmium-polluted soil, some crops, such as rice, can accumulate considerable amounts (more than 1000 μg/Kg). The average daily intake of Cd via food in European countries and North America is 15-25 μg but there may be large variations depending on age and dietary habits. In Japan, the average intake is generally 40-50 μg but may be much higher in cadmium-polluted areas (Bernard and Lauwerys, 1980; WHO, 1992; Alonso et al. 2001; Leroyer et al. 2001).

6.3. Absorption, Distribution and Excretion

In mammals, Cd is virtually absent at birth but accumulates with age, especially in the liver and kidneys. Rapid renal concentration occurs mainly during the early years of life (Henke et al. 1970), and 50–75% of the total body burden is found in these two organs.

6.3.1. Absorption

The factors that affect the absorption of ingested Cd include the animal species, type of compound, dose, frequency of administration, age or stage of development, pregnancy and lactation, presence or absence of drugs, nutritional status, and interactions with various nutrients. Studies in experimental animals have shown that 0.5–8% of Cd nitrate or chloride is absorbed after a single exposure (Friberg et al. 1974). In humans given radioactive Cd, the average amount absorbed is 5% (Flanagan et al. 1978).

Metallothioneins (MT) are metal-binding proteins of low relative molecular mass with a high content of cysteine residues that have a particular affinity for Cd and affect its toxicity. The regulation, degradation, and biological significance of mammalian MT's have been reviewed in detail (Miles et al. 2000).

Several functions have been assigned to MT. The essential metals Zn and Cu may induce its synthesis, and it is involved in the storage of these metals. Zinc–MT can detoxify free radicals. Administration of Zn and induction of MT can inhibit the toxicity of agents such as carbon tetrachloride, ethanol, and ionizing radiation, which act in part through oxidative injury. Cd also induces MT, and intracellular binding of Cd to MT protects against its toxicity. Cd is transported in the plasma as a complex with MT and may be toxic to the kidney when excreted in the glomerular filtrate. Most of the Cd in urine is bound to MT (Olsvik et al. 2000; Wimmer et al. 2005).

MT occurs as at least four genetic variants or isomers in humans, I, II, III, and IV. The two major forms, I and II, are ubiquitous in most organs, particularly in liver and kidney but also in brain (Wimmer et al. 2005).

MT-III is found in human brain and differs from I and II by having six glutamic acid residues near the terminal part of the protein. MT-III is thought to be a growth inhibitory factor, and metals do not regulate its expression; however, it does bind Zn. Its expression is down regulated in Alzheimer disease (AD). Another proposed role for MT-III is the participation in the utilization of Zn as a neuromodulator, since MT-III is present in the neurons that store Zn in their terminal vesicles. MT-IV occurs during differentiation of stratified squamous epithelium, but it is not known to have a role in the absorption or toxicity of Cd (Olsvik et al. 2000; Wimmer et al. 2005).

Cd bound to MT in food does not appear to be absorbed or distributed in the same way as inorganic Cd compounds (see WHO, 1992). Mice exposed to cadmium–MT had lower concentrations of Cd in blood and liver but a higher concentration in kidney than mice exposed to the same amount of Cd chloride (Cherian et al. 1978; Wimmer et al. 2005).

Low dietary concentrations of calcium (Ca^{2+}) and protein promote absorption of Cd from the intestinal tract of experimental animals (Friberg et al. 1974). A low iron (Fe) status in laboratory animals and humans has also been shown to result in greater absorption of Cd (Flanagan et al. 1978). In particular, women with low body Fe stores, as reflected by low serum ferritin concentrations, had an average gastrointestinal absorption rate twice as high as that of a control group of women, of the order of 10%.

Studies in rats showed that Fe status is an important predictor of Cd bioavailability. A high Fe status resulted in decreased total and fractional Cd accumulation from wheat endosperm and bran diets. In rats with reduced Fe status, the inclusion of wheat bran in the diet increased Cd uptake (Wing at al. 1992). Similar effects have been found in other studies of marginal Fe status. The bioavailability of Cd in rats can be increased significantly even in the absence of overt deficiency (Reeves and Chaney, 2001).

6.3.3. Distribution

Once absorbed, Cd is transported in the blood, mainly in erythrocytes, and is bound intracellularly to protein fractions of low and high relative molecular mass (Nordberg, 1972). The fraction with a low relative molecular mass is similar to MT. Plasma MT has an important role in the transport of Cd. It can contain up to 11% of Cd by weight, bound to sulfhydryl groups (Elinder and Nordberg, 1985), and occurs in large quantities in the liver, particularly after exposure to this metal. MT occurs in varying amounts in other tissues, such as the kidneys, and its concentration correlates with that of Cd in these tissues. The low relative molecular mass of free MT in plasma allows it to filter through the glomeruli and subsequently be reabsorbed in the proximal tubules, which in turn results in selective accumulation of Cd in the cortex (Nordberg, 1972). Transport of Cd bound to MT from blood to renal tubular cells is rapid and virtually complete, while the kidneys do not take up free Cd to a similar extent (Johnson and Foulkes, 1980).

While Cd can reach the embryo or fetus early in gestation, little transfer occurs across the fully developed placenta (Ahokas and Dilts, 1979). Cd can induce MT in the placenta, and the placenta retains Cd after exposure to low concentrations. The essential elements Zn and Cu are also bound to MT in the placenta but, for reasons not understood, Cd is retained and the essential metals are transported to the fetus (Goyer, 1995). The concentrations of Cd in the organs of embryos, fetuses and neonates are three orders of magnitude lower than the corresponding concentrations in adult women (Chaube et al. 1973).

Long-term exposure to Cd leads to selective accumulation in the liver and renal cortex, such that up to 75% of the total body burden is found in these organs (Friberg et al. 1985). The accumulation of Cd in the liver and its subsequent redistribution to the kidney is due to efficient MT synthesis in the liver; cadmium–metallothionein is slowly released into the plasma, filtered through the glomeruli and reabsorbed in the proximal tubules. After exposure to normal dietary concentrations of Cd (10–30 μg/day), about 50% of the body burden is found in kidneys, about 15% in the liver, and about 20% in muscle (Kowal et al. 1979). Lower concentrations are found in brain, bone, and fat (Cherry, 1981). Accumulation in the kidneys continues up to 50–60 years of age and falls thereafter, possibly due to age-related changes in kidney function. The hepatic and renal concentrations may fall subsequent to renal damage and increased leakage of bound Cd into the urine. Differences in the population mean concentrations of Cd in the renal cortex in various countries have been attributed to differences in daily dietary intake (Friberg et al. 1985).

6.3.4. Excretion

Little Cd is normally excreted in the urine. The rate of excretion increases slowly with increasing body burden, but, as renal dysfunction develops, it increases sharply and the hepatic and renal Cd concentrations fall (Nomiyama and Nomiyama, 1976). In the general population, the mean urinary Cd concentration, mainly bound to MT, ranges from < 0.5 to 2.0 pg/L, representing about 0.01% of the body burden, and increases with age (Nordberg, 1972). Estimates of biliary and gastrointestinal excretion after oral administration of Cd are somewhat uncertain in that most fecal Cd is unabsorbed material. The mechanism of fecal excretion may involve both sloughed mucosal cells and excretion in the bile. After an initial rapid phase, biliary excretion represents 0.02–0.04% of the body burden, and most is associated with a fraction of low relative molecular mass (Elinder and Nordberg, 1985). After low or moderate doses, the amount excreted in the feces is about the same as that excreted in urine. Minor routes of excretion include hair, breast milk, and pancreatic fluid, but collectively these routes make little contribution to the total excretion or biological half time of Cd. The slow excretion of Cd results in extremely long biological half times in animals, lasting from 200 days to 22 years (Friberg et al. 1985). The retention functions are multiphasic, involving several compartments with different half times. The half time of the slowest compartment is usually greater than 20% of the life span of the animal.

6.4. Cadmium Toxicity

Much of the information that has become available since publication of Environmental Health Criteria 134 (WHO, 1992) on the effects of long-term exposure to low doses of Cd on human health is from a cross-sectional, population-based epidemiological study conducted between 1985 and 1989 in Belgium, known as the Cadmibel, or Cadmium in Belgium Study (Lauwerys et al. 1990).

The main findings of the Cadmibel Study with regard to exposure are: (1) the body burden of Cd increases with age, at least until 60 years; (2) higher alcohol consumption is associated with less excretion of Cd in urine; (3) smokers have a higher burden of Cd than do nonsmokers; (4) premenopausal women have a higher burden of Cd than do similarly aged men (at least among nonsmokers); (5) urinary excretion of Cd increases in women after the menopause; and (6) residence in areas where there is heavy environmental pollution with Cd is associated with higher body burdens, including a 30% greater urinary excretion (Lauwerys et al. 1991; Sartor et al. 1992).

Cd can induce several cellular dysfunctions including cell death, decreased DNA repair and increased mutagenesis. They appear to be mainly mediated by an indirect production of Reactive Oxygen Species (ROS), which at least in part, is due to the inhibition of cellular antioxidants and constitutes oxidative stress. As other toxic metals, Cd can interfere with proteins that contain a Zn finger motif, which are implicated in the maintenance of genome stability or in DNA repair and DNA damage signaling (Valverde et al. 2000; Hartwig et al. 2002).

In cadmium-exposed cells, DNA strand breaks and chromosomal aberrations are found only at high, cytotoxic concentrations of the metal (Hartwig et al. 2002). A more specific effect of Cd with potential genetic consequences is the inhibition of DNA repair mechanism, mainly, base excision repair (Beyersmann and Hechtenberg, 1997; Hartwig et al. 2002).

6.4.1. Access of Cd to the brain

The nervous system has not received as much attention as a potential target organ for Cd as have the kidney and the skeletal system, possibly because Cd does not easily cross the blood–brain barrier, at least in adults, and any neurotoxic effect must be secondary to its other effects, such as interference with Zn metabolism and the expression of metallothionein-III (Jin et al. 1998). When Cd is administered to adult animals, only a small portion of it distributes to brain (Cantilena and Klaassen, 1981; Choudhuri et al. 1996). Two factors probably play important roles in preventing accumulation of Cd in brains of adult animals: the preferential accumulation of Cd in liver and kidney (Kotsonis and Klaassen, 1977; Chart and Cherian, 1993), and the presence of a well-developed blood-brain barrier (Lucis et al. 1972).

Cd is more toxic to newborn and young rats than to adult rats; this age difference in susceptibility might be due to differences in the blood–brain barrier integrity (Wong and Klaassen, 1982).

In vertebrates, Cd enters the central nervous system either through the olfactory bulb (Sunderman, 2001) or through the blood–brain barrier. Penetration of Cd into the brain can be facilitated by ethanol (Pal et al. 1993), chronic exposure to Cd (Shukla et al. 1996) or by forming complexes with other substances (O'Callaghan and Miller, 1986). Once Cd is present in the intercellular spaces, it interferes with a number of cellular and physiological processes.

The blood–brain barrier and circumventricular epithelial cells with tight junctions limit the entrance of Cd into the central nervous system. In addition, the choroid plexus epithelium accumulates highly toxic metals from the blood or cerebrospinal fluid (CSF) (Murphy, 2000). Injected Cd cannot easily enter the brain as it is blocked not only by the blood–brain barrier and blood cerebrospinal surfaces but also by the ependymal and pial surfaces (Takeda et al. 1999). Studies with radiotracers have shown the greatest concentration of Cd in circumventricular areas, including the choroid plexus, hypophysis, meninges, pineal gland, and olfactory bulb in rats (Arivison, 1980).

In a study of 7–14-day-old mice, the tracer moved from the blood vessels to the brain parenchyma and particularly the granular layer of the cerebellum (ATSDR, 1999). In other studies, no radiolabel was found in brain parenchyma or spinal cord, but some was found in sensory ganglia and spinal roots. Therefore, the peripheral nervous system may be particularly sensitive to Cd (Murphy, 2000). It seems that the administration of Cd initially affects the integrity and permeability of the vascular endothelium and the necrotic changes in nerve cells are only secondary to this effect. Some studies show that the choroid plexus –the principal site for the formation of CSF, and its ability to sequestrate and concentrate heavy metals, may protect the brain from the heavy metals coming from the blood (Zheng et al. 1991).

As a general choroid plexus toxicant, Cd directly destroys the plexus ultrastructure. In both chronic (22 weeks) and acute (1–24 days) exposure models, the levels of Cd in the choroid plexus found were high, while Cd in the CSF fell below the detection limit (Zheng et al. 1991). A postmortem human study revealed that the Cd concentration in the choroid plexus was about 2–3 times higher than that found in the cerebral cortex (Manton et al. 1984). Cadmium-produced deterioration of the plexus structure can be characterized by the loss of microvilli, a rupture of the apical surface, and an increased number of blebs (Arvidson and Tjalve, 1986).

Cadmium-induced injury in the cerebral microvessels is thought to be associated with oxidative stress. Following *in vivo* Cd exposure, there was an early increase followed by a later decrease in microvessel enzymes involved in cellular redox reactions such as superoxide dismutase (SOD), glutathione peroxidase (GSH) and catalase. Thus, a depletion of microvessel antioxidant defense systems and a resultant increase in lipid peroxidation (LPO) may provoke microvessel damage (Shukla et al. 1996).

In male rats given Cd chloride at 10 mg/L of drinking water, corresponding to a dose of Cd of 2 mg/Kg per day, *ad libitum* for 90 days, the concentration of Cd in the brain was increased by 76% after 30 days and by 165% after 90 days. The permeability of the blood–brain barrier to fluorescein dye and the concentration of malondialdehyde in brain microvessels were increased at the end of exposure, whereas the activity of SOD was decreased. The authors concluded that cadmium-induced dysfunction of the blood–brain barrier may be related to depletion of microvessel antioxidants and increased LPO (Shukla et al. 1996).

Accumulation in the brain seems to be dependent on the administration route. Oral treatments failed to induce any significant increase in trout (Melgar et al. 1997), but Cd has been shown to be taken up by olfactory epithelium and transported to the brain in pikes (Tallkvist et al. 2002).

On the other hand, MT in glial cells and ependymal cells near circumventricular organs also minimizes the diffusion of Cd into other parts of the brain. The use of immunohistochemical staining techniques showed different concentrations of MT in rat and mouse brains, but as the concentration appeared to be higher in adult mouse brain than in young or adult rat brain it may be difficult to compare the results of studies in the two species (Nishimura et al. 1992). Cd accumulation and expression of mRNA of MT-I, -II, and -III were observed in the brains of adult mice and also during different stages of development. There appeared to be an inverse relationship between expression of MT and Cd accumulation in the brain (Choudhuri et al. 1996). The effect of MT on the neurotoxicity of Cd is thus not clear.

6.4.2. Neurotoxicity

Several studies of workers exposed to Cd showed increased frequencies of subjective symptoms such as fatigue, headache, sleep disturbances, sensory and motor dysfunctions, anorexia, and anosmia (Murphy, 2000), reduced visuomotor performance and difficulties of concentration and postural balance were observed (Viaene et al. 2000). A case report of two individuals exposed to Cd during a fire suggested that it had caused persistent neurophysiological deficits (e.g. slowed reaction times, abnormal balance, constricted visual fields, decreased vibration sensitivity) and persistent neuropsychological deficits (e.g. delayed recall, dexterity, coordination, decision-making) (Kilburn and McKinley, 1996). No controls were included in this investigation, nor could the deficits be ascribed solely to Cd, as exposure to other potentially neurotoxic agents was involved.

In children exposed to Cd, it has been reported effects on higher nervous functions, such as lowered IQ (Thatcher et al. 1982; Jiang et al. 1990) or behavioral abnormalities (Marlowe, 1986).

Hart et al. (1993) investigated the potential neurobehavioral effects of occupational exposure to Cd in 31 workers with a mean duration of exposure of 14 years. The workers were classified by the concentration of Cd in a 24-h urine sample as low (mean, 6.0; range, 1–11 [units not stated]) and high (mean, 43; range, 25–110). No group of unexposed persons was included. Efforts were made to assess the potential confounding effects of exposure to known neurotoxicants. The group with high Cd concentrations achieved significantly lower scores for attention, psychomotor speed (symbol–digit modalities, digit symbol) and memory (paired-associate learning, digit symbol incidental recall) but not in measures of general intelligence, vigilance, conceptual reasoning, mental flexibility, motor speed, or mood state.

The only well-controlled epidemiological study of adult exposure to Cd and neurobehavioral effects was reported by Viaene et al. (2000). A sample of 42 workers with a mean of 12.6 years of exposure to Cd and 47 controls were given a questionnaire (the neurotoxicity symptom checklist-60), a standard neurological examination, and computer-administered neurobehavioral tests of visomotor performance, memory, and concentration. The statistical analyses were adjusted for a variety of potential confounders, including age, reported alcohol consumption, smoking habits, exposures to other neurotoxicants such as lead and organic solvents, education, and use of hypnotic medications. The exposure indices were the maximum concentration of Cd measured in urine during each person's work years (mean, 13 μg/g of creatinine) and the urinary concentration at the time of the examination (mean, 4.6 μg/g of creatinine). The exposed workers reported significantly more signs and symptoms of peripheral neuropathy, particularly in the categories of sensorimotor ability, equilibrium, and concentration. The only finding in the neurological examination that was related to exposure to Cd was a higher frequency of a positive forehead reflex. In the neurobehavioral tests, the exposed workers showed significant slowing of simple reaction time and significantly worse scores in the symbol–digit substitution test. The scores in the two tests were also significantly related to the maximum urinary concentration of Cd. These neurological effects were seen in workers without microproteinuria, suggesting that the effects on the nervous system are unlikely to be secondary to the effects of Cd on renal function.

Moreover, olfactory dysfunction is a well-known symptom after chronic exposure to Cd. Rose et al. (1992) reported that factory workers exposed to Cd fumes from a brazing operation performed more poorly than control subjects on a standardized smell function test. CNS symptoms such as headache and vertigo were also observed in the cadmium-exposed workers (Shukla and Singhal, 1984).

Cd-induced alterations of the spontaneous cortical activity have been published in the literature. Some studies have shown that repeated doses of Cd to rats for 9 days cause epileptiform EEG activity (Vataev et al. 1994). Cd modifies the function of Ca^{2+} channels and is used as a Ca^{2+} channel blocker *in vitro* experiments (Calabresi et al. 1990; Soliakov and Wonnacott, 1996); compared to the controls, the effect of Cd treatment on the spontaneous cortical activity was similar in the three centers with the somatosensory area showing the most marked alterations (Papp et al. 2003).

Vataev et al. (1994) reported significant changes of EEG recorded from the somatosensory, visual and auditory cortical areas and from the hippocampus after single or repeated administration of Cd to rats in doses comparable (or lower) than those applied by Papp et al. (2003). In spite of its low permeability across the blood-brain barrier, Cd was described to accumulate in various parts of the rat brain after two months oral exposure by ≈/50mg/Kg body weight, a dose comparable to Clark et al. (1985). Sensory systems seem generally to show a specific sensitivity to Cd, evidenced by the high accumulation of the metal in the thalamus and olfactory bulb in treated rats (Clark et al. 1985) or the reduced brainstem auditory response (Whitworth et al. 1999).

It also has been demonstrated that the neurological pathologic effect of Cd in experimental animals includes cerebral bleeding and cerebral edema in the neonatal mouse (Gabbiani et al. 1967; Webster and Valois, 1981), hyperactivity (Smith et al. 1982), and increased aggressiveness in juvenile rats (Holloway and Thor, 1988).

Behavioral effects that are not apparent clinically can be detected in the offspring of animals exposed to small repeated doses of Cd during pregnancy (Murphy, 2000). The offspring of pregnant rats given 50 mg/L in drinking water, corresponding to 4–5 mg/Kg per day, throughout gestation had the same body weight as controls but lower brain weights after 7 and 14 days but not 21 days (Gupta et al. 1993). Brain protein, DNA, and RNA contents were the similar in control and treated rats, although there was a twofold increase in the concentration of Cd in the tissues. The activities of several enzymes were decreased at various times after birth: succinate dehydrogenase at 7 days, cyclic nucleotide phosphorylase at 14 days, and acetylcholine at 21 days.

When Cd was given during gestation and lactation at 5–6 mg/Kg in drinking water, the brain Zn and DNA contents and DNA synthetase and thymidine kinase activities were reduced. The brains weighed less at 7, 14, and 21 days after birth (Gupta et al. 1993).

Cd chloride administered by gavage to female rats on 5 days a week for 5 weeks and then during mating and gestation at a dose of Cd of 0.04, 0.4, or 4 mg/Kg per day significantly reduced locomotor activity in offspring tested at 2 months of age (Baranski et al. 1983). Administration of 60 mg/L (corresponding to 5 mg/kg per day) in drinking water during days 1–20 of gestation caused alterations in locomotor activity and behavior in the open field in adult offspring of each sex. Adult female offspring showed decreased acquisition of avoidance behavior. The Cu content of the brains of 2-week old offspring of each sex and of 16-week-old females was decreased, but the concentration of Zn in brain was decreased only in 16-week-old animals (Baranski et al. 1983). These results suggest that changes in the concentrations of essential metals may also play a role in the apparent effects of Cd during development.

The offspring of rats given Cd at a dose of 0.20, 0.62, or 2.0 mg/Kg per day on days 7 and 15 of gestation showed decreased horizontal motor activity, increased immobility after treatment with amphetamine during a 5-min stress swim (at all doses), increased social interactions (at 0.62 and 2.0 mg/Kg per day), and extended latency of reacquisition (at 0.62 and 2.0 mg/Kg per day). The weights of these animals at birth and weaning were not different from those of controls, and the dams showed no toxic effects (ATSDR, 1999).

6.4.3. Cadmium Neuropathology

Recent reports are dealing with the influence of Cd in synaptic neurotransmission in the brain of rats (*in vivo* microdialysis in rat amygdala) (Minami et al. 2001), rat brain neurotransmitter levels (Andersson et al. 1997; Carageorgiou et al. 2000), and brain antioxidant status (Carageorgiou et al. 2004).

A variety of neurobehavioral and biochemical effects are produced on the nervous system of rodents given repeated doses of Cd (Murphy, 2000). As Cd and Zn are metabolically competitive, Cd may replace Zn in a number of metalloenzymes, proteins, and ion channels. Exposure to Cd generally increases the concentrations of noradrenaline and dopamine (DA) and impairs enzymes involved in the synthesis of neurotransmitters. Serotonin production may be altered, depending on age, brain region, and duration of exposure (Hobson et al. 1986; Lafuente et al. 2001). Nation et al. (1989) demonstrated variations in DA content in the striatum and hippocampus of rats exposed to Cd alone, or in combination with lead. Besides, Andersson et al. (1997) showed differences in DA, and their metabolic rates (DA/3,4-dihydoxyfenyl acetic acid, DOPAC) in the frontal cortex, nucleus accumbens, olfatory tubercule, and the striatum of rats exposed to Cd. Lafuente et al. (2001) demonstrate that after Cd exposure DA content increased in posterior hypothalamus as compared to the control group. Besides, DA turnover, measured as DOPAC/DA ratio, diminished in the median eminence and posterior hypothalamus; and Desole et al. (1991) found that after Cd exposure, DA turnover decreased in both median eminence and posterior hypothalamus. This inhibitory effect on DA turnover may be expected considering that this effect was previously shown in other brain areas (i.e. striatum).

Administration of Cd orally to weanling rats for 30 days caused increased secretion of DA, noradrenaline, and serotonin in certain regions of the brain at 1.0 mg/Kg per day but not at 0.1 mg/Kg per day; however, locomotor activity was increased in both groups. Neonatal animals given repeated doses of Cd showed decreased exploration in open fields and inner squares and improved performance in T-mazes (Ando et al. 1998).

The administration of Cd to either developing or adult animals increases both 5-hydroxytryptamine (5-HT) and 5-Hydroxyindoleacetic (5-HIAA) concentrations in different brain regions of adult rats, while 5-HT and 5-HIAA brain regional contents were decreased in developing rats (Gupta et al. 1993).

Cd has also shown to influence the acetylcholine metabolism (Devi and Fingerman, 1995; Carageorgiou et al. 2005), the metabolism of monoamines and acetylcholine were more affected in rat amygdale that excitatory transmitters by Cd administration (Minami et al. 2001). Carageorgiou et al. (2004) demonstrate that Acetylcholinesterase (AChE) activity increases in the rat brain following long-term Cd administration. AChE is a very important enzyme for cholinergic neurotransmission. AChE can potentiate Ca^{2+} influx into cells (Webb et al. 1996). Areas of higher AChE expression generally correlate with brain regions that degenerate early in AD.

Through its well-known blocking effect on Ca^{2+} channels, Cd can affect transmitter release (Soliakov and Wonnacott, 1996) or other Ca^{2+} regulated phenomena (Waisberg et al. 2003). In this way, Cd is used as a blocker for voltage-dependent Ca^{2+} channels. Because Ca^{2+} ions are essential for neuronal activity, a reduced plasma Ca^{2+} concentration could induce a drop in brain Ca^{2+} levels that could modify neuronal biosynthesis activity (Vetillard and Bailhache, 2005). Likewise, Cd has been shown to inhibit gamma aminobutyric acid-A (GABA-A) receptor activity in snail neurons by increasing intracellular Ca^{2+} (Molnar et al. 2004).

On the other hand, for several years, cadmium/drug interactions have been issues of central interest for some researchers. Previous studies have shown that chronic oral Cd exposure antagonizes behavioral effects common to acute administrations of ethanol (Nation et al. 1988; Burkey et al. 1994).

The finding of cadmium/cocaine antagonism is of particular interest in that, coupled with the aforementioned alcohol data; it suggests the metal may alter behavioral adaptations normally produced by a variety of psychoactive agents (Nation et al. 1995). The results of this investigation showed a pronounced augmentation of the stimulatory effects of cocaine in control animals. Behavioral sensitization to the motor-activating properties of cocaine was also observed in cadmium-treated animals, but the emergence of the effect was delayed and less pronounced than was the case for controls. Nation et al. (1995) argue that adult male rats chronically exposed to a water supply containing 100 ppm Cd chloride were less responsive than their control counterparts to the stimulatory effects of repeated administrations of cocaine. Thus, focal concentration of transmitter availability in discrete brain regions is predictive of responsiveness to cocaine, and ultimately a determinant of addictive (self-administration) behavior (Glick et al. 1994).

Cocaine does not directly release presynaptic DA, but rather prevents DA uptake by inhibiting the DA transporter (Carroll et al. 2002). Cd has been demonstrated to both increase and decrease the basal concentration of DA in brain (Lafuente et al. 2003), actions that are likely mediated through a calcium-mediated process.

One of the more popular conceptualizations of the initiation of behavioral sensitization to cocaine involves a sequence of cellular events in the ventral tegmental area (VTA) (Kalivas, 1995). A converging line of research points to the following cascade: (i) in the VTA, somatodendritic DA release is increased by cocaine administration; (ii) DA stimulates DA receptors which promotes glutamate release from cortical afferents; (iii) elevation of extracellular glutamate in the region of the VTA activates NMDA glutamate sub-type receptors on dopaminergic projections to the nucleus accumbens; (iv) extracellular DA levels rise in the region of the nucleus accumbens, a condition which serves as the putative substrate for many of the psychoactive effects peculiar to cocaine treatment, including behavioral sensitization (Petit et al. 1990; Keller et al. 1992).

6.5. Mechanisms of Cd Neurotoxicity

Cd is a transition metal that preferentially remains in a 2^+ oxidative state and therefore is unable to participate in classical redox chemistry such as the Fenton reaction; it does not generate ROS directly (Stohs and Bagchi, 1995; HaMai et al. 2001). Instead, divalent Cd, as well as other single-state heavy metals, such as mercury, is believed to generate ROS through selective inhibition of the mitochondrial electron transport chain (Stohs and Bagchi, 1995; Wang et al. 2004). Cd has been shown to inhibit electron transport through complex III (Wang et al. 2004). Mitochondria, under physiological conditions, are the major intracellular source of ROS because they perform about 90% of cellular aerobic metabolism and, in the process, release 1–5% of that oxygen as ROS (Agar and Durham, 2003). Therefore agents that impair mitochondrial efficiency serve to increase ROS production. Oxidative stress affects numerous cellular components, such as DNA, lipids, and proteins, through oxidation reactions (Stohs and Bagchi, 1995; Dröge, 2002; Beal, 2002). These alterations in structure produce significant changes in function and may result in pathogenesis. Increased oxidative stress is a prominent feature of many neurodegenerative disorders such as AD, Parkinson disease (PD), and amyotrophic lateral sclerosis (ALS), and increased levels of oxidative stress are observed in motor neurons of individuals with familial ALS (Beal, 2002; Agar and Durham, 2003).

Also, it has been reported that Cd can induce ROS accumulation indirectly by inhibiting antioxidant enzymes, such as, SOD, catalase, GSHPx, and glutathione reductase (Shukla et al. 1989; Acan and Tezcan, 1995). The ability of Cd to induce oxidative stress in brain cells has been reported as the induction of ROS, after the interaction of Cd^{2+} with mitochondrial sites, leading to the breakdown of the mitochondrial potentials, a consequent reduction of intracellular glutathione (GSH) levels and increase in catalase and SOD activities (López et al. 2006). Monroe and Halvorsen (2006) provided evidence that $CdCl_2$ increases oxidative stress in nerve cells, effect that is completely blocked by exogenous neurotrophic factors, such as ciliary neurotrophic factor and leukemia-inhibitory factor. Those actions involve the inhibition of the transphosphorylation of receptor associated to Janus kinases, Jak1 and Jak2. The neuronal inhibition of Jak kinase activated selectively in neurons by increasing intracellular levels of oxidative stress offers a new mechanism by which heavy metals may exert their neurotoxic effects. Likewise, *in vivo* studies have shown that the exposure of adult rats to low to moderate doses of Cd induced LPO in all tissues, mainly in lung and brain, and that the inhibition of GSHPx should be considered as a potentially significant event in the generation of free radicals by Cd (Manca et al. 1991).

Saturation of the lipid bilayer of several areas of the brain regions as a result of LPO in Cd-exposed animals causes disturbances in membrane fluidity and intracellular Ca^{2+} concentrations (Kumar et al. 1996).

It has been reported that in adult rats co-exposed to ethanol and Cd, there is a Cd-induced increase of LPO in the corpus striatum and cerebral cortex (Pal et al. 1993).

Moreover, ubiquitin binding to protein substrates is often signaled by post-translational modifications, such as glycosylation or phosphorylation. These processes may be activated by Cd. Oxidative modification of proteins occurring in the presence of Cd by perturbation of redox homeostasis can also target oxidant-sensitive proteins for ubiquitin-dependent degradation (Martelli et al. 2006).

Cd accumulation in the brain damages both neurons and glial cells. Chronic treatment of mice with Cd causes LPO and GSH depletion in the brain (Kumar et al. 1996), and, in HT4, a hippocampal cell line, and in rat primary mesencephalic cultures, Cd was found to disrupt intracellular sulfhydryl homeostasis by a mechanism involving oxidative stress and protein ubiquitination (Figueiredo-Pereira et al. 2002; Rockwell et al. 2004). Cd was found to cause cell death in primary oligodendrocytes via ROS generation and glutathione depletion (Almazan et al. 2000) and in microglia and dopaminergic neuron-glia cultures (Liu and Hong, 2003; Block et al. 2007). These studies demonstrate that Cd neurotixcity in cultured brain cells involves oxidative stress from activated microglia in Cd toxicity to neurons, at least to dopaminergic neurons.

Activated microglia is present in the vicinity of degenerating neurons in the substantia nigra, and is likely a major contribution factor in parkinsonism (Langston et al. 1999; Block et al. 2007). Dopaminergic neurons in the substantia nigra are relatively deficient in oxidative defenses (Yoo et al. 2003), rendering it more susceptible to ROS-induced damage.

Likewise, the injury to the CNS produced by Cd appears to be related to a largely loss of oxidative phosphorylation, together with a variety of conditions that produce CNS damage after inhibition of oxidative phosphorylation, all of which selectively damage the brain white matter (Fern et al. 1996). Changes in the homeostasis of cytosolic Ca^{2+} concentration can affect the regulation of many neuronal functions (Alshuaib and Byerly, 1996). Neuronal excitation causes a transient increase in intracellular Ca^{2+}, which in turn mediates a neuronal response. The increase in intracellular Ca^{2+} is dependent on voltage-dependent channels and on its release from intracellular Ca^{2+} stores. The intracellular Ca^{2+} is rapidly restored to resting levels by extrusion from the neuron by ATP-driven Ca^{2+} pumps, the Na^{2}/Ca^{2+}-exchanger, Ca^{2+}-binding proteins, and sequestration into the endoplasmic reticulum and mitochondria (Alshuaib et al. 2003). Cd inhibits all of the known pathways of cellular Ca^{2+} influx, and acts as a competitive ion to Ca^{2+}at the voltage-dependent Ca^{2+} channels and it is a potent blocker of the Ca^{2+}-dependent neurotransmitter release (Guan et al. 1988). This effect on Ca^{2+} influx is probably due to the interaction of the heavy metal ion with thiol groups of proteins involved in intracellular Ca^{2+} sequestration (Beyersmann and Hechtenberg, 1997). On the other hand, Cd has been reported to elevate the intracellular Ca^{2+} concentrations. That sustained increase is believed to be the main cause for cellular death (Orrenius and Nicotera, 1994). The uptake and intercellular distribution of Ca^{2+} is especially susceptible to interference by Cd^{2+}. Yoshida (2001) found a dose-dependent, elevated concentration of cytoplasmic and nuclear Ca^{2+} in Cd-exposed neurons.

It has been reported that Cd may block the influx of Ca^{2+} through membrane channels into the nerve terminal following the action potential, these decrease in Ca^{2+} influx caused by Cd would be associated with an altered transmitter release (Antonio et al. 1999). Several studies have shown that Cd is a potent inhibitor of the brain (Na+/K+)-ATPase (Antonio et al. 2002), Mg^{+} ATPase and that Cd inhibits choline transport in synaptosomes (Chandra et al. 1994). Cd interacts by either inhibiting or stimulating the activity of adenylate cyclase, depending on the concentration of the cation, presumably through its interaction with an enzymatic site closely related to the allosteric site of the regulatory unit of the Cd–ATP complex (Lundberg et al. 1987). Alterations in the mechanisms of neurotransmitters release have also been implicated in Cd neurotoxicity. The levels of excitatory neurotransmitters (glutamate and aspartate) were found decreased, while the inhibitory neurotransmitters (glycine and GABA) were increased in the amygdala of Cd-exposed animals, suggesting that Cd affects the balance excitation/inhibition of the synaptic neurotransmission (Minami et al. 2001).

This cadmium-dependent signaling cascade triggers Ca^{2+} release from internal stores, most likely via IP3-sensitive Ca^{2+} stores. Therefore, in spite of its inhibitory action on many types of Ca^{2+} channels and pumps (Berridge et al. 2000), Cd can upregulate the internal concentration of Ca^{2+} and play the role of an alternative signaling molecule controlling various transduction pathways (Misra et al. 2002).

Another mechanism by which Cd interferes with Ca^{2+} homeostasis is by its ability to modulate extracellular calcium-sensing receptors (CaSR). Either acutely or chronically applied, Cd potently modulates the activity of CaSR (Chang and Shoback, 2004).

The cell death induced by Cd in neurons is mediated by two mechanisms, apoptotic and necrotic. The apoptosis produced by Cd was reversed by vitamin C while this antioxidant molecule did not affect the necrotic-like cell death. It also appears that in the apoptotic mechanism mediated by Cd, but not in the necrotic mechanisms, oxidative stress could be implicated. The induction of ROS in these cells type could be mediated by mitochondria alterations because Cd produces a breakdown of the mitochondrial membrane potential (López et al. 2006). On the other hand, the fall in the ATP levels could be due to a breakdown in the mitochondrial electron transport as Wang et al. (2004) suggest.

Several reports have shown that Cd induces apoptosis in many tissues and cells both *in vivo* in Anterior Pituitary Cells (Poliandri et al. 2003) and *in vitro* in HL-60 Cells (Kondoh etal. 2002), cortical neurons (López et al. 2003) and C6 glioma cells (Wätjen et al. 2002). However, the mechanism responsible for the apoptotic action of Cd is poorly understood.

Three apoptotic pathways have been described: (1) mitochondria-dependent pathway; (2) death receptor dependent pathway; (3) endoplasmic reticulum (ER) pathway. There is an individual initiator caspase in each apoptotic pathway: caspase-9, in the mitochondrial pathway, caspase-8, in the death receptor-dependent pathway and caspase-12, in the ER pathway (Mehmet, 2000). Cd probably induces apoptosis through the mitochondrial pathway.

Other reports (Manca et al. 1991; Kumar et al. 1996; Shukla et al. 1996; Méndez-Armenta et al. 2003; El-Demerdash et al., 2004) have also shown that Cd can induce an elevation of LPO, and intracellular Ca^{2+} (Annunziato et al. 2003; Alshuaib et al. 2003), two well-known processes activating neuronal apoptosis. The increase of free radicals and Ca^{2+} levels associated to Cd exposure may induce mitochondrial disruption (Fern et al. 1996) and therefore the release of cytochrome c.

Furthermore, evidence suggests that the elevation of intracellular free Ca^{2+} concentration is associated with the development of apoptosis (Hu et al. 2005; Rekasi et al. 2005). Cd can interfere with essential metal ions including Ca^{2+}, Zn and Cu for uptake, and especially affect Ca^{2+} signaling in hepatic cells (Baker et al. 2003).

Robertson and Orrenius (2000) demonstrated that cells treated with $CdCl_2$ showed distinct changes with lower dose (60μM) showing mainly apoptosis whereas higher dose (120 μM) revealed mainly necrosis. These results are consistent with previous studies (Hamada et al. 1997), showing again that Cd may induce either apoptosis or necrosis in mammalian cells depending on the exposure conditions and the model used.

López et al. (2006) found that in cortical neurons, Cd exposure induced cellular death, which was, in part, reversed by vitamin C. This indicated that oxidative stress could be implicated in the mechanism by which Cd induces death in cortical neurons. The death induced by Cd in these cells, is mediated by two mechanisms, apoptotic and necrotic (Figure 1). Cortical neurons treated with Cd ions at concentrations between 1 and 100 μM, in either the absence or in the presence of serum in the treatment medium, generated ROS. The induction of ROS in these cells type could be mediated by mitochondria alterations because, as we mentioned above Cd produces a breakdown of the mitochondrial membrane potential. The decreases in ATP levels and in the mitochondria membrane potential began at 10 and 50-μM Cd ion, respectively, while the ROS formation was detected at lower doses (100 nM or 1 μM).

Cd also induces LPO in cortical neurons but this does not seem to be mediated by Cd per se, but as consequence of the ROS increment induced by this ion, since in their liposome preparation, a membrane model (Bangham et al. 1965), the only Cd treatment did not induce LPO. However, Fe^{2+}, a lipid peroxidation inductor, and SIN-1, a peroxynitrite donor that induces LPO, produce LPO in liposomes. These results agree with those found by Casalino et al. (2002) who found that vitamin E, an antioxidant agent, reversed the LPO induced by Cd in rat kidney mitochondrial fractions.

Several studies describe that, as a consequence of the Cd treatment, a toxic effect appears which correlates with GSH depletion (Figueiredo-Pereira et al. 1998; 1999). Thus, the results of López et al. (2006) are in accordance with the results found by Figueiredo-Pereira et al.

(1998) in HT4 and mesencephalic cells. These authors found that, in these cells, Cd treatment promotes a dose-dependent decrease of reduced GSH.

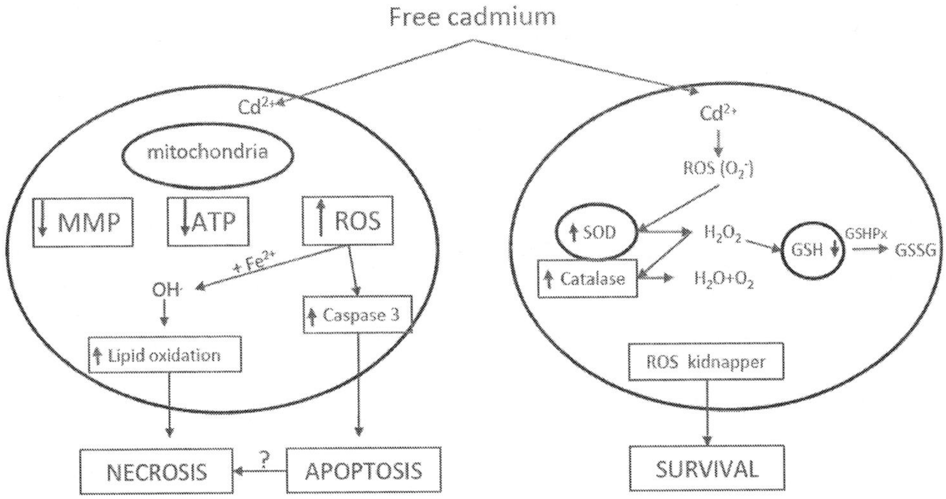

Modified from López et al. 2006.

Figure 1. Representation of a summary related to the mechanism by which (a) cadmium induces neuronal death in cortical neurons and (b) in neurons which are resistant to cadmium toxicity.

In summary, the cellular mechanisms underlying Cd toxicity are complex and still debated: Cd^{2+} affects intracellular Ca^{2+} binding proteins, including tubulin, troponin C, and calmodulin, binds human serum transferrin, disrupts mitochondrial respiration, and heavily interferes with the Ca^{2+} influx/efflux machinery. Cd is a potent inhibitor of voltage-dependent Ca^{2+} channels. This inhibition involves competition with Ca^{2+} for the cation binding site(s) in the channel pore (Yang et al. 1993); as the site affinity for Cd^{2+} is higher than that for Ca^{2+}, Cd^{2+} inhibits the Ca^{2+} current.

6.6. Cadmium and Neurodegeneration

Environmental contaminants such as Cd, Fe, Pb and oxidative stress (free radical formation) have been linked to AD and PD by many authors, with much work performed over the last two decades.

Few papers have discussed the concentrations and the role of Cd in the brain, in any detail. Spyrou et al. (1999) found decreased levels of Cd, compared to 'normals', in the frontal lobe of Alzheimer subjects, and Ward and Mason (1987) found no significant differences between 'normals' and AD subjects for the cortex and hippocampus.

These results show for the 'normal' ageing subjects significantly higher concentrations of Cd in the thalamus compared to the temporal lobes and hippocampus. The thalamus, which is mainly grey matter, has a higher concentration of MT than white matter. This is because the MT, which binds Cd and Zn is mainly found in the cytosol of cells.

Also, Cd movement in the brain may disrupt Ca^{2+} dependent mechanisms, causing an increase in oxidative stress and Ab peptides themselves are capable of spontaneously forming radicals that damage enzymes (Markesbery, 1999).

One possible contributory role of Cd toxicity in the development of neurological or psychiatric disorders is associated with dysfunction of central dopaminergic pathways, such as schizophrenia or PD (Remy and Samson, 2003; Evans and Lees, 2004). Some findings suggest that an environmentally relevant body burden of Cd is associated with alteration in the number and/or function of striatal DA neurons. As such, those exposed to Cd via tobacco smoke (Yue, 1992) or industrial pollution (Thun et al. 1985) may be at greater risk for the development of dopamine-related disorders or may display a decreased sensitivity to antipsychotic medications or drugs used in PD.

In this way Okuda et al. (1997) reported a case of a 64-year-old man suffered from acute exposure to Cd, followed by multiple organ failure. Three months after exposure, the patient developed parkinsonian features. The case suggests that Cd intoxication may damage the basal ganglia, resulting in parkinsonism.

According to Chen et al. (2008) the mechanism of Cd-induced neurodegeneration in PC12 and SH-SY5Y cells exposed to Cd is that Cd-induced apoptosis in these cells in a time- and concentration-dependent manner. Cd rapidly activated the mitogen-activated protein kinases (MAPK) including extracellular signal-regulated kinase 1/2 (Erk1/2), c-Jun N-terminal kinase (JNK) and p38. Inhibition of Erk1/2 and JNK, but not p38, partially protected the cells from Cd-induced apoptosis. Consistently, over-expression of dominant negative c-Jun or down-regulation of Erk1/2, but not p38 MAPK, partially prevented Cd-induced apoptosis. Cd also activated mammalian target of rapamycin (mTOR)-mediated signaling pathways. Treatment with rapamycin, an mTOR inhibitor, blocked Cd-induced phosphorylation of S6K1 and eukaryotic initiation factor 4E binding protein 1, and markedly inhibited Cd-induced apoptosis. Down-regulation of mTOR by RNA interference also in part, rescued cells from Cd-induced death. These findings indicate that activation of the signaling network of MAPK and mTOR is associated with Cd-induced neuronal apoptosis. These authors strongly suggest that inhibitors of MAPK and mTOR may have a potential for prevention of Cd-induced neurodegeneration. Misra et al. (2003) also found that Cd increases the level of kinases of the RAS pathway: RAF-1, MAPK, MEK1/2, ERK1/2, p38 MAPK and JNK. Furthermore, Cd treatment induces translocation of NF-kappaB from cytosolic to the membrane fraction and increases DNA binding activity of the latter. In this way, Cd stimulates the expression of the intercellular adhesion molecule-1 (ICAM-1) (Jeong et al. 2004).

Concerning to ALS, which is a progressive neurodegenerative disorder, 10% of the ALS patients are congenital (familial ALS), and the other 90% are sporadic ALS (SALS). It has been shown that mutations found in the Cu, Zn-SOD cause 20% of the familial ALS due to its low enzyme activity. It has been reported that Cd exposure causes ALS by reducing the SOD activity (Bar-sela et al. 2001). However, the mechanism of Cd decrease the activity of Cu, Zn-SOD protein is not clear. We hypothesize that Cd may cause the misfolding of Cu, Zn-SOD protein to alter its catalytic mechanism.

Huang et al. (2006) hypothesized that heavy metals may interfere the structure of Cu, Zn-SOD protein to suppress its activity in some of the SALS. These authors expressed and characterized the recombinant human Cu, Zn-SOD under various concentrations of Cu^{2+}, Zn^{2+}, and Cd^{2+}. By atomic absorption spectrophotometry, they demonstrated that adding Cd significantly increased the content of Cd ion, but reduced its Zn^{2+} content and enzyme activity of the Cu, Zn-SOD protein. The data of circular dichroism spectra demonstrated that the secondary structure of Cu, Zn-SOD/Cd is different from Cu, Zn-SOD, but close to apo-SOD. In addition to the effect of Cd on Cu, Zn-SOD, Cd was also shown to induce neural cell apoptosis. To further investigate the mechanism of neural cell apoptosis induced by Cd, they used proteomics to analyze the altered protein expressions in neural cells treated with Cd. The altered proteins include cellular structural proteins, stress-related and chaperone proteins, which are involved in ROS, enzyme proteins, and proteins that mediated cell death and survival signaling. Taken together, their results demonstrate that Cd decreases the content of Zn^{2+}, changes the conformation of Cu, Zn-SOD protein to decrease its enzyme activity, and causes oxidative stress-induced neural cell apoptosis.

Final Considerations

In this chapter, we have summarized the findings regarding to cadmium-induced neuronal damage, and the possible mechanisms involved in that damage. Oxidative stress, interference with Ca^{2+}, and zinc-dependent processes and apoptosis induction are the key processes involved in Cd neurotoxicity.

While there is ample data involving a wide variety of endpoints to suggest that Cd produces neurotoxicity, the nature of the toxicity as well as the primary mechanisms are still not well understood. However, it has been reported important neurobehavioral deficits, changes in the neurotransmission due to Ca^{2+} modifications, generation of necrotic or apoptotic cell death through LPO and other mechanisms, mitochondrial and energy failure, etc., alterations frequently found in neurodegenerative disorders. Human epidemiological studies have shown that many people are constantly exposed to Cd, either occupationally or environmentally, by drinking water and food and especially by smoking. The affected persons may be at higher risk of behavioral and functional neurotoxic disorders. Thus further investigations must deal with those neuronal mechanisms in detail in order to understand cadmium-induced neurotoxicity.

References

Acan, N.L. and Tezcan, E.F. (1995) Inhibition kinetics of sheep brain glutathione reductase by cadmium ion. *Biochem Mol Med.* 54:33–37.

Agar, J. and Durham, H. (2003) Relevance of oxidative injury in the pathogenesis of motor neuron disease. *Amyotroph Lateral Scler Other Mot Neuron Disord.* 4:232–242.

Agency for Toxic Substances and Disease Registry (ATSDR). (1999) Toxicological Profile for Cadmium. US Department of Health and Human Services, Washington DC, US Public Health Service, US Department of Health and Human Services.

Ahokas, R. and Dilts, P. (1979) Cadmium uptake by the rat embryo as a function of gestational age. *Am J Obstet Gynecol.* 135: 219–222.

Almazan, G., Liu, H.N., Khorchid, A., Sundararajan, S., Martinez-Bermudez, A.K. and Chemtob, S. (2000) Exposure of developing oligodendrocytes to cadmium causes HSP72 induction, free radical generation, reduction in glutathione levels, and cell death. *Free Radic Biol Med.* 29:858–869.

Alonso, E., Cambra, K. and Martinez, T. (2001) Lead and cadmium exposure from contaminated soil among residents of a farm area near an industrial site. *Arch Environ Health*, 56: 278-282.

Alshuaib, W.B. and Byerly, L. (1996) Modulation of membrane currents by cyclic AMP in cleavage arrested Drosophila neurons. *J Exp Biol.* 199: 537–548.

Alshuaib, W.B., Cherian, P.S., Hasan, Y.M. and Fahim, A.M. (2003) Drug effects on calcium homeostasis in mouse CA1 hippocampal neurons. *Intern J Neurosci.* 113: 1317–1332.

Andersson, H., Petersson-Grawe, K., Lindquist, E., Luthman, J., Oskarsson, A. and Olson, L. (1997) Low-level cadmium exposure of lactating rats causes alterations in brain serotonin levels in the offsprings. *Neurotoxicol Teratol.* 19: 105–115.

Ando, M., Hiratsuka, H., Nakagawa, J, Sato, S., Hayashi, Y. and Mitsumori, K. (1998) Cadmium accumulation in rats treated orally with cadmium chloride for 8 months. *J Toxicol Sci.* 23: 243–248.

Antonio, M.T., Corpas, I. and Leret, M.L. (1999) Neurochemical changes in newborn rats brain after gestational cadmium and lead exposure. *Toxicol Lett.* 104: 1-9.

Antonio, M.T., Lopez, N. and Leret, M.L. (2002) Pb and Cd poisoning during development alters cerebellar and striatal function in rats. *Toxicology*, 176: 59–66.

Annunziato, L., Amoroso, S., Pannaccione, A., Cataldi M, Pignataro G, D'Alessio A, Sirabella R, Secondo A, Sibaud, L. and Di Renzo, G.F. (2003) Apoptosis induced in neuronal cells by oxidative stress: role placed by caspases and intracellular calcium ions. *Toxicol Lett.* 139: 125–133.

Arivison, B. (1980) Regional differences in the severity of cadmium-induced lesions in the peripheral nervous system in mice. *Acta Neuropathol.* 49: 213–224.

Arvidson, B. and Tjalve, H. (1986) Distribution of cadmium-109 in the nervous system of rats after intravenous injection. *Acta Neuropathol.* 69:111–116.

Baker, T.K., Vanvooren, H.B., Smith, W.C. and Carfagna, M.A. (2003) Involvement of calcium channels in the sexual dimorphism of cadmium-induced hepatotoxicity. *Toxicol Lett.* 137: 185–192.

Bangham, A.D., Standish, M.M. and Watkinns, J.C. (1965) Diffusion of univalent ions across the lamellae of swollen phospholipids. *J Mol Biol.* 13: 238–252.

Baranski, B., Stetkiewicz, I., Sitarek, K. and Szymczak, W. (1983) Effects of oral, subchronic cadmium administration on fertility, prenatal and postnatal progeny development in rats. *Arch Toxicol.* 54: 297–302.

Bar-sela, S., Reingold, S. and Richter, E.D. (2001) Amyotrophic lateral sclerosis in a battery-factory worker exposed to cadmium. *Int J Occup Environ Health*, 7: 109–112.

Beal, M.F. (2002) Oxidatively modified proteins in aging and disease. *Free Radic Biol Med.* 32:797–803.

Bernard, A. and Lauwerys, R. (1980) Effects of cadmium exposure in man. Handbook of experimental pharmacology, Vol. 80. Berlin, Springer-Verlag, pp. 135–177.

Berridge, M.J., Lipp, P. and Bootman, M,D. (2000) The versatility and universality of calcium signaling. Nat Rev Mol Cell Biol, 1: 11–21.

Beyersmann, D. and Hechtenberg, S. (1997) Cadmium, gene regulation, and cellular signaling in mammalian cells. *Toxicol Appl Pharmacol.* 144: 247–261.

Block, M.L., Zecca, L. and Hong, J.S. (2007) Microglia-mediated neurotoxicity: uncovering the molecular mechanisms. *Nat Rev Neurosci.* 8:57-69.

Burkey, R.T., Nation, J.R., Grover, C.A. and Bratton, G.R. (1994) Chronic cadmium exposure attenuates ethanol-induced hypoalgesia in the adult rat. *Alcohol Clin Exp Res.* 17: 423-427.

Calabresi, P., Mercuri, N.B., Stefani, A. and Bernardi, G. (1990) Synaptic and intrinsic control of membrane excitability of neostriatal neurons. I. An in vivo analysis. *J Neurophysiol.* 63: 651-662.

Cantilena, L.R. and Klaassen, C.D. (1981) Comparison of the effectiveness of several chelators after single administration on the toxicity, distribution and excretion of cadmium. *Toxicol Appl Pharmacol.* 58:452-460.

Carageorgiou, H., Boviatsis, St., Carageorgiou-Kassaveti, M., Pantos, C., Messari, I. and Papadopoulou-Daifoti, Z. (2000) Proceedings of the ''2nd International symposium on trace elements in human: New perspectives''. In: Dopamine, serotonin and their metabolite levels in certain rat brain areas after acute and chronic administration of cadmium. Eds: S. Ermidou-Pollet, S. Pollet, Athens, Greece 7–9, pp. 723–730.

Carageorgiou, H., Tzotzes, V., Pantos, C., Mourouzis, C., Zarros, A. and Tsakiris, S. (2004) In vivo and in vitro effects of cadmium on adult rat brain total antioxidant status, acetylcholinesterase, (Na+,K+)-ATPase and Mg2π-ATPase activities: protection by Lcysteine. *Basic Clin Pharmacol Toxicol.* 94: 112–118.

Carageorgiou, H, Tzotzes, V., Sideris, A., Zarros, A. and Tsakiris, S. (2005) Cadmium Effects on Brain Acetylcholinesterase Activity and Antioxidant Status of Adult Rats: Modulation by Zinc, Calcium and L-Cysteine Co-Administration. *Basic Clini Pharmacol Toxicol.* 97: 320–324.

Carroll, F.I., Lewin, A.H. and Mascarella, S.W. (2002) Dopamine-transporter uptake blockers: structure–activity relationships. In: Reith ME, editor. Neurotransmitter transporters: structure, function and regulation. Totowa, NJ7 Humana Press; p. 381–432.

Casalino, E., Calzaretti, G., Sblano, C. and Landriscin, C. (2002) Molecular inhibitory mechanism of antioxidant enzymes in rat liver and kidney by cadmium. *Toxicology*, 179: 37–50.

Chandra, S.V., Musthy, R.C., Husain, T. and Bansal, S.K. (1994) Effect of interaction of heavy metals on (Na^+K^+) ATPase and uptake of H-DA and H-NA in rat brain synaptosomes. *Acta Pharmacol Toxicol.* 54: 210–213.

Chang, W. and Shoback, D. (2004) Extracellular Ca2+-sensing receptors—an overview. *Cell Calcium*, 35: 183–196.

Chart, H.M. and Cherian, M.G. (1993) Mobilization of hepatic cadmium in pregnant rats. *Toxicol Appl Pharmacol.* 120: 308-314.

Chen, L., Liu, L., Luo, Y. and Huang, S. (2008) MAPK and mTOR pathways are involved in cadmium-induced neuronal apoptosis. *J Neurochem.* 105:251-61.

Cherian, M.G., Goyer, R.A. and Valberg, L. (1978) Gastrointestinal absorption and organ distribution of oral cadmium chloride and cadmium metallothionein in mice. *J Toxicol Environ Health,* 4: 861–868.

Chaube, S., Nishimura, H. and Swinyard, C. (1973) Zinc and cadmium in normal human embryos and fetuses. *Arch Environ Health*, 26: 237–240.

Cherry, W. (1981) Distribution of cadmium in human tissues. In: Nrigu, J., ed., Cadmium in the Environment, Vol. II, Chichester, John Wiley, pp. 111–122.

Choudhuri, S., Liu, W.L., Berman, N.E. and Klaassen, C.D. (1996) Cadmium accumulation and metallothionein expression in brain of mice at different stages of development. *Toxicol Lett.* 84: 127–133.

Clark, D.E., Nation, J.R., Bourgeois, A.J., Hare, M.F, Baker, D.M. and Hindeberger, E.J. (1985) The regional distribution of cadmium in the brains of orally exposed adult rats. *Neurotoxicology*, 6: 109-114.

Desole, M.S., Miele, M., Esposito, G., Fresu, L., Enrico, P., De Natale, G., Anania, V. and Miele, E. (1991). Cadmium-induced changes in the activity of the dopaminergic and purinergic systems and in ascorbic acid catabolism in the rat striatum. *Clin Ther.* 31: 229-234.

Devi, M. and Fingerman, M.T.I. (1995) Inhibition of acetylcholinesterase activity in the central nervous system of the red swamp crayfish, Procambarus clarkii, by mercury, cadmium, and lead. *Bull Environ Contam Toxicol.* 55: 746-750.

Dröge, W. (2002) Free radicals in the physiological control of cell function. *Physiol Rev.* 82:47–95.

El-Demerdash, F.M., Yousef, M.I., Kedwany, F.S. and Baghdadi, H.H. (2004) Cadmium-induced changes in lipid peroxidation, blood hematology, biochemical parameters and semen quality of male rats: protective role of Vitamin E and beta-carotene. *Food Chem. Toxicol.* 42: 1563–1571.

Elinder, C.G. (1992) Cadmium as an environmental hazard. *IARC* 1118: 123-132.

Elinder, C.G. and Nordberg, M. (1985) In: Friberg, L., Elinder, C.-G., Kjellstrom, T. and Nordberg, G.F. Eds., Cadmium and Health: A Toxicological and Epidemiological Appraisal, Boca Raton, FL: CRC Press.

Evans, A.H. and Lees, A.J. (2004) Dopamine dysregulation syndrome in Parkinson's disease. *Curr Opin Neurol.* 17: 393– 8.

Fern, R., Black, J.A., Ransom, B.R. and Waxman, S.G. (1996) Cd^{+}-induced injury in CNS white matter. *J Neurophysiol.* 76: 3264–3273.

Figueiredo-Pereira, M.E., Yakushin, S. and Cohen, G. (1998) Disruption of the intracellular sulfhydyl homeostasis by cadmium-induced oxidative stress leads to protein thiolation and ubiquitination on neuronal cells. *J Biol Chem.* 273:12703–12709.

Figueiredo-Pereira, M.E., Yakushin, S. and Cohen, G. (1999) The ubiquitinin/proteosome pathway: friend or foe in zinc- cadmium, and H2O2-induced neuronal oxidative stress. *Mol Biol Rep.* 26:65–69.

Figueiredo-Pereira, M.E., Li, Z., Jansen, M. and Rockwell, P. (2002) N-acetylcysteine and celecoxib lessen cadmium cytotoxicity which is associated with cyclooxygenase-2 up-regulation in mouse neuronal cells. *J Biol Chem.* 277:25283–25289.

Flanagan, P., McLellan, J., Haist, J., Cherian, M., Chamberlaijn, M. and Valberg, L. (1978) Increased dietary cadmium absorption in mice and human subjects with iron deficiency. *Gastroenterology*, 74: 841–846.

Friberg, L., Elinder, C., Kjellstrom, T. and Nordberg, G. (1985) Cadmium and Health, A Toxicological and Epidemiological Appraisal, Vol. I, Exposure, Dose and Metabolism, Cleveland, OH, CRC Press.

Friberg, L., Piscator, M., Nordberg, G. and Kjellstrom, T. (1974) Cadmium in the Environment, 2nd Ed., Cleveland, Ohio, CRC Press.

Gabbiani, G., Baic, D. and Dezxel, C. (1967) Toxicity of cadmium for the central nervous system. *Exp Neurol.* 18: 154-160.

Glick, S.D., Raucci, J., Wang, S., Keller, R.W. and Carlson, J.N. (1994) Neurochemical predisposition to self-administer cocaine in rats: Individual differences in dopamine and its metabolites. *Brain Res.* 653: 148-154.

Goyer, R.A. (1995) Transplacental transfer of lead and cadmium. In: Goyer, R.A. & Cherian, M.G., Eds., Toxicology of Metals: Biochemical Aspects, Heidelberg: Springer-Verlag, pp. 1–13.

Guan, Y.Y., Quastel, D.M.J. and Saint, D.A. (1988) Single Ca^{2+} entry and transmitter release system at the neuromuscular synapse. *Synapse*, 2: 558–564.

Gupta, A., Murthy, R., Thakur, S., Dubey, M. and Chandra, S. (1993) Neurochemical changes in developing rat brain after pre- and postnatal cadmium exposure. *Bull Environ Contam Toxicol.* 51; 12–17.

Hamada, T., Tanimoto, A. and Sasaguri, Y. (1997) Apoptosis induced by cadmium. *Apoptosis*, 2: 359–367.

HaMai, D., Bondy, S.C., Becaria, A. and Campbell, A. (2001) The chemistry of transition metals in relation to their potential role in neurodegenerative processes. *Curr Top Med Chem.* 1:541–551.

Hart, R.P., Rose, C.S. and Hamer, R.M. (1993) Neuropsychological effects of occupational exposure to cadmium. *J Clin Exp Neuropsychol.* 11: 933–943.

Hartwig, A., Asmuss, M., Blessing, H., Hoffmann, S., Jahnke, G., Khandelwal, S., Pelzer, A. and Burkle, A. (2002) Interference by toxic metal ions with zinc-dependent proteins involved in maintaining genomic stability. *Food Chem Toxicol.* 40: 1179–1184.

Henke, G., Sachs, H. and Bohn, G. (1970) Cadmium determination by neutron activation analysis of liver and kidneys from children and young people. *Arch Toxikol.* 26: 8–16.

Hobson, M., Milhouse, M. and Rajanna, B. (1986) Effects of cadmium on the uptake of dopamine and norepinephrine in rat brain synaptosomes. *Bull Environ Contam Toxicol.* 37: 421-426.

Holloway, R.W. and Thor, H.D. (1988) Social memory deficits in adult male rats exposed to cadmium in infancy. *Neurotoxicol Teratol.* 10: 193:197.

Hu, Q.L., Chang, J.L., Tao, L.T., Yan, G.L., Xie, M.C. and Wang, Z. (2005) Endoplasmic reticulum mediated necrosis-like apoptosis of HeLa cells induced by Ca2+ oscillation. *J Biochem Mol Biol.* 38: 709–716.

Huang, Y.H., Shih, C.M., Huang, C.J., Lin, C.M., Chou, C.M., Tsai, M.L., Liu, T.P., Chiu, J.F. and Chen, C.T. (2006) Effects of cadmium on structure and enzymatic activity of Cu,Zn-SOD and oxidative status in neural cells. *J Cell Biochem*. 98: 577–589.

Jensen, A. and Bro-Rasmussen, F. (1991). Environmental cadmium in Europe. *Rev Environ Contamination and Toxicol.* 124: 101–181.

Jeong, E.M., Moon, C.H., Kim, C.S., Lee, S.H., Baik, E.J., Moon, C.K. and Jung, Y.S. (2004) Cadmium stimulates the expression of ICAM-1 via NFkappaB activation in cerebrovascular endothelial cells. *Biochem Biophys Res Commun.* 320: 887–892.

Jiang, H.M., Han, G.A. and He, Z.L. (1990) Clinical significance of hair cadmium content in the diagnosis of mental retardation of children. *Chin Med J* (Engl.), 103: 331-334.

Jin, T., Lu, J. and Nordberg, M. (1998) Toxicokinetics and biochemistry of cadmium with special emphasis on the role of metallothionein. *Neurotoxicology*, 19, 529–535.

Johnson, D. and Foulkes, E. (1980) On the proposed role of metallothionein in the transport of cadmium. *Environ Res*. 21: 360–365.

Kalivas, P.W. (1995) Interactions between dopamine and excitatory amino acids in behavioral sensitization to psychostimulants. *Drug Alcohol Depend*. 37: 95-100.

Keller, R.W., Maisonneuve, I.M., Carlson, J.N. and Glick, S.D. (1992) Within-subject sensitization of striatal dopamine release after a single injection of cocaine: an in vivo microdialysis study. *Synapse*, 11: 28-34.

Kilburn, K.H. and McKinley, K.L. (1996) Persistent neurotoxicity from a battery fire: Is cadmium the culprit? *South Med J.* 89: 693–698.

Kondoh, M., Araragi, S., Sato, K., Higashimoto, M., Takiguchi, M. and Sato, M. (2002) Cadmium induces apoptosis partly via caspase-9 activation in HL-60 cells. *Toxicology*, 170: 111–117.

Kotsonis, F.N. and Klaassen, C.D. (1977) Toxicity and distribution of cadmium administered to rats at sublethal doses. *Toxicol Appl Pharmacol*. 41: 667-680.

Kowal, N., Johnson, D., Kraemer, D. and Pahren, H. (1979) Normal levels of cadmium in diet, urine, blood and tissues of inhabitants of the Unites States. *J Toxicol Environ Health*, 5: 995–1014.

Kumar, R., Agarwal, A.K. and Seth, P.K. (1996) Oxidative stress-mediated neurotoxicity of cadmium. *Toxicol Lett*. 89: 65–69.

Lafuente, A., Gonzalez-Carracedo, A., Romero, A. and Esquifino, A.I. (2003) Effect of cadmium on 24-h variations in hypothalamic dopamine and serotonin metabolism in adult male rats. *Exp Brain Res.* 149:200–2006.

Lafuente, A., Márquez, N., Pazo, D. and Esquifino, A.I. (2001) Cadmium effects on dopamine turnover and plasma levels of prolactin, GH and ACTHJ. *Physiol Biochem*. 57: 231-236.

Langston, J.W., Forno, L.S., Tetrud, J., Reeves, A.G., Kaplan, J.A. and Karluk, D. (1999) Evidence of active nerve cell degeneration in the substantia nigra of humans years after 1- methyl-4-phenyl-1,2,3,6-tetrahydropyridine exposure. *Ann Neurol.* 46: 598-605.

Lauwerys, R., Amery, A., Bernard, A., Bruax P, Buchet J-P, Claeys F, De Plaen P, Ducoffre G, Fagard R, Lijnen P, Nick L, Roels, H., Rondia, D., Saint-Remy, A, Sartor, F. and Staessen, J. (1990) Health effects of environmental exposure to cadmium: Objectives, design and organization of the Cadmibel Study: A cross-sectional morbidity study carried out in Belgium from 1985 to 1989. *Environ Health Perspectives*, 87: 283–289.

Lauwerys, R., Bernard, A., Buchet, J.P., Roels, H., Bruaux, P., Claeys, F., Ducoffre, G., De Plaen, P., Staessen, J., Amery, A., et al. (1991) Does environmental exposure to cadmium represent a health risk? Conclusions from the Cadmibel Study. *Acta Clin Belg*. 46: 219–225.

Leroyer, A., Hemon, D., Nisse, C., Auque, G., Mazzuca, M. and Haguenoer, J.M. (2001) Determinants of cadmium burden levels in a population of children living in the vicinity of nonferrous smelters. *Environ Res*. 87: 147-159.

Liu, B. and Hong, J.S. (2003) Primary rat mesencephalic neuron-glia, neuron-enriched, microglia-enriched, and astroglia-enriched cultures. *Methods Mol Med*. 79:387-395.

López, E., Arce, C., Oset-Gasque, M.J., Cañadas, S. and Gonzalez, M.P. (2006) Cadmium induces reactive oxygen species generation and lipid peroxidation in cortical neurons in culture. *Free Radic Biol Med*. 40: 940–951.

López, E., Figueroa, S., Oset-Gasque, M.J. and Gonzalez, M.P. (2003) Apoptosis and necrosis: two distinct events induced by cadmium in cortical neurons in culture. *Br J Pharmacol*. 138: 901–911.

Lucis, O.J., Lucis, R. and Shaikh, Z.A. (1972) Cadmium and zinc in pregnancy and lactation. *Arch Environ Health*, 25: 14-22.

Lundberg, U., Milanes, C.L., Pernalete, N., Weisinger, J., Contreras, E.I., Pas, M.V. and Bellurin-Font, E. (1987) Effect of cadmium of canine renal cortical renal cortical adenylate cyclase. *Am J Phisiol*. F401–F407.

Manca, D., Ricard, A.C., Trottier, B. and Chevalier, G. (1991) Studies on lipid peroxidation in rat tissues following administration of low and moderate doses of cadmium chloride. *Toxicology*, 67: 303-323.

Manton, W.I., Kirkpatrick, J.B. and Cook, J.D. (1984) Does the choroid plexus really protect the brain from lead? *Lancet* II:351.

Markesbery, W.R. (1999) The role of oxidative stress in Alzheimer's Disease. *Arch Neurol*. 56:1449– 52.

Marlowe, M. (1986) Exposure to metal pollutants and behavioral disorders in children: a review of the evidence. *Rev Environ Health*, 6: 85-117.

Martelli, A., Rousselet, E., Dycke, C., Bouron, A. and Moulis, J.M. (2006) Cadmium toxicity in animal cells by interference with essential metals. *Biochimie*, 88: 1807–1814

Mehmet, H. (2000) Caspases find a new place to hide. *Nature*, 403: 29–30.

Melgar, M.J., Perez, M., Garcia, M.A., Alonso, J. and Miguez, B. (1997) The toxic and accumulative effects of short-term exposure to cadmium in rainbow trout (Oncorhynchus mykiss). *Vet Hum Toxicol*. 39:79–83.

Méndez-Armenta, M., Barroso-Moguel, R., Villeda-Hernández, J., Nava-Ruız, C. and Rios, C. (2001) Histopathological alterations in the brain regions of rats after perinatal combined treatment with cadmium and dexamethasone. *Toxicology*, 161: 189-199.

Méndez-Armenta, M., Villeda-Hernandez, J., Barroso-Moguel, R., Nava-Ruız C, Jimenez-Capdeville, M.E. and Rıos, C. (2003) Brain regional lipid peroxidation and metallothionein levels of developing rats exponed to cadmium and dexamethasone. *Toxicol Lett*. 144: 151–157.

Miles, A.T., Hawksworth, G.M., Beattie, J.H. and Rodilla, V. (2000) Induction, regulation, degradation and biological significance of mammalian metallothioneins. *Crit Rev Biochem Mol Biol.* 35: 35–75.

Minami, A., Takeda, A., Nishibaba, D., Takefuta, S. and Oku, N. (2001) Cadmium toxicity in synaptic neurotransmission in the brain. *Brain Res.* 894: 336-339.

Misra, U.K., Gawdi, G., Akabani, G. and Pizzo, S.V. (2002) Cadmium-induced DNA synthesis and cell proliferation in macrophages: the role of intracellular calcium and signal transduction mechanisms. *Cell Signal*, 14: 327–340.

Misra, U.K., Gawdi, G., Akabani, G. and Pizzo, S.V. (2003) Induction of mitogenic signalling in the 1LN prostate cell line on exposure to submicromolar concentrations of cadmium^{+}. *Cell Signal*, 15: 1059–1070.

Molnar, G., Salanki, J. and Kiss, T. (2004) Cadmium inhibits GABA-activated ion currents by increasing intracellular calcuim level in snail neurons. *Brain Res*. 1008:205–211.

Monroe, K.R. and Halvorsen, W.S. (2006) Cadmium blocks receptor mediated Jak/STAT signaling in neurons by oxidative stress. *Free Radic Biol Med.* 41: 493–502.

Morselt, A.F.W. (1991) Environmental pollutants and disease: a cell biological approach using chronic cadmium exposure in the animal model as a paradigm case. *Toxicology*, 70: 1–84.

Murphy, V. (2000) Cadmium: Acute and chronic neurological disorders. In: Yasui, M., Strong, M., Ota, K. & Verity, M.A., eds, Mineral and Metal Neurotoxicology, Boca Raton, FL: CRC Press, pp 229–240.

Nation, J.R., Frye, G.D., Von Stultz, J. and Bratton, G.R. (1989) Effects of combined lead and cadmium exposure: Changes in schedule-controlled responding and in dopamine, serotonin, and their metabolites. *Behav Neurosci.* 103: 1108-1114.

Nation, J.R., Livermore, C.L. and Bratton, G.R. (1995) Cadmium exposure attenuates the initiation of behavioral sensitization to Cocaine. *Brain Res.* 702: 223-232.

Nation, J.R., Wellman, P.J., Von Stultz, J., Taylor, B., Clark, D.E. and Bratton, G.R. (1988) Cadmium exposure results in decreased responsiveness to ethanol. *Alcohol.* 5: 99-102.

Nishimura, N., Nishimura, H., Ghaffar, A. and Tohyama, C. (1992) Localization of metallothionein in the brain of rat and mouse. *J Histochem Cytochem.* 40: 309–315.

Nomiyama, K. and Nomiyama, H. (1976) Mechanisms of urinary excretion of cadmium: Experimental studies in rabbits. In: Noordberg, G., ed., Effects and Dose–Response Relationships of Toxic Metals, Amsterdam, Elsevier Science, pp. 371–379.

Nordberg, G. (1972) Cadmium metabolims and toxicity. *Environ Physiol Biochem.* 2: 7–36.

Nordberg, F.G. (2004) Cadmium and health in the 21st century-historical remarks and trends for the future. *BioMetals*, 17: 485–489.

O'Callaghan, J.P. and Miller, D.B. (1986) Diethyldithiocarbamate increases distribution of cadmium to brain but prevents cadmium-induced neurotoxicity. *Brain Res.* 370: 354–358.

Okuda, B., Iwamoto, Y., Tachibana, H. and Sugita, M. (1997) Parkinsonism after acute cadmium poisoning. *Clin Neurol Neurosurg.* 99:263–265.

Olsvik, P.A., Gundersen, P., Andersen, R.A. and Zachariassen, K.E. (2000) Metal accumulation and metallothionein in two populations of brown trout, Salmo trutta, exposed to different natural water environments duringa run-off episode. *Aquat Toxicol.* 50:301–316.

Orrenius, S. and Nicotera, P. (1994) The calcium ion and cell death. *J Neural Trans.* 43: 1–11.

Pal, R., Nath, R. and Gill, K.D. (1993) Influence of ethanol on cadmium accumulation and its impact on lipid peroxidation and membrane bound functional enzymes (Na+,K+-ATPase and acetylcholinesterase) in various regions of adult rat brain. *Neurochem Int.* 23: 451–458.

Papp, A., Nagymajtényi, L. and Dési, I. (2003) A study on electrophysiological effects of subchronic cadmium treatment in rats. *Environ Toxicol Pharmacol.* 13: 181-186.

Petit, H.O., Pan, H., Parsons, L.H. and Justice, J.B. (1990) Extracellular concentrations of cocaine and dopamine are enhanced during chronic cocaine administration. *J Neurochem.* 55: 798-804.

Poliandri, B.H.A., Cabilla, P.J., Velardez, O.M., Bodo, A.C.C. and Duvilanski, H.B. (2003) Cadmium induces apoptosis in anterior pituitary cells that can be reversed by treatment with antioxidants. *Toxicol Appl Pharmacol.* 190: 17–24.

Reeves, P. and Chaney, R.L. (2001) Mineral Status of Female Rats Affects the Absorption and Organ Distribution of Dietary Cadmium Derived from Edible Sunflower Kernels (Helianthus annuus L.). *Environ Res.* 85: 215-225.

Rekasi, Z., Czompoly, T., Schally, A.V., Boldizsar, F., Varga, J.L., Zarandi, M., Berki, T., Horvath, R.A. and Nemeth, P. (2005) Antagonist of growth hormonereleasing hormone induces apoptosis in LNCaP human prostate cancer cells through a Ca2+-dependent pathway. *Proc Natl Acad Sci.* 102: 3434–3440.

Remy, P. and Samson, Y. (2003) The role of dopamine in cognition: evidence from functional imaging studies. *Curr Opin Neurol.* 16: S37– S41.

Robertson, J.D. and Orrenius, S. (2000) Molecular mechanisms of apoptosis induced by cytotoxic chemicals. *Crit Rev Toxicol.* 30: 609–627.

Rockwell, P., Martinez, J., Papa, L. and Gomes, E. (2004) Redox regulates COX-2 upregulation and cell death in the neuronal response to cadmium. *Cell Signal*, 16:343–353.

Rose, C.S., Heywood, P.G. and Costanzo, R.M. (1992) Olfactory impairment after chronic occupational cadmium exposure. *J Occup Med.* 34: 600–605.

Sartor, F.A., Rondia, D.J., Claeys, F.D., Staessen, J.A., Lauwerys, R.R., Bernard, A.M., Buchet, J.P., Roels, H.A., Bruaux, P.J., Ducoffre, G.M., Lijnen, P.J., Thijs, L.B. and Amery, A.K. (1992) Impact of environmental cadmium pollution on cadmium exposure and body burden. *Arch Environ Health*, 47: 347–353.

Shukla, G.S., Hussain, T., Srivastava, R.S. and Chandra, S.V. (1989) Glutathione peroxidase and catalase in liver, kidney, testis and brain regions of rats following cadmium exposure and subsequent withdrawal. *Ind Health*, 27:59–69.

Shukla, A., Shukla, G.S. and Srimal, R.C. (1996) Cadmium-induced alterations in blood–brain barrier permeability and its possible correlation with decreased microvessel antioxidant potential in rat. *Hum Exp Toxicol.* 15: 400–405.

Shukla, S.G. and Singhal, R.L. (1984) The present status of biological effects of toxic metals in the environment: lead, cadmium and manganese. *Can J Physiol Pharmacol.* 62: 1015-1031.

Smith, M.J., Pihl, R.O. and Garber, B. (1982) Postnatal cadmium exposure and long-term behaviour changes in the rat. *Neurobehav Toxicol Teratol.* 4:283– 287.

Soliakov, L. and Wonnacott, S. (1996) Voltage-sensitive Ca2+/channels involved in nicotinic receptor-mediated [3H]dopamine release from rat striatal synaptosomes. *J Neurochem.* 67: 163-170.

Spyrou, N.M., Stedman, J.D. and Cutts, D.A. (1999) Trace elements in Alzheimer's Disease. *Res Pract Alzheimer's Dis.* 2: 125 –129.

Stohs, S.J. and Bagchi, D. (1995) Oxidative mechanisms in the toxicity of metal ions. *Free Radic Biol Med.* 18:321–336.

Sunderman, F.W. (2001) Nasal toxicity, carcinogenicity, and olfactory uptake of metals. *Ann Clin Lab Sci.* 31: 3 –24.

Takeda, A., Takefuta, S., Ijiro, H., Okada, S. and Oku, N. (1999) 109-Cadmium transport in rat brain. *Brain Res Bull*, 49: 453–457.

Tallkvist, J., Persson, E., Henriksson, J. and Tjalve, H. (2002) Cadmium-metallothionein interactions in the olfactory pathways of rats and pikes. *Toxicol Sci.* 67:108–113.

Thatcher, R.W., Lester, M.L., McAlaster, R. and Horst, R. (1982) Effects of low levels of cadmium in lead on cognitive functioning in children. *Arch Environ Health*, 37: 159-166.

Thun, M.J., Schnorr, T.M., Smith, A.B., Halperin, W.E. and Lemen, R.A. (1985) Mortality among a cohort of US cadmium production workers—an update. *J Natl Cancer Inst.* 74:325–33.

Valverde, M., Fortoul, T.I., Diaz-Barriga, F., Mejia, J. and del Castillo, E.R. (2000) Induction of genotoxicity by cadmium chloride inhalation in several organs of CD-1 mice. *Mutagenesis*, 15: 109–114.

Vataev, S.I., Malgina, N.A. and Oganesian, G.A. (1994) The effect of cadmium on the structure of the circadian cycle of waking/sleep and on the EEG in Wistar rats. *Zh Evol Biokhim Fiziol.* 30: 408-419.

Vetillard, A. and Bailhache, T. (2005) Cadmium: An Endocrine Disrupter That Affects Gene Expression in the Liver and Brain of Juvenile Rainbow Trout. *Biology of Reproduction*, 72: 119–126.

Viaene, M.K., Masschelein, R., Leenders, J., De Groof, M., Swerts, L.J.V.C. and Roels, H.A. (2000) Neurobehavioral effects of occupational exposure to cadmium: A cross sectional epidemiological study. *Occup Environ Med.* 57: 19–27.

Waisberg, M., Joseph, P., Hale, B. and Beyermann, D. (2003) Molecular and cellular mechanisms of cadmium carcinogenesis. *Toxicology*, 192:95–117.

Wang, Y., Fang, J., Leonard, S.S. and Rao, K.M. (2004) Cadmium inhibits the electron transfer chain and induces reactive oxygen species. *Free Radic Biol Med.* 36:1434–1443.

Ward, N.I. and Mason, J.A. (1987) Neutron activation analysis techniques for identifying elemental status in Alzheimer's Disease. *J Radioanal Nucl Chem.* 113(2):515–26.

Wätjen, W., Cox, M., Biagioli, M. and Beyermann, D. (2002) Cadmium-induced apoptosis in C6 glioma cells: mediation by caspase 9-activation. *BioMetals*, 15: 15–25.

Webb, C.P., Nedergaard, S., Giles, K. and Greenfield, S.A. (1996) Involvement of the NMDA receptor in a non-cholinergic action of acetylcholinesterase in guinea-pig substantia nigra pars compacta neurons. *Eur J Neurosci.* 8: 837–841.

Webster, W.S. and Valois, A.A. (1981) The toxicity effects of cadmium on the neonatal mouse central nervous system. *J Neuropathol Exp Neurol.* 40:247– 257.

WHO (1992) Cadmium (Environmental Health Criteria 134), Geneva.

WHO (2000) Cadmium Chapter 6.3, Air Quality Guidelines - Second Edition. Geneva.

Wimmer, U., Wang, Y., Georgiev, O. and Schaffner, W. (2005) Two major branches of anti-cadmium defense in the mouse: MTF-1/metallothioneins and glutathione. *Nucleic Acids Res.* 33: 5715–5727.

Wing, A., Wing, K., Tidehag, P., Hallmans, G. and Sjostrom, R. (1992) Cadmium accumulation from diets in rats with different iron status. *Nutr Res.* 12: 1205–1215.

Whitworth, C., Hudson, T.E. and Rybak, L.P. (1999) The effect of combined administration of cadmium and furosemide on auditory function in the rat. *Hearing Res.* 129: 61-70.

Wong, K.L. and Klaassen, D.C. (1982) Neurotoxic effects of cadmium in Young rats. *Toxicol Appl Pharmacol.* 63: 330–337.

Yang, J., Ellinor, P.T., Sather, W.A., Zhang, J.I.F. and Tsien, R.W. (1993) Molecular determinants of Ca21 selectivity and ion permeation in L-type Ca21 channels. Nature, 366: 158–161.

Yates, E.M. (1992) The world needs cadmium – a miner's viewpoint. In: Cook, M.E. et al., ed. Cadmium 92. London, Cadmium Association, 1992, pp.1–7.

Yoshida, S. (2001) Re-evaluation of acute neurotoxic effects of Cd2+ on mesencephalic trigeminal neurons of adult rat. *Brain Res.* 892: 102–110.

Yoo, M.S., Chun, H.S., Son, J.J., DeGiorgio, L.A., Kim, D.J., Peng, C. and Son, J.H. (2003) Oxidative stress regulated genes in nigral dopaminergic neuronal cells: correlation with the known pathology in Parkinson's disease. *Brain Res Mol Brain Res.* 110:76- 84.

Yue, L. (1992) Cadmium in tobacco. *Biomed Environ Sci.* 5:53–56.

Zheng, W., Perry, F.D., Nelson, L.D. and Aposhian, V.H. (1991) Choroid plexus protects cerebrospinal fluid against toxic metals. *FASEB J.* 5: 2188–2193.

In: Progress in Neurodegeneration
Editors: M.R. Avila-Costa and V.A. Martinez
ISBN 978-1-60741-317-2

Chapter 7

Zinc

Ricardo García Ruiz
Department of Neuroscience, Neuromorphology Lab. UNAM, Mexico

7.1. General Description

Zinc (Zn) is the 30th element of the Periodic table of elements. The most common and most stable oxidation number of Zn is +2 (Zn^{2+}). Zn is a ubiquitous trace element found in plants and animals. The adult human body contains approximately 1.5 to 2.5 grams present in all organs, tissues, fluids and secretions. The level of free intracellular Zn^{2+} is as low as 0.5nM, as estimated from measurements of the zinc-specific ^{19}F-NMR signal of a fluorinated metal chelating probe (Benters et al. 1997). About 90% of total body Zn is found in skeletal muscle and bone, with much smaller amounts in the liver, gastrointestinal tract, skin, kidney, brain, lung, prostate and other organs (Frederickson and Bush, 2001).

Zn plays an essential role in cell membrane integrity; it helps manage insulin action and blood glucose concentration and has an essential role in the development and maintenance of the body's immune system. Zn is also required for bone and teeth mineralization, normal taste and wound healing. Zn is a component of more than 70 different enzymes that function in many aspects of cellular metabolism, involving metabolism of proteins, lipids and carbohydrates (Parkin, 2004). The functions of Zn comprise the stabilization of conformation in transcription factors. Zn also has been found to modulate cellular signal transduction processes and even to function as a modulator of synaptic neurotransmission in the case of the zinc-containing neurons in the forebrain (Valko et al. 2005).

The significance of Zn in human nutrition and public health was recognized relatively recently. Clinical Zn deficiency in humans was first described in 1961, when the consumption of diets with low Zn bioavailability due to high phytic acid content was associated with "adolescent nutritional dwarfism" in the Middle East (Prasad et al. 1961). Since then, Zn insufficiency has been recognized by a number of experts as an important public health issue, especially in developing countries (Prasad, 1998).

In the brain Zn is essential for synthesis of coenzymes that mediate biogenic-amine synthesis and metabolism. Zn from vesicles in presynaptic terminals of certain glutaminergic neurons modulates postsynaptic N-methyl-D-aspartate (NMDA) receptors for glutamate. Large amounts of Zn released from vesicles by seizures or ischemia can kill postsynaptic neurons. Acute Zn deficiency impairs brain function of experimental animals and humans. Zn deficiency in experimental animals during early brain development causes malformations, whereas deficiency later in brain development causes microscopic abnormalities and impairs subsequent function. A limited number of studies suggest that similar phenomena can occur in humans (Sandstead et al. 2000).

7.1.1. Sources

Zn is a very common substance that occurs naturally. Many food products contain certain concentrations of Zn. Drinking water also contains certain amounts of Zn, which may be higher when it is stored in metal tanks. Industrial sources or toxic waste sites may cause the Zn amounts in drinking water to reach levels that can cause health problems (ATSDR, 2005).

Zn occurs naturally in air, water and soil, but its concentrations are rising unnaturally, due to addition of Zn through human activities. Most Zn is added during industrial activities, such as mining, coal and waste combustion and steel processing. Some soils are heavily contaminated with Zn, and these are to be found in areas where Zn has to be mined or refined, or were sewage sludge from industrial areas has been used as fertilizer (ATSDR, 2005).

Zn is the 23rd most abundant element in the Earth's crust. The dominant ore is Zn blende, also known as sphalerite. Other important Zn ores are wurzite, smithsonite and hemimorphite. The main Zn mining areas are Canada, Russia, Australia, USA and Peru. World production exceeds 7 million tons a year and commercially exploitable reserves exceed 100 million tons. More than 30% of the world's need for Zn is met by recycling.

Ores containing Zn are widespread geologically and geographically and many ore bodies are still awaiting development when sufficient demand occurs (ATSDR, 2005).

The best sources of Zn are oysters (richest source) red meats, poultry, cheese (ricotta, Swiss, gouda), shrimp, crab, and other shellfish. Other good, though less easily absorbed sources of Zn include legumes (especially lima beans, black-eyed peas, pinto beans, soybeans, peanuts), whole grains, miso, tofu, brewer's yeast, cooked greens, mushrooms, green beans, tahini, and pumpkin and sunflower seeds.

7.2. Absorption

7.2.1. Inhalation Exposure

Quantitative studies regarding absorption of Zn and Zn compounds after inhalation exposure in humans are limited. The absorption of inhaled Zn depends on the particle size and solubility, both of which may greatly influence the deposition and clearance of Zn aerosols, particularly insoluble Zn oxide (a review of the role of particle size in the deposition of particles is found in Witschi and Last, 2001). Elevated levels of Zn have been found in the blood and urine of workers exposed to zinc oxide fumes (Hamdi, 1969).

The rates or percentages of absorption of inhaled Zn in animals are not available; however, studies provide data on Zn retention in the lungs. Zn retention values were 19.8, 11.5, and 4.7% in the lungs of guinea pigs, rats, and rabbits, respectively, after inhalation exposure (nose-only) to 3.5–9.1 mg zinc/m^3 as Zn oxide aerosol for 2–3 hours (Gordon et al. 1992). The aerosol had a mass median diameter of 0.17 μm. The retention of Zn in lungs was dose related in male Wistar rats administered a single intratracheal instillation of 0.07–3.7 mg zinc/m^3 as zinc oxide (Hirano et al. 1989). A half-life of 14 hours was calculated.

No studies were located regarding distribution in humans after inhalation exposure to Zn. However, occupational studies provided indirect evidence that zinc might distribute to tissues to produce systemic effects (Brown, 1988; Malo et al. 1990).

Zn levels in the lungs of cats peaked immediately after acute exposure to 12–61 mg zinc/Kg/day as Zn oxide for approximately 3 hours and remained high for 2 days postexposure, then dropped significantly thereafter (Drinker and Drinker, 1928). Levels in pancreas, liver, and kidneys increased slowly.

7.2.2. Oral Exposure

Several studies have measured oral absorption rates of Zn in humans. Absorption ranged from 8 to 81% following short-term exposures to Zn supplements in the diet; differences in absorption are probably due to the type of diet (amount of Zn ingested, amount and kind of food eaten) (Aamodt et al. 1983; Hunt et al. 1991; Sandström and Sandberg 1992). For example, dietary protein facilitates Zn absorption; fractional Zn absorption ranged from 8% for low-protein rolls to 26% for high-protein rolls 3 days after individuals ingested 0.05 mg zinc/Kg (Hunt et al. 1991).

The body's natural homeostatic mechanisms control Zn absorption from the gastrointestinal tract (Davies, 1980). Persons with adequate nutritional levels of Zn absorb approximately 20–30% of all ingested Zn. Those who are zinc-deficient absorb greater proportions of administered Zn (Johnson et al. 1988).

Like several other metals, Zn can induce metallothionein (MT) production in intestinal mucosal cells (Richards and Cousins, 1975). Zn binds to MT, which remains in the mucosal cells lining the gastrointestinal tract, and the bound metal is excreted from the body upon sloughing off of these cells. Although the affinity of Zn for MT is relatively low, the protein may serve to prevent absorption of excess Zn in the body (Foulkes and McMullen, 1987).

Absorption of Zn in rats is increased when MT levels are lower (Flanagan et al. 1983). It is hypothesized that Zn entering luminal cells is associated with cysteine-rich intestinal protein (CRIP), and a small amount is bound to MT; however, as the luminal Zn concentration increases, the proportion of cytosolic Zn associated with CRIP is decreased with a concomitant increase in Zn binding to MT (Hempe and Cousins, 1992).

In humans, dairy products that contain both calcium (Ca^{2+}) and phosphorus decrease Zn absorption and plasma Zn concentration (Pecoud et al. 1975). Zn binds to phosphate, which results in coprecipitation of Zn with Ca^{2+} phosphate in the intestines (Nelson et al. 1985). Dietary phytate also reduces Zn absorption. The addition of 400μmol phytate to the diet decreased Zn absorption from 43.3±17.9% in females fed bread containing 0.02 mg zinc/Kg (zinc-65 isotope) to 14.3±3.2% (Sandström and Sandberg, 1992). Rats given diets supplemented with radiolabeled Zn and phytate excreted significantly more Zn in the feces than rats given diets supplemented with radiolabeled Zn but without phytate (Davies and Nightingale, 1975). The authors suggested that the decrease in absorption was due to the formation of zinc-phytate complexes in the intestines. Phytate also reduced reabsorption of Zn secreted into the gastrointestinal tract of humans (Sandström and Sandberg, 1992).

Endogenous substances, such as amino acids, can also influence the absorption of Zn (ATSDR, 2005). Complexing of Zn with amino acids generally enhances its absorption in all segments of the intestine (Wapnir and Stiel, 1986). Although neither Zn nor the amino acid proline are readily absorbed in the colon, complexing of Zn with proline during an *in vivo* intestinal perfusion in rats resulted in increased Zn absorption.

It has been reported that a single oral dose of 0.7 mg zinc/Kg as Zn sulfate given to 11 individuals resulted in peak Zn levels in the plasma at 2–3 hours (Sturniolo et al. 1991). Similarly, Neve et al. (1991) reported peak serum Zn concentration at 2.3 hours with 0.7 mg zinc/Kg as Zn sulfate.

7.2.3. Dermal Exposure

Dermal absorption of Zn occurs, but its mechanism is not clearly defined. Studies are very limited regarding the absorption of Zn through the skin (ATSDR, 2005). Historically, Zn oxide has been used clinically to promote the healing of burns and wounds (Gordon et al. 1981). Absorption has been observed in burn patients treated with gauze dressings containing Zn oxide (Hallmans, 1977). The pH of the skin, the amount of Zn applied, and the vehicles administered with zinc all affect its absorption (Agren 1990, 1991).

7.3. Distribution

Zn is one of the most abundant trace metals in humans. It is found normally in all tissues and tissue fluids and is a cofactor in over 300 enzyme systems. Together, muscle and bone contain approximately 90% of the total amount of Zn in the body (≈60 and 30%, respectively) (Wastney et al. 1986). Organs containing sizable concentrations of Zn are the liver, gastrointestinal tract, kidney, skin, lung, brain, heart, and pancreas (Bentley and Grubb, 1991; He et al. 1991). High concentrations of Zn were also detected in the prostate (Forssen, 1972), retina, and sperm (Bentley and Grubb, 1991). Zn levels may vary considerably from one individual to another (Forssen, 1972).

To some degree, the distribution of Zn in some tissues appears to be regulated by age (Schroeder et al. 1967). Zn concentrations increase in the liver, pancreas, and prostate and decrease in the uterus and aorta with age. Levels in the kidneys and heart peak at approximately 40–50 years of age and then decline.

Hormones, such as the adrenocorticotrophic hormone (ACTH), appear to regulate the concentration of Zn in the liver. ACTH, secreted by the anterior pituitary gland, stimulates the secretion of glucocorticoids. Glucocorticoids, or hormones with glucocorticoid activity, have been shown *in vitro* to stimulate the net Zn uptake in cultured liver cells and at the same time activate the gene that regulates MT synthesis (Failla and Cousins, 1978). However, there are no *in vivo* data to support these *in vitro* findings. MT in the cells of the intestinal mucosa binds Zn, thus regulating its release into the blood (ATSDR, 2005).

7.3.1. Metabolism

Plasma provides a metabolically active transport compartment for Zn (Cousins, 1985). Zn is most often complexed to organic ligands (existing in loosely or firmly bound fractions) rather than free in solution as metallic ion (Gordon et al. 1981). Zn is found in diffusible or nondiffusible forms in the blood (NAS/NRC, 1979). In the diffusible form, approximately two-thirds of plasma Zn is freely exchangeable and loosely bound to albumin (Cousins, 1985); the zinc-albumin complex has an association constant of about 106 (NAS/NRC, 1979). The diffusible form of Zn also includes Zn bound to amino acids (primarily histidine and cysteine). The zinc-albumin complex is in equilibrium with the zinc-amino acid complex (Henkin, 1974). The zinc-amino acid complex can be transported passively across tissue membranes to bind to proteins. As we mentioned before, an important binding protein in the kidney and liver is MT, although other tissue-binding proteins may be present.

7.3.2. Excretion

The principal route of excretion of ingested Zn in humans is through the intestine (Davies and Nightingale, 1975; Reinhold et al. 1991; Wastney et al. 1986). Zn loss in the body is by secretion via the gut, and the remainder occurs in the urine (Wastney et al. 1986). Fecal excretion of zinc increases as intake increases (Spencer et al. 1976). Excretion of Zn in the urine also reflects zinc intake (Wastney et al. 1986). Minor routes of elimination are saliva secretion, hair loss, and sweat (Prasad, 1998).

Other factors may affect Zn excretion. For example, low dietary intake of Zn or malnutrition can increase the urinary excretion of Zn. This release of Zn is a result of tissue breakdown and catabolism during starvation; and elevated urinary excretion of Zn may persist after intake levels return to normal (Spencer et al. 1976). Administration of histidine or high-protein diet may increase urinary Zn excretion; however, a corresponding increase in Zn absorption may maintain Zn balance in the body (Hunt et al. 1991).

7.4. Zinc in the Brain

Sheline et al. (1943) first reported ^{65}Zn uptake by brain in dogs and mice. Uptake was slower and the amount retained was less than that in other tissues. A decade later, Maske

(1955) serendipitously discovered that diphenylthiocarbizone (dithizone) stains a pool of Zn that is strikingly localized. Staining of hippocampal mossy fibers was intense. Subsequently, Hu and Friede (1968) measured Zn in 24 regions of human brain by atomic absorption spectroscopy. Concentrations in hippocampus were highest, but gray matter of the cortex was nearly as rich. White matter had the lowest concentrations. Concentrations of Zn in newborn brain were lower than in adults (Sandstead et al. 2000).

Soon after Maske, McLardy (1960) found Zn in mossy-fiber "giant" boutons. Later, he reported several other Zn-containing fiber systems (McLardy, 1970). After McLardy (1960), von Euler (1962) found that bathing the surface of the hippocampus with H_2S-saturated saline removed Zn and changed the evoked potential response after electrical stimulation. He noted that the H_2S caused a variety of changes and therefore was cautious in concluding that removal of Zn caused the changes.

Nearly simultaneously with the above anatomical studies, Zeigler et al. (1964) measured the effect of Zn deficiency on the kinetics of Zn in chick brain. Zn deficiency increased the ^{65}Zn uptake but had no apparent effect on the concentration of stable Zn. Cox et al. (1969) used rats to confirm that Zn deficiency had little effect on the concentration of Zn in brain. He also showed that high intakes of Zn increased its concentration in brain. About a decade later, Wallwork et al. (1983) used weanling rats to confirm that Zn deficiency has little effect on brain Zn, with the exception of a decreased concentration of Zn in the olfactory bulb. In addition, he found that brain copper was increased by Zn deficiency.

Studies by Haug (1967) built on the work of McLardy (1960). With the use of electron microscopy and a modified silver-sulfide stain (Timm, 1958), Haug showed electron-dense silver particles that were located within the mossy-fiber giant boutons, evenly distributed, and not in mitochondria. Later, Haug et al. (1971) showed that transection of mossy-fiber axons caused a rapid disappearance of Zn from the vesicles in the presynaptic boutons.

Nearly two decades after von Euler, Hesse et al. (1979) confirmed that Zn status can affect synaptic responses in the hippocampus. Using Zn-deprived rats, he showed decreases in evoked responses after repeated low frequency stimulation of the dentate gyrus. In contrast, repeated stimulation of commissural axons did not result in decreased evoked responses. Hesse suggested that his findings were caused by a decrease in vesicle Zn. More recent findings suggest that this is unlikely. Commissural axon terminals were shown to contain as much vesicle Zn as mossy-fiber terminals (Frederickson et al. 1992; Long et al. 1995).

Shortly after Hesse et al. (1979), Frederickson et al. (1982; 1983), with the use of stable-isotope dilution mass spectrometry, found that ~8% of Zn in the hippocampus is in vesicles. Soon after, three groups showed that Zn is released from axon terminals during electrophysiologic activity. Howell et al. (1984) showed that electrical stimulation *in vitro* caused uptake of ^{65}Zn tracer by presynaptic terminals of mossy-fiber axons, and that previously incorporated ^{65}Zn was released. Assaf and Chung (1984) reported similar findings on the basis of the chemical analysis of poststimulation superfusate, and Sloviter (1985) showed by electron microscopy and modified silver stain that electrical stimulation decreased vesicle Zn in mossy-fiber axon terminals. About the same time Perez-Clausell and Danscher (1985) showed by electron microscopy and modified silver stain that Zn is present in ~10% of the clear round vesicles of Gray's Type I (excitatory) synaptic boutons. These authors subsequently showed by *in vivo* sulfide binding that Zn released from vesicles can move from the synaptic cleft to the extracellular space (Perez-Clausell and Danscher, 1985).

Peters et al. (1987) and Westbrook and Mayer (1987) showed that vesicle Zn that is released into the synaptic cleft during neurotransmission modulates NMDA-specific postsynaptic receptors for glutamate in a rapid, dose-dependent and reversible manner. Consistent with Zn having a modulator role, Fukahori et al. (1988) found lower Zn concentrations in the dentate area of the hippocampus of a strain of mice with a high propensity for seizures. Zn deficiency decreased hippocampal Zn and increased seizures, whereas high intakes of Zn increased hippocampal Zn and decreased seizures (Fukahori and Itoh, 1990). Mitchell et al. (1990) confirmed that Zn status can affect seizure susceptibility. *In vivo* chelation of Zn with dithizone increased the sensitivity of rats to kainic acid–induced seizures. Morton et al. (1990) also found that Zn status affected seizure threshold. Subcutaneous administration of Zn decreased noise-induced seizures in DBA/2J mice, but had no effect on seizures caused by kainic acid.

Findings of Frederickson et al. (1990) were consistent with vesicle Zn affecting cognition. Reversible chelation of Zn *in vivo* "produced a time-locked and selective disruption of hippocampal-dependent spatial-working memory." Subsequently, Browning and O'Dell (1994; 1995) found in Guinea pigs that Zn deficiency decreased the concentration of postsynaptic NMDA-specific glutamate-mediated Ca^{2+} channels in cortical synaptosomes.

On the other hand, high concentrations of extracellular Zn can kill neurons. Yokoyama et al. (1986) found that 30 μmol/L or more of Zn in tissue culture killed neurons. Soon after, Frederickson et al. (1988; 1989) reported the toxicity of Zn for neurons *in vivo*. With the use of a quinoline fluorescence technique, they found that kainic acid–induced seizures caused loss of Zn from the presynaptic axon terminals of hippocampal mossy fibers and that, coincidentally, the postsynaptic neurons showed intense fluorescence for Zn and signs of degeneration. Following, Tonder et al. (1990) found similar abnormalities in rats that had been subjected to cerebral ischemia. Sensi et al. (1997), Yin and Weiss (1995) and Yin et al. (1998) suggested mechanisms whereby Zn enters postsynaptic neurons. They include passage through voltage-gated Ca^{2+} channels, transporter-mediated exchange with intracellular sodium, passage through NMDA receptor–gated channels and penetration through calcium-permeable -amino-3-hydroxy-5-methyl-4-isoxazole-propionate (AMPA)- or kainate receptor–gated channels.

Other *in vitro* evidence of the toxicity of Zn was provided by Bush et al. (1993, 1994a, b, and c). At physiologic concentrations and pH, Zn complexed with "amyloid protein precursor," and "A-β-1–40," a component of cerebral amyloid that is present in spinal fluid. A-β-1–40 solubility was decreased and resistance of the resulting amyloid to tryptic digestion was increased. These authors suggested that a similar *in vivo* phenomenon might contribute to dementia.

The turnover of Zn in the brain is slow. The maximum uptake of Zn is probably 6–10 days after parenteral injection into rats (Takeda, 2000) and the half time for elimination from the rat brain is in the range of 16–43 days (Takeda et al. 1995).

7.4.1. Zinc and Brain Function

Zinc's roles in brain function are poorly understood. However, it is well known that this metal is essential pre- and postnatally for growth, maturation, and function. In early pregnancy, Zn is essential for cell multiplication and implantation of the embryo and for cell differentiation and organ formation. Deficiency causes teratology in all tissues. Zn deficiency in later pregnancy impairs neuronal replication and migration (as observed in cerebellar external granular cells). Synaptogenesis is impaired (as observed in Purkinje cells). It has been proposed that Zn deficiency impairs Ca^{2+} channels causing a decrease in intracellular Ca^{2+} that suppresses gene expression of growth factors and synthesis of nucleic acids and proteins. Whatever the mechanism, effects in experimental animals include poorly reversible impairments in learning and memory later in life, which appears associated with decreased neuronal survival. It is unknown if similar phenomena occur in humans. It is known however that low fetal growth, a process caused by maternal Zn deficiency, is a risk factor for coronary heart disease, type 2 diabetes mellitus, chronic lung disease, and obesity. One wonders if fetal growth is related to later neuronal health and function (Sandstead, 2003).

Approximately 90% of the total brain Zn exists as Zn MT. Zn has three functions in Zn MT: catalytic, coactive (or co-catalytic) and structural (Vallee and Falchuk, 1993).

Zn deficiency from dietary inadequacy was first described among poor Iranian farm boys by Prasad et al. (1963). Subsequently, the condition was identified among poor Egyptian farm boys who displayed dwarfism, hypogonadism, iron deficiency, hookworm and schistosomiasis (Prasad et al. 1963; Sandstead et al. 1967). Abnormal behaviors occurred in some of the people studied. In the second decade of this century, the International Health Board of the Rockefeller Foundation (1919) reported an association between hookworm infection and low cognitive performance in U.S. Army recruits and in children from Southeastern mill towns. The same year Waite and Nelson (1919) found a direct association between the severity of hookworm infection and impaired mental development in children from North Queensland, Australia. Thus it is suspected that Zn deficiency contributed to the cognitive abnormalities described.

Thirty-three years ago Henkin et al. (1975) discovered that severe Zn deficiency impaired neuromotor and cognitive performance of adults. They induced Zn deficiency by administration of large doses of histidine, which caused high urinary excretion of Zn. All subjects developed abnormal taste and smell perception. Some were ataxic, some were depressed, some hallucinated and some developed paranoia. Soon after Henkin's report Moynahan (1976) described abnormal behavior in a patient with acrodermatitis enteropathica, and Kay et al. (1976) found abnormal behaviors in patients with Zn deficiency as a result of inadequate parenteral feeding.

Effects of postnatal Zn nutriture on infant development were reported by Friel et al. (1993). Linear growth and motor development were higher in newborns <1500 g that were given 11 mg Zn/L of formula from birth to 6 months compared with infants given 6.7 mg Zn/L. Later, Sazawal et al. (1996) reported that repletion with 10 mg Zn/d simultaneously with potentially limiting vitamins increased activity and energy expenditure of low-income urban Indian children, aged 12–23 months. Similarly, Bentley et al. (1997) found that Guatemalan infants given 10 mg Zn/d for 7 months sat up and played more than infants given placebo. Ashworth et al. (1998) also found that Zn repletion improved behavioral ratings. His subjects were low-birth-weight Brazilian infants, aged 12 months, who were given 5 mg Zn/d 6 d/wk during the first 8 postnatal weeks. Controls given 1 mg Zn/d lagged behind.

In other children study Thatcher et al. (1984) found a direct association between an index of Zn status (hair Zn concentration) and reading performance on a standardized test. In addition, coherence of the frontal lobe EEG was related directly to the concentration of Zn in hair. Consistent with Thatcher, Wachs et al. (1995) found that certain preadolescent behaviors of Egyptian children were associated with the consumption of foods that were derived from animals and are rich in Zn.

In adults, Henrotte et al. (1977) found that low concentrations of Zn were associated with lower frequency of the EEG during hyperventilation. Later these authors reported an association between Type A personality, high Zn concentration and low urinary Zn concentration, as contrasted with Type B personality (Henrotte et al. 1985). When Type A subjects were exposed to stress, they excreted more Zn in their urine than did type B subjects. Goldstein and Pfeiffer (1978) reported that treatment of schizophrenic patients with Zn was followed by a decrease in EEG amplitude (toward normal), in contrast to the effect of placebo. The change was consistent with a decrease in cortical excitability. Subsequently, Tang (1991) reported lower concentrations of Zn in hair from female epileptic patients than from controls. In addition, the occurrence of seizures was associated with low plasma Zn concentrations (Sandstead et al. 2000).

Relevant to Zn nutrition of the elderly, Burnet (1981) suggested that low Zn nutriture increases the risk of dementia. He based his thesis on the requirement of Zn for DNA synthesis and repair (Lieberman and Ove 1962; Lieberman et al. 1963). The more recent findings of Tully et al. (1995) appear to support Burnet's idea. They found a negative association between the serum Zn concentration 1 year before dead and the frequency of "senile" and "diffuse" plaques in the brains of 12 elderly women who were examined *postmortem*.

7.4.2. Zinc Transport into the Brain

The zinc-binding affinity for ligands in serum is important for understanding the mechanism of Zn transport into the brain via the brain barrier system, the blood–brain and the blood–cerebrospinal fluid (CSF) barriers. Serum Zn (approximately 15 mM) is partitioned among three fractions: the protein-bound form, the low molecular weight ligand-bound form and free Zn^{2+} (Takeda, 2000). The Zn bound to serum proteins and low molecular weight ligands is about 98 and 1–2%, respectively, of total serum Zn (Prasad and Oberleas, 1970). On the other hand, the concentration of free Zn^{2+} in mammalian serum has been estimated to be as low as 10^{-9} –10^{-10} M (Magneson et al. 1987); free Zn is a very small fraction of the exchangeable Zn in serum.

The largest component of exchangeable Zn in the serum is bound to albumin. The bovine albumin molecule is believed to have two zinc-binding sites of equally high affinity (Henkin, 1974). A brain autoradiogram with ^{65}Zn in the Nagase analbuminemic rat, which has a genetic mutation affecting albumin mRNA processing and lacks serum albumin, demonstrates that albumin is not essential for Zn transport into the brain (Takeda et al. 1997a). However, serum albumin may participate in Zn transport as a large pool of exchangeable Zn in normal animals (Takeda, 2000).

The next largest component of exchangeable Zn in the serum is bound to amino acids, i.e. histidine and cysteine (Hempe and Cousins, 1992). In this way, Aiken et al. (1992) report that ^{65}Zn uptake in the brain as well as in other tissues, expressed relative to the plasma ^{65}Zn level, was enhanced by L- histidine infusion. Buxani-Rice et al. (1994) report that ^{65}Zn transport into the brain during a short cerebrovascular perfusion was enhanced by addition of 100 mM L-histidine. These results suggest that L-histidine is involved in Zn transport into the brain via the brain barrier system. Zn may be present as $Zn(His)_2$ and $Zn(His)^+$ in the serum, and other complexes such as $Zn(Cys)(His)^-$ may also be present (Giroux and Henkin, 1972). Thus, it is possible that histidine might serve to transfer Zn to plasma membrane proteins such as DMT1, a divalent metal transporter (Gunshin et al. 1997), and/or an unidentified Zn transporter (Colvin, 1998) on the brain capillary endothelial cells and choroidal epithelial cells.

The mechanism by which Zn is taken up in the brain parenchyma cells is poorly understood. The Zn concentration in the extracellular fluid in the brain was estimated to be approximately 0.15 μM, judging from the CSF concentration (Palm et al. 1986). The intracellular Zn concentration in the brain was estimated to be approximately 150 μM, judging from the average total brain Zn concentration (Takeda et al. 2000). There is a 1000 times difference in Zn concentration between the brain parenchymal cells and the extracellular fluid. This may indicate the existence of energy-dependent Zn uptake in neurons and glial cells (Takeda, 2000).

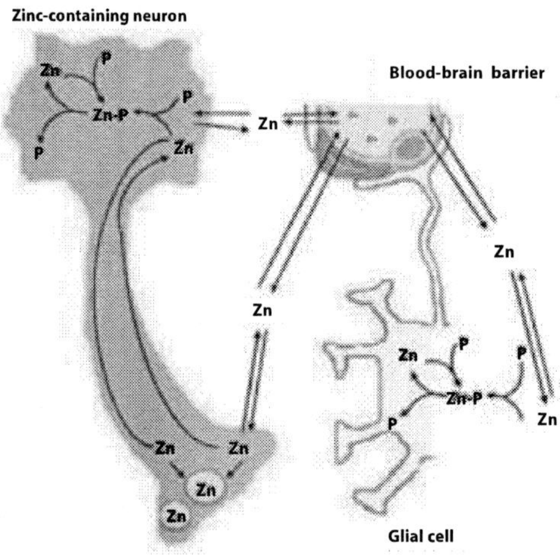

Modified from Takeda, 2000.

Figure 1. Transportation and utilization of Zn in the brain. Zn is transported into the brain via not only the blood–brain barrier but also the blood– CSF barrier. Zn is taken up by neurons, which may have two Zn uptake sites, i.e. the cell body and the neuron terminal, and also by glial cells, and it is then incorporated into zinc-binding proteins. In Zn-containing glutaminergic neurons, the Zn taken up is transported into the presynaptic vesicles and utilized as a neuromodulator. P, protein; shaded P, degraded protein

Genes that are involved in mammalian Zn transport have recently been cloned. Four putative Zn transporters, known as ZnT-1 through ZnT-4, are now known (McMahon and Cousins, 1998). ZnT-1 is ubiquitously expressed and associated with Zn efflux (Tsuda et al. 1997). However, a Zn transporter involved in Zn uptake in cells has not yet been cloned.

High-affinity Zn uptake was observed in hippocampal slices (Howell et al. 1984). DMT1 is present in the hippocampal pyramidal and granule cells, cerebellar granule cells, the preoptic nucleus and pyramidal cells of the piriform cortex in high densities (Gunshin et al. 1997). This transporter appears to be involved in Zn uptake in neurons (Colvin et al. 2000). On the other hand, it is not known where the preferential uptake of Zn in neurons occurs (Figure 1).

7.4.3. Zinc Release from Neuron Terminals

Zn is transported anterogradely and retrogradely via the axonal transport system. This transport is important for the supply of Zn to zinc-requiring sites. Injection of ^{65}Zn into either the striatum or the substantia nigra shows that it is transported in reciprocal projections between the striatum and the substantia nigra via the medial forebrain bundle (Takeda, 2000). Zn transport is inhibited by injection of colchicine, an inhibitor of axonal transport, into the medial forebrain bundle (Takeda et al. 1998). Zn transport after injection of ^{65}Zn into the olfactory bulb demonstrates that Zn taken up in the mitral and/or tufted cells (the secondary olfactory neuron) is transported along the olfactory tract (Takeda et al. 1997b).

Localization of Zn in the presynaptic area was observed in the hippocampal formation (Kim et al. 1999). Zn concentrations in the vesicles in the giant boutons of hippocampal mossy fibers were estimated to be 300–350 μM (Frederickson et al. 1983). The presence of zinc-containing neurons has been demonstrated in the brain, especially the telencephalon (Figure 2) (Perez-Clausell and Danscher, 1985). The hippocampal and amygdalar regions may possess zinc-containing neuron terminals in high densities (Frederickson et al. 2000). All zinc-containing neurons reported so far are considered to be glutaminergic (Crawford and Connor, 1973). However, not all glutaminergic neurons are zinc-containing (Frederickson and Moncrieff, 1994).

In hippocampal slices, electrical stimulation enhances the uptake of exogenous Zn in the mossy fiber neuropil (Howell et al. 1984). This may mean re-uptake of Zn by presynaptic neurons. However, the mechanism of re-uptake of Zn is unknown. An unidentified Zn transporter and/or receptor-mediated endocytosis might be involved in the reuptake (Frederickson, 1989). On the other hand, Zn transporter-3 (ZnT-3) is highly expressed in brain regions such as the hippocampus, amygdala and entorhinal cortex which are rich in vesicular Zn and serves to transport cytosolic Zn into the synaptic vesicles (Wenzel et al. 1997).

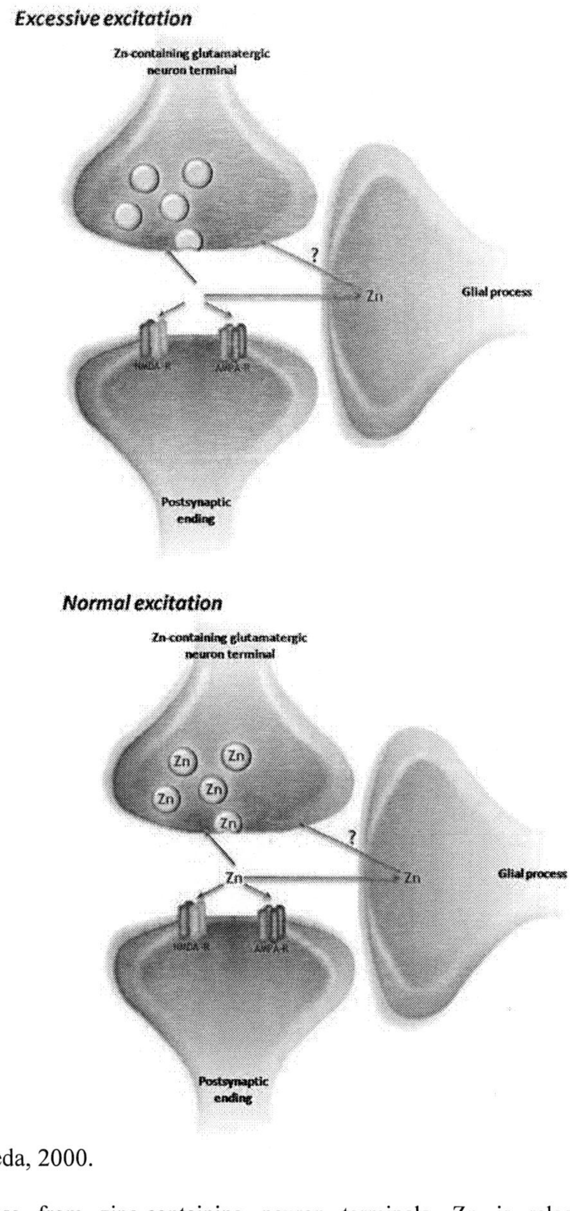

Modified from Takeda, 2000.

Figure 2. Zn release from zinc-containing neuron terminals. Zn is released from zinc-containing glutaminergic neuron terminals. However, the mechanism of removal of zinc released into the synaptic cleft is unclear, as described in the text. During excessive excitation of zinc-containing glutaminergic neurons, there is a decrease of Zn concentration in the presynaptic vesicles, followed by translocation of Zn to postsynaptic neurons, although the translocation of Zn might also occur from sites other than the presynaptic vesicles. Shaded Zn represents low levels of Zn.

Metallothionein-III (MT-III), a brain-specific member of the MT family of metal-binding proteins (Ebadi, 1986), is also highly expressed in glutaminergic neurons that contain vesicular Zn (McMahon and Cousins, 1998). MT-III may be an important regulator of Zn in the brain and its absence has been implicated in the development of Alzheimer disease (Uchida et al. 1991). However, the vesicular Zn concentration is not changed by disruption of the MT-III gene (Erickson et al. 1997), thus MT-III might not be involved in Zn transport into the synaptic vesicles (Takeda, 2000).

Glutamate released into the synaptic cleft is primarily removed by transport into glial cells via a glutamate transporter (Zerangue and Kavanaugh, 1996), whereas the mechanism of removal of Zn released into the synaptic cleft is unknown. Although glial cells appear to be a candidate (Takeda et al. 1999), a Zn transporter is not identified in glial cells. A portion of the Zn is taken up by presynaptic neurons as described above, and also taken up by postsynaptic neurons; the Zn may be taken up by mechanisms via the voltage-gated Ca^{2+} channel (Weiss et al. 1993), NMDA receptor (Koh and Choi, 1994) and AMPA/ kainate receptor (Yin and Weiss, 1995).

7.5. Zinc Neurotoxicity

Zn lacks redox activity and is quite nontoxic compared to other transition metals such as manganese and iron (Bertholf, 1988). However, it has been demonstrated that excess Zn in the extracellular fluid is neurotoxic in the brain (Cuajungco and Lees, 1997; Frederickson et al. 2000). Zn promotes *in vitro* aggregation of β-amyloid protein, the main component of the senile plaques typically observed in Alzheimer disease (AD) brains (Bush et al. 1994c). A high concentration of extracellular Zn could trigger the formation of amyloid plaques in the brain (Lee et al. 1999; Huang et al. 2000). Suh et al. (2000) observed histochemically reactive Zn in the amyloid deposits of dense-core (senile) plaques, in the amyloid angiopathy surrounding diseased blood vessels, and in the somata and dendrites of neurons showing the characteristic neurofibrillary tangles of AD brains.

Perforant path stimulation induces a concurrent loss of Timm stain in the mossy fibers and damage of hilar interneurons and CA3 pyramidal cells innervated by the mossy fibers (Simons, 1993). Synaptically released Zn enters postsynaptic neurons in toxic excess during seizures, resulting in a loss of Zn in the presynaptic vesicles. Excessive influx of Zn into postsynaptic neurons leads to degeneration (Frederickson et al. 1989; Lee et al. 2000a). The translocation of Zn from presynaptic neuron terminals into postsynaptic neurons also occurs after cerebral ischemia (Park and Koh, 1999) and traumatic brain injury (Gilly and Armstrong, 1984). After transient forebrain ischemia, Zn released into the synaptic clefts accumulates in neuronal cell bodies followed by degeneration of the neurons (Koh et al. 1996).

On the other hand, prevention of Zn translocation by a Zn chelator has been shown to be neuroprotective in both seizures and ischemia (Koh et al. 1996).

Preferential Zn influx through the calcium-permeable AMPA/kainate receptor after exposure to NMDA, kainite or high K^+ in the presence of 300 μM Zn^{2+} causes prolonged mitochondrial superoxide production (Sensi et al. 1999b). Zn^{2+}-dependent production of reactive oxygen species may be associated with degeneration of postsynaptic neurons. In the case of exposure of cortical cultures to Zn^{2+}, synaptically released zinc-induced neuronal death exhibits features of both apoptosis and necrosis and is probably mediated by free radical injury (Kim et al. 1999), although the findings might imply neuronal necrosis. Therefore, vesicular Zn may be a key modulator of the neuronal death associated with transient global ischemia and sustained seizures, as well as in other neurological disease states (Takeda, 2000).

Thus, disruption of Zn homeostasis has been implicated in several neurodegenerative diseases including AD (Huang et al. 2000; Suh et al. 2000), prion disease (Watt and Hooper, 2003), amyotrophic lateral sclerosis (ALS) (Valentine and Hart, 2003) and Wilson Disease (Brewer, 2000) and some brain alterations such as ischemia and seizures.

7.5.1. Mechanism of Zn-induced Neurotoxicity

The multiple pathways by which intracellular Zn^{2+} promotes injury are just beginning to be understood (See Figure 3). Given the cation's multimodal effects on cell physiology, these injury pathways are likely to be complex, with probable overlaps and/or synergisms between mechanisms (Capasso et al. 2005). It is now clear that the two classical models of neuronal death, apoptosis and necrosis, are coexisting processes which may occur in parallel (Leist and Jaattela, 2001), with the intensity of the insult and its effect on the status of cellular energy levels determining which process is more prominent in a given cell. In that respect, mitochondria, by controlling ATP production, appear to function as the checkpoint between an either more predominantly necrotic or apoptotic demise (Capasso et al. 2005).

For instance, a subacute ischemic insult may result in still partially-functioning neuronal mitochondria able to generate sufficient ATP to allow apoptosis to fully progress. In contrast, a fulminant insult would result in rapidly compromised mitochondria and energy levels, forcing the neurons to a necrotic death (Sensi et al. 1999a). Given Zn^{2+}'s well-known effects on mitochondrial function, it is therefore not surprising that the cation could be involved in the induction of both necrotic and apoptotic processes (Aizenman et al. 2000; Jiang et al. 2000; Bossy-Wetzel et al. 2004).

Several studies have indicated that mitochondria are important sites for the neurotoxic effects triggered by elevated $[Zn^{2+}]$i levels. Mitochondria serve as the major cellular source of ROS, and following rapid and massive cytosolic Zn^{2+} rises, a great disruption of mitochondrial function occurs in the form of a dramatic loss of mitochondrial membrane potential and increased ROS generation (Yin and Weiss, 1995; Sensi et al. 1999a; 2000). These Zn^{2+}-triggered ROS have the potential to diffuse out of mitochondria, promoting neuronal demise.

On the other hand, recent findings indicate that submicromolar cytosolic Zn^{2+} loads (as, for example, would be expected to be seen in neurons in the ischemic penumbra) are able to induce a disruption of mitochondrial function sufficient to set in motion the apoptotic cell death pathways. Submicromolar Zn^{2+} levels promote mitochondrial swelling and increased respiration in isolated mitochondria. These effects are consistent with the induction of mitochondrial permeability transition pore (mPTP), a key step in the activation of the apoptotic cascade (Sensi et al. 2003). In addition, by inducing oxidative stress, Zn^{2+} can also indirectly cause further enhancement of the mPTP induction (Wudarczyk et al. 1999).

In intact cortical neurons, submicromolar $[Zn^{2+}]$i rises are adequate to trigger substantial mitochondrial swelling and release of Cytochrome-c, while pharmacological blockade of mPTP opening results in attenuation of both Zn^{2+}-triggered release of these factors as well as Zn^{2+}-dependent neuronal loss (Jiang et al. 2001).

Zn^{2+} also acts as a potent inhibitor of energy production. Moderate intracellular Zn^{2+} loads can inhibit multiple enzymes in the glycolytic pathway, leading to reduced levels of glyceraldehyde-3-phosphate dehydrogenase (GAPDH), phosphofructokinase, and NAD^+ glycohydrolase (Krotkiewska and Banas, 1992; Kukimoto et al. 1996), and ultimately to ATP depletion. GAPDH inhibition, in particular, is linked to depletion of cytosolic NAD^+, which can be prevented by addition of pyruvate: supplying pyruvate reverses Zn^{2+}-inducedNAD^+ depletion and is potently neuroprotective against Zn^{2+}-dependent toxicity (Sheline et al. 2000).

7.6. Diseases Associated with Zinc-Mediated Neuronal Death

7.6.1. Ischemia

Interruption of blood flow after transient global ischemia induces delayed neuronal death, the development of an infarct, and subsequent cognitive dysfunction, are believed to be the hallmark of pathogenesis of vascular dementia in elderly people (Lee et al. 2000b). In response to ischemia glutamate is released from nerve terminals and accumulates in synaptic clefts. Excess of glutamate causes over-stimulation of its receptors. Then, it induces the entry of large quantities of Ca^{2+} to responding neurons through NMDA-type glutamate receptors or voltage-dependent Ca^{2+} channels, and thereafter, the increased intracellular Ca^{2+} triggers various pathways of apoptotic neuronal death (Choi, 2005).

As described above, Zn is co-released with glutamate to synaptic clefts by membrane depolarization in the ischemic condition. Koh and Choi (1994) reported that Zn caused apoptotic death of primary cultured cortical neurons. They also revealed the accumulation of chelatable Zn in degenerating neurons of the hippocampus after transient global ischemia (Koh et al. 1996). This Zn translocation occurred in vulnerable neurons in the hippocampus after global ischemia but before the onset of the delayed neuronal death, and enhanced the infarct (Lee et al. 2002). Administration of Ca^{2+} EDTA, a Zn-selective membrane-impermeable chelator, inhibited zinc-induced death of cultured cortical neurons (Koh et al. 1996), blocked the accumulation of Zn, protected the hippocampal neurons after transient global ischemia and reduced the infarct voume (Lee et al. 2002). These results indicate that Zn is a key factor in delayed neuronal death after ischemia, which might be involved in the pathogenesis of vascular dementia (Weiss et al. 2000).

7.6.2. Seizures

The role of Zn^{2+} in epilepsy remains controversial. Excessive exogenous Zn^{2+} can act as a potent convulsant, and in fact intracerebral Zn injection has been used since the 1980's to experimentally induce epilepsy (Itoh and Ebadi, 1982; Pei et al. 1983; Pei and Koyama,

1986). However, a more recent set of studies has indicated that by modulating synaptic activity, the lower levels of Zn^{2+} found endogenously may serve instead as an effective anticonvulsant. Indeed, reduction or loss of vesicular Zn^{2+} via dietary Zn deficiency, chelation, or Znt-3 knockout has consistently been found to enhance seizure susceptibility (Takeda et al. 2003; Blasco-Ibanez et al. 2004). In addition, Zn^{2+}'s selective inhibition of synaptic GABA transporters, recently described by Cohen-Kfir (2005) may be another mechanism that can promote an anticonvulsant effect.

Besides its role as a modulator of seizure activity, Zn^{2+} has also proven to be a mediator of neuronal death in status epilepticus. Zn^{2+} translocation occurs in kainate-induced status epilepticus (Sloviter, 1985; Frederickson et al. 1988; 1989; Suh et al. 2001), and the neuronal loss associated with this condition is attenuated by chelation of extracellular Zn^{2+} (Yi et al. 2003). As in ischemia, Zn^{2+} accumulation in epilepsy may be the result not only of translocation, but also of intracellular mobilization. This hypothesis is supported by studies in Znt-3 KO mice, where kainate-induced seizures were found to produce toxic $[Zn^{2+}]i$ accumulation in degenerating hippocampal neurons (Lee et al. 2000c). The origin of these seizure-driven $[Zn^{2+}]i$ rises seems to be region specific; a recent study on Znt-3 and MT-III double-knockout mice indicates that the main mechanism of intracellular Zn^{2+} accumulation in the CA3 hippocampal subregion is translocation, whereas $[Zn^{2+}]i$ rises in CA1, dentate gyrus, and the thalamus are mainly the result of intracellular mobilization from MT-III (Lee et al. 2003).

7.7. Neurodegenerative Diseases

By extrapolation from its likely role in neuronal death after ischemic or epileptic brain insults, the speculation can be entertained that Zn might also be a factor contributing to more indolent neuronal death in certain neurodegenerative conditions. The demonstrated ability of lower levels of Zn exposure to induce apoptosis aids this speculation, although direct evidence for apoptosis in human neurodegenerative diseases is itself limited (Thompson, 1995). Some specific links to Zn can be found with regard to AD (Uchida et al. 1991; Bush et al. 1993; Bush et al. 1994c; and others), and the amyotrophic lateral sclerosis-parkinsonism-dementia complex found in the Western Pacific (Duncan et al. 1992). In particular, Zn has been found to promote the aggregation of Aβ1-40 peptide into amyloid plaque (Bush et al. 1994c), possibly a key step in AD pathogenesis. It is intriguing that cortical and hippocampal amyloid plaque formation in AD occurs in association with zinc-containing glutamatergic terminals; in cerebellum, glutamatergic terminals contain little Zn and only diffuse amyloid deposits occur.

7.7.1. Alzheimer Disease

Increasing evidence has suggested the implication of Zn in the pathogenesis of AD. However, it remains still controversial. Although the precise etiology of AD is still not yet clear, it is widely believed that the abnormal deposition of β-amyloid protein (AβP) a major component of senile plaques, in the brain and its neurotoxicity may be based on the molecular mechanism of AD (See chapter 2). AβP is a 39-43 amino acid residue peptide derived from a large precursor protein amyloid precursor protein (APP). AβP has an intrinsic tendency to form insoluble aggregates with β-pleated sheet structures. Interestingly, the aggregation and the subsequent conformational change of AβP strongly correlate with its neurotoxicity. Therefore, factors that promote the aggregation of AβP may be involved in the pathogenesis of AD (Konoha et al. 2006). Bush et al. (1994c) found that Zn remarkably enhance the aggregation of AβP *in vitro*. Zn also binds to APP and modulates the binding of APP to extracellular matrix (Bush et al. 1993); APP also binds to copper (Hesse et al. 1994) and regulates copper homeostasis (White et al. 1999). Moreover, clioquinol, a copper/zinc-sensitive chelator, was reported to inhibit the accumulation of AβP in brains of experimental animals (Cherny et al. 2001).

However, considering that Zn is profusely present in the brain and that low concentration of Zn is enough to initiate the aggregation of AβP, the adverse role of Zn in AD is still disputable (Konoha et al. 2006). It is possible that Zn influences the homeostasis of other trace metals and contributes to the pathogenesis of this disease, because other metals including aluminum, iron and copper also accelerate the aggregation of AβP (Kawhara, 2003). Meanwhile, the protective role of Zn in the pathogenesis of AD has been suggested. Arispe et al. (1993) found that AβP forms cation-selective (including Ca^{2+}) ion channels on artificial lipid membranes. Kawhara et al. (1977) have revealed that AβP forms ion channels on neuronal cell membranes and caused the abnormal increase of intracellular Ca^{2+} level (Kawhara, 2003). Therefore, it is possible that channel-formation by AβP and the subsequent increase in Ca^{2+} level may trigger the apoptotic neurodegeneration and finally promotes the pathogenesis of AD (Kawhara, 2004). Some studies demonstrate that Zn inhibits AβP-channel in lipid bilayer membranes as well as on membranes of neuronal cells (Kawhara et al. 1977; Arispe et al. 1993; Lin et al. 1999). Considering that both of Zn APP coexists in synapses and secrets during neuronal excitation, it is possible that Zn may play a role as an endogenous blocker of AβP channel. It was also reported that Zn has concentration-dependent dual effects in AβP-neurotoxicity: low concentrations of Zn protects neurons, but high concentration enhance AβP-neurotoxicity (Moreira et al. 2000).

Furthermore, there are several zinc-related MT related with AD. Uchida et al. (1991) found the protein GIF, which has the ability of inhibiting the outgrowth of neuronal processes. GIF is abundantly present in the normal brain, however, remarkably depleted in the brain of AD patients. They studied the structure of GIF and found that it is MT-III, which binds to Zn and copper. Zinc-binding to S100β, a Ca^{2+} binding protein, influences it's binding with hyperphosphorylated tau protein in AD patients (Yu and Fraser, 2001). Therefore, there is no doubt about the implication of Zn in the pathogenesis of AD; however, its role is complex and still controversial.

7.7.2. Other Neurodegenerative Diseases

Prion disease is a transmissible amyloidogenic disease including Creutzfeldt-Jacob disease and Kuru disease in human, and bovine spongiform encephalopathy. The conversion of normal cellular form of prion protein (PrP^c) to the pathogenic protease-resistant form (PrP^{sc}) is believed that it is based on the transmission and the pathogenesis of prion disease (Brown et al. 1997; Konoha et al. 2006). It is hypothesized that PrP^c is a copper-metalloprotein and modulates copper homeostasis. As copper and Zn are competitive in the binding with MT, Zn also binds to the copper-bindig domain of PrP^c. Zn as well as copper induces the endocytosis of PrP^c (Perera and Hooper, 2001) and causes the aggregation of PrP fragment peptide (PrP106-126) and enhances its neurotoxicity (Jobling et al. 2001). Considering that Zn concentration in the brain is much higher than that of copper, Zn might contribute to the functions of PrP^c.

In 1994, linkage analysis of gene of familial ALS patients demonstrated that mutations in the gene of copper-zinc superoxide dismutase (Cu-Zn SOD) was responsible for the disease (Aoki et al. 1993), Cu-Zn SOD is a primary antioxidant enzyme which enables the regulation of superoxide. Thus, the link between Zn and other neurodegenerative diseases should not be discarded (Konoha et al. 2006).

7.7.3. Brain Aging

Brain aging is characterized by a gradual and progressive loss in the number of nerve cells and decreased cerebral blood flow, along with an increase in advanced glycation end-products, free radical production, and markers of oxidative stress (Morrison and Hof, 1997; Mrak et al. 1997). One new area of investigation in brain aging concerns the role of Zn^{2+} homeostasis (Doraiswamy and Finefrock, 2004; Mocchegiani et al. 2005). Evidence for a Zn^{2+} role in aging-related neurodegeneration comes from studies in animal models of accelerated aging (SAMP 10 mice). SAMP10 mice have been found to display reduced levels of synaptic Zn^{2+} in the mossy fiber pathway (Saito et al. 2000), a phenomenon that may be attributed to the reduced expression of the Zn^{2+} transporter Znt-3. Interestingly, these mice also appear to be more sensitive to excitotoxic damage in the hippocampus. These apparently paradoxical findings should be interpreted in light of the known property of synaptically released Zn^{2+} to indirectly prevent excitotoxicity via stimulation of GABA release together with inhibition of GABA reuptake from interneurons (Takeda et al. 2004; Cohen-Kfir et al. 2005), and by the cation's action as a tonic inhibitor of NMDAR activity (Vogt et al. 2000).

Thus, it is conceivable that Zn^{2+} deficiency in SAMP10 mice might induce a state of chronic excitotoxicity. Several studies have shown age-related reduction in the mossy fiber density and Zn^{2+} staining along with a decrease in synaptic density in the CA2 and CA3 hippocampal subregions (Geinisman et al. 2004; Mocchegiani et al. 2005).

As mentioned above, MTs may play a pivotal role in maintaining Zn^{2+} homeostasis, and recent studies have shown that these proteins tend to be expressed in higher concentrations within the brain of aging rats (Mocchegiani et al. 2001). Astrocytes from aging rats show increased MT-I and MT-II gene expressions (Mocchegiani et al. 2001), while MT-III gene expression has been found to be increased in the neurons of aging hippocampi (Giacconi et al. 2003). MTs are potent antioxidants and protective factors against stress conditions and thus, their abnormal expression in the aging brain may simply reflect a protective endogenous response to a subchronic state of inflammatory and or oxidative stress (Capasso et al. 2005).

On the other hand, as MT presence becomes increased as a consequence of persistently high levels of inflammatory cytokines (IL-1 and IL-6), it is possible that their protective actions can be overwritten by concomitant increase in ROS-driven [Zn2+]i accumulation (Aizenman et al. 2000; Mocchegiani et al. 2001; Bossy-Wetzel et al. 2004). In the aging brain, a persistent increase in the levels of inflammatory cytokines may provoke a chronic up-regulation of MT gene expression (Mocchegiani et al. 2001), leading to higher availability of intracellularly releasable Zn^{2+} (Mocchegiani et al. 2004). It is intriguing to envision a bimodal action for MTs: early in life they enhance neuronal survival and functioning by counteracting oxidative stress, but later on in the aging brain, their overexpression becomes disadvantageous by negatively interfering with Zn^{2+} homeostasis (Mocchegiani et al. 2005).

Finally, one additional aspect of the aging brain concerns the issue of Zn^{2+}-triggered perturbation of mitochondrial function. As mentioned previously, Zn^{2+} potently interferes with mitochondrial behavior, and in older neurons, such action would almost certainly have a greater impact on cell viability compared to young, healthy nerve cells. The capability of the organelles to cope with death signaling, cationic loads, and/or oxidative stress dramatically decrease with aging (Balaban et al. 2005; Short et al. 2005). Mutations in mitochondrial DNA (mtDNA) induced by increased oxidative stress inside and outside of the organelles are perpetuated and accumulate in aging cells (Wallace, 1999). mtDNA mutations may result in the expression of defective mitochondrial proteins and therefore less-efficient metabolism, as well as ways to handle Zn^{2+}. Altered mitochondrial metabolism may in turn disrupt the critical recycling of ROS, leading to increased mitochondrial oxidative stress and further effects on mtDNA, producing a vicious, feed-forward cycle (Capasso et al. 2005).

Final Considerations

In light of the evidences presented in this chapter, we resume the possible pathways of Zn neurotoxicity in Figure 3.

Some neurological disorders are associated with alteration of Zn homeostasis, fact that might influence vesicular Zn levels. Extra excitation of zinc-containing glutaminergic neurons causes a decline in vesicular Zn followed by damage of postsynaptic neurons. The decrease appears to act adversely on brain function. The mechanism of Zn homeostasis, not only in the entire brain but also in neurons and glia cells, is important for understanding Zn homeostasis in the pre-synaptic vesicles and for clarification of vesicular Zn function.

It is clear that Zn is a powerful toxic metal implicated in the neuronal injury observed in cerebral ischemia, epilepsy, and in some neurodegenerative processes. The mechanisms by which Zn exerts its neurotoxicity include the production of reactive oxygen species and the disruption of metabolic enzymatic activity, eventually leading to activation of apoptotic and/or necrotic processes, leading to neuronal death. Beside acute neuronal injury, it has been revealed the role of Zn dysmetabolism in Alzheimer disease since this metal is believed to trigger the Aβ aggregation and senile plaque formation.

Finally, recent findings suggest that alteration of Zn homeostasis might also be a critical contributor to aging-related neurodegenerative processes.

Multiple evidences suggest that the modulation of intracellular and extracellular Zn might be an important therapeutical target for the treatment for many neurological conditions ranging from stroke to Alzheimer disease. Thus, further research about the role of Zn and its toxicity might lead to the development of new treatments for neurodegenerative diseases.

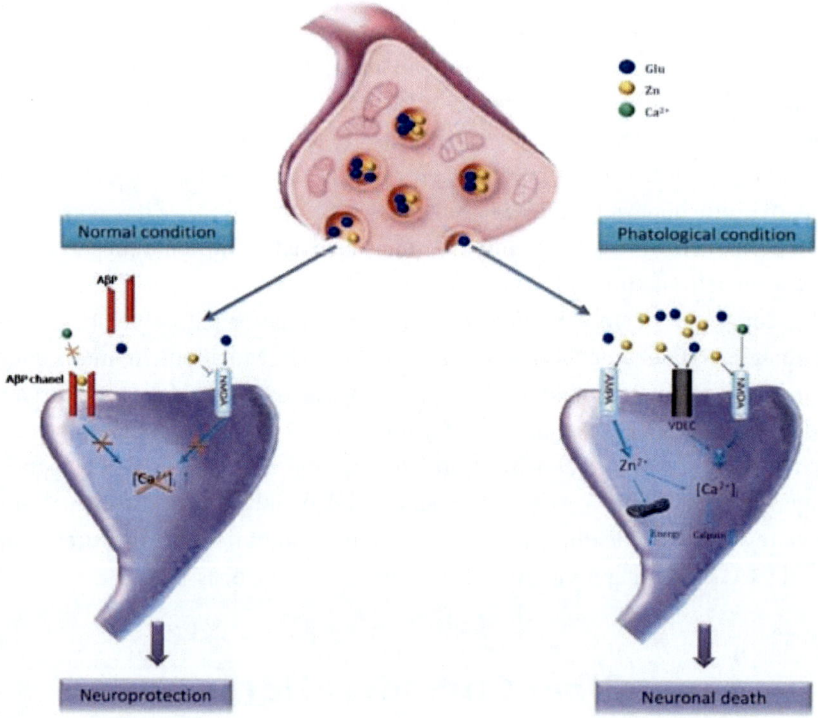

Modified from Konoha et al. 2006.

Figure 3. Scheme of Zinc Neurotoxicity. Zn coexists with Glutamate (Glu) in presynaptic vesicles and is secreted with neuronal excitation. In normal conditions, secreted Zn binds to NMDA type glutamate receptor (NMDA-R) and modulates postsynaptic excitability. AβP, which also is secreted in synapses, aggregates and forms channels on membranes. Zn binds to AβP channel and inhibits Ca^{2+} influx through the channel. Therefore, Zn plays a protective role. However, in the pathological conditions, great amounts of Zn are released in the synapses and translocated into postsynaptic target neurons through the AMPA-type glutamate receptor (AMPA-R)-related Ca^{2+} channel or other pathways such as voltage-dependent L-type Ca^{2+} channel (VDLC). That Zn then inhibits numerous enzymes including mitochondria respiratory enzymes and causes energy depletion. Furthermore, Zn increases intracellular Ca^{2+} levels and enhances the effects of glutamate. Increased Ca^{2+} triggers various apoptotic pathways. Dyshomeostasis of Zn and Ca^{2+} eventually produce delayed neuronal death.

References

Aamodt, R.L., Rumble, W.F. and Henkin, R.I. (1983). Zinc absorption in humans: Effects of age, sex, and food. In: Inglett G, ed. The nutritional bioavailability of zinc. Washington, DC: The American Chemical Society, 61-82.

Agren, M.S. (1990) Percutaneous absorption of zinc from zinc oxide applied topically to intact skin in man. *Dermatologica*, 180:36-39.

Agren, M.S. (1991) Influence of 2 vehicles for zinc oxide on zinc absorption through intact skin and wounds. *Acta Derm Venereol (Stockh)*, 71:153-156.

Aiken, S.P., Horn, N.M. and Saunders, N.R. (1992) Effect of histidine on tissue zinc distribution in rats. *BioMetals*, 5: 235–243.

Aizenman, E., Stout, A.K., Hartnett, K.A., Dineley, K.E., Mclaughlin, B. and Reynolds, I.J. (2000) Induction of neuronal apoptosis by thiol oxidation: putative role of intracellular zinc release. *J Neurochem.* 75: 1878–1888.

Aoki, M., Ogasawara, M., Matsubara, Y., Narisawa, K., Nakamura, S., Itoyama, Y. and Abe, K. (1993) Mild ALS in Japan associated with novel SOD mutation. *Nat Genet.* 5: 323-324.

Arispe, N., Rojas, E. and Pollard, H.B. (1993) Alzheimer disease amyloid β protein forms calcium channels in bilayer membranes: Blockade by tromethamine and aluminum. *Proc Natl Acad Sci USA*, 90: 567-571.

Ashworth, A., Morris, S.S., Lira, P.I. and Grantham-McGregor, S.M. (1998) Zinc supplementation, mental development and behavior in low birth weight term infants in northeast Brazil. *Eur J Clin Nutr*. 52:223-227.

Assaf, S.Y. and Chung, S.H. (1984) Release of endogenous Zn2+ from brain tissue during activity. *Nature*, 308:734-736.

ATSDR (Agency for Toxic Substances and Disease Registry). (2005) Toxicological Profile ofr Zinc. US Department of Health and Human Services, Washington DC, US Public Health Service, US Department of Health and Human Services.

Balaban, R.S., Nemoto, S. and Finkel, T. (2005) Mitochondria, oxidants, and aging. *Cell*, 120: 483–495.

Benters, J., Flögel, U., Schäfer, T., Leibfritz, D., Hechtenberg, S. and Beyersmann, D. (1997) Study of the interactions of cadmium and zinc ions with cellular calcium homoeostasis using 19F-NMR spectroscopy. *Biochem J.* 322: 793-799.

Bentley, M.E., Caulfield, L.E., Ram, M., Santizo, M.C., Hurtado E, Rivera JA, Ruel, M.T. and Brown, K.H. (1997) Zinc supplementation affects the activity patterns of rural Guatemalan infants. *J Nutr.* 127: 1333-1338.

Bentley, P.J. and Grubb, B.R. (1991) Experimental dietary hyperzincemia tissue disposition of excess zinc in rabbits. *Trace Elements in Medicine*, 8:202-207.

Bertholf, R.L. (1988) Zinc, in: H.G. Seiler, H. Sigel (Eds.), Handbook on Toxicity of Inorganic Compounds, Dekker, New York, pp. 788–800.

Blasco-Ibanez, J.M., Poza-Aznar, J., Crespo, C., Marques-Mari, A.I., Gracia-Llanes, F.J. and Martinez-Guijarro, F.J. (2004) Chelation of synaptic zinc induces overexcitation in the hilar mossy cells of the rat hippocampus. *Neurosci Lett.* 355: 101–104.

Bossy-Wetzel, E., Talantova, M.V., Lee, W.D., Scholzke, M.N., Harrop, A., Mathews, E., Gotz, T., Han, J., Ellisman, M.H., Perkins, G.A. and Lipton, S.A. (2004) Crosstalk between nitric oxide and zinc pathways to neuronal cell death I volving mitochondrial dysfunction and p38-activated K+ channels. *Neuron*, 41: 351–365.

Brewer, G.J. (2000) Recognition, diagnosis and management of Wilson's disease. *Proc Soc Exp Biol Med.* 223: 39-46.

Brown, J.J. (1988) Zinc fume fever. *Br J Radiol.* 61:327-329.

Brown, D.R., Qin, K., Herms, J.W., Madlung, A., Manson, J., Strome, R., Fraser, P.E., Kruck, T., von Bohlen, A., Schulz-Schaeffer, W., Giese, A., Westaway, D. and Kretzschmar, H. (1997) The cellular prion protein binds copper in vivo. *Nature*, 390: 684-687.

Browning, J.D. and O'Dell, B.L. (1994) Low zinc status in guinea pigs impairs calcium uptake by brain synaptosomes. *J Nutr.* 124:436-443.

Browning, J. and O'Dell, B. (1995) Zinc deficiency decreases the concentration of N-methyl-D-aspartate receptors in guinea pig cortical synaptic membranes. *J Nutr.* 125:2083-2089.

Burnet, F.M. (1981) A possible role of zinc in the pathology of dementia. *Lancet*, 1:186-188.

Bush, A.I., Multhaup, G., Moir, R.D., Williamson, T.G., Small, D.H., Rumble, B., Pollwein, P., Beyreuther, K. and Masters, C.L. (1993) A novel zinc(II) binding site modulates the function of the beta A4 amyloid protein precursor of Alzheimer's disease. *J Biol Chem*, 268:16109-16112.

Bush, A.I., Pettingell, W.H., de Paradis, M.D. and Tanzi, R.E. (1994a) Modulation of A beta adhesiveness and secretase site cleavage by zinc. *J Biol Chem.* 269:12152-12158.

Bush, A.I., Pettingell, W.H., de Paradis, M., Tanzi, R.E. and Wasco, W. (1994b) The amyloid beta-protein precursor and its mammalian homologues. Evidence for a zinc-modulated heparin-binding superfamily. *J Biol Chem.* 269:26618-26621.

Bush, A.I., Pettingell, W.H., Multhaup, G., de Paradis, M., Vonsattel, J.P., Gusella, J.F., Beyreuther, K., Masters, C.L. and Tanzi, R.E. (1994c) Rapid induction of Alzheimer A beta amyloid formation by zinc. *Science*, 265:1464-1467.

Buxani-Rice, S., Ueda, F. and Bradbury, M.W.B. (1994) Transport of zinc-65 at the blood–brain barrier during short cerebrovascular perfusion in the rat: its enhancement by histidine. *J Neurochem.* 62: 665–672.

Capasso, M., Jeng, J.M., Malavolta, M., Mocchegiani, E. and Sensi, S.L. (2005) Zinc dyshomeostasis: a key modulator of neuronal injury. *J Alzheimer's Dis.* 8:93-108.

Cherny, R.A., Atwood, C.S., Xilinas, M.E., Gray, D.N., Jones, W.Dm., McLean, C.A., Barnham, K.J., Volitakis, I., Fraser, F.W., Kim, Y., Huang, X., Goldstein, L.E., Moir, R.D., Lim, J.T., Beyreuther, K., Zheng, H., Tanzy, R.E., Masters, C.L. and Bush, A.I. (2001) Treatment with copper-zinc chelator markedly and rapidly inhibits beta-amyloid accumulation in Alzheimer's disease transgenic mice. *Neuron*, 30: 665-676.

Choi, D.W. (2005) Neurodegeneration: cellular defenses destroyed. *Nature*, 433: 696-698.

Cohen-Kfir, E., Lee, W., Eskandari, S. and Nelson, N. (2005) Zinc inhibition of {γ}-aminobutyric acid transporter 4 (GAT4) reveals a link between excitatory and inhibitory neurotransmission. *Proc Natl Acad Sci USA*, 102: 6154–6159.

Colvin, R.A. (1998) Characterization of a plasma membrane zinc transporter in rat brain. *Neurosci Lett.* 247:147–150.

Colvin, R.A., Davis, N., Nipper, R.W. and Carter, P.A. (2000) Zinc transport in the brain: routes of zinc influx and efflux in neurons. *J Nutr*. 130: 1484S–1487S.

Cousins, R.J. (1985) Absorption, transport, and hepatic metabolism of copper and zinc: Special reference to metallothionein and ceruloplasmin. *Physiol Rev*. 65:238-309.

Cox, D.H., Chu, R.C. and Schlicker, S.A. (1969) Zinc deficiency in the maternal rat during gestation, and zinc, iron, copper, and calcium content and enzyme activity in maternal and fetal tissues. *J Nutr*. 98:449-458.

Crawford, I.L. and Connor, J.D. (1973) Localization and release of glutamic acid in relation to the hippocampal mossy fiber pathway. *Nature*, 244: 422-423.

Cuajungco, M.P. and Lees, G.J. (1997) Zinc metabolism in the brain: relevance to human neurodegenerative disorders. *Neurobiol Dis*. 4: 137-169.

Davies, N.T. (1980) Studies on the absorption of zinc by rat intestine. *Br J Nutr*. 43:189-203.

Davies, N.T. and Nightingale, R. (1975) The effects of phytate on intestinal absorption and secretion of zinc, and whole body retention of Zn, copper, iron and manganese in rats. *Br J Nutr*. 34:243-258.

Doraiswamy, P.M. and Finefrock, A.E. (2004) Metals in our minds: therapeutic implications for neurodegenerative disorders *Lancet Neurol*. 3: 431–434.

Drinker, K. and Drinker, P. (1928) Metal fume fever: V. Results of the inhalation by animals of zinc and magnesium oxide fumes. *J Ind Hyg*. 10:56-70.

Duncan, M.W., Marini, A.M., Watters, R., Kopin, I.J. and Markey, S.P. (1992) Zinc, a neurotoxin to cultured neurons, contaminates cycad flour prepared by traditional guamanian methods. *J Neurosci*. 12: 1523– 37.

Ebadi, M. (1986) Biochemical characterization of a metallothionein-like protein in rat brain. *Biol Trace Elem Res*. 11: 101–116.

Erickson, J.C., Hollopeter, G., Thomas, S.A., Froelick, G.J. and Palmiter, R.D. (1997) Disruption of the metallothionein-III gene in mice: analysis of brain zinc, behavior, and neuron vulnerability to metals, aging, and seizures. *J Neurosci*. 17: 1271–1281.

Failla, M.L. and Cousins, R.J. (1978) Zinc accumulation and metabolism in primary cultures of adult rat liver cells: Regulation by glucocorticoids. *Biochem Biophys Acta*, 543:293-304.

Flanagan, P.R., Haist, J. and Valberg, L.S. (1983) Zinc absorption, intraluminal zinc and intestinal metallothionein levels in zinc-deficient and zinc-repleted rodents. *J Nutr*. 113:962-972.

Foulkes, E.C. and McMullen, D.M. (1987) Kinetics of transepithelial movement of heavy metals in rat jejunum. *Am J Physiol*. 253:G134-G138.

Forssen, A. (1972) Inorganic elements in the human body: I. Occurrence of Ba, Br, Ca, Cd, Cs, Cu, K, Mn, Ni, Sn, Sr, Y, and Zn in the human body. *Ann Med Exp Biol Fenn*, 50:99-162.

Frederickson, C.J. (1989) Neurobiology of zinc and zinc-containing neurons. *Int Rev Neurobiol*. 31:145–238.

Frederickson, C.J. and Bush, A.I. (2001) Synaptically released zinc: physiological functions and pathological effects. *Biometals*, 14: 353-366.

Frederickson, R.E., Frederickson, C.J. and Danscher, G. (1990) In situ binding of bouton zinc reversibly disrupts performance on a spatial memory task. *Behav Brain Res*. 38:25-33.

Frederickson, C.J., Hernandez, M.D., Goik, S.A., Morton, J.D. and McGinty, J.F. (1988) Loss of zinc staining from hippocampal mossy fibers during kainic acid induced seizures: a histofluorescence study. *Brain Res.* 446:383-386.

Frederickson, C.J., Hernandez, M.D. and McGinty, J.F. (1989) Translocation of zinc may contribute to seizure-induced death of neurons. *Brain Res.* 480:317-321.

Frederickson, C.J., Klitenick, M.A., Manton, W.I. and Kirkpatrick, J.B. (1983) Cytoarchitectonic distribution of zinc in the hippocampus of man and the rat. *Brain Res.* 273:335-339.

Frederickson, C.J., Manton, W.I., Frederickson, M.H., Howell, G.A. and Mallory, M.A. (1982) Stable-isotope dilution measurement of zinc and lead in rat hippocampus and spinal cord. *Brain Res.* 246:338-341.

Frederickson, C.J. and Moncrieff, D.W. (1994) Zinc-containing neurons. *Biol Signals*, 3: 127–139.

Frederickson, C.J., Suh, S.W., Silva, D. and Thompson, R.B. (2000) Importance of zinc in the central nervous system: the zinc-containing neuron. *J Nutr.* 130: 1471S–1483S.

Frederickson, C.J., Rampy, B.A., Reamy-Rampy, S. and Howell, G.A. (1992) Distribution of histochemically reactive zinc in the forebrain of the rat. *Chem Neuroanat.* 5:521-530.

Friel, J.K., Andrews, W.L., Matthew, J.D., Long, D.R., Cornel, A.M., Cox, M., McKim, E. and Zerbe, G.O. (1993) Zinc supplementation in very-low-birth-weight infants. *J Pediatr Gastroenterol Nutr.* 17:97-104.

Fukahori, M. and Itoh, M. (1990) Effects of dietary zinc status on seizure susceptibility and hippocampal zinc content in the El (epilepsy) mouse. *Brain Res.* 529:16-22.

Fukahori, M., Itoh, M., Oomagari, K. and Kawasaki, H. (1988) Zinc content in discrete hippocampal and amygdaloid areas of the epilepsy (El) mouse and normal mice. *Brain Res.* 455:381-384.

Geinisman, Y., Ganeshina, O., Yoshida, R., Berry, R.W., Disterhoft, J.F. and Gallagher, M. (2004) Aging, spatial learning, and total synapse number in the rat CA1 stratum radiatum. *Neurobiol Aging*, 25: 407–416.

Giacconi, R., Cipriano, C., Muzzioli, M., Gasparini, N., Orlando, F. and Mocchegiani, E. (2003) Interrelationships among brain, endocrine and immune response in aging and successful aging: role of metallothionein III isoform. *Mech Aging Dev.* 124: 371–378.

Gilly, W.F. and Armstrong, C.M. (1984) Threshold channels — a novel type of sodium channel in squid giant axon. *Nature*, 309: 448–452.

Giroux, E.L. and Henkin, R.I. (1972) Competition for zinc among serum albumin and amino acids. *Biochim Biophys Acta*, 273: 64-72.

Goldstein, L. and Pfeiffer, C. (1978) Effects of single oral doses of trace elements on the quantitated EEG of normal subjects and chronic schizophrenics. *Psychopharm Bull*, 14:55-57.

Gordon, T., Chen, L.C., Fine, J.M., Schlesinger, R.B., Su, W.Y., Kimmel, T.A. and Amdur, M.O. (1992) Pulmonary effects of inhaled zinc oxide in human subjects, guinea-pigs, rats, and rabbits. *Am Ind Hyg Assoc J.* 53:503-509.

Gordon, E.F., Gordon, R.C. and Passal, D.B. (1981) Zinc metabolism: Basic, clinical, and behavioral aspects. *J Pediatr.* 99:341-349.

Gunshin, H., Mackenzie, B., Berger, UV, Gunshin Y, Romero MF, Boron WF, Nussberger, S., Gollan, J.L. and Hediger, M.A. (1997) Cloning and characterization of a mammalian protein-coupled metal-ion transporter. *Nature*, 388: 482–488.

Hallmans, G. (1977) Treatment of burns with zinc tape: A study of local absorption of zinc in humans. *Scand J Plast Reconstr Surg*. 11:155-161.

Hamdi, E.A. (1969) Chronic exposure to zinc of furnace operators in a brass foundry. *Br J Ind Med.* 26:126-134.

Haug, F.M. (1967) Electron microscopical localization of the zinc in hippocampal mossy fibre synapses by a modified sulfide silver procedure. *Histochemie*, 8:355-368.

Haug, F.M., Blackstad, T.W., Simonsen, A.H. and Zimmer, J. (1971) Timm's sulfide silver reaction for zinc during experimental anterograde degeneration of hippocampal mossy fibers. *J Comp Neurol.* 142:23-31.

He, L.S., Yan, X.S. and Wu, D.C. (1991) Age-dependent variation of zinc-65 metabolism in LACA mice. *Int J Radiat Biol.* 60:907-916.

Hempe, J.M. and Cousins, R.J. (1992) Cysteine-rich intestinal protein and intestinal metallothionein: An inverse relationship as a conceptual model for zinc absorption in rats. *J Nutr*. 122:89-95.

Henkin, R.I. (1974). Metal-albumin-amino acid interactions: Chemical and physiological interrelationships. In: Friedman M, ed. Chemical and physiological interrelationships in protein-metal interactions. New York, NY: Plenum Press, 299-328.

Henkin, R.I., Patten, B.M., Re, P.K. and Bronzert, D.A. (1975) A syndrome of acute zinc loss. Cerebellar dysfunction, mental changes, anorexia, and taste and smell dysfunction. *Arch Neurol.* 32:745-75.

Henrotte, J.G., Plouin, P.F., Levy-Leboyer, C., Moser G, Sidoroff-Girault N, Franck, G., Santarromana, M. and Pineau, M. (1985) Blood and urinary magnesium, zinc, calcium, free fatty acids, and catecholamines in type A and type B subjects. *J Am Coll Nutr.* 4:165-172.

Henrotte, J.G., Verdeaux, G., Monod, H., Audebert, A.A. and Renard, G. (1977) Relationship between blood zinc level and EEG changes under the influence of hyperpnea in normal subjects. *Encephale*, 3:159-164.

Hesse, G.W., Hesse, K.A. and Catalanotto, F.A. (1979) Behavioral characteristics of rats experiencing chronic zinc deficiency. *Physiol Behav*. 22:211-215.

Hesse, L., Beher, D., Masters, C.L. and Multhaup, G. (1994) The beta A4 amyloid precursor protein binding to copper. *FEBS Lett.* 349: 109-116.

Hirano, S., Higo, S., Tsukamoto, N., Kobayashi, E. and Suzuki, K.T. (1989) Pulmonary clearance and toxicity of zinc oxide instilled into the rat lung. *Arch Toxicol.* 63:336-342.

Howell, G.A., Welch, M.G. and Frederickson, C.J. (1984) Stimulation-induced uptake and release of zinc in hippocampal slices. *Nature*, 308:736-738.

Hu, K.H. and Friede, R.L. (1968) Topographic determination of zinc in human brain by atomic absorption spectrophotometry. *J Neurochem*. 15: 677-685.

Huang, X., Caujungco, M.P., Atwood, C.S., Moir, R.D., Tanzi, R.E. and Bush, A.I. (2000) Alzheimer's disease, β-amyloid protein and zinc. *J Nutr*. 130: 1488S–1492S.

Hunt, J.R., Lykken, G.I. and Mullen, L.K. (1991) Moderate and high amounts of protein from casein enhance human absorption of zinc from whole-wheat or white rolls. *Nutr Res.* 11:413-418.

International Health Board Effects of hookworm infection. (1919) Report of the International Health Board, vol. 5:40-79 Rockefeller Foundation New York, NY.

Itoh, M. and Ebadi, M. (1982) The selective inhibition of hippocampal glutamic acid decarboxylase in zinc-induced epileptic seizures. *Neurochem Res.* 7: 1287–1298.

Jiang, D., Sullivan, P.G., Sensi, S.L., Steward, O. and Weiss, J.H. (2001) Zn(2+) induces permeability transition pore opening and release of pro-apoptotic peptides from neuronal mitochondria. *J Biol Chem.* 276: 47524–47529.

Jiang, L.J., Vasak, M., Vallee, B.L. and Maret, W. (2000) Zinc transfer potentials of the alpha – and beta-clusters of metallothionein are affected by domain interactions in the whole molecule. *Proc Natl Acad Sci USA*, 97: 2503–2508.

Jobling, M.F., Huang, X., Stewart LR, Barnham, K.J., Curtain C, Volitakis I, Perugini M, White AR, Cherny, R.A., Masters, C.L., Barrow CJ, Collins SJ, Bush, A.I. and Cappai, R. (2001) Copper and zinc binding modulates the aggregation and neurotoxic properties of the prion peptide PrP106-126. *Biochemistry*, 40:8073-8084.

Johnson, P.E., Hunt, J.R. and Ralston, N.V. (1988) The effect of past and current dietary Zn intake on Zn absorption and endogenous excretion in the rat. *J Nutr.* 118: 1205-1209.

Kawhara, M. (2003) Aluminum-induced conformational change and the pathogenesis of Alzheimer's disease. *J Health Sci.* 49: 341-347.

Kawhara, M., Arispe, N., Kuroda, Y. and Rojas, E. (1977) Alzheimer's disease amyloid beta-protein forms Zn2+-sensitive, cation-selective channels across excised membrane parches from hypothalamic neurons. *Biophys J.* 73: 67-75.

Kawhara, M. (2004) Disruption of calcium homeostasis in Alzheimer's disease and other conformational disease. *Curr Alzheimer Res.* 1: 87-95.

Kay, R.G., Tasman-Jones, C., Pybus, J., Whiting, R. and Black, H. (1976) A syndrome of acute zinc deficiency during total parenteral alimentation in man. *Ann Surg.* 183:331-340.

Kim, E.Y., Koh, J.Y., Kim, Y.H., Sohn, S., Joe, E. and Gwag, B.J. (1999) Zn^{2+} entry produces oxidative neuronal necrosis in cortical cell cultures. *Eur J Neurosci.* 11: 327–334.

Koh, J.Y. and Choi, D.W. (1994) Zinc toxicity on cultured cortical neurons: involvement of N-methyl-D-aspartate receptors. *Neuroscience*, 60: 1049–1057.

Koh, J.Y., Suh, S.W., Gwag, B.J., He, Y.Y., Hsu, C.Y. and Choi, D.W. (1996) The role of zinc in selective neuronal death after transient global cerebral ischemia. *Science*, 272: 1013–1016.

Konoha, K., Sadakane, Y. and Kawahara, M. (2006) Zinc Neurotoxicity and its Role in Neurodegenerative Diseases. *J Health Sci.* 52: 1-8.

Krotkiewska, B. and Banas, T. (1992) Interaction of Zn2+ and Cu2+ ions with glyceraldehyde-3-phosphate dehydrogenase from bovine heart and rabbit muscle. *Int J Biochem.* 24: 1501–1505.

Kukimoto, I., Hoshino, S., Kontani, K., Inageda, K., Nishina, H., Takahashi, K., and Katada, T. (1996) Stimulation of ADP-ribosyl cyclase activity of the cell surface antigen CD38 by zinc ions resulting from inhibition of its NAD+ glycohydrolase activity. *Eur J Biochem.* 239: 177–182.

Lee, J.Y., Kim, J.H., Palmiter, R.D. and Koh, J.Y. (2003) Zinc released from metallothionein-III may contribute to hippocampal CA1 and thalamic neuronal death following acute brain injury. *Exp Neurol.* 184: 337–347.

Lee, J.Y., Mook-Jung I. and Koh, J.Y. (1999) Histochemically reactive zinc in plaques of the Swedish mutant beta-amyloid precusor protein transgenic mice. *J Neurosci.* 19: 10.

Lee, J.Y., Park, J., Kim, Y.H., Kim, D.H., Kim, C.G. and Koh, J.Y. (2000a) Induction by synaptic zinc of heat shock protein-70 in hippocampus after kainate seizures. *Exp Neurol.* 161: 433–441.

Lee, J.M., Grabb, M.C., Zipfel, G.J. and Choi, D.W. (2000b) Brain tissue responses to ischemia. *J Clin Invest.* 106: 723-731.

Lee, J.Y., Cole, T.B., Palmiter, R.D. and Koh, J.Y. (2000c) Accumulation of zinc in degenerating hippocampal neurons of ZnT3-null mice after seizures: evidence against synaptic vesicle origin. *J Neurosci.* 20: 1–5.

Lee, J.M., Zipfel, G.J., Park, K.H., He, Y.Y., Hsu, C.Y. and Choi, D.W. (2002) Zinc translocation accelerates infarction after mild transient focal ischemia. *Neuroscience*, 115: 871-878.

Leist, M. and Jaattela, M. (2001) Four deaths and a funeral: from caspases to alternative mechanisms, *Nat Rev Mol Cell Biol.* 2: 589–598.

Lieberman, I., Abrams, R., Hunt, N. and Ove, P. (1963) Levels of enzyme activity and deoxyribonucleic acid synthesis in mammalian cells cultured from the animal. *J Biol Chem.* 238:3955-3962.

Lieberman, I. and Ove, P. (1962) Deoxyribonucleic acid synthesis and its inhibition in mammalian cells-cultured from the animal. *J Biol Chem.* 237:1634-1642.

Lin, H., Zhu, Y.J. and Lal, R. (1999) Amyloid β protein (1-40) forms calcium-permeable, Zn2+-sensitive channel in reconstituted lipid vesicles. *Biochemistry*, 38:11189-11196.

Long, Y., Hardwick, A.L. and Frederickson, C.J. (1995) Zinc-containing innervation of the subicular region in the rat. *Neurochem Int.* 27:95-103.

Magneson, G.R., Puvathingal, J.M. and Ray, W.J. (1987) The concentrations of free Mg and free Zn in equine blood plasma. *J Biol Chem.* 262: 11140–11148.

Malo, J.L., Malo, J., Cartier, A. and Dolovich, J. (1990) Acute lung reaction due to zinc inhalation. *Eur Res J.* 3:111-114.

Maske, H. (1955) Uber den topchemischen Natchweis von Zink in Ammonshorn verschieder Säugetieve. *Naturwissenschaften*, 42:424.

McLardy, T. (1960) Neurosyncytial aspects of the hippocampal mossy fiber system. *Confin Neurol.* 20:1-19.

McLardy, T. (1970) Anatomical rationale of ablative surgery for temporal lobe seizures and dyscontrol: suggested stereo-chemode chelate-blockade alternative. *Acta Neurochir.* 23:119-124.

McMahon, R.J. and Cousins, R.J. (1998) Regulation of the zinc transporter ZnT-1 by dietary zinc. *Proc Natl Acad Sci USA*, 95: 4841–4846.

Mitchell, C.L., Barnes, M.I. and Grimes, L.M. (1990) Diethyldithiocarbamate and dithizone augment the toxicity of kainic acid. *Brain Res.* 506:327-330.

Mocchegiani, E., Bertoni-Freddari, C., Marcellini, F. and Malavolta, M. (2005) Brain ageing and neurodegeneration: role of zinc ion availability. *Progress in Neurobiology*, 75: 367-390.

Mocchegiani, E., Giacconi, R., Cipriano, C., Gasparini, N., Bernardini, G., Malavoltam, M., Menegazzi, M., Cavalieri, E., Muzzioli, M., Ciampa, A.R. and Suzuki, H. (2004) The variations during the circadian cycle of liver CD1d-unrestricted NK1.1+TCR gamma/delta+ cells lead to successful aging. Role of metallothionein/IL-6/gp130/PARP-1 interplay in very old mice. *Exp Gerontol.* 39: 775–788.

Mocchegiani, E., Giacconi, R., Cipriano, C., Muzzioli, M., Fattoretti, P., Bertoni-Freddari, C., Isani, G., Zambenedetti, P. and Zatta, P. (2001) Zinc-bound metallothioneins as potential biological markers of aging. *Brain Res Bull*, 55: 147–153.

Moreira, P., Pereira, C., Santos, M.S. and Oliveira, C. (2000) Effect of zinc ions on the cytotoxicity induced by amyloid beta-peptide. *Antioxid Redox Signal*, 2: 317-325.

Morrison, J.H. and Hof, P.R. (1997) Life and death of neurons in the aging brain. *Science*, 278: 412–419.

Morton, J.D., Howell, G.A. and Frederickson, C.J. (1990) Effects of subcutaneous injections of zinc chloride on seizures induced by noise and by kainic acid. *Epilepsia*, 31:139-144.

Moynahan, E.J. (1976) Letter: Zinc deficiency and disturbances of mood and visual behavior. *Lancet*, 1:91.

Mrak, R.E., Griffin, S.T. and Graham, D.I. (1997) Aging-associated changes in human brain. *J Neuropathol Exp Neurol.* 56: 1269–1275.

NAS/NRC. (1979) Zinc. Subcommittee on Zinc, Committee on Medical and Biologic Effects of Environmental Pollutants, Division of Medical Sciences, National Academy of Sciences/National Research Council. Baltimore, MD: University Park Press.

Nelson, L.S.J., Jacobs, F.A. and Brushmiller, J.G. (1985) Solubility of calcium and zinc in model solutions based on bovine and human milks. *J Inorg Biochem.* 24:255-265.

Neve, J., Hanocq, M., Peretz, A., Abi Khalil, F., Pelen, F., Famaey, J.P. and Fontaine, J. (1991) Pharmacokinetic study of orally administered zinc in humans: Evidence for an enteral recirculation. *Eur J Drug Metab Pharmacokinet*, 16:315-323.

Palm, R., Strand, T. and Hallmans, G. (1986) Zinc, total protein, and albumin in CSF of patients with cerebrovascular diseases, *Acta Neurol Scand.* 74: 308–313.

Park, J.A. and Koh, J.Y. (1999) Induction of an immediate early gene egr-1 by zinc through extracellular signal-regulated kinase activation in cortical culture: its role in zinc-induced neuronal death. *J Neurochem*. 73:450–456.

Parkin, G. (2004) Synthetic analogues relevant to the structure and function of zinc enzymes. *Chem Rev*. 104: 699-767.

Pecoud, A., Donzel, P. and Schelling, J.L. (1975) Effects of foodstuffs on the absorption of zinc sulfate. *Clin Pharmacol Ther.* 17:469-474.

Pei, Y.Q. and Koyama, I. (1986) Features of seizures and behavioral changes induced by intrahippocampal injection of zinc sulfate in the rabbit: a new experimental model of epilepsy. *Epilepsia*, 27: 183–188.

Pei, Y., Zhao, D., Huang, J. and Cao, L. (1983) Zinc-induced seizures: a new experimental model of epilepsy. *Epilepsia,* 24:169–176.

Perera, W.S.S. and Hooper, N.M. (2001) Ablation of the metal ion-induced endocytosis of the prion protein by disease-associated mutation of the octarepeat region. *Current Biol.* 11: 519 – 523.

Perez-Clausell, J. and Danscher, G. (1985) Intravesicular localization of zinc in rat telencephalic boutons. A histochemical study. *Brain Res*. 337:91-98.

Peters, S., Koh, J. and Choi, D.W. (1987) Zinc selectively blocks the action of N-methyl-D-aspartate on cortical neurons. *Science*, 236:589-593.

Prasad, A.S. (1998) Zinc deficiency in humans: a negle cted problem. *J Am Coll Nutr.* 17: 542-543.

Prasad, A.S., Halsted, J.A. and Nadimi, M. (1961) Syndrome of iron deficiency anemia, hepatosplenomegaly, hypogonadism, dwarfism, and geophagia. *Am J Med*. 31:532-546.

Prasad, A.S. and Oberleas, D. (1970) Binding of zinc to amino acids and serum proteins in vitro, *J Lab Clin Med.* 76: 416–425.

Prasad, A.S., Miale, A. Jr., Farid, Z., Sandstead, H.H. and Schulert, AR. (1963) Zinc metabolism in patients with the syndrome of iron deficiency anemia, hepatosplenomegaly, dwarfism, and hypognadism. *J Lab Clin Med*. 61:537-49.

Richards, M.P. and Cousins, R.J. (1975) Mammalian zinc homeostasis: Requirement for RNA and metallothionein synthesis. *Biochem Biophys Res Commun*. 64:1215-1223.

Reinhold, J.G., Faradji, B., Abadi, P. and Ismail-Beigi, F. (1991) Decreased absorption of calcium, magnesium, and phosphorous by humans due to increased fiber and phosphorous consumption as wheat bread. *Nutr Rev*. 49: 204-206.

Saito, T., Takahashi, K., Nakagawa, N., Hosokawa T, Kurasaki, M., Yamanoshita O, Yamamoto Y, Sasaki, H., Nagashima, K. and Fujita, H. (2000) Deficiencies of hippocampal Zn and ZnT3 accelerate brain aging of rat. *Biochem Biophys Res Commun*. 279: 505–511.

Sandstead, H.H. (2003) Zinc is essential for brain development and function. *J Trace Elements Exp Med*. 16: 165 – 173.

Sandstead, H.H., Frederickson, C.J. and Penland, J.G. (2000) History of zinc as related to brain function. *J Nutr*. 130:496S–502S.

Sandstead, H.H., Prasad, A.S., Schulert, A.R., Farid, Z., Miale, A., Bassilly, S. and Darby, W.J. (1967) Human zinc deficiency, endocrine manifestations and response to treatment. *Am J Clin Nutr*. 20:422-442.

Sandström, B. and Sandberg, A.S. (1992). Inhibitory effects of isolated inositol phosphates on zinc absorption in humans. *J Trace Elements Electrol Health Dis*. 6:99-103.

Sazawal, S., Bentley, M., Black, R.E., Dhingra, P., George, S. and Bhan, M.K. (1996) Effect of zinc supplementation on observed activity in low socioeconomic Indian preschool children. *Pediatrics*, 98:1132-1137.

Schroeder, H.A., Nason, A.P. and Tipton, I.H. (1967) Essential trace metals in man: Zinc: Relation to environmental cadmium. *J Chronic Dis*. 20:179-210.

Sensi, S.L., Canzoniero, L.M., Yu, S.P., Ying, H.S., Koh, J.Y., Kerchner, G.A. and Choi, D.W. (1997) Measurement of intracellular free zinc in living cortical neurons: routes of entry. *J Neurosci*. 17:9554-9564.

Sensi, S.L., Ton-That, D., Sullivan, P.G., Jonas, E.A., Gee, K.R., Kaczmarek, L.K. and Weiss, J.H. (2003) Modulation of mitochondrial function by endogenous Zn2+ pools. *Proc Natl Acad Sci USA*, 100: 6157–6162.

Sensi, S.L., Ying, H.S., Carriedo, S.G., Rao, S.S. and Weiss, J.H. (1999a) Preferential Zn2+ influx through Ca2+-permeable AMPA/kainate channels triggers prolonged mitochondrial superoxide production. *Proc Natl Acad Sci USA*, 96: 2414–2419.

Sensi, S.L., Yin, H.Z. and Weiss, J.H. (1999b) Glutamate triggers preferential Zn^{2+} flux through Ca2+ permeable AMPA channels and consequent ROS production. *Neuroreport*, 10: 1723–1727.

Sensi, S.L., Yin, H.Z. and Weiss, J.H. (2000) AMPA/kainate receptortriggered Zn2+ entry into cortical neurons induces mitochondrial Zn2+ uptake and persistent mitochondrial dysfunction. *Eur J Neurosci.* 12: 3813–3818.

Sheline, C.T., Behrens, M.M. and Choi, D.W. (2000) Zinc induced cortical neuronal death: contribution of energy failure attributable to loss of NAD(+) and inhibition of glycolysis. *J Neurosci.* 20: 3139–3146.

Sheline, G., Chaikoff, I., Jones, H. and Montgomery, M. (1943) Studies on the metabolism of zinc with the aid of its radioactive isotope. *J Biol Chem.* 149:139-151.

Short, K.R., Bigelow, M.L., Kahl, J., Singh, R., Coenen-Schimke, J., Raghavakaimal, S. and Nair, K.S. (2005) Decline in skeletal muscle mitochondrial function with aging in humans. *Proc Natl Acad Sci USA*, 102: 618–5623.

Simons, T.J.B. (1993) Measurement of free Zn ion concentration with the fluorescent probe mag-fura-2 (furaptra). *J Biochem Biophys Methods*, 27: 25–37.

Sloviter, R.S. (1985) A selective loss of hippocampal mossy fiber Timm stain accompanies granule cell seizure activity induced by perforant path stimulation. *Brain Res.* 330:150-153.

Spencer, H., Osis, D., Kramer, L. and Norris, C. (1976) Intake, excretion, and retention of zinc in man. In: Prasad AS, ed. Trace elements in human health and disease. Vol. 1: Zinc and copper. New York, NY: Academic Press, 345-361.

Sturniolo, G.C., Montino, C, Rossetto, L., Martin, A., D'Inca, R., D'Odorico, A. and Naccarato, R. (1991) Inhibition of gastric acid secretion reduces zinc absorption in man. *J Am Coll Nutr.* 10:372-375.

Suh, S.W., Jensen, K.B., Jensen, M.S., Silva, D.S., Kesslak, P.J., Danscher, G. and Frederickson, C.J. (2000) Histochemically-reactive zinc in amyloid plaques, angiopathy, and degenerating neurons of Alzheimer's diseased brains. *Brain Res.* 852: 274–278.

Suh, S.W., Thompson, R.B. and Frederickson, C.J. (2001) Loss of vesicular zinc and appearance of perikaryal zinc after seizures induced by pilocarpine. *Neuroreport*, 12: 1523–1525.

Takeda, A. (2000) Movement of zinc and its functional significance in the brain. *Brain Res Rev.* 34:137–48.

Takeda, A., Hirate, M., Tamano, H., Nisibaba, D. and Oku, N. (2003) Susceptibility to kainate-induced seizures under dietary zinc deficiency. *J Neurochem.* 85: 1575–1580.

Takeda, A., Kawai, M. and Okada, S. (1997a) Zinc distribution in the brain of Nagase analbuminemic rat and enlargement of the ventricular system. *Brain Res.* 769: 193–195.

Takeda, A., Kodama, Y., Ohnuma, M. and Okada, S. (1998) Zinc transport from the striatum and substantia nigra. *Brain Res Bull*, 47: 103-106.

Takeda, A., Minami, A., Seki, Y. and Oku, N. (2004) Differential effects of zinc on glutamatergic and GABAergic neurotransmitter systems in the hippocampus. *J Neurosci Res*. 75: 225–229.

Takeda, A., Ohnuma, M., Sawashita, J. and Okada, S. (1997b) Zinc transport in the rat olfactory system. *Neurosci Lett.* 225: 69–71.

Takeda, A., Sawashita, J. and Okada, S. (1995) Biological half-lives of zinc and manganese in rat brain. *Brain Res*. 695: 53-58.

Takeda, A., Takefuta, S., Okada, S. and Oku, N. (2000) Relationship between brain zinc and transient learning impairment of adult rats fed zinc-deficient diet. *Brain Res*. 859: 352–357.

Takeda, A., Sawashita, J., Takefuta, S., Ohnuma, M. and Okada, S. (1999) Role of zinc released by stimulation in rat amygdala. *J Neurosci Res*. 57: 405–410.

Tang, S.M. (1991) Hair zinc and copper concentration in patients with epilepsy. *Zhonghua Shen Jing Jing Shen Ke Za Zhi*, 24: 94–97, 124–125.

Thatcher, R.W., McAlaster, R., Lester, M.L. and Cantor, D.S. (1984) Comparisons among EEG, hair minerals and diet predictions of reading performance in children. *Ann NY Acad Sci*. 433:87-96.

Timm, F. (1958) Zur Histochemie der Schwermetalle. Das Sulfid-Silber-Verfahren. *Dtsch Z Gesamte Gerichtl Med*. 46:706-712.

Thompson, C.B. (1995) Apoptosis in the pathogenesis and treatment of disease. *Science*, 267: 1456– 62.

Tonder, N., Johansen, F.F., Frederickson, C.J., Zimmer, J. and Diemer, N.H. (1990) Possible role of zinc in the selective degeneration of dentate hilar neurons after cerebral ischemia in the adult rat. *Neurosci Lett*. 109:247-252.

Tsuda, M., Imaizumi, K., Katayama, T., Kitagawa, K., Wanaka, A., Tohyama, M. and Takagi, T. (1997) Expression of zinc transporter gene, ZnT-1, is induced after transient forebrain ischemia in the gerbil. *J Neurosci.* 17: 6678–6684.

Tully, C.L., Snowdon, D.A. and Markesbery, W.R. (1995) Serum zinc, senile plaques, and neurofibrillary tangles: findings from the Nun Study. *Neuroreport*, 6:2105-2108.

Uchida, Y., Takiko, K., Titani, K.K., Ihara, Y. and Tomonaga, M. (1991) The growth inhibitory factory that is deficient in the Alzheimer's disease brain is a 68 amino acid metallothionein-like protein. *Neuron*, 7: 337–347.

Valentine, J.S. and Hart, P.J. (2003) Misfolded CuZnSOD and amyotrophic lateral sclerosis. *Proc Natl Acad Sci USA*, 100: 3617-3622.

Valko, M., Morris, H. and Cronin, M.T.D. (2005) Metals, Toxicity and Oxidative Stress. *Curr Med Chem*. 12: 1161-1208.

Vallee, B.L. and Falchuk, K.H. (1993) The biological basis of zinc physiology. *Physiol Rev*. 73:79–118.

Vogt, K., Mellor, J., Tong, G. and Nicoll, R. (2000) The actions of synaptically released zinc at hippocampal mossy fiber synapses. *Neuron*, 26: 187–196.

von Euler, C. (1962) On the significance of the high zinc content in the hippocampal fomation. Physiologie de Hippocampus, 135-143 CNRS Paris, France.

Wachs, T., Bishry, Z., Moussa, W., Younis, F., McCabe, G., Harrison, G., Sweifi, E., Kirksey, A., Galal, O., Jerome, O. and Shaheen, F. (1995) Nutritional intake and context as predictors of cognition and adaptive behaviors of Egyptian school-age children. *Int J Behav Dev*. 18:425-450.

Waite, J. and Nelson, I. (1919) A study of the effects of hookworm infection upon mental development of North Queensland school children. *Med J Aust.* 1:1-8.

Wallace, D.C. (1999) Mitochondrial Diseases in Man and Mouse. *Science*, 283: 482–1488.

Wallwork, J.C., Milne, D.B., Sims, R.L. and Sandstead, H.H. (1983) Severe zinc deficiency: effects on the distribution of nine elements (potassium, phosphorus, sodium, magnesium, calcium, iron, zinc, copper and manganese) in regions of the rat brain. *J Nutr.* 113:1895-1905.

Wapnir, R.A. and Stiel, L. (1986) Zinc intestinal absorption in rats: Specificity of amino acids as ligands. *J Nutr*. 116: 2171 2179.

Wastney, M.E., Aamodt, R.L., Rumble, W.F. and Henkin, R.I. (1986) Kinetic analysis of zinc metabolism and its regulation in normal humans. *Am J Physiol.* 251:R398-R408.

Watt, N.T. and Hooper, N.M. (2003) The prion protein and neuronal zinc homeostasis. *Trends Biochem Sci.* 28: 406-410.

Weiss, J.H., Hartley, D.M., Koh, J. and Choi, D.W. (1993) AMPA receptor activation potentiates zinc neurotoxicity. *Neuron*. 10: 43–49.

Weiss, J.H., Sensi, S.L. and Koh, J.Y. (2000) Zn (2+): a novel ionic mediator of neural injury in brain disease. *Trends Pharmacol Sci.* 21: 395-401.

Wenzel, H.J., Cole, T.B., Born, D.D., Schwartzkroin, P.A. and Palmiter, R.D. (1997) Ultrastructural localization of zinc transporter-3 (ZnT-3) to synaptic vesicle membranes within mossy fiber boutons in the hippocampus of mouse and monkey. *Proc Natl Acad Sci USA*, 94: 12676–12681.

Westbrook, G.L. and Mayer, M.L. (1987) Micromolar concentrations of Zn2+ antagonize NMDA and GABA responses of hippocampal neurons. *Nature*, 328:640-643.

White, A.R., Multhaup, G., Maher, F., Bellingham, S., Camakaris, J., Zheng, H., Bush, A.I., Beyreuther, K., Masters, C.L. and Cappai, R. (1999) The Alzheimer's disease amyloid precursor protein modulates copper-induced toxicity and oxidative stress in primary neuronal cultures. *J Neurosci.* 19: 9170-9179.

Witschi, H.R. and Last, J.A. (2001) Toxic responses of the respiratory system. In: Klaassen CD, ed. Casarett & Doull's toxicology: The basic science of poisons. 6th Ed. New York, NY: McGraw-Hill, 515–534.

Wudarczyk, J., Debska, G. and Lenartowicz, E. (1999) Zinc as an inducer of the membrane permeability transition in rat liver mitochondria. *Arch Biochem Biophys.* 363: 1–8.

Yi, J.S., Lee, S.K., Sato, T.A. and Koh, J.Y. (2003) Co-induction of p75(NTR) and the associated death executor NADE in degenerating hippocampal neurons after kainate-induced seizures in the rat. *Neurosci Lett.* 347: 126–130.

Yin, H.Z., Ha, D.H., Carriedo, S.G. and Weiss, J.H. (1998) Kainate-stimulated Zn2+ uptake labels cortical neurons with Ca2+- permeable AMPA/kainate channels. *Brain Res.* 781:45-55.

Yin, H.Z. and Weiss, J.H. (1995) Zn(2+) permeates Ca(2+) permeable AMPA/kainate channels and triggers selective neural injury. *Neuroreport*, 6:2553-2556.

Yokoyama, M., Koh, J. and Choi, D.W. (1986) Brief exposure to zinc is toxic to cortical neurons. *Neurosci Lett.* 71:351-355.

Yu, W.H. and Fraser, P.E. (2001) S100beta interaction with tau is promoted by zinc and inhibited by hyperphosphorylation in Alzheimer's disease. *J Neurosci.* 21: 2240-2246.

Zeigler, T., Leach, R., Scott, M., Huegin, F., Mcevoy, R. and Strain, W. (1964) Effect of zinc nutrition upon uptake and retention of zinc-65 in the chick. *J Nutr.* 82:489-494.

Zerangue, N. and Kavanaugh, M.P. (1996) Flux coupling in a neuronal glutamate transporter. *Nature*, 383: 634–637.

In: Progress in Neurodegeneration
Editors: M.R. Avila-Costa and V.A. Martinez
ISBN 978-1-60741-317-2

Chapter 8

Vanadium

Maria Rosa Avila-Costa
Department of Neuroscience, Neuromorphology Lab
UNAM, Mexico

8.1. General Description

Vanadium (V) is a transition element, is a member of Group 5, and is the 23rd element of the Periodic table. Andres Manuel Del Rio was the first chemist to provide the idea of this new element in 1801. But it was discovered by Nils Sefstrom, a Swedish chemist in 1830 (Mukherjee et al. 2004). Pure V is a bright sliver-white, soft and ductile metal and 22nd most abundant element in the earth's crust. V has become subject of interest amongst nutritionists since the discovery that various marine species have this metal as an essential element (Almedeida et al. 2001). Although most food contains low amount of V (<1 ng/g), food is the major source of exposure to V for general population (Barceloux, 1999).

V may be beneficial and possibly essential in humans, but certainly essential for some living organisms (Kustin et al. 1983; Michibata and Sakurai, 1990; Nielsen and Uthus, 1990; Rehder, 1999; Michibata et al. 2003a,b; Crans et al. 2004). The usual oxidation states of the vanadium are V (III), V (IV) and V (V). The V ions play a role in biology as counter ions for various proteins, DNA and RNA.

V is widely distributed on Earth, however its role as a micronutrient in humans is not yet established. Humans are exposed to V mainly through the polluted atmosphere from combustion products of vanadium-bearing fuel oils, fumes and dust. Food contains a very low content of V usually below 1 ng/g. V enters the organism by inhalation, skin and gastrointestinal tract and accumulates mainly in the liver, kidney, spleen, bones and to a lesser extent in lungs and brain (Hansen et al. 1982; Bracken et al. 1985; Barceloux, 1999).

The structural role is often manifested by the maintenance of various biological structures, whereas a functional role is to bring key reactivity to a reaction center for a protein (Crans et al. 2004). V ions have many structural roles reflected by its structural and electronic analogy to phosphorus (Rehder, 1999). In addition, the V ion is an enzyme cofactor, (Van Pee et al. 2000; Rehder, 2003) and is found in certain tunicates (Michibata et al. 2003b) and possibly mammals (Nielsen and Uthus, 1990).

During the last few decades, the facade of V as a 'slightly' toxic and carcinogenic element eventually ratified to an essential trace element with anti–diabetic and anti-carcinogenic properties (Mukherjee et al. 2004).

Neurotoxic effects of V are not well understood yet, but it is known that acute poisoning in animals by ingestion or inhalation of V compounds leads to nervous disturbances, paralysis of legs, respiratory failure, convulsions, bloody diarrhea, and death (Wenning and Kirsch, 1988). V penetrates the blood–brain barrier (Berman, 1980; Avila-Costa et al. 2005) and induces neurochemical alterations: noradrenaline, dopamine, and 5-hydroxytriptamine altered levels and an inhibitory effect on the uptake and release of noradrenaline were found in the rat brain during V poisoning (Witkowska and Brzezinski, 1979; 1983; Avila-Costa et al. 2004).

It is well established that vanadate (V^{5+}) and vanadyl (V^{4+}), the V oxidation states of biological interest, may cause a number of adverse toxic effects in mammals depending on circulating V levels.

Industrial exposure of metal workers to vanadium pentoxide (V_2O_5) leads to cardiovascular diseases and a variety of symptoms involving central nervous system (CNS), gastrointestinal and respiratory systems (WHO, 1988). Moreover, it has been suggested that raised tissue levels of V may be of etiological importance in manic-depressive illness (Nechay, 1984; Naylor et al. 1984; Conri et al. 1986; Campbell et al. 1988). Serum V levels in adults with depression were significantly greater than those of controls, regardless of the type of depression, tremor and impaired conditioned reflexes, as well as congestion of brain spinal cord (WHO, 1988). Moreover, reduced cognitive abilities in humans chronically exposed to this element were found (Barth et al. 2002).

8.2. Sources

Metallic V does not occur in nature. Over 70 V minerals are known, carnatite and vanadinite being the most important from the point of view of mining. V is also found in phosphate rock and certain iron ores, and is present in some crude oils in the form of organic complexes. It is also found in small percentages in meteorites.

Production of V is linked with that of other metals such as iron, uranium, titanium, and aluminum. As rich minerals rarely occur in large deposits, ores with low V content, which exist in large amounts, are important. Extraction of V from fossil fuels, including vanadium-rich oil and coal, tars, bitumens, and asphaltites is important in several countries (WHO, 1988).

During the first half of the 1980s, the global production of V (as V_2O_5) ranged from 34 to 45 million Kg, China, Finland, South Africa, the USA, and the USSR being the biggest producers (WHO, 1988).

V is mainly (75 - 85%) used in ferrous metallurgy as an alloy additive in various types of steel. Its use in nonferrous metals and is important for the atomic energy industry, aircraft construction, and space technology. V is also widely used as a catalyst in the chemical industry, where V_2O_5 and metavanadates are especially important for the production of sulfuric acid and plastics. Small quantities of V are used in a variety of other applications. Emissions of V may be high in the vicinity of large plants producing steel alloys. V is also released into the air: during the re-smelting of scrap steel and the transformation of titaniferrous and vanadic magnetite iron ores into steel; from the roasting of V slags; from V_2O_5 smelting furnaces; and from electric furnaces in which ferrovanadium is smelted (WHO, 1988).

The global recycling of V involves the release of V from natural and anthropogenic sources to the air, water, and soil, the transport of particles of V in water and air, wet and dry deposition, absorption, and complexing (ATSDR, 1997).

8.2.1. Air

Exposure of the general population to V in air results primarily from the combustion of petroleum, coal, and heavy oils during the generation of electricity and heat. During the burning of oil fields in Kuwait, air particulate matter in Bahrain contained 11–42 ng V/m^3 (Madany and Raveendran, 1992). Concentrations in coal, crude oil, and carbon fossils average approximately 100 μg V/g, 600 μg V/g, and 1500 μg V/g, respectively (Meish and Bielig, 1980). Very little V is present in natural gas. Additional industrial sources of V in the metallurgy industry include discharges from furnaces roasting V slag, V_2O_5 smelting furnaces, and crucibles, which melt ferrovanadium alloys. Anthropogenic sources account for about two thirds of the atmospheric V, which occurs in the form of V oxides (ATSDR, 1997). Continental dust, marine aerosols, and, to a lesser extent, volcanic emissions account for the rest of the V in the atmosphere.

Vanadium enters the air as an aerosol of simple or complex V oxides from anthropogenic sources and as mineral particles of less soluble trivalent forms from natural sources (Barceloux, 1999).

8.2.2. Soil

The actual content of V depends on the type of parent rock. The soil types, which contain the highest concentration of V, are Tundra podzols and clays. Elemental V does not occur in nature because of its propensity to react with other elements, particularly oxygen (ATSDR, 1997). The largest releases of V to the soil result from the weathering of rocks rather than from anthropogenic sources, such as super-phosphate fertilizers (0.05–2 g V/kg) and basic slag (1–5 g V/kg) (Byerrum et al. 1974).

Basic rocks contain higher concentrations of V compared with neutral and acidic rocks.

8.2.3. Water

Drinking water is not an important source of exposure to V for the general population with typical V concentrations <1 μg/L drinking water (WHO, 1988). Natural releases of V from the erosion of soil and the weathering of rock far exceed the deposition of V from anthropogenic sources in the air. Most samples from drinking water supplies contain levels of V below detectable limits. When detected, the concentration in drinking water usually is 1–6 μg V/L with a range up to about 20 μg V/L (WHO, 1988). The level of V in freshwater depends on geographical difference in leachates and effluents from both natural and anthropogenic sources (Barceloux, 1999).

8.2.4. Food

Even though most foods contain low concentrations (<1 ng V/g) (Byrne and Kosta, 1978), food is the major source of exposure to V for the general population. The daily requirement for V probably is small (<10 μg/d) with the highest source of V present in black pepper, dill seed, mushrooms (0.05–2 μg/g), parsley (1.8 μg/g), shellfish, spinach (0.5–0.8 μg/g), and some prepared foods (Nielsen, 1991). In general, seafood contains higher concentrations of V compared with terrestrial animal sources. Beverages, fresh fruit and vegetables, cereals, liver, fats, and oil contain smaller amounts (<1–10 ng V/g). The processing of food tends to raise the concentrations of V (Myron et al. 1977). High concentrations of V exist in tobacco, and smoke from tobacco contains 1–8 ppm V (Byrne and Kosta, 1978). The mushroom, *Amanita muscaria* contains about 100 times the concentration (100 ppm V) found in other plants and mushrooms (Bertrand, 1950). The estimated daily dietary intake for the US population is about 10–60 μg V (Harland and Harden-Williams, 1994).

8.2.5. Occupational Exposure

V levels near metallurgical industries usually average about 1 mg V/m^3 whereas ambient air near industries, which produce V metal or compounds, contains concentrations of V of a few mg V/m^3 (WHO, 1988). Very high levels of V result from boiler-cleaning operations as a result of the high concentration (approximately 10–25%) of V oxides in the dust. During these cleaning operations, concentrations of 50–100 mg V/m^3 are frequent with concentrations ranging up to 500 mg V/m^3 (Barceloux, 1999).

In terms of occupational exposure, the most important V compounds are V_2O_5, vanadium trioxide, ferrovanadium, vanadium carbide, and vanadium salts, such as sodium and ammonium vanadate. The oxides and salts are commonly used in industry in powder form, giving rise to the possibility of dust and aerosol formation, when the substances are crushed or ground. Many metallurgical processes involve the production of vapor containing V_2O_5, which condenses to form respirable aerosols. Boiler-cleaning operations generate dusts containing the pentoxide and trioxide compounds. Combustion of residual fuels with high V content is likely to produce aerosols of the pentoxide as well as oxide complexes of V with other metals (WHO, 1988).

8.3. Absorption, Distribution and Excretion

Absorption, excretion and storage mechanisms of V in living system are not thoroughly understood. Several reports show that V is poorly (only about 10%) absorbed from gastrointestinal tract (Nriagu, 1998; Poucheret et al. 1998). These reports suggests that most of the ingested V is transformed into the cationic vanadyl form in stomach before being absorbed in the duodenum through an unidentified mechanism (Hirano and Suzuki, 1996). Again, V in its anionic vanadate form has been found to be absorbed at much higher quantities (about five times more than vanadyl form) through anionic transport system (Hirano and Suzuki, 1996). It has also been reported that vanadyl undergoes spontaneous oxidation to vanadate *in vivo* (Li et al. 1996). Multivalent existence of V in nature and in living systems put forth the chemical complexity of this element. This multifaceted chemical character of V in turn echoes in its biological and biochemical properties, especially in metabolism and in absorption. Again vanadate after reaching the blood stream is converted into vanadyl ion, although the vanadate form also exists. Thus, vanadate (by transferrin) and vanadyl (by albumin and transferrin) are rapidly transported by blood proteins to various tissues (Fantus et al. 1995). Blood parameters showed little or no reflection of toxicity after a long-term supplementation of V compounds (Fawcett et al. 1997; Guidotti et al. 1997), which might be due to the brisk transport of V from blood to the tissues. Upon supplementation, V is incorporated in various organs and tissues including liver, kidney, brain, heart, muscles and bone. Kidney, spleen, bone and liver tissues of rat have been shown to accumulate distinctly high amounts of V in chronically treated animals through oral administration (Hamel and Duckworth, 1995; Ramandham et al. 1991).

The effects of V administration persist even after V has been withdrawn for several days (Cam et al. 2000). Unabsorbed V is excreted in feces. When V was administrated through parenteral route, 10% of the V was found in the feces of humans and rats (Barceloux, 1999; Setyawati et al. 1998; Alimonti et al. 2000). V is excreted through bile and through urine (Alimonti et al. 2000). It is, thus, the bile route through which a significant amount of V may be ultimately excreted through feces. Moreover, it may be suggested that V content in feces does not reflect the amounts of V absorbed or unabsorbed (Mukherjee et al. 2004). Figure 1 summarizes the kinetics of Vanadium.

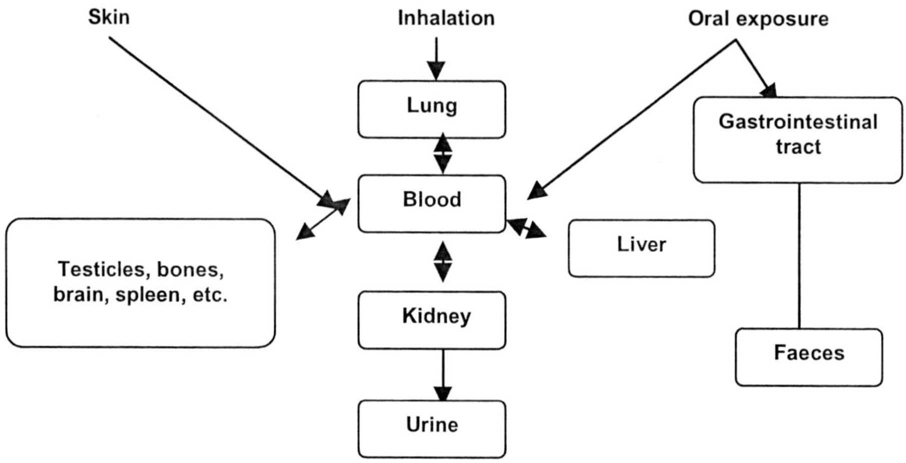

Figure 1. Vanadium kinetics.

8.4. General Toxicity

The toxicity of V compounds depends on a variety of factors including the route of administration and the inherent toxicity of the particular compound. In general, the toxicity of V compounds is low, and the toxicity is least following ingestion and greatest following parenteral administration. Inhalation is a route of exposure that produces intermediate toxicity (WHO, 1988; Barceloux, 1999; National Toxicology Program, 2002). The toxicity of V increases with higher valences and the pentavalent compounds are usually the most toxic. Acidification tends to reduce the toxicity of V compounds (Mitchell, 1953).

Studies in animals have shown that equivalent doses of V_2O_5 are better tolerated by small animals, including rats and mice, than by larger animals, such as rabbits and horses (Hudson, 1964). The LD50 of V_2O_5 is highly species-dependent. Differences in diet and route of V administration may contribute to these discrepancies (IARC, 2006).

In humans, acute V poisoning can manifest itself in a number of symptoms including eye irritation and tremors of the hands (Lewis, 1959). In addition, a greenish coloration of the tongue has been observed in humans exposed to high concentrations of V_2O_5 and is probably due to the formation of trivalent and tetravalent V complexes (Wyers, 1946). The green color disappears within 2–3 days of cessation of exposure (Lewis, 1959).

8.4.1. Mechanism of Action

The systemic effects of V probably result from its oxidizing ability (Kiviluoto et al. 1981). V also inhibits oxidative phosphorylation (Hathcock et al. 1966), but the role of this mechanism in toxicity is unclear. Vanadate interferes with phosphate-containing enzymes (ATP phosphohydrolases, adenylate kinase, glyceraldehyde-3-phosphate dehydrogenase, ribonuclease) at relatively high (μmol-mmol) extracellular concentrations (Cantley et al. 1977). Based on *in vitro* data, vanadate probably is a noncompetitive inhibitor of Na K-ATPase (K1= 880 nM). This species of V is a more effective inhibitor of ATPases in concentrations >50 μM, whereas vanadyl is an effective inhibitor in very low concentrations (10 nM) (Janiszewska et al. 1994).

8.5. Nervous System Effects

Neurophysiological effects have been reported following acute exposure (by oral administration and subcutaneous injection) of dogs and rabbits to V oxides and salts (vanadium trioxide, V_2O_5, vanadium trichloride and ammonium metavanadate) (IARC, 2006).

These effects included disturbances of the CNS, such as impaired conditioned reflexes and neuromuscular excitability (Roshchin, 1967). The animals behaved passively, refusing to eat, and lost weight. In cases of severe poisoning, diarrhea, paralysis of the hind limbs and respiratory failure were followed by death (Roshchin, 1967).

In a study reported by Seljankina (1961 cited by WHO, 1988), solutions of V_2O_5 were administered orally to rats and mice at doses of 0.005–1 mg/Kg per day for periods ranging from 21 days at the higher concentrations to 6 months at the lower concentrations. A dose of 0.05 mg/Kg was found to be the threshold dose for functional disturbances in conditioned reflex activity in both mice and rats. Repeated exposure to aqueous solutions (0.05–0.5 mg/Kg per day, for 80 days) of V_2O_5 impaired conditioned reflex mechanisms in rats.

In male CD-1 mice exposed by inhalation to 0.02 M V_2O_5 2h twice a week for 4 weeks, Golgi staining revealed a drastic reduction in dendritic spines in the striatum compared with controls, showing that the inhalation of V_2O_5 causes severe neuronal damage in the corpus striatum (Avila-Costa et al. 2004). Using the same inhalation model, after 12 weeks of exposure, a decrease in dendritic spines of granule cells of the olfactory bulb was observed (Mondragón et al. 2003).

In addition, ultrastructural modifications in nuclear morphology of these cells were evident, Golgi apparatus was dilated and an increase in lipofucsin granules was observed, as well as necrosis of some cells (Colin-Barenque et al. 2003). In the cerebellum, necrosis and apoptosis of the Purkinje and granule cell layers were seen (Meza et al. 2003).

There is scant information about V and neurodegeneration, for that reason our group have developed an inhalation model in mice to evaluate different brain areas, starting with the olfactory bulb, which is one of the main entrances to the brain, followed by the cerebral ventricles and the consequences in striatum, substantia nigra and hippocampus, structures highly related with neurodegenerative disorders.

8.5.1. Olfactory Bulb

The ability to smell, like other primary sensory abilities, diminishes with age. Olfactory decrement is especially prominent in the elderly and in neurodegenerative diseases. This decrement is correlated with neuronal and synaptic loss in the olfactory bulb, which in some way correlates with exposure to airborne toxins (Mesholam et al. 1998; Ponsen et al. 2004; Jankovic, 2008).

In this way, after V inhalation, we found that mice show evident olfactory bulb alterations, which include: reduction of dendritic spines density of granule cells (Figure 2 A and B) and apoptotic-like cell death (Figure 3).

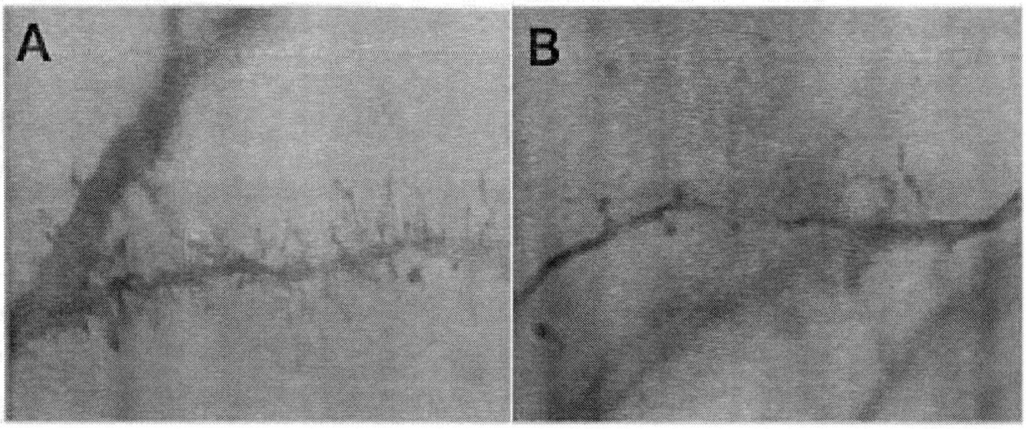

Figure 2. A. Control granule cell dendrite of the olfactory bulb in which is notorious the presence of spines. B. Evidences the dendritic spine loss after V inhalation. 4000 X.

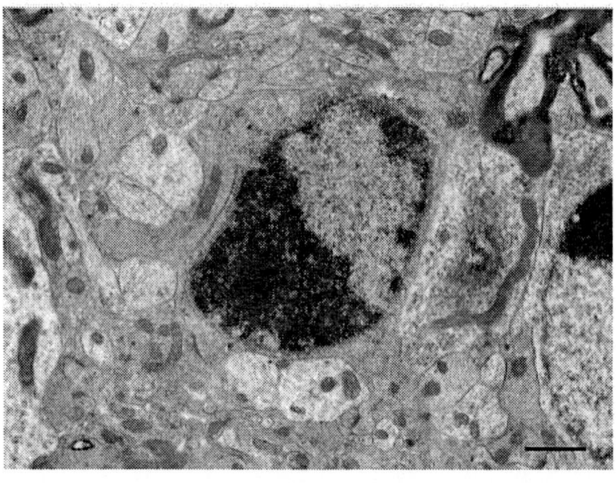

Figure 3. Apoptotic granule cell of exposed mouse evidenced by condensation and margination of the chromatin. Scale bar = 1.5 μm.

8.5.2. Cerebral Ventricles

The blood–brain barrier (BBB) plays a pivotal role in the removal of endogenous and exogenous toxins. Regulation of blood–brain tissue exchange is accomplished by ependymal cells, which possess intercellular tight junctions. Loss of BBB function is an etiologic component of many neurological disorders. Cumulative evidence has revealed that the brain barriers are subject to toxic insults from heavy metal exposure (Zhen et al. 2002).

In our laboratory we investigate the effects of V_2O_5 inhalation on mouse fourth ventricle ependymal epithelium (Avila-Costa et al. 2005). Scanning electron microscopy analysis demonstrates loss of cilia from the area around the floor of the fourth ventricle, a tendency for the remaining cilia to conglomerate, sloughed cells and disintegrated structure compared to that of control mice (Figure 4 A and B); and transmission electron microscopy showed ependymal cell layer detachment from the basal membrane and dissolution of the tight junctions between ependymal cells (Figure 5 A and B).

We conclude that this damage can allow toxicants to modify the permeability of the epithelium and promote access of inflammatory mediators to the underlying neuronal tissue causing injury and neuronal death. Since Berman (1980) establish that vanadate anion penetrates BBB and leads to CNS dysfunctions.

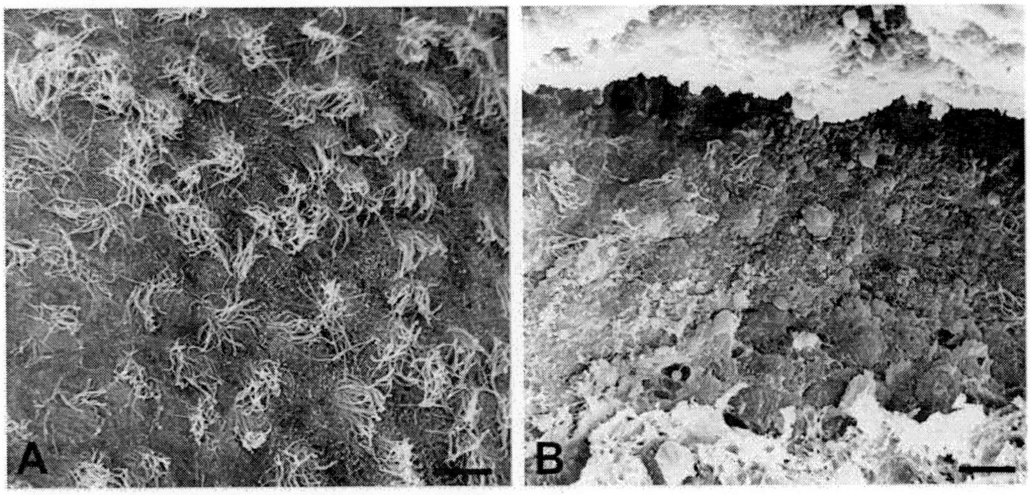

Figure 4. Scanning electron microscopy micrographs of the floor of the fourth ventricle. A. Control mouse and B. Vanadium exposed mouse; the micrograph evidences cilia loss. Bar 10μm.

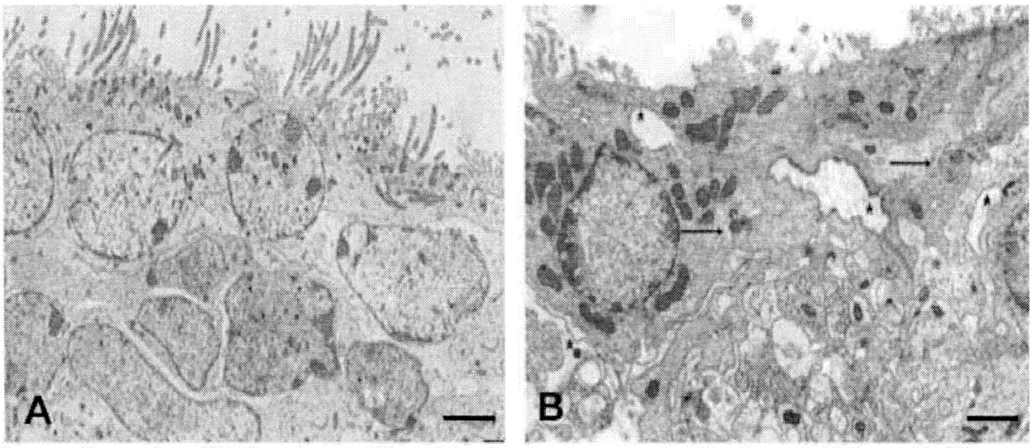

Figure 5. Transmission electron microscopy micrographs of control mouse (A) and V exposed mouse (B). Panel B shows the opening of the intracellular junctions (arrows), and ependymal cells detachment (asterisks) (ruthenium red). Bar 10μm.

8.5.3. Basal Ganglia

Using the same model, we demonstrated that mice, which inhaled V_2O_5 had a significant dopaminergic neuronal loss in the substantia nigra and as a consequence, developed morphological alterations of the striatum medium size spiny neurons (Figure 6) (Avila-Costa et al. 2004).

The relevance of our basal ganglia findings resides in the facts that demonstrate that V produces free radicals (Sasi et al. 1994) and interferes with the catecholaminergic brain levels (Sharma et al. 1986). As it is well established, the basal ganglia are brain structures especially vulnerable to oxidative stress as a consequence of its lower antioxidants concentrations, and the elevated presence of iron in the substantia nigra. The dopamine catabolism implies the generation of a great amount of free radicals, which in turn produces an oxidative stress state (Luo et al. 1998). Furthermore, there is evidence that oxidative stress is implicated with neurodegenerative diseases such as Parkinson (Simonian and Coyle, 1996).

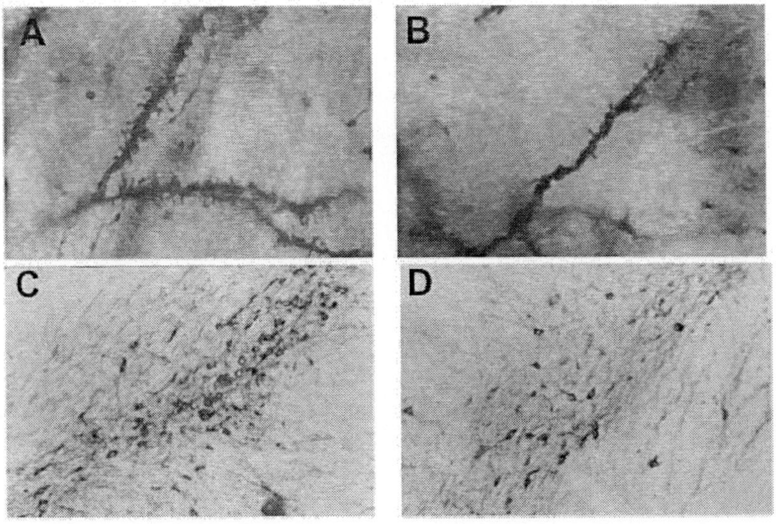

Figure 6. A. Control striatum medium size spiny neuron in which is notorious the presence of spines on the dendrites. B. Evidences the dendritic spine loss after V inhalation. C and D. Representative TH-immunostained coronal sections of substantia nigra of (C) control mouse and (D) exposed mouse. The cell loss due to the V inhalation was significantly different from control mice. 4000 X.

8.5.4. Hippocampus

Recent findings from our group, demonstrate that mice which inhaled V, also manifest spatial memory deterioration as a consequence of hippocampal neuronal death (Avila-Costa et al. 2006). Results showed that V inhalation produces a time dependent loss of dendritic spines (Figure 7), necrotic-like cell death (Figure 8), and notorious alterations of the hippocampus CA1 neuropile, which correlate with spatial memory impairment (data not shown).

It has been proposed that temporary inactivation or lesions of the dorsal hippocampus cause impairments in the acquisition and retrieval of spatial memory in tasks such as those evaluated in the Morris water maze. Substantial bodies of evidences indicate that hippocampal neuron loss is widely viewed as a hallmark of normal aging and to contribute directly to age-related deficits in learning and memory (Grady et al. 1995).

On the other hand, it has been agreed that many conditions lead to a decreased number of dendritic spines (Fiala et al. 2002). These conditions also demonstrate decreased neuronal number, suggesting that spines are lost as a result of a severe decline in the number and availability of axonal inputs to dendritic spines (Fiala et al. 2002). According with our results we assume, that the decreased number of dendritic spines and the number of necrotic cells are the attainable explanation for the memory impairments identified after V inhalation, since behavioral deficiency highly correlated with those hippocampal cell changes.

Furthermore, memory impairments is related with the phosphorylation of some signaling molecules, which have been proposed as one of the most prevalent mechanisms for modulating neuronal functions, including important mechanisms of memory acquisition

(Kandel, 2001). It appears that V activates extracellular signal–regulated kinases (ERKs) trough competitive inhibition of protein tyrosine phosphatase (Zhao et al. 1996) that acts by irreversible oxidation of the catalytic site of these enzymes (Huyer et al. 1997), whose oxidation results in their inhibition (Augereau et al. 2005) affecting the tyrosine phosphorylation leading to cell alterations. Moreover, protein phosphorilation is involved in the regulation of neurotransmitter receptors such as NMDA, which mediates whole cell currents and intracellular Ca^{2+} and plays an essential role in memory mechanisms (Tlan Wang and Salter, 1994).

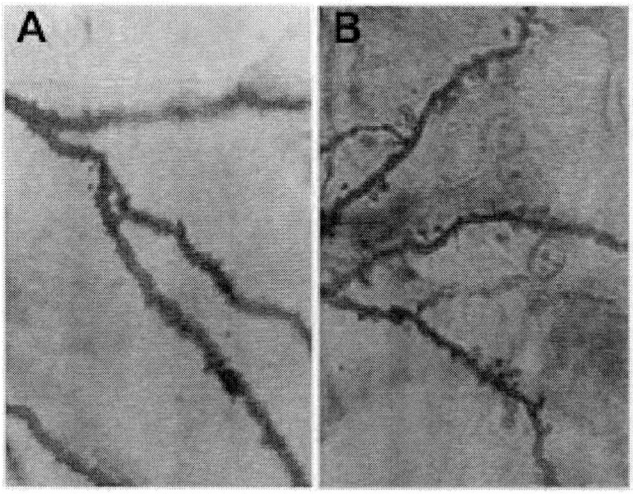

Figure 7. A control pyramidal cell of hippocampus CA1, it is notorious the presence of spines on the dendrites. B evidences the dendritic spine loss after V inhalation. 1000 X.

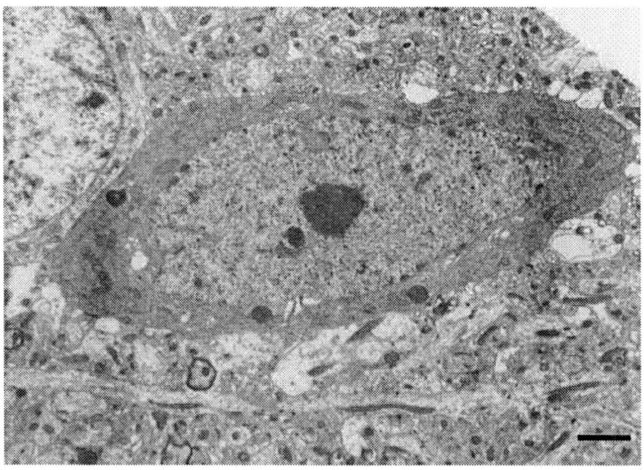

Figure 8. Necrotic pyramidal cell of exposed mouse, displaying electron-dense cytoplasm, dilated Golgi apparatus and endoplasmic reticulum. Scale bar = 1 μm.

Final Considerations

The facts mentioned above provide evidences that V interferes with important cerebral mechanisms that are directly involved with the pathogenesis of neurodegenerative disorders. First, the olfactory bulb alterations which are considered as the first symptom of neurodegeneration, and also serves as the main entrance to the brain; second, the ependymal epithelium disruption, that allows toxicants to modify the permeability of the epithelium and promote access of inflammatory mediators to the underlying neuronal tissue causing injury and neuronal death; third, the fact that V interferes with dopaminergic brain concentrations leading to cell alterations in those structures related with Parkinson disease; and finally that V inhibits important enzymes which modulate neuronal functions, that leads to changes that include the generation of ROS which in turn produces cell death, and as a consequence functional alterations, such as memory deterioration, which is related to Alzheimer disease. Those findings would allow us planning strategies to protect the brain from toxicants such as V, which have increased in the atmosphere during the last decades and constitute an important health problem since metal pollution has been related with neurodegenerative diseases.

References

Alimonti, A., Petrucci, F., Krachler, M., Bocca, B. and Caroll, S. (2000) Reference values for chromium, nickel and vanadium in urine of youngsters form the urban area of Rome. *J Environ Monit.* 2: 351–354.

Almedeida, M., Filipe, S., Hunianes, M., Mala, M.F., Melo, R., Severin, N., Silva, J.A., Frueesto da Silva, J.J. and Wever, R. (2001) Vanadium haloperoxidases from brown algae of the Laminariaceae family. *Phytochemistry*, 57: 633–642.

ATSDR Agency for Toxic Substances and Disease Registry. (1997) ATSDR's Toxicological Profiles: Vanadium. Boca Raton, Florida: Lewis Publishers, CRC Press, Inc.

Avila-Costa, M.R., Colín-Barenque, L., Zepeda-Rodríguez, A., Antuna, S., Pasos, F., Saldivar, O.L., Espejel-Maya, G., Mussali-Galante, P., Avila-Casado, M del C. and Fortoul, T.I. (2005) Ependymal epithelium disruption after vanadium pentoxide inhalation: A mice experimental model. *Neurosci Lett.* 381: 21–25.

Avila-Costa, M.R., Fortoul, T.I., Niño-Cabrera, G., Colín-Barenque, L., Bizarro-Nevares, P., Gutiérrez-Valdez, A.L., Ordóñez-Librado, J.L., Rodríguez-Lara, V., Mussali-Galante, P., Díaz-Bech, P. and Anaya-Martínez, V. (2006) Hippocampal cell alterations induced by the inhalation of vanadium pentoxide (V_2O_5) promote memory deterioration. *Neurotoxicolgy*, 27: 1007-1012.

Avila-Costa, M.R., Montiel Flores, E., Colin-Barenque, L., Ordóñez, J.L., Gutiérrez, A.L., Niño-Cabrera, H.G., Mussali-Galante, P. and Fortoul, T.I. (2004). Nigrostriatal modifications After Vanadium Inhalation: An Immunocytochemical and Cytological Approach. *Neurochem Res.* 29: 1365–1369.

Augereau, O., Claverol, S., Boudes, N., Basurko, M.J., Bonneu, M., Rossignol, R., Mazat, J.P., Letellier, T. and Dachary-Prigent, J. (2005) Identification of tyrosine-phosphorylated proteins of the mitochondrial oxidative phosphorylation machinery. *Cel and Mol Life Sci*. 62: 1-11.

Barceloux, D.G. (1999) Vanadium. *J Toxicol Clin Toxicol*. 37:265-78.

Barth, A., Schaffer, A.W., Konnaris, C., Blauensteiner, R., Winker, R., Osterode, W. and Rudiger, H.W. (2002). Neurobehavioral effects of vanadium. *J Toxicol Environ Health*, 65:677–683.

Berman, E. (1980). Toxic Metals and Their Analysis. Philadelphia: Hayden.

Bertrand, D. (1950) Survey of contemporary knowledge of biochemistry; II. The biochemistry of vanadium. *Bull Am Mus Natl Hist*. 94: 407–455.

Bracken, W.M., Sharma, R.P. and Elsner, Y.Y. (1985) Vanadium accumulation and subcellular distribution in relation to vanadate induced cytotoxicity in vitro. *Cell Biol Toxicol*. 1: 259–268.

Byerrum, R.U., Eckardt, R.E., Hopkins, L.L., Libsch, J.F., Rostoker, W., Zenz, C., Gordon, W.A., Mountain, J.T., Hicks, S.P. and Boaz, T.D. (1974) Vanadium. Washington DC, National Academy of Sciences, pp. l–117.

Byrne, A.R. and Kosta, L. (1978) Vanadium in foods and in human body fluids and tissues. *Sci Total Environ*. 10:17–30.

Cam, M.C., Brownsey, R.W. and McNeill, J.H. (2000 o 1997) Mechanisms of vanadium action: insulin-mimetic and insulin-enhancing agensts? *Can J Physiol Pharmacol*. 78: 829–847.

Campbell, C.A., Peet, M. and Ward, N.I. (1988) Vanadium and other trace elements in patients taking lithium. *Biol Psychiat*. 24:775–781.

Cantley, L.C.Jr., Josephson, L., Warner, R., Yanagisawa, M., Lechene, C. and Guidotti, G. (1977) Vanadate is a potent (Na, K)-ATPase inhibitor found in ATP derived from muscle. *J Biol Chem*. 252:7421–7423.

Colin-Barenque, L., Avila-Costa, M.R., Sánchez, I., López, I., Niño-Cabrera, G., Pasos, F., Delgado, V. and Fortul, T.I. (2003) Muerte neuronal en el bulbo olfatorio de ratón inducida por exposición subaguda y crónica de vanadio. In: Proceedings of the XLVI Congreso Nacional de la Sociedad de Ciencias Fisiológicas, A.C., Aguascalientes, Ags. México.

Conri, C., Simonff, M., Fleury, B. and Moreau, F. (1986). Variations in serum vanadium levels during treatment of mental depression. *Biol Psychiatry*, 21:1335–1339.

Crans, D.C., Smee, J.J., Gaidamauskas, E. and Yang, L. (2004) The chemistry and biochemistry of vanadium and the biological activities exerted by vanadium compounds. *Chem Rev*. 104: 849-902.

Fantus, I.G., Deragon, G., Lai, R. and Tang, S. (1995) Modulation of insulin action by vanadate: evidence of a role for phosphotyrosine phosphatase activity to alter cellular signaling. *Mol Cell Biochem*. 153: 103–112.

Fawcett, J.P., Farquhar, S.J., Thou, T. and Shand, B.I. (1997) Oral vanadyl sulphate does not affect blood cells, viscosity or biochemistry in humans. *Pharmacol Toxicol*. 80: 202–206.

Fiala, J.C., Spacek, J. and Harris, K.M. (2002) Dendritic Spine Pathology: Cause or Consequence of Neurological Disorders? *Brain Res Rev*. 39: 29–54.

Grady, C.L., McIntosh, A.R., Horwitz, B., Maisog, J.M., Ungerleider, L.G., Mentis, M.J., Pietrini, P., Schapiro, M.B. and Haxby, J.V. (1995) Age-Related Reductions in Human Recognition Memory Due to Impaired Encoding. *Science*, 269: 218-221.

Guidotti, T.L., Audette, R.J. and Martin, C.J. (1997) Interpretation of the trace metal analysis profile for patients occupationally exposed to metals. *Occup Med* (*Lond.*). 47: 497–503.

Hamel, F.G. and Duckworth, W.C. (1995) The relationship between insulin and vanadium metabolism in insulin target tissues. *Mol Cell Biochem.* 153: 95–102.

Hansen, T.V., Aaseth, J. and Alexander, J. (1982) The effect of chelating agents on vanadium distribution in the rat body and on uptake by human erythrocytes. *Arch Toxicol.* 50: 195-202.

Harland, B.F. and Harden-Williams, B.A. (1994) Is vanadium of human nutritional importance yet? *J Am Diet Assoc.* 94:891–894.

Hathcock, J.N., Hill, C.H. and Tore, S.B. (1966) Uncoupling of oxidative phosphorylation by vanadate. *Can J Biochem.* 44: 983–988.

Hirano, S. and Suzuki, K.T. (1996) Exposure metabolism and toxicity for rare earths and related compounds. *Environ Health Perspect.* 104: 85–95.

Hudson, T.G.F. (1964) Vanadium: Toxicology and biological significance. In: Browning, E., ed. Elsevier Monographs on Toxic Agents, Amsterdam, Elsevier Publishing Company, pp. 67–78.

Huyer, G., Liu, S., Kelly, J., Moffat, J., Payette, P., Kennedy, B., Tsaprailis, G., Gresser, M.J. and Ramachandran, C. (1997) Mechanism of inhibition of proteintyrosine phosphatases by vanadate and pervanadate. *J Biol Chem.* 272: 843– 851.

IARC Monographs on the Evaluation of Carcinogenic Risks to Humans. (2006) Vanadium Pentoxide. Volume 86: 227-292.

Janiszewska, G., Lachowicz, L., Jaskolski, D. and Gromadzinska, E. (1994) Vanadium inhibition of human parietal lobe ATPases. *Int J Biochem.* 26: 551–553.

Jankovic, J. (2008) Parkinson's disease: clinical features and diagnosis. *J Neurol Neurosurg Psychiatry*, 79:368-376.

Kandel, E.R. (2001) The molecular biology of memory storage: a dialogue between genes and synapses. *Science*, 294: 1030 –1038.

Kiviluoto, M., Pyy, L. and Pakarinen, A. (1981) Clinical laboratory results of vanadium exposed workers. *Arch Environ Health*, 36: 109–113.

Kustin, K., McLeod, G.C., Gilbert, T.R. and Briggs, L.B.R.T. (1983) Vanadium and other metal ions in the physiological ecology of marine organisms. *Struct Bonding*, 53: 139-160.

Lewis, C.E. (1959) The biological effects of vanadium. II. The signs and symptoms of occupational vanadium exposure. *Arch Ind Health*, 19: 497–503.

Li, J., Elberg, G., Crans, D.C. and Shechter, Y. (1996) Evidences for the distinct vanadyl(+4)-dependent activating system for manifesting insulin-like effects. *Biochemistry*, 35: 8314–8328.

Luo, Y., Umegaki, H., Wang, X., Abe, R. and Roth, G.S. (1998) Dopamine induces apoptosis through an oxidation-involved SAPK/JNK activation pathway. *J Biol Chem.* 273:3756–3764.

Madany, I.M. and Raveendran, E. (1992) Polycyclic aromatic hydrocarbons, nickel and vanadium in air particulate matter in Bahrain during the burning of oil fields in Kuwait. *Sci Total Environ.* 116: 281–289.

Meish, H.U. and Bielig, H.J. (1980) Chemistry and biochemistry of vanadium. *Basic Res Cardiol.* 75:413–417.

Mesholam, R.I., Moberg, P.J., Mahr, R.N. and Doty, R.L. (1998) Olfaction in neurodegenerative disease : A meta-analysis of Olfactory functioning in Alzheimer's and Parkinson's diseases. *Arch Neurol.* 55: 84-90.

Meza, P., Colin-Barenque, L., Mondragón, A., Avila-Costa, M.R., Sánchez I, López I, Niño-Cabrera G, Pasos F, Delgado V. and Fortoul, T. (2003) Efecto de la inhalación subaguda y crónica de vanadio sobre la ultraestructura del cerebelo de ratón. In: Proceedings of the XLVI Congreso Nacional de la Sociedad de Ciencias Fisiológicas, A.C., Aguascalientes, Ags. México.

Michibata, H. and Sakurai, H. (1990) Vanadium in Biological Systems; N.D. Chasteen, Ed.; Kluwer Academic Publishers: Boston.

Michibata, H., Uyama, T., Ueki T. and Kanamori, K. (2003a) Vanadocytes, cells hold the key to resolving the highly selective accumulation and reduction of vanadium in ascidians. *Microsc Res Technol.* 56: 421-434.

Michibata, H., Yamaguchi, N., Uyama, T. and Ueki, T. (2003b) Molecular biological approaches to the accumulation and reduction of vanadium by ascidians. *Coord Chem Rev.* 237:41-51.

Mitchell, W.G. (1953) Influence of pH on toxicity of vanadium in mice. *Proc Soc Exp Biol Med.* 84:404–405.

Mondragón, A., Colin-Barenque, L., Meza, P., Avila-Costa, M.R., Sánchez, I., López, I., Niño-Cabrera, G., Pasos, F., Delgado, V., Ordóñez, J., Gutiérrez, A., Aley, P. and Fortoul, T. (2003) Efectos de la inhibición crónica de vanadio sobre la citología del bulbo olfatorio de ratón. In: Proceedings of the XLVI Congreso Nacional de la Sociedad de Ciencias Fisiológicas, A.C., Aguascalientes, Ags. México.

Mukherjee, B., Patra, B., Mahapatra, S., Banerjee, P., Tiwari, A. and Chatterjee, M. (2004) Vanadium--an element of atypical biological significance. *Toxicol Lett.* 150: 135-143.

Myron, D.R., Givand, S.H. and Nielsen, F.H. (1977) Vanadium content of selected foods as determined by flameless atomic absorption spectroscopy. *J Agric Food Chem.* 25: 297–299.

National Toxicology Program (2002) Toxicology and Carcinogenesis Studies of Vanadium Pentoxide (CAS No. 1314-62-1) in F3344/N rats and B6C3F1 mice (Inhalation Studies) (Technical Report series No. 507; NIH Publication No. 03-4441) Research Triangle Park, NC.

Naylor, G.J., Smith, A.H.W., Bryce-Smith, D. and Ward, N.I. (1984) Tissue vanadium levels in manic-depressive psychosis. *Psychol Med.* 14: 767–772.

Nechay, R. (1984). Mechanisms of action of vanadate. *Annu Rev Pharmacol Toxicol.* 24:510–524.

Nielsen, F.G. (1991) Nutritional requirements for boron, silicon, vanadium, nickel, and arsenic: Current knowledge and speculation. *FASEB J.* 5: 2661–2667.

Nielsen, F.H. and Uthus, E.O. (1990) The essentiality and metabolism of vanadium. In: Chasteen ND. Ed. Vanadium in Biological Systems. Kluwer Academic Publishers: Boston.

Nriagu, J.P. (1998) Vanadium in the Environment, Part 2: Health Effects, JohnWiley and sons, New York, Chichester,Weinheim, Brisbane, Singapore, Toronto.

Ramandham, S., Heyliger, C., Gresser, M.J., Tracey, A.S. and McNeil, J.H. (1991) The distribution and half life for retention of vanadium in organs of normal and diabetic rats orally fed vanadium(IV) and vanadium(V). *Biol Trace Elem Res.* 130: 119–124.

Ponsen, M.M., Stoffers, D., Booij, J., van Eck-Smit, B.L., Wolters, E.Ch. and Berendse H.W. (2004) Idiopathic hyposmia as a preclinical sign of Parkinson's disease. *Ann Neurol.* 56: 173–81.

Poucheret, P., Verma, S., Grynpas, M.D. and McNeill, J.H. (1998) Vanadium and diabetes. *Mol Cell Biochem.* 188: 73–80.

Rehder, D. (1999) The coordination chemistry of vanadium as related to its biological functions. *Coord Chem Rev.* 182: 297-322.

Rehder, D. (2003) Biological and medicinal aspects of vanadium. *Inorg Chem Commun.* 6: 604-617.

Roshchin, A.V. (1967) Toxicology of vanadium compounds used in modern industry. *Hyg Sanit.* 32: 345–352.

Sasi, M.M., Haider, S.S., Fakhri, M-El. and Ghwarsha, K.M. (1994) Microchromatographic analysis of lipids, protein and occurrence of lipid peroxidation in various brain areas of vanadium exposed rats: A possible mechanism of vanadium neurotoxicity. *Neurotoxicity*, 15: 413-420.

Setyawati, I.A., Thompson, K.H., Yeun, V.G., Sun, Y., Battell, M., Lyster, D.M., Vo, C., Ruth, T.J., Zeisler, S., McNeill, J.H. and Orvig, C. (1998) Kinetic analysis and comparison of uptake, distribution, and excretion of 48 V-labeled compounds in rats. *J Appl Physiol.* 84: 569–575.

Sharma, R.P., Coulombe, R.A. and Srisuchart, B. (1986). Effects of dietary vanadium exposure on levels of regional brain neurotransmitters and their metabolites. *Biochem Pharmacol.* 35: 461–465.

Simonian, N.A. and Coyle, J.T (1996) Oxidative stress in neurodegenerative disease. *Annu Rev Pharmacol Toxicol.* 36: 83-106.

Tlan Wang, Y. and Salter, M.W. (1994) Regulation of NMDA receptors by tyrosine kinases and phosphatases. *Nature*, 369: 233-235.

Van Pee, K.H., Keller, S., Wage, T., Wynands, I., Schnerr, H. and Zehner, S. (2000) Enzymatic halogenation catalyzed via a catalytic triad and by oxidoreductases. *Biol Chem.* 381: 1-5.

Wenning, R. and Kirsch, N. (1988). Vanadium. In: Seiler HG, Sigel H, Sigel A. eds. Handbook on Toxicity of Inorganic Compounds. New York: Marcel Dekker, pp. 749–765.

WHO (1988). Vanadium. In: Environmental Health Criteria. No. 81. Technical Report Series, Geneva: WHO.

Witkowska, D. and Brzezinski, J. (1979). Alteration of brain noradrenaline, dopamine and 5-hydroxytryptamine levels during vanadium poisoning. *Pol J Pharmacol Pharm.* 31: 393–398.

Witkowska, D. and Brzezinski, J. (1983). Effect of metavanadate on the uptake and release of noradrenaline in rat brain cerebral cortex slices. *Toxicol Lett.* 17: 223–231.

Wyers, H. (1946) Some toxic effects of vanadium pentoxide. *Br J Ind Med.* 3: 177–182.

Zhao, Z., T. Zhongjia, C.D., Diltz, M., You, and Fischer, E.H. (1996) Activation of mitogen-activated protein (MAP) kinase pathway by pervanadate, a potent inhibitor of tyrosine phosphatases. *J Biol Chem.* 271: 22251–22255.

Zhen, X., Torres, C., Cai, G. and Friedman, E. (2002) Inhibition of protein tyrosine/mitogen-activated protein kinase phosphatase activity is associated with d2 dopamine receptor supersensitivity in a rat model of parkinson's disease. *Mol Pharmacol.* 62: 1356–1363.

In: Progress in Neurodegeneration
Editors: M.R. Avila-Costa and V.A. Martinez
ISBN 978-1-60741-317-2

Chapter 9

Lead

Ana Luisa Gutierrez-Valdez
Department of Neuroscience, Neuromorphology Lab
UNAM, Mexico

9.1. General Description

Lead (Pb) is a naturally occurring bluish-gray metal found in small amounts in the earth's crust. Pb can be found in all parts of our environment. Much of it comes from human activities including burning fossil fuels, mining, and manufacturing. It is rarely found naturally as a metal. It is usually found combined with two or more other elements to form Pb compounds.

Metallic Pb is resistant to corrosion (i.e., not easily attacked by air or water). When exposed to air or water, thin films of Pb compounds are formed that protect the metal from further attack. Pb is easily molded and shaped. It can be combined with other metals to form alloys. Pb and Pb alloys are commonly found in pipes, storage batteries, weights, shot and ammunition, cable covers, and sheets used to shield us from radiation. The largest use for Pb is in storage batteries in cars and other vehicles (ATSDR, 2007).

Environmental levels of Pb have increased more than 1,000-fold over the past three centuries as a result of human activity. The greatest increase occurred between the years 1950 and 2000, and reflected increasing worldwide use of leaded gasoline. Pb can enter the environment through releases from mining Pb and other metals, and from factories that make or use Pb, lead alloys, or lead compounds. Pb is released into the air during burning coal, oil, or waste. Before the use of Pb gasoline was banned, most of the Pb released into the U.S. environment came from vehicle exhaust. In 1979, cars released 94.6 million kilograms (208.1 million pounds) of Pb into the air in the United States. In 1989, when the use of Pb was limited but not banned, cars released only 2.2 million Kg (4.8 million pounds) to the air. Since EPA banned the use of leaded gasoline for highway transportation in 1996, the amount of Pb released into the air has decreased further. Before the 1950s, Pb was used in pesticides applied to fruit orchards (ATSDR, 2007).

Once Pb gets into the atmosphere, it may travel long distances if the Pb particles are very small. Pb is removed from the air by rain and by particles falling to land or into surface water.

Short-term exposure to high levels of Pb can cause vomiting, diarrhea, convulsions, coma or even death. However, even small amounts of Pb can be harmful, especially to infants, young children and pregnant women. Symptoms of long-term exposure to lower Pb levels may be less noticeable but are still serious. Anemia is common and damage to the nervous system may cause impaired mental function. Other symptoms are appetite loss, abdominal pain, constipation, fatigue, sleeplessness, irritability and headache. Continued excessive exposure, as in an industrial setting, can affect the kidneys (WHO, 1995).

Studies of Pb workers suggest that long-term exposure may be associated with increased mortality due to cerebrovascular disease. The same was found in a study of adults from the general population who were hospitalized for Pb poisoning during childhood. Population studies suggest that there is a significant association between bone-lead levels and elevated blood pressure. Blood Pb levels (PbBs) also have been associated with small elevations in blood pressure. Between the two biomarkers, bone Pb appears to be the better predictor. Pb also affects kidney functions; glomerular filtration rate appears to be the function affected at the lowest PbBs. Decreased glomerular filtration rate has been consistently observed in populations with mean PbB <20 μg/dL and two studies have reported effects at PbB <10 μg/dL (ATSDR, 2007).

9.2. Sources of Lead Exposure

Native Pb is rare in nature. Currently Pb is usually found in ore with zinc, silver and copper and it is extracted together with these metals. The main Pb mineral is in Galena and there are also deposits of cerrussite and anglesite which are mined. Galena is mined in Australia, which produces 19% of the world's new Pb, followed by the USA, China, Peru' and Canada. Some is also mined in Mexico and West Germany. World production of new Pb is 6 million tons a year, and workable reserves total are estimated 85 million tons, which is less than 15 year's supply (WHO, 1995).

The Pb salts enter the environment through the exhausts of cars. The larger particles will drop to the ground immediately and pollute soils or surface waters, the smaller particles will travel long distances through air and remain in the atmosphere. Part of this Pb will fall back on earth when it is raining. This lead-cycle caused by human production is much more extended than the natural lead-cycle. It has caused Pb pollution to be a worldwide issue (ATSDR, 2007).

9.2.1. Air

According to the Toxics Release Inventory (TRI04, 2006), in 2004, a total of 215,216 pounds of Pb were released to air from 4,337 reporting facilities (TRI04, 2006). In addition, a total of 923,449 pounds of Pb compounds were released to air from 4,294 reporting facilities (TRI04, 2006). Releases of Pb and Pb compounds to air constitute, respectively, 1.78 and 40.23% of all on-site releases.

The emissions of Pb and Pb compounds to the atmosphere reported to TRI has declined from 2.8 million pounds in 1988 to about 1.1 million pounds in 2004 as new industries were added to TRI reporting requirements (TRI04, 2006). In 2000, before the reporting thresholds were drastically reduced, air emissions were 1.5 million pounds. In the past, transportation, particularly automotive sources, were the major contributor to air emissions of Pb. Today, industrial processes, especially metal processing, are the major sources of Pb emissions to the atmosphere with the highest concentrations found around smelters and battery manufacturers (EPA, 2003).

9.2.2. Water

Of the known aquatic releases of Pb, the largest ones are from the steel and iron industries and Pb production and processing operations (EPA, 1982). Urban runoff and atmospheric deposition are significant indirect sources of Pb found in the aquatic environment. Pb reaching surface waters is sorbed to suspended solids and sediments (EPA, 1982).

Although aquatic releases of Pb from industrial facilities are expected to be small with respect to emissions to land and air, Pb may be present in significant levels in drinking water. In areas receiving acid rain the acidity of drinking water may increase; this increases the corrosivity of the water, which may, in turn, result in the leaching of Pb from water systems, particularly from older systems during the first flush of water through the pipes (McDonald, 1985). In addition, the grounding of household electrical systems to the plumbing can increase corrosion rates and the subsequent leaching of Pb from the Pb solder used for copper pipes (ATSDR, 2007).

9.2.3. Soil

While the majority of Pb releases are to land, they constitute much lower exposure risks than releases to air and water. In 1997, before new industries were added to TRI, 95% of Pb and Pb compound releases to land reported to TRI were from the primary metals industrial sector, primarily metal smelters. In 2004, metal mining, coal mining, electrical utilities, and Resource Conservation and Recovery Act (RCRA)/solvent recoveries (hazardous waste facilities), as well as primary metals, are the industrial sectors contributing most heavily to releases to land. Many of these facilities with large releases, such as metal mines, are located in sparsely populated areas. Hazardous waste facilities are highly regulated. Most of the Pb released to land becomes tightly bound and immobile (ATSDR, 2007).

9.2.4. Paint

Although the sale of residential lead-based paint was banned in, flaking paint, paint chips, and weathered powdered paint, which are most commonly associated with deteriorated housing stock in urban areas, remain major sources of Pb exposure for young children residing in these houses, particularly for children afflicted with pica (the compulsive, habitual consumption of nonfood items) (Bornschein et al. 1986; EPA 2003). Pb concentrations of 1–5 mg/cm^2 have been found in chips of lead-based paint (Billick and Gray, 1978), suggesting that consumption of a single chip of paint would provide greater short-term exposure than any other source of lead (EPA, 2003). An estimated 40–50% of occupied housing may contain lead-based paint on exposed surfaces (Chisolm, 1986).

9.2.5. Other Sources

Pb has been detected in a variety of foods. Pb may be introduced into food through uptake from soil into plants or atmospheric deposition onto plant surfaces, during transport to market, processing, and kitchen preparation (EPA 1986). In the FDA Total Diet Study (TDS) 1991–1996, food was purchased 4 times/year and a market basket consisting of about 260 foods from three representative cities within the geographical region analyzed for different elements, including Pb (Capar and Cunningham, 2000). Pb was below the limit of quantization in all TDS food in the following food categories: milk and cheese, eggs, meat, poultry, and fish; legumes and nuts, grain and cereal products, vegetables, mixed dishes and meals, desserts, snacks, fats and dressings, and infant and junior foods. Only five products had quantifiable concentrations of Pb, namely: canned peaches (0.032 mg/kg), canned pineapple (0.013 mg/kg), canned fruit cocktail (0.031 mg/kg), sweet cucumber pickles (0.036 mg/kg), and dry table wine (0.023 mg/kg) (ATSDR, 2007).

Other factors such as absorption of Pb from cooking water and cookware can influence the amount of Pb in cooked vegetables. Ceramic dishes may contain Pb in their glazes, and Pb in glass has been shown to leach into wine. The degree to which Pb is released from food once it is consumed also influences a person's uptake of Pb (ATSDR, 2007).

9.3. Absorption, Distribution, Metabolism, and Excretion

The current provisional tolerable weekly intake (PTWE) is 25/μg Pb/Kg body weight. These values are reached or exceeded, for example, in Cuba, India, and Thailand. The intakes reported for Italy, Germany, and New Zealand were half or more of the PTWI; lower intakes were found in Sweden, Denmark, and in the United States (Schäfer et al. 1999).

Up to 50% of inhaled inorganic Pb may be absorbed in the lungs. Adults take up 10–15% of Pb in food, whereas children may absorb up to 50% via the gastrointestinal tract. Pb in blood is bound to erythrocytes, within a few minutes, where it is bound to the cell membranes and to hemoglobin. Pb shows 16-fold enrichment in erythrocytes compared to plasma, where it mainly binds to albumin. The uptake into erythrocytes seems to be saturable, the elimination is slow and principally via urine, glomerular filtration makes up 76% of total Pb excretion; 16% is eliminated with the feces via the bile or via pancreatic and intestinal secretion. Less than 8% is eliminated through hairs, nails, and perspiration (Schäfer et al. 1999).

From the blood, Pb distributes into the soft tissues, such as the liver (about 1 mg Pb/Kg), the kidney (about 0.8 mg Pb/Kg), and the brain (about 0.1 mg Pb/Kg). In the cells, Pb mainly binds to the membranes and mitochondria. Pb is accumulated in the skeleton, and is only slowly released from this body compartment. Half-life of Pb in blood is about 1 month and in the skeleton 20–30 years (WHO, 1995).

Inorganic Pb enters the body mainly through inhalation and ingestion. Abrasive blasting, sanding, removal with ``heat guns", scaling, grinding, welding, cutting, and torch burning-all can produce Pb particulates in the respirable range. Heat or burning Pb or lead-based paint generates Pb vapors. These forms of Pb are readily inhaled and absorbed (Fischbein, 1998). Between 30 and 5% of inhaled Pb is deposited in the respiratory tract and absorbed. Pb deposited in the lower respiratory tract is virtually completely absorbed.

In adults, inorganic Pb does not penetrate the blood–brain barrier, whereas this barrier is less developed in children. The high gastrointestinal uptake and the permeable blood–brain barrier make children especially susceptible to Pb exposure and to subsequent brain damage. Organic Pb compounds penetrate body and cell membranes. Tetramethyl lead and tetraethyl lead penetrate the skin easily. These compounds may also cross the blood–brain barrier in adults, and thus adults may suffer from Pb encephalopathy related to acute poisoning by organic Pb compounds (Järup, 2003).

9.4. Health Effects

Pb exerts its toxic effects on many organ systems, across a wide range of exposure levels. Beginning with subtle, subclinical disturbances of enzyme function and biochemical aberrations, the toxic effects of Pb can progress to severe, clinically evident disease with disruption of multiple organ functions (Fischbein, 1998). Disruption of subcellular, mitochondrial energy metabolism by interference with enzymatic function, competition with essential metals for binding sites, and disturbance of ion transport mechanisms are likely to account for many of the subclinical and clinical manifestations of Pb toxicity (EPA, 1986).

Acute exposure to Pb is known to cause proximal renal tubular damage (WHO, 1995). Long-term Pb exposure may also give rise to kidney damage and, in a recent study of Egyptian policemen, urinary excretion of NAG (N-acetyl-β-D-glucosaminidase, Markers of tubular damage) was positively correlated with duration of exposure to Pb from automobile exhaust, blood Pb and nail Pb (Mortada et al. 2001).

In adults, the effects of high-dose exposure to Pb have long been recognized and described (Landrigan et al. 1990). The clinical picture is characterized by anemia, abdominal colic and peripheral neuropathy (extensor weakness, ``wrist/ankle drop").

The symptoms of acute Pb poisoning also are represented by headache, irritability, abdominal pain and various symptoms related to the nervous system. Pb encephalopathy is characterized by sleeplessness and restlessness. Children may be affected by behavioral disturbances, learning and concentration difficulties. In severe cases of Pb encephalopathy, the affected person may suffer from acute psychosis, confusion and reduced consciousness. People who have been exposed to Pb for a long time may suffer from memory deterioration, prolonged reaction time and reduced ability to understand. Individuals with average blood PB levels under 3 μmol/l may show signs of peripheral nerve symptoms with reduced nerve conduction velocity and reduced dermal sensibility. If the neuropathy is severe the lesion may be permanent. The classical picture includes a dark blue Pb sulphide line at the gingival margin. In less serious cases, the most obvious sign of Pb poisoning is disturbance of hemoglobin synthesis, and long-term Pb exposure may lead to anemia (Järup, 2003).

A report has shown that long-term low-level Pb exposure in children may also lead to diminished intellectual capacity (WHO, 1995).

Workers with lower-level, chronic or recurrent exposure to Pb may remain asymptomatic or develop vague, non-specific symptoms (e.g., myalgias, fatigue, irritability, headaches) reflecting subtler organ system damage and impairment than is seen in acute intoxications. Such individuals often continue to work and come to clinical attention only as the result of blood Pb screening or monitoring programs (Levin and Goldberg, 2000).

9.5. Lead Neurotoxicity

J. Lockhart Gibson of Brisbane, Australia made one of the earliest reports of Pb neurotoxicity in 1892 (Gibson et al. 1892). The major symptom associated with the diagnosis of Pb neurotoxicity during this era was the overt encephalopathy that is known to occur at blood Pb levels of 70 μg/dL or greater (Chisolm, 2001). In 1943, Byers and Lord suggested that Pb had lasting effects on the central nervous system (CNS) (Byers and Lord, 1943). A definitive demonstration of the lasting, subclinical effects of Pb was initially made in 1979 by Dr. Herb Needleman, who observed that children who were previously, but not currently, exposed to Pb exhibited lasting neurobehavioral and cognitive deficits (Needleman et al. 1979). Since the elucidation of the lasting, subclinical effects of Pb on the developing human nervous system, there have been many studies that have attempted to characterize, in detail, the specific components of behavior and cognition that are affected by this metal. The most consistent finding among these studies is that low level of Pb intoxication modestly alters the performance on standardized cognitive assessments (Ruff and Bijur, 1989). Studies on cognitive ability have demonstrated a deficit between 0 and 5 points on the IQ scale for every 10 μg/dL increase in blood Pb level (Bellinger, 1995). In addition, consistent alterations are observed on attention, visual-motor reasoning skills, social behavior, mathematics and reading abilities (Lanphear et al. 2000; Canfield et al. 2003).

In severe, acute Pb poisoning, dramatic disturbances of CNS function may occur, more frequently seen in lead-poisoned children with blood Pb levels >100 mg/dL. The clinical picture may include convulsions, delirium, and coma (Fischbein, 1998). More common among workers is the toxic encephalopathy associated with longer-term, lower level exposure to Pb, accompanied by non-specific, often subtle symptoms, including headache, dizziness, sleep disturbances, short-term memory deficits, depression, fatigue, irritability, joint and/or muscle pain, and loss of libido (Lilis et al. 1977).

Impaired cognitive performance has been demonstrated in relation to chronic low-level exposure to Pb in a number of studies of occupational groups (Stollery et al. 1989; Balbus-Kornfeld et al. 1995). Recently, neuropsychological decrements have been found to persist among Pb battery workers with a history of blood Pb levels ranging from 50 to 100 mg/dL with subsequent low-level exposure to Pb, indicating long-lasting and perhaps permanent functional impairment (Hanninen et al. 1998). CNS impairment with intellectual and cognitive dysfunction can occur at blood Pb levels in the 40±70 mg/dL range, and there is evidence that impaired short-term memory and difficulty in concentrating may occur when blood Pb levels reach 40 mg/dL (EPA, 1986). Emotional lability, with irritability, sleep disturbance, fatigue, and depression may be among the earliest symptoms experienced by workers with excessive absorption of Pb and have been shown to occur when blood Pb concentrations exceed 40 mg/dL (Baker et al. 1983). In other study of elderly women, blood Pb levels as low as 8 mg/dL were significantly associated with decrements in cognitive function measured by neuropsychological testing (Muldoon et al. 1996).

In this way, it has been demonstrated that the learning impairment, and other behavioral disturbances in affected children and laboratory animals, suggests that hippocampus might be one region adversely affected during early life. Indeed, recent findings demonstrated that early Pb exposure disrupts expression and phosphorylation of the cAMP-responsible element binding protein (CREB), a transcription factor directly related to the neuronal plasticity in the hippocampus of juvenile rats (Toscano et al. 2003). Furthermore, early Pb exposure altered the N-methyl-Daspartate receptor subunit composition in favor of the prevalence NR2B receptor subunit and decreased expression of the NR 2A subunit, which might be important in hippocampal development and maturation (Toscano et al. 2002) Figure 1 summarizes the effects of Pb in the CNS.

9.5.1. Mechanism of Pb-induced Neurotoxicity

In the CNS, Pb is selectively accumulated in astroglia, both *in vivo* (Holtzman et al. 1984; Struzynska et al. 2005) and *in vitro* (Lindahl et al. 1999). Three mechanisms for Pb entry into cells have been identified in various cell cultures and isolated cell models, usually under nonphysiological conditions (e.g., depolarized cells). These mechanisms include uptake into erythrocyte ghosts via an anion exchange (Simons, 1993), uptake into depolarized and nondepolarized bovine chromatin cells via L-type calcium channels (Legare et al. 1998), and transport into several cell types, including C6 rat glioma cells, by a cation channel that is activated by depletion of intracellular calcium stores (Kerper and Hinkle, 1997a). None of these mechanisms for Pb transport has been validated *in vivo* (Qian and Tiffany-Castiglioni, 2003).

The mechanism by which Pb disturbs CNS function has been the subject of some investigation. Effects of Pb on neurotransmitter system functions in specific areas of the brain have been demonstrated in rats following low-level exposure in rats (Kala and Jadhav, 1995a;

b). Brain mitochondrial respiration and ADP phosphorylation is affected *in vitro* by this metal (Dumas et al. 1985).

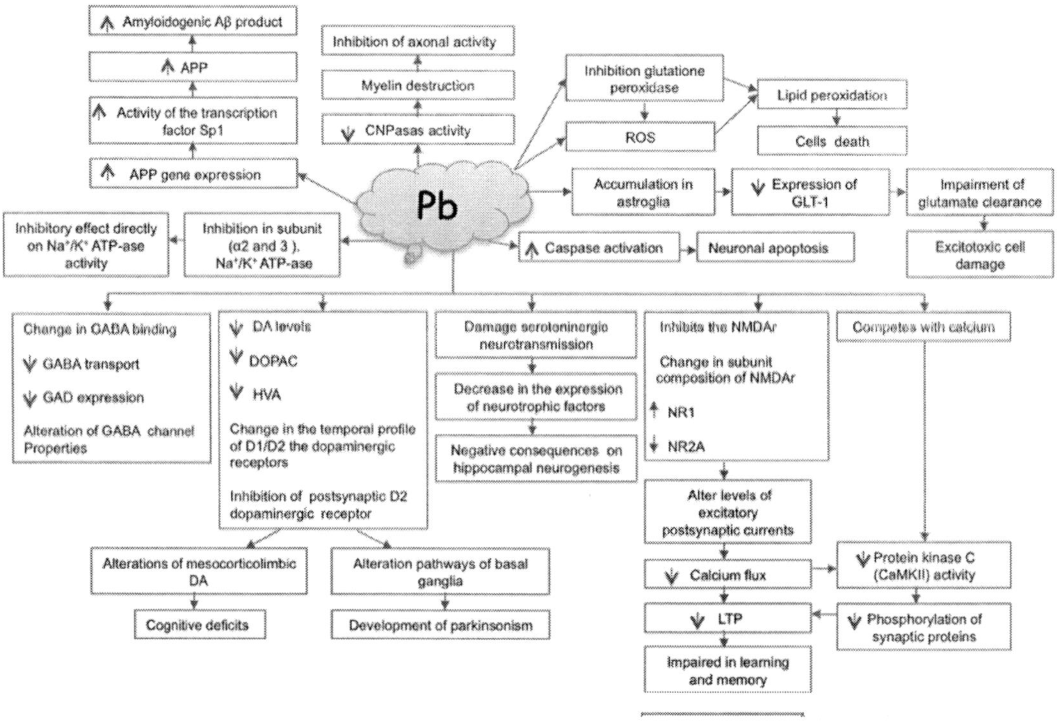

Figure 1. Effects of Pb exposure in the CNS. Abbreviations: APP (the-amyloid precursor protein), Aβ (β-amyloid -peptides are the primary constituents of amyloid deposits-), Sp1 (activity of specificity protein 1 -a transcription factor involved in the regulation of the *APP* gene-), LTP (Long-term potentiation), CNPase (2´,3´cyclic nucleotide 3-phosphodiesterase - enzyme preferentially located in myelin and was shown to be an integral protein-), ROS (Reactive Oxygen Species), GLT-1 (Astrocytic glutamate/aspartate transporters - which regulate extracellular glutamate concentration-), GABA (g-aminobutyric acid), GAD (glutamic acid decarboxylase -an enzyme regulating the synthesis of GABA-), DOPAC (3,4-dihydroxyphenilacetic acid - the metabolites of dopamine -), DA (Dopamine), HVA (Homovanillic acid - the metabolites of dopamine -), CaMKII (calcium/calmodulin-dependent protein kinase II).

The direct neurotoxic actions of Pb include apoptosis, excitotoxicity, influences on neurotransmitter storage and release processes, mitochondria, second messengers, cerebrovascular endothelial cells, and both astroglia and oligodendroglia (Lidsky and Schneider, 2003). Although all of lead's toxic effects cannot be tied together by a single unifying mechanism, lead's ability to substitute for calcium (Ca^{2+}) [and perhaps zinc (Bressler and Goldstein, 1991)] is a factor common to many of its toxic actions. For example, lead's ability to pass through the blood–brain barrier (BBB) is due in large part to its ability to substitute for Ca^{2+} ions (Zheng et al. 2003). Experiments with metabolic inhibitors suggest that back-transport of Pb via the Ca-ATPase pump plays an important role in this process (Bradbury and Deane, 1993). More direct evidence for the role of the Ca-ATPase pump in the transport of Pb into the brain has been provided by *in vitro* studies of brain capillary endothelial cells, the primary constituent of the BBB (Kerper and Hinkle, 1997a, b; Zheng et al. 2003). At higher blood levels, Pb disrupts the function of endothelial cells in the BBB. This may lead to hemorrhagic encephalopathy, characterized by seizures and coma (NourEddine et al. 2005).

It has been demonstrated that Na^+/K^+ ATP-ase activity decreased in the cerebral cortex of pups intoxicated with Pb, whereas in cerebellum both ATP-ases (Na^+/K^+ and Mg^+ -dependent) activities decreased significantly. On the other hand, the energy metabolism of discrete brain areas was significantly altered by lead-treatment of rats. The more affected area was the striatum, where ATP, ADP and AMP levels decreased significantly in the intoxicated pups. Adenylate energy charge, an additional index of cell energy state was decreased also in the striatum of intoxicated pups (-25%). Moreover, Pb treatment diminished significantly the ATP/ADP ratio in striatum and hippocampus (Antonio and Leret, 2000).

Pb accumulation has also been suggested to generate lipid peroxidation and therefore to affect antioxidant enzymes, including catalase (Somashekaraiah et al. 1992; Acharya and Acharya, 1997; Villeda-Hernandez et al. 2006; Kumar Bokara et al. 2007). There have been demonstrated the effect of acute Pb acetate administration on brain catalase activity. In all these studies, the induction of brain catalase increased progressively with time following Pb administration (Valenzuela et al. 1989; Somashekaraiah et al. 1992).

A possible mechanism through which Pb may alter brain catalase activity has been suggested previously (Valenzuela et al. 1989; Somashekaraiah et al. 1992). According to this suggestion, the selenoenzyme glutathione peroxidase catalyses the detoxification of lipid peroxides. There are several reports of this enzyme being inhibited in different brain regions after Pb exposure. The endogenous inhibition of that enzyme could therefore be a major cause of both oxidative stress and catalase induction as a compensatory mechanism to eliminate hydroperoxides (Correa et al. 2004). In that regard, Valenzuela et al. (1989) speculated that an increase in the formation of lipid hydroperoxides in the lead-intoxicated cerebellum of rats may have served as a signal to maintain higher levels of catalase in order to enhance the detoxification process. The delayed induction of catalase has been reported in previous studies. For example, Somashekaraiah et al. (1992) concluded that the administration of Pb depletes antioxidant enzymes in chick embryos 9 hrs after injection, and suggested that this effect provided a suitable condition for the enhancement of lipid peroxidation. However, after 72 h, levels of lipid peroxidation decreased to normal with a significant increase in the levels of antioxidant enzymes such as catalase (Somashekaraiah et al. 1992).

On the other hand, the mechanism by which Pb disrupts normal physiological processes is based on the similarity of ionized lead (Pb^{2+}) to Ca^{2+}. Both are divalent cations; however, Pb^{2+} can disrupt the physiological effects of Ca^{2+} at concentrations several orders of magnitude lower than the concentration of Ca^{2+} (NourEddine et al. 2005). In the developing brain, Pb^{2+} causes an inappropriate release of neurotransmitters at rest and competes with Ca^{2+} to interfere with evoked neurotransmitter release. This increase in basal release and decrease in evoked release may interfere with selective pruning of synaptic connections in the brain during the first few years of brain development (Zawia, 2003).

Moreover, growing evidence indicates that some of the toxic effects of Pb are localized to the endoplasmic reticulum (ER). In addition to serving as a site for synthesis and folding of proteins destined for transport to the Golgi apparatus, the ER is the major intracellular Ca^{2+}

storage site and regulating organelle. The Ca^{2+} concentration in the lumen of the ER is 3–4 orders of magnitude that of the cytosol, a gradient that is maintained by sarco/endoplasmic reticulum calcium-ATPases, or SERCA proteins (Mendolesi and Pozzan, 1998). As shown in a variety of cell types, Pb inhibits Ca^{2+}-ATPase (Mas-Oliva, 1989; Hechtenberg and Beyersmann, 1991), stimulates Ca^{2+}-activated depletion of internal Ca21 stores (Kerper and Hinkle, 1997a), impairs store refilling (Wiemann et al. 1999), and binds to the ER-localized 78-kD molecular chaperone glucose regulated protein (GRP78) (Qian et al. 2000) also known as immunoglobulin heavy chain–binding protein. Under conditions of oxidative or chemical stress, the ER undergoes a stress response termed the unfolded (or misfolded) protein response (Kozutsumi et al. 1988; Wooden et al. 1991).

Bressler et al. (1999) demonstrate that Pb is a potent activator of protein kinase C (PKC) in enzymatic assays and, in intact cells; Pb activates downstream events through a mechanism that is dependent on PKC. The underlying hypothesis is that Pb impairs learning by interfering with activity-dependent mechanisms that reinforce synapses during a critical period of development when postnatal experiences are crucial for cognitive development. The activity-dependent mechanisms that are most likely disrupted by Pb are those that require Ca^{2+}, particularly PKC. Models such as LTP that enable to study synaptic plasticity have emphasized the importance of PKC. Inhibitors of PKC block the induction of LTP and the phosphorylation of synaptic proteins by PKC is associated with LTP (Ramakers et al. 1995). Processes important for synaptic transmission such as the biosynthesis of neurotransmitters (Haycock, 1993), ligand-receptor interactions (Patel et al. 1995), conductance of ion channels (Hofmann et al. 1994) and dendritic branching (Hundle et al. 1995) are regulated, in part, by PKC and, therefore, may be affected by Pb.

9.5.2. Pb and Neurotransmission

It has been demonstrated that Pb causes activation of PKC and binds to PKC more avidly than Ca^{2+}, its physiologic activator. Alteration of PKC function compromises transmitter and second-messenger systems within the cell leading to further changes in gene expression and protein synthesis (NourEddine et al. 2005).

There was a general decrease in neurotransmitter levels in several brain areas of lead-treated rats. The most affected area was the hippocampus, where dopamine (DA), 5-hydroxytryptamine (5-HT) as well as their main respective metabolites, DOPAC and 5-HIAA decreased significantly. In the hypothalamus, noradrenaline (NA), DA and 5-HT decreased significantly, whereas in the cerebellum of lead-treated rats, NA and 5-HT showed a significant reduction (-77.71% and -23.56% respectively). The striatum was the area less affected, since only DA levels showed a strong reduction (-31.65%) in the intoxicated pups (Antonio and Leret, 2000).

As we mentioned above, Antonio and Leret (2000) also confirmed that Pb might exert an inhibitory effect directly on Na^+/K^+ ATP-ase activity in cerebellum and cerebral cortex of intoxicated pups. It is known that Na^+/K^+ ATPase, a key enzyme for maintaining the ion distribution about the cellular membrane which is required for neuronal activity, is among the enzymes particularly affected by Pb (Carfagna et al. 1996). This fact may relate to the different alpha subunit composition of Na^+/K^+ ATP-ase of different tissues. Enzyme extracted from brain has mostly α2 and α3 isozymes present in significant amounts (Gerbi et al. 1993). All the subunits of the enzyme (α1, 2 and 3) differ with the number of reactive sulphydryl groups in the catalytic site (Brodsky and Guidotti, 1990) and α2 and 3 forms of the enzyme are much more sensitive to the inhibitory effect of Pb than the α1 form is (Fox et al. 1991). Thus it can be suggested that one of the possible mechanisms of the inhibition may be the interaction of Pb with –SH groups in the catalytic site of the enzyme.

The general decrease in neurotransmitters observed in the study of Antonio and Leret (2000) could be related with a decreased synthetic capacity, because it has been described that Pb could inhibit tyroxine hydroxylase activity (Ramin et al. 1993) or alternatively with Pb interference with cellular energy metabolism, inhibiting ATP synthesis that leads to the dysfunction of energy-dependent neuronal events such as neurotransmitter uptake (Boykin et al. 1991).

It also has been demonstrated that Pb affects GABAergic neurons in a study measured [3H]-GABA specific binding in the rat brain, and found binding to be increased in the cerebellum, but decreased in the sriatum, following chronic Pb exposure (Memo et al. 1980). Latter studies in synaptosomal preparations of adult rat brains found a diminished transport of [14C]-GABA following Pb treatment with a decrease in both the uptake and depolarization-evoked release of the neurotransmitter (Struzynska and Sulkowski, 2004). This was paralleled by a decrease in the expression of GAD (Glutamic acid decarboxylase), though also by an increase in a GABA transporter protein (GAT-1), possibly to compensate for the extracellular increase in GABA (Struzynska and Sulkowski, 2004). Others have observed that synaptosomes preloaded with [3H]-GABA demonstrate a Pb induced alterations, but spontaneous GABA release (Minnema et al. 1991).

Some studies have suggested that the release of GABA by Pb is related to modifications of various channels. For example, acute Pb treatment reduced both GABA B affinity (K_d) by 30%, and receptor density (B_{max}) by 15%, while chronic Pb treatment increased receptor capacity by about 20% in spite of decreased receptor affinity (Waskiewicz, 1996). These results suggest that Pb can affect GABA B binding in two ways: by reducing the binding affinity and by altering the binding capacity. Others have examined the effect of Pb on Ca^{2+}-channels and GABA release. Using the whole-cell patch clamp technique it was determined that Pb concentrations >10 nM reversibly blocked the tetrodotoxin-sensitive release of GABA, as evidenced by a reduction of the amplitude and frequency of GABA-mediated postsynaptic currents evoked by spontaneous neuronal firing (Braga et al. 1999a). Further experiments suggested that Pb exerted its effect directly by binding to voltage-gated Ca^{2+} channels (Braga et al. 1999a; b).

Pb also reduces the stimulated release of glutamate (Glu) in various areas of the brain, including cerebellar granule, hippocampal and cerebrocortical neurons. But the reasons for this phenomenon are still unclear (Fitsanakis and Aschner, 2005).

Experiments involving various combinations of recombinant NMDA subunits have been conducted in *Xenopus laevis*. Using this approach, it was determined that Pb induced inhibition of Glu-activated currents is indeed dependent on the subtype expressed. Interestingly, chronic exposure to low levels of Pb during development alters the type of NMDA receptors expressed in the developing and young adult rat brain. Thus, rats exposed to Pb showed a greater number of MK-801 binding sites, and these sites were associated with an increase in the NR1/NR-2B subunits composition (Toscano et al. 2002).

NourEddine et al. (2005) demonstrated that oral administration of 1000 ppm of Pb acetate to young rats for 30 days caused a reduction in locomotor activity and stereotypic exploratory behavior during a 20 min testing period. This locomotor hypoactivity induced by Pb was accompanied by a reduction in stereotypic behavior (sniffing, lickings, biting and grooming). These outcomes suggested that Pb might interfere with catecholaminergic and particularly dopaminergic neurotransmission. Therefore, these authors examined the effect of the Pb acetate on the uptake of DA in striatal synaptosomal preparations. The collected data showed a clear inhibition of the uptake of 3H-DA with an IC50 of 3.5×10^{-5} M. This inhibition of the uptake of DA suggests that the behavioral effects of Pb may be involved in dopaminergic neurotransmission. Likewise, chronic post weaning low-level Pb exposure produces cognitive deficits associated with Pb-induced alterations of mesocorticolimbic DA function. This study examined Pb-induced changes in the temporal profile of D1/D2 receptor protein and DA levels in the nucleus accumbens, hippocampus, and the frontal cortex (Gedeon et al. 2001).

The dopaminergic receptors (D1 and D2) have been implicated in the stereotypic behavior (Al tadjir et al. 1990; Tursky et al. 1998), thus, the results reported suggest that Pb induces a reduction in the catecholaminergic transmission, either by an inhibition of the synthesis of DA and its release to the synaptic site, or by an inhibition of the postsynaptic D2 receptors (NourEddine et al. 2005).

Autoradiographic studies also showed that Pb exposure decreased the density of D2 receptors in the cerebral cortex (Ma et al. 1999), and it also has been shown a diminished evoked release of DA in synaptosomes isolated from the striatum of animals exposed to Pb (Sulkowski et al. 1999).

It is likely that Pb intoxication might also produce cognitive deficits in adult animals, since there is evidence showing that sub-chronic (14 days) exposure of adult mice inhibits constitutive nitric oxide synthase (cNOS) activity in brain synaptosomes (García-Arenas et al. 2004) and it has been hypothesized that nitric oxide is important in synaptic plasticity (Holscher, 1997). There are evidences showing that Pb readily crosses the BBB and that, in the adult rats, enters the brain with rapid kinetics (Kerper and Hinkle, 1997a).

It is well known that cNOS activity is close related to the activation of NMDA receptor (NMDAr) (Arancio et al. 1996; Holscher, 1997) thus, the effect of Pb on cNOS activity observed in this study can be associated with functional effects on NMDAr. *In vitro* neurochemical studies have shown that Pb has a marked inhibitory effect on the activation of the NMDAr ion channel complex (Uteshev et al. 1993), which is also involved in synaptic plasticity subserving LTP (Alkondon et al. 1990). The interactions of Pb with the NMDAr channel complex may be mediated by its interaction with zinc regulatory site in the receptor complex, since Pb presents a higher affinity for this binding site than zinc (Bressler et al. 1999). The regional specificity on cNOS inhibition might be related to its ability to bind to the zinc allosteric site. This is particularly important in hippocampus and cerebellum because zinc is highly concentrated in those regions (Sawashita et al. 1997).

9.6. Lead and Neurodegenerative Diseases

Parkinson disease (PD) and Alzheimer disease (AD) are the two most common neurodegenerative diseases of the older population. PD affects more than 500,000 persons (National Institute of Neurological Disorders and Stroke 2004; Siderowf and Stern, 2003). Causation of both PD and AD is complex. In a minority of cases, particularly in early onset AD and PD, etiology appears to be primarily genetic (Tanner et al. 1999). But in most cases, causation appears to involve interactions among multiple genetic and environmental factors (Foster, 2002; Kennedy et al. 2003). Landrigan et al. (2005) hypothesize that exposure of the developing brain to still undefined toxic environmental agents during stages of vulnerability in early life—*in utero* and in early postnatal life—may be an important contributor to causation.

Through detailed reconstructions of neonatal and medical histories of birth cohorts in the United Kingdom, David Barker of the University of Southampton proposed what is now termed "the Barker hypothesis" (Osmond and Barker, 2000), the concept that parameters of fetal, infant, and childhood growth may be predictors of disease in later life. At the 2003 Mount Sinai Conference on Early Environmental Origins of Neurological Degeneration, Landrigan et al. (2005) explored the plausibility of extending the Barker hypothesis to encompass brain development and to explore the impacts of toxic chemicals on brain development.

Conferees generally supported the hypothesis that early exposures to environmental toxicants could later affect the brain and that such associations are biologically conceivable (De la Fuente-Fernandez and Calne, 2002). This consensus was based on experimental studies of associations between early-life exposures to pesticides and PD (Thiruchelvam et al. 2000a; b), as well as on epidemiologic studies of the toxic and apparently irreversible effects on the developing brain of *in utero* exposures to Pb, methylmercury, and polychlorinated biphenyls (Grandjean et al. 1997). A mechanistic hypothesis proposed by Langston et al. (1999) that early exposures to neurotoxic chemicals reduce the number of neurons in critical areas of the brain such as the substantia nigra (SN) to levels below to those needed to sustain function in the face of the neuronal deterioration associated with advancing age.

Childhood exposure to Pb, even at relatively low levels (Canfield et al. 2003), results in a decline of cognitive function that persists into adulthood and that manifests as a persistent lowering of IQ score plus alteration in behavior (Needleman et al. 1990). Each increase of 10 μg/dL in the lifetime average blood Pb concentration was found to be associated with a 4.6-point decrease in IQ (Schwartz et al. 2000). There appears to be no minimum threshold level below which Pb does not cause brain injury (Canfield et al. 2003). In addition, elevated Pb levels in childhood have been associated with lower class standing in high school, lower vocabulary and grammatical reasoning scores, poorer hand–eye coordination, and self-reports of minor delinquent activity (Needleman et al. 1990) and may predict dementia (Landrigan et al. 2005).

Occupational exposure to Pb among adults is associated with poorer neurobehavioral test scores and with deficits in manual dexterity, executive ability, verbal intelligence, and verbal memory (Schwartz et al. 2000). Some data suggest that cognitive function can decline progressively in older Pb workers in relation to cumulative past occupational exposure to Pb

(Stewart et al. 1999). Susceptibility to the persistent effect of Pb on the CNS may be enhanced in persons who have at least one apolipoprotein E-4 allele (Stewart et al. 2002).

It has also been reported that both, biochemical and behavioral data implicate dopaminergic neurotransmitter systems in the neurotoxicity of Pb (Pokora et al. 1996). Reported effects are consistent with the hypothesis that Pb exposure, by some as yet undetermined mechanism, depletes dopamine (DA) availability. Pb exposure decreases DA turnover (Lasley et al. 1984), synaptosomal DA release (Minnema et al. 1986), and synaptic transmission in peripheral nerve (Cooper et al. 1984). It also impairs autoreceptor-mediated regulation of DA release, an effect accompanied by decreased levels of DA metabolites (Lasley, 1992). In other studies, chronic Pb exposure decreased levels of DA and metabolites in nucleus accumbens (Kala and Jadhav, 1995a) and, in microdialysis studies, decreased both basal and K^+-stimulated DA release in nucleus accumbens (Kala and Jadhav, 1995a; b). Pokora et al. (1996) examined the hypotheses that low level Pb exposure could increase DA binding sites, would do so preferentially in nucleus accumbens, and that such effects would be modified by concurrent DA agonist treatment in mice.

These authors conclude that chronic low-level Pb exposure alters DA system function, and by so doing alters its response to chronic DA agonist treatments, presenting substantial potential implications not only for DA-mediated neurodegenerative diseases such as PD, but also for the efficacy of DA-based therapeutic treatments. That Pb exposure alone, at very low levels in rodents (a species considerably less sensitive to Pb than humans), can decrease DA binding sites raises the possibility that the Pb burden sustained by some segment of the world's populations could be sufficient, in particular given its accumulation in the body over time, to eventually serve as a predisposing factor for neurodegenerative diseases that involve disruptions of DA system functions. Moreover, the therapeutic efficacy of chronic administration of DA agonists, such as L-dopa in the case of PD or methylphenidate in children with attention deficit disorder, could conceivably be altered by an elevated Pb burden, as indicated by the marked interactions between Pb exposure alone and chronic DA agonist treatments seen in the Pokora et al. (1996) study.

In this way Gorell et al. (1997) and Coon et al. (2006) demonstrated a 2-fold increase in the risk of PD among workers with > 20 years of occupational exposure to Pb. Associations of such chronic occupational exposure to combinations of Pb–iron and Pb–copper were even more robust. These studies support the hypothesis that Pb plays a role in the etiology of PD in exposed individuals. Although the biochemical mechanism of Pb neurotoxicity is not completely understood, a growing body of evidence suggests that metal cations of Pb, iron, and aluminum stimulate free radical formation, which results in neurodegeneration via peroxidative damage to the cell membrane. Sandhir et al. (1994) observed that increasing Pb concentration in rat brain produces heightened levels of lipid peroxidation and decreased activity of neuroprotective antioxidant enzymes and acetylcholinesterase. The authors suggested that lipid peroxidation eventually could lead to neuronal cell death through deterioration of the cell membrane. Uversky et al. (2001), using *in vitro* models of human brain cells, found that increasing levels of heavy metal cations stimulate the conformational changes that can lead to fibrillation of recombinant α-synuclein. The authors argued that the aggregation and fibrillation of α-synuclein provoked by the presence of heavy metal cations could directly cause the intracellular protein inclusions that are observed in the SN of PD patients. Quinlan et al. (1988) observed that although Pb ions alone did not induce peroxidation, they did accelerate the rate of peroxidation caused by iron ions. The study of Coon et al. (2006) provides additional objective evidence to support the hypothesis that long-term exposure to heavy metals, such as Pb, contributes to the accumulation of peroxidative damage and neurodegenerative cell death that is observed in PD.

Early exposure to Pb has also been associated with dysfunction of the hippocampus, an area of the brain important in memory function (Petit et al. 1983), and also with the subsequent adult appearance of neurofibrillary tangles (Nicklowitz and Mandybur, 1975).

The sporadic nature of AD argues for an environmental link that may drive AD pathogenesis; however, the triggering factors and the period of their action are unknown. Recent studies in rodents have shown that exposure to Pb during brain development predetermined the expression and regulation of the amyloid precursor protein (APP) and its amyloidogenic β-amyloid (Aβ) product in old age. Wu et al. (2008) reported that the expression of AD-related genes [APP, BACE1 (β-site APP-cleaving enzyme 1)] as well as their transcriptional regulator (Sp1) was elevated in aged (23-year-old) monkeys exposed to Pb as infants. Furthermore, developmental exposure to Pb altered the levels, characteristics, and intracellular distribution of Aβ staining and amyloid plaques in the frontal association cortex. These effects were accompanied by a decrease in DNA-methyltransferase activity and higher levels of oxidative damage to DNA, indicating that epigenetic imprinting in early life influenced the expression of AD-related genes and promoted DNA damage and pathogenesis. These data suggest that AD pathogenesis is influenced by early life Pb exposure and argue for both an environmental trigger and a developmental origin of AD.

Furthermore, the pro-oxidant effects of heavy metals such as Pb can exacerbate the age-related increase in oxidative stress that is related to the decline of the antioxidant defense systems. Brain inflammatory reactions also generate oxidative stress. Chronic inflammation can contribute to the formation of the senile plaques that are typical for AD (Monnet-Tschudi et al. 2006). In agreement with this view, nonsteroidal anti-inflammatory drugs and antioxidants suppress early pathogenic processes leading to AD, thus decreasing the risk of developing the disease. The effects of Pb were also tested in aggregating brain-cell cultures of fetal rat telencephalon, a three-dimensional brain-cell culture system. The continuous application for 10 to 50 days of non-cytotoxic concentrations of Pb resulted in their accumulation in brain cells and the occurrence of delayed toxic effects. When applied at non-toxic concentrations, Pb becomes neurotoxic under pro-oxidant conditions. Furthermore, this metal induces glial cell reactivity, a hallmark of brain inflammation. Pb increases the expression of the APP; and stimulates the formation of insoluble Aβ, which plays a crucial role in the pathogenesis of AD and causes oxidative stress and neurotoxicity *in vitro* (Monnet-Tschudi et al. 2006). A considerable body of evidence suggests that the heavy metals such as Pb contribute to the etiology of neurodegenerative diseases and emphasizes the importance of taking preventive measures in this regard (Basha et al. 2005; Monnet-Tschudi et al. 2006).

Basha et al. (2005) concluded that the long latency of AD suggests that this neurodegenerative disease remains asymptomatic for decades before a progressive accumulation of damage becomes clinically detectable. Their findings suggest that the initial events that trigger this disease begin very early in life and may be worsened by re-exposure to environmental agents late in life. Therefore, they propose that amyloidogenesis is promoted by a latent response to developmental reprogramming of the expression of the APP gene by early exposure to Pb, as well as enhancement of Aβ aggregation in old age. In rodents, these events occur without Pb-induced disturbances to the enzymatic processing of APP. The above-described findings provide further evidence for the developmental basis of amyloidogenesis and late-life disturbances in AD-associated proteins by environmental agents.

Finally, the etiology of amyotrophic lateral sclerosis (ALS) likely involves an environmental component; Kamel et al. (2005; 2008) assessed literature on ALS and Pb exposure, they found that ALS was associated with self-reported occupational Pb exposure, with a dose response for cumulative days of exposure. ALS was also associated with blood and bone Pb levels, with a 1.9-fold increase in risk for each μg/dL increment in blood Pb and a 2.3- to 3.6-fold increase for each doubling of bone Pb. A polymorphism in the delta-aminolevulinic acid dehydratase gene was associated with a 1.9-fold increase in ALS risk. These results, together with previous studies, suggest that Pb exposure plays a role in the etiology of ALS. An increase in mobilization of Pb from bone into blood may play a role in the acute onset of disease.

Moreover, Kamel et al. (2008) found that the variant allele (ALAD 2) of the polymorphism denoted as ALAD K59N was positively associated with an approximate twofold increase in risk of ALS after adjustment for age, sex, region, education, and physical activity. In the course of their analysis, they also identified a previously unknown polymorphism, denoted as ALAD IVS2+299G>A. The researchers theorize that although ALAD alleles did not modify the relationship of ALS to Pb in this cross-sectional study, genetic susceptibility conferred by these polymorphisms might still affect risk through a mechanism related to internal Pb exposure. ALAD 2 appears to promote retention of Pb in blood and migration of Pb from bone to blood. The current findings are consistent with the hypothesis that this increased retention of Pb in blood relative to bone increases its availability to target tissues and hence its toxicity. The authors speculate that alterations in Pb toxicokinetics conferred by the presence of the ALAD 2 allele may subtly increase exposure to Pb throughout a person's lifetime, thereby elevating risk. They conclude that because the study is small and the observation unique, further research is necessary to confirm or refute the hypothesis. Considering that the frequency of the ALAD 2 allele is approximately 10% in Caucasian populations, if this study's conclusions are confirmed, it will be an important contribution to identifying a large number of people who could be at elevated risk for developing a devastating, incurable disease (Hood, 2003).

Final Considerations

The exposition to Pb have been shown to interfere with a large amount of intracellular targets, thereby contributing to several pathogenic processes typical of neurodegenerative disorders, including oxidative stress, deregulation of protein turnover, mitochondrial dysfunction, and brain inflammation. Exposure to Pb early in development can predispose the brain for developing a neurodegenerative disease later in life. Alternatively, heavy metals can exert their harmful effects through acute neurotoxicity or through slow accumulation during prolonged periods of life.

The world population continues to sustain lifetime exposures to Pb due to residual contamination of dust, soil, food, and water supplies from the many years of use of Pb-based paints and gasoline. It is expected that 1 of 6 children in the world population still has Pb-Blood levels above the currently level of concern. Clearly, additional efforts are warranted for fully understanding the basis of Pb-induced changes in CNS function per se as well as its interactions with neurodegenerative disorders.

References

Acharya, S. and Acharya, U.R. (1997) In vivo lipid peroxidation responses of tissues in lead-treated Swiss mice. *Ind Health*, 35: 542–544.

Alkondon, M., Costa, A.C., Radhakrishnan, V., Aronstam, R.S. and Albuquerque, E.X. (1990) Selective blockade of NMDA-activated channel currents may be implicated in learning deficits caused by lead. *FEBS Lett.* 261:124–30.

Al tadjir, G., Starr, M.S. and Starr, B.S. (1990) Procon vulsan effect of SKF 38393 mediated by migral D1 receptor. *Eur J Pharmacol.* 182: 45–51.

Antonio, M.T. and Leret, M.L. (2000) Study of the neurochemical alterations produced in discrete brain areas by perinatal low-level lead exposure. *Life Sciences*, 67:635-642.

Arancio, O., Kiebler, M. and Lee, C.J. (1996) Nitric oxide directly in the presynaptic neuron to produce long-term potentiation in cultured hippocampal neurons. *Cell*, 87:1025–35.

ATSDR (Agency for Toxic Substances and Disease Registry). (2007) Toxicological Profile for Lead. US Department of Health and Human Services, Washington DC, US Public Health Service, US Department of Health and Human Services.

Balbus-Kornfeld, J.M., Stewart, W., Bolla, K.I. and Schwartz, B.S. (1995) Cumulative exposure to inorganic lead and neurobehavioral test performance in adults: an epidemiological review. *Occup Environ Med.* 52:2-12.

Baker, E.L., Feldman, R.G., White, R.F. and Harley, J.P. (1983) The role of occupational lead exposure in the genesis of psychiatric and behavioral disturbances. *Acta Psychiatr Scand Suppl.* 67:38-48.

Basha, M.R., Murali, M., Siddiqi, H.K., Ghosal, K., Siddiqi, O.K., Lashuel, H.A., Ge, Y.W., Lahiri, D.K. and Zawia, N.H. (2005) Lead (Pb) exposure and its effect on APP proteolysis and Abeta aggregation. *FASEB J.* 19:2083-2084.

Bellinger, D. (1995) Lead and neuropsychological function in children: progress and problems in establishing brain– behavior relationships. *Adv Child Neuropsychol.* 3: 12–45.

Billick, I.H. and Gray, V.E. (1978) Lead based paint poisoning research: Review and evaluation 1971-1977. Washington, DC: U.S. Department of Housing and Urban Development.

Bornschein, R.L., Succop, P.A., Krafft, K.M., Clark, C.S., Peace, B. and Hammond, P.B. (1986) Exterior surface dust lead, interior house dust lead and childhood lead exposure in an urban environment. In: Hemphil DD, ed. Trace substances in environmental health. Vol. 20. Columbia, MO: University of Missouri 322-332.

Boykin, M.J., Chetty, C.S. and Rajanna, B. (1991) Effects of lead on kinetics of 3H-dopamine uptake by rat brain synaptosomes. *Ecotoxicol Environ Saf.* 22: 88-93.

Bradbury, M.W. and Deane, R. (1993) Permeability of the blood–brain barrier to lead. *Neurotoxicology*, 14: 131–136.

Braga, M.F., Pereira, E.F. and Albuquerque, E.X. (1999a) Nanomolar concentrations of lead inhibit glutamatergic and GABAergic transmission in hippocampal neurons. *Brain Res.* 826: 22– 34.

Braga, M.F., Pereira, E.F., Marchioro, M. and Albuquerque, E.X. (1999b) Lead increases tetrodotoxin-insensitive spontaneous release of glutamate and GABA from hippocampal neurons. *Brain Res.* 826: 10– 21.

Bressler, J.P. and Goldstein, G.W. (1991) Mechanisms of lead neurotoxicity. *Biochem Pharmacol.* 41: 479–484.

Bressler, J., Kim, K.A., Chakraborti, T. and Goldstein, G. (1999) Molecular mechanisms of lead neurotoxicity. *Neurochem Res.* 24:595–600.

Brodsky, J.L. and Guidotti, G. (1990). Sodium affinity of brain Na(+)-K(+)-ATPase is dependent on isozyme and environment of the pump. *Am Physiol Soc.* 258: C803-C811.

Byers, E.R. and Lord, E.E. (1943) Late effects of lead poisoning on mental development. *J Dis Child*, 66: 471– 494.

Canfield, R.L., Henderson, C.R., Cory-Slechta, D.A., Cox, C., Jusko, T.A. and Lanphear, B.P. (2003) Intellectual impairment in children with blood lead concentrations below 10 microg per deciliter. *N Engl J Med.* 348: 1517–1526.

Capar, S.G. and Cunningham, W.C. (2000) Element and radionuclide concentrations in food: FDA total diet study 1991-1996. *J AOAC Int.* 83:157-177.

Carfagna, M.A., Ponsler, G.D. and Muhoberac, B.B. (1996) Inhibition of ATPase activity in rat synaptic plasmamembranes by simultaneous exposure to metals. *Chem Biol Int.* 100: 53-65.

Chisolm, J.J. (1986) Removal of lead paint from old housing: The need for a new approach. *Am J Public Health.* 76:236-237.

Chisolm, J.J. (2001). Evolution of the management and prevention of childhood lead poisoning: dependence of advances in public health on technological advances in the determination of lead and related biochemical indicators of its toxicity. *Environ Res.* 86: 111 –121.

Coon, S., Stark, A., Peterson, E., Gloi, A., Kortsha, G., Pounds, J., Chettle, D. and Gorell, J. (2006). Whole-body lifetime occupational lead exposure and risk of Parkinson's disease. *Environ Health Perspect.* 114: 1872-1876.

Cooper, G.P., Suszkiw, J.B. and Manalis, R.S. (1984) Heavy metals: effects on synaptic transmission . *Neurotoxicology*, 5: 247-266.

Correa, M., Roig-Navarro, A. and Aragon, C. (2004) Motor behavior and brain enzymatic changes after acute lead intoxication on different strains of mice. *Life Sciences*, 74: 2009–2021.

De la Fuente-Fernandez, R. and Calne, D. (2002) Evidence for environmental causation of Parkinson's disease. *Parkinsonism Relat Disord.* 8:235–241.

Dumas, P., Gueldry, D., Loireau, A., Chomard, P., Buthieau, A.M. and Autissie,r N. (1985) Effects of lead poisoning on properties of brain mitochondria in young rats. *C R Soc Biol.* 179:175-183.

EPA. (1982) Standards of performance for lead-acid battery manufacturing plants. U.S. Environmental Protection Agency. Code of Federal Regulations. 40 CFR 60. Subpart KK.

EPA. (1986) Air quality criteria for lead. Research Triangle Park, NC: U.S. Environmental Protection Agency, Office of Research and Development, Office of Health and Environmental Assessment, Environmental Criteria and Assessment Office. EPA600883028F.

EPA. (2003) National air quality and emissions trends report. 2003 Special studies edition. Research Triangle Park, NC: U.S. Environmental Protection Agency. EPA454R03005.

Fitsanakis, A.V. and Aschner, M. (2005) The importance of glutamate, glycine, and g-aminobutyric acid transport and regulation in manganese, mercury and lead neurotoxicity. *Toxicol App Pharmacol.* 204: 343– 35.

Fischbein, A. (1998) Occupational and environmental exposure to lead. In: Rom W, editor. Environmental and occupational medicine. 3rd ed. Philadelphia: Lippincott-Raven, p 973-996.

Foster, H. (2002) Why the preeminent risk factor in sporadic Alzheimer's disease cannot be genetic. *Med Hypoth.* 59: 57–61.

Fox, D.A., Rubinstein, S.D. and Hsu, P. (1991) Developmental lead exposure inhibits adult rat retinal, but not kidney, Na^+,K^+-ATPase. *Toxicol Appl Pharmacol.* 109: 482-493.

García-Arenas, G., Ramírez-Amaya, V., Balderas, I., Sandoval, J., Escobar, M., Ríos, C. and Bermúdez-Rattoni, F. (2004) Cognitive deficits in adult rats by lead intoxication are related with regional specific inhibition of cNOS. *Behav Brain Res.* 149: 49–59.

Gedeon, Y., Ramesh, G.T., Wellman, P.J. and Jadhav, A.L. (2001) Changes in mesocorticolimbic dopamine and D1/D2 receptor levels after low level lead exposure: a time course study. *Toxicol Lett.* 2–3: 217–226.

Gerbi, A., Debray, M., Maixent, J.M., Chanez, C. and Bourre, J.M.J. (1993) Effect of dietary -linolenic acid on functional characteristic of Na,K-ATPase isoenzymes in whole brain membrane of weaned rats. *Biochim Biophys Acta.* 1165:291-298.

Gibson, J., Love, W., Hardie, D., Bancroft, P. and Turner, A. (1892). Notes on lead poisoning as observed among children in Brisbane, Intercolonial Medical Congress of Australia, Third Session.

Gorell, J.M., Johnson, C.C., Rybicki, B.A., Peterson, E.L., Kortsha, G.X., Brown, G.G. and Richardson, R.J. (1997) Occupational exposures to metals as risk factors for Parkinson's disease. *Neurology*, 48:650–658.

Grandjean, P., Weihe, P., White, R.F., Debes, F., Araki, S., Yokoyama, K., Murata, K., Sørensen, N., Dahl, R. and Jørgensen, P.J. (1997) Cognitive deficit in 7-year old children with pre-natal exposure to methylmercury. *Neurotoxicol Teratol.* 19:417–428.

Hanninen, H., Aitio, A., Kovala, T., Luukkonen, R., Matikainen, E., Mannelin, T., ErkkilaÈ, J. and RiihimaÈki, V. (1998) Occupational exposure to lead and neuropsychological dysfunction. *Occup Environ Med.* 55:202-209.

Haycock, J.W. (1993) Multiple signaling pathways in bovine chromaffin cells regulate tyrosine hydroxylase phosphorylation at Serl9, Ser31, and Ser40. *Neurochem Res.* 18: 15-26.

Hechtenberg, S. and Beyersmann, D. (1991) Inhibition of sarcoplasmic reticulum Ca(21)-ATPase activity by cadmium, lead and mercury. *Enzyme*. 45:109 115.

Hofmann, F., Biel, M. and Flockerzi, V. (1994) Molecular basis for Ca2+ channel diversity. *Annu Rev Neurosci.* 17: 399-418.

Holscher, C. (1997) Nitric oxide, the enigmatic neuronal messenger: its role in synaptic plasticity. *Trends Neurosci*. 20:298–303.

Holtzman, D., DeVries, C., Nguyen, H., Olson, J. and Bensch, K. (1984) Maturation of resistance to lead encephalopathy: Cellular and subcellular mechanisms. *Neurotoxicology*, 5: 97–124.

Hood, E. (2003) Toxic Oil Timeline. Diagnosing Effects Decades Later. *Environ Health Persp*. 111: 538-539.

Hundle, B., McMahon, T., Dadgar, J. and Messing, R.O. (1995) Overexpression of epsilon-protein kinase C enhances nerve growth factor-induced phosphorylation of mitogen-activated protein kinases and neurite outgrowth. *J Biol Chem.* 270: 30134-30140.

Järup, L. (2003) Hazards of heavy metal contamination. *British Med Bull*, 68:167-182.

Kala, S.V. and Jadhav, A.L. (1995a) Region-specific alterations in dopamine and serotonin metabolism in brains of rats exposed to low levels of lead. *Neurotoxicology*, 16:297-308.

Kala, S.V. and Jadhav, A.L. (1995b) Low level lead exposure decreases in vivo release of dopamine in the rat nucleus accumbens : a microdialysis study. *J Neurochem.* 65: 1631-1635.

Kamel, F., Umbach, D.M., Hu, H., Munsat, T.L., Shefner, J.M., Taylor, J.A. and Sandler, D.P. (2005) Lead exposure as a risk factor for amyotrophic lateral sclerosis. *Neurodegener Dis.* 2: 195-201.

Kamel, F., Umbach, D.M., Stallone, L., Richards, M., Hu, H. and Sandler, D.P. (2008) Association of lead exposure with survival in amyotrophic lateral sclerosis. *Environ Health Perspect*. 116: 943-947.

Kennedy, J.L., Farrer, L.A., Andreason, N.C., Mayeux, R. and St. George-Hyslop, P. (2003) The genetics of adult-onset neuropsychiatric disease: complexities and conundra? *Science*. 302:822–826.

Kerper, L.E. and Hinkle, P.M. (1997a) Lead uptake in brain capillary endothelial cells: activation by calcium store depletion. *Toxicol Appl Pharmacol.* 146: 127–33.

Kerper, L.E. and Hinkle, P.M. (1997b) Cellular uptake of lead is activated by depletion of intracellular calcium stores. *J Biol Chem*. 272: 8346–52.

Kozutsumi, Y., Segal, M., Normington, K., Gething, M.J. and Sambrook, J. (1988) The presence of malfolded proteins in the endoplasmic reticulum signals the induction of glucose-regulated proteins. *Nature*, 332: 462–464.

Kumar Bokara, K., Brown, E., McCormick, R., Rao Yallapragada, P., Rajanna, S. and Bettaiya, R. (2007) Lead-induced increase in antioxidant enzymes and lipid peroxidation products in developing rat brain. *BioMetals*, 21: 9-16.

Langston, W., Forno, L.S., Tetrud, J., Reeves, A.G., Kaplan, J.A. and Karluk, D. (1999) Evidence of active nerve cell degeneration in the substantia nigra of humans years after 1-methyl-4-phenyl-1,2,3,6-tetrahydropyridine exposure. *J Ann Neurol.* 46:598–605.

Landrigan, P.J., Silbergeld, E., Froines, J.R. and Pfeffer, R.M. (1990) Lead in the modern workplace (Editorial). *Am J Public Health*, 80:907-908.

Landrigan, P.J., Sonawane, B., Butler, R.N., Trasande, L., Callan, R. and Droller, D. (2005) Early environmental origins of neurodegenerative disease in later life. *Environ Health Perspect.* 113: 1230-1233.

Lanphear, B.P., Dietrich, K., Auinger, P. and Cox, C. (2000) Cognitive deficits associated with blood lead concentrations <10 microg/dL in US children and adolescents. *Public Health Rep.* 115: 521– 529.

Lasley, S.M. (1992) Regulation of dopaminergic activity, but not tyrosine hydroxylase, is diminished after chronic inorganic lead exposure. *Neurotoxicology*, 13: 625-636.

Lasley, S.M., Greenland, R.D., Minnema, D.J., and Michaelson, A. (1984) Influence of chronic inorganic lead exposure on regional dopamine and 5-hydroxytryptamine turnover in rat brain. *Neurochem Res.* 9: 1675-1688.

Legare, M.E., Barhoumi, R., Hebert, E., Bratton, G.R., Burghardt, R.C. and Tiffany-Castiglioni, E. (1998) Analysis of Pb21 entry into cultured astroglia. *Toxicol Sci.* 46: 90–100.

Levin, S.M. and Goldberg, M. (2000). Clinical Evaluation and Management of Lead-Exposed Construction Workers. *Am J Ind Med.* 37:23-43.

Lidsky, T.I. and Schneider, J.S. (2003) Lead neurotoxicity in children: basic mechanisms and clinical correlates. *Brain*, 126: 5-19.

Lilis, R., Fischbein, A., Eisinger J, Blumberg, W.E., Diamond, S., Anderson, H.A., Rom, W., Rice, C., Sarkozi, L., Kon, S. and Selikoff, I.J. (1977) Prevalence of lead disease among secondary lead smelter workers and biological indicators of lead exposures. *Environ Res.* 14:255-285.

Lindahl, L.S., Bird, L., Legare, M.E., Mikeska, G., Bratton, G.R. and Tiffany-Castiglioni, E. (1999) Differential ability of astroglia and neuronal cells to accumulate lead: Dependence on cell type and on degree of differentiation. *Toxicol Sci.* 50: 236–243.

Ma, T., Chen, H.H. and Ho, I.K. (1999) Effects of chronic lead (Pb) exposure on neurobehavioral function and dopaminergic neurotransmitter receptors in rats. *Toxicol Lett.* 105:111–121.

Mas-Oliva, J. (1989) Effect of lead on the erythrocyte (Ca21,Mg21)-ATPase activity: Calmodulin involvement. *Mol Cell Biochem.* 89:87–93.

McDonald, M.E. (1985) Acid deposition and drinking water. *Environ Sci Technol.* 19:772-776.

Mendolesi, J. and Pozzan, T. (1998) The endoplasmic reticulum Ca21 store: A view from the lumen. *Trends Biochem Sci.* 23:10–14.

Memo, M., Lucchi, L., Spano, P.F. and Trabucchi, M. (1980) Effect of chronic lead treatment on GABA-ergic receptor function in rat brain. *Arch Toxicol.* 46: 249– 256.

Minnema, D.J., Cooper, G.P. and Schamer, M.M. (1991). Differential effects of triethyllead on synaptosomal [3H]dopamine vs. [3H]acetylcholine and [3H]gamma-aminobutyric acid release. *Neurotoxicol Teratol.* 13: 257–265.

Minnema, D.J., Greenland, R.D. and Michaelson, I.A. (1986) Effect of in vitro inorganic lead on dopamine release from superfused rat striatal synaptosomes. *Toxicol Appl Pharmacol.* 84: 400-411 .

Monnet-Tschudi, F., Zurich, M.G., Boschat, C., Corbaz, A. and Honegger, P. (2006) Involvement of environmental mercury and lead in the etiology of neurodegenerative diseases. *Rev Environ Health*, 21: 105-17.

Mortada, W.I., Sobh, M.A., El-Defrawy, M.M. and Farahat, S.E. (2001) Study of lead exposure from automobile exhaust as a risk for nephrotoxicity among traffic policemen. *Am J Nephrol*, 21: 274–279.

Muldoon, S.B., Cauley, J.A., Kuller, L.H., Morrow, L., Needleman, H.L., Scott, J. and Hooper, F.J. (1996) Effects of blood lead levels on cognitive function of older women. *Neuroepidemiology*, 15:62-67.

National Institute of Neurological Disorders and Stroke. (2004) Parkinson's Disease Backgrounder. Bethesda, MD:National Institute of Neurological Disorders and Stroke. Available:http://www.ninds.nih.gov/health_and_medical/pubs/parkinson's_disease_backgrounder.htm.

Needleman, H.L., Gunnoe, C., Leviton, A., Reed, R., Peresie, H., Maher, C. and Barrett, P. (1979) Deficits in psychologic and classroom performance of children with elevated dentine lead levels. *N Engl J Med.* 300: 689– 695.

Needleman, H.L, Schell, A., Bellinger, D., Leviton, A. and Allred, E.N. (1990) The long-term effects of exposure to low doses of lead in childhood. An 11-year follow-up report. *N Engl J Med.* 322: 83–88.

Nicklowitz, W.J. and Mandybur, T.I. (1975) Neurofibrillary changes following childhood lead encephalopathy. *J Neuropathol Exp Neurol.* 34: 445-455.

NourEddine, D., Miloud, S. and Abdelkader, A. (2005) Effect of lead exposure on dopaminergic transmission in the rat brain. *Toxicology*, 207:363-368.

Osmond, C. and Barker, D. (2000) Fetal, infant, and childhood growth are predictors of coronary heart disease, diabetes, and hypertension in adult men and women. *Environ Health Perspect.* 108: 545–553.

Patel, A.J., Hunt, A., Jacquesberg, W., Kiss, J. and Rodriguez, J. (1995) Effects of protein kinase C modulation on NMDA receptor mediated regulation of neurotransmitter enzyme and c-fos protein in cultured neurons. *Neurochem Res.* 20: 561-569.

Petit, T.L., Alfano, D.P. and LeBoutillier, J.C. (1983) Early lead exposure and the hippocampus: a review and recent advances. *Neurotoxicology*, 4: 79-94.

Pokora, M.J., Richfield, E.K. and Cory-Slechta, D.A. (1996) Preferential Vulnerability of Nucleus Accumbens Dopamine Binding Sites to Low-Level Lead Exposure: Time Course of Effects and Interactions with Chronic Dopamine Agonist Treatments. *J Neurochem.* 67: 1540-1550.

Qian, Y., Harris, E.D., Zheng, Y. and Tiffany-Castiglioni, E. (2000) Lead targets GRP78, a molecular chaperone, in C6 rat glioma cells. *Toxicol Appl Pharmacol.* 163: 260–266.

Qian, Y. and Tiffany-Castiglioni, E. (2003) Lead-induced endoplasmic reticulum (ER) stress responses in the nervous system. *Neurochem Res.* 28: 153-62.

Quinlan, G.J., Halliwell, B., Moorhouse, C.P. and Gutteridge, J.M. (1988) Action of lead(II) and aluminium (III) ions on iron-stimulated lipid peroxidation in liposomes, erythrocytes and rat liver microsomal fractions. *Biochim Biophys Acta.* 962:196–200.

Ramakers, G.M.J., Degraan, P.N.E., Urban, I.J.A., Kraay D, Tang T., Pasinelli, P., Oestreicher, A.B. and Gispen, W.H. (1995) Temporal differences in the phosphorylation state of pre-and postsynaptic protein kinase C substrates B-50/GAP-43 and neurogranin during long term potentiation. *J Biol Chem.* 270: 13892-13898.

Ramin, S.M., Kedzierski, W. and Porter, J.C. (1993) Action of catecholamine secretory processes of fetal hypothalamic and adrenal cells. *Mol Cell Neurosci.* 4: 449-454.

Ruff, H.A. and Bijur, P.E. (1989) The effects of low to moderate lead levels on neurobehavioral functioning in children: toward a conceptual model. *J Dev Behav Pediatr.* 10: 103– 109.

Sandhir, R., Julka, D. and Gill, K.D. (1994) Lipoperoxidative damage on lead exposure in rat brain and its implications on membrane bound enzymes. *Pharmacol Toxicol.* 74:66–71.

Sawashita, J., Takeda, A. and Okada, S. (1997). Change of zinc distribution in rat brain with increasing age. *Brain Res Dev Brain Res.* 102:295–298.

Schäfer, S.G., Dawes, R.L.F., Elsenhans, B., Forth, W. and Schomann, K. (1999) Metals. In: Toxicology. Hans Marquardt, Siegfried G. Schäfer, Roger O. McClellan, and Frank Welsch (Eds). Academic Press, San Diego, Ca. Pp. 755-804.

Schwartz, B.S., Stewart WF, Bolla, K.I., Simon, P.D., Bandeen-Roche, K., Gordon, P.B., Links, J.M. and Todd, A.C. (2000) Past adult lead exposure is associated with longitudinal decline in cognitive function. *Neurology.* 55: 1144–1150.

Siderowf, A. and Stern, M. (2003) Update on Parkinson Disease. *Ann Intern Med.* 138:651–658.

Simons, T.J.B. (1993) Lead transport and binding by human erythrocytes in vitro. *Eur J Physiol.* 423: 307–313.

Somashekaraiah, B.V., Padmaja, K. and Prasad, R.K. (1992) Lead induced lipid peroxidation and antioxidant defense components of developing chick embryos. *Free Rad Biol Med.* 13: 107–114.

Stewart, W.F., Schwartz, B.S., Simon, D., Bolla, K.I., Todd, A.C. and Links, J. (1999) Neurobehavioral function and tibial and chelatable lead levels in 543 former organo-lead workers. *Neurology.* 52: 1610–1617.

Stewart, W.F., Schwartz, B.S., Simon, D., Kelsey, K. and Todd, A.C. (2002) ApoE genotype, past adult lead exposure, and neurobehavioral function. *Environ Health Perspect.* 110:501–505.

Stollery, B.T., Banks, H.A., Broadbent, D.E. and Lee, W.R. (1989) Cognitive functioning in lead workers. *Br J Ind Med.* 46:698-707.

Struzynska, L. and Sulkowski, G. (2004) Relationships between glutamine, glutamate, and GABA in nerve endings under Pb-toxicity conditions. *J Inorg Biochem.* 98: 951– 958.

Struzynska, L., Chalimoniuk, M. and Sulkowski, G. (2005) The role of astroglia in Pb-exposed adult rat brain with respect to glutamate toxicity. *Toxicology*, 212: 185–194.

Sulkowski, G., Dabrowska-Bouta, B., Waskiewicz, J. and Rafalowska, U. (1999) Inhibition of dopamine tansport and binding of (3H) spiperone to dopamine D2 receptor by acute Pb-toxicity in vivo. *Folia Neuropathol.* 37: 205–209.

Tanner, C.M., Ottman, R., Goldman, S.M., Ellenberg, J., Chan, P., Mayeux, R. and Langston, J.W. (1999) Parkinson disease in twins: an etiologic study. *JAMA*. 281:341–346.

Thiruchelvam, M., Brockel, B.J., Richfield, E.K., Baggs, R.B. and Cory-Slechta, D.A. (2000a) Potentiated and preferential effects of combined paraquat and maneb on nigrostriatal dopamine systems: environmental risk factors for Parkinson's disease? *Brain Res.* 873:225–234.

Thiruchelvam, M., Richfield, E.K., Baggs, R.B., Tank, A.W. and Cory-Slechta, D.A. (2000b) The nigrostriatal dopaminergic system as a preferential target of repeated exposures to combined paraquat and maneb: implications for Parkinson's disease. *J Neurosci.* 20:9207–9214.

Toscano, C.D., Hashemzadeh-Gargari, H., McGlothan, J.L. and Guilarte, T.R. (2002) Developmental Pb^{2+} exposure alters NMDAR subtypes and reduces CREB phosphorylation in the rat brain. *Brain Res Dev Brian Res.* 139: 217– 226.

Toscano, C.D., McGlothan, J.L. and Guilarte, T.R. (2003) Lead exposure alters cyclic-AMP response element binding protein phosphorylation and binding activity in the developing brain. *Dev Brain Res.* 145: 219–228.

TRI04. (2006). TRI explorer: Providing access to EPA's toxics release inventory data. Washington, DC: Office of Information Analysis and Access. Office of Environmental Information. U.S. Environmental Protection Agency. Toxics Release Inventory. http://www.epa.gov/triexplorer/. October 23, 2006.

Tursky, L., Calvalheiro, E.A. and Bortolotto, Z.A. (1998) Dopamine sensitive anticonvulsion site in the rat stratium. *J Neurosci.* 8:4027–4031.

Uteshev, V., Busselberg, D. and Hass, H.L. (1993) Pb^{2+} modulates the NMDA–receptor-channel complex. *Arch Pharmacol.* 347:209–13.

Uversky, V.N., Li, J. and Fink, A.L. (2001) Metal-triggered structural transformations, aggregation, and fibrillation of human alphasynuclein. A possible molecular NK between Parkinson's disease and heavy metal exposure. *J Biol Chem*. 276:44284–44296.

Valenzuela, A., Lefauconnier, J., Chaudiere, J. and Bourre, J. (1989) Effects of lead acetate on cerebral glutathione peroxidase and catalase in the suckling rat. *Neurotoxicology*. 10: 63–70.

Villeda-Hernandez, J., Mendez Armenta, M., Barroso-Moguel, R., Trejo-Solis, M.C., Guevara, J. and Rios, C. (2006). Morphometric analysis of brain lesions in rat fetuses prenatally exposed to low-level lead acetate: correlation with lipid peroxidation. *Histol Histopathol.* 21: 609–617.

Waskiewicz, J. (1996) Alterations of GABA-B binding caused by acute and chronic lead administration. *Acta Neurobiol Exp*. 56: 227–231.

Wiemann, M., Schirrmacher, K. and Büsselberg, D. (1999) Interference of lead with the calcium release activated calcium flux of osteoblast-like cells. *Calcif Tissue Int*. 65:479–485.

WHO. (1995) Lead. Environmental Health Criteria, vol. 165. Geneva: World Health Organization, Technical Report Series, Geneva: WHO.

Wooden, S.K., Li, L.J., Navarro, D., Qadri, I., Pereira, L. and Lee, A.S. (1991) Transactivation of the grp78 promoter by malfolded proteins, glycosylation block, and calcium ionophore is mediated through a proximal region containing a CCAAT motif which interacts with CTF/NF-I. *Mol Cell Biol.* 11: 5612–5623.

Wu, J., Basha, M.R., Brock, B., Cox, D.P., Cardozo-Pelaez, F., McPherson, C.A., Harry, J., Rice, D.C., Maloney, B., Chen, D., Lahiri, D.K. and Zawia, N.H. (2008) Alzheimer's disease (AD)-like pathology in aged monkeys after infantile exposure to environmental metal lead (Pb): evidence for a developmental origin and environmental link for AD. *J Neurosci.* 28: 3-9.

Zawia, N.H. (2003) Transcriptional involvement in neurotoxicity. *Toxicol Appl Pharmacol.* 306: 37–48.

Zheng, W., Aschner, M. and Ghersi-Egea, J.F. (2003) Brain barrier systems: a new frontier in metal neurotoxicological research. *Toxicol Appl Pharmacol.* 192: 1-11.

In: Progress in Neurodegeneration
Editors: M.R. Avila-Costa and V.A. Martinez

ISBN 978-1-60741-317-2

Chapter 10

Iron

Enrique Montiel-Flores, José Luis Ordoñez-Librado, Ana Luisa Gutierrez-Valdez and Maria Rosa Avila-Costa
Department of Neuroscience, Neuromorphology Lab. UNAM, Mexico

10.1. General Description

Iron (Fe) is the 26th element of the Periodic table. The electronic structure of Fe and its capacity to drive one-electron reactions predetermines Fe as a major component in the production and metabolism of free radicals in biological systems. In the organism Fe is commonly found in three oxidation states: Fe(II), Fe(III), and to a much lesser extent Fe(IV). At physiological pH, while Fe(III) precipitates as oxyhydroxide polymers, Fe(II) is soluble. On the other hand, Fe(II) is unstable in aqueous media and tends to react with molecular oxygen to form Fe(III) and superoxide (Valko et al. 2005).

Fe is an essential element required for growth and survival of almost every organism. The average-weight human contains approximately 4-5 g of Fe. Fe deficiency is a widely spread condition that affects approximately 500 million people around the world. The consequences of Fe deficiency can range from anemia to mental retardation in growing children.

Fe is a component of several metalloproteins and plays a crucial role in vital biochemical activities, such as oxygen sensing and transport, electron transfer, and catalysis (Aisen et al. 2001). Fe is thus indispensable for life. The biological functions of Fe are based on its chemical properties, e.g., its capacity to form a variety of coordination complexes with organic ligands in a dynamic and flexible mode, and its favorable redox potential to switch between the ferrous, [Fe(II)], and ferric, [Fe(III)], states. The bioavailability of Fe is generally limited, because under aerobic conditions, Fe(II) is readily oxidized in solution to Fe(III), which is virtually insoluble at physiological pH [K_{free} Fe(III) = 10^{-18} M] (Papanikolaou and Pantopoulos, 2005).

Fe overload is a less frequent condition, and involves defects in Fe absorption, transport and secondary Fe disorders. Toxicity and chronic Fe toxicity is a status that can be associated with: (1) primary hemochromatosis, a genetic disorder related to increased intestinal absorption of Fe, (2) high dietary Fe intake and (3) frequent blood transfusions. Cases of acute Fe toxicity are relatively rare and mostly related to hepatoxicity (Cargnoni et al. 2002). A high content of Fe has been associated with several pathological conditions, including liver and heart disease (Rasmussen et al. 2001), cancer (Valko et al. 2001), neurodegenerative disorders (Berg et al. 2001), diabetes (Lao and Ho, 2004), hormonal (Fraga and Oteiza, 2002) and immune system abnormalities (Walker and Walker, 2000).

10.2. Sources of Iron Exposure

10.2.1. Air

In remote areas, Fe levels in air are about 50–90 ng/m^3; at urban sites, levels are about 1.3 μg/m^3. Concentrations up to 12 μg/m^3 have been reported in the vicinity of iron- and steel-producing plants (National Research Council, 1979).

10.2.2. Water

The median Fe concentration in rivers has been reported to be 0.7 mg/L. In anaerobic ground water where Fe is in the form of Fe(II), concentrations will usually be 0.5–10 mg/L, but concentrations up to 50 mg/L can sometimes be found (National Research Council, 1979). Concentrations of Fe in drinking-water are normally less than 0.3 mg/L but may be higher in countries where various Fe salts are used as coagulating agents in water-treatment plants and where cast Fe, steel, and galvanized iron pipes are used for water distribution (WHO, 1996).

10.2.3. Food

Fe occurs as a natural constituent in plants and animals. Liver, kidney, fish, and green vegetables contain 20–150 mg/kg, whereas red meats and egg yolks contain 10–20 mg/kg. Rice and many fruits and vegetables have low Fe contents (1–10 mg/kg) (WHO, 1996). Reported daily intakes of Fe in food — the major source of exposure — range from 10 to 14 mg (National Research Council, 1989). Drinking water containing 0.3 mg/L would contribute about 0.6 mg to the daily intake. Intake of Fe from air is about 25 μg/day in urban areas.

10.3. Absorption, Distribution, Metabolism, and Excretion

Fe is an essential metal for almost all living organisms due to its involvement in a large number of iron-containing enzymes and proteins, yet it is also toxic. The mechanisms involved in Fe absorption across the intestinal tract, its transport in serum and delivery to cells and Fe storage within cells is briefly reviewed (Crichton et al. 2002).

Dietary free Fe, on reduction from the Fe(III) to Fe(II) state on the luminal surface of the proximal small intestine, is transported into the enterocytes by the apical transporter DMT1 (also known as DCT1, Nramp2) (Schümann et al. 1999; Nemeth et al. 2004). Dietary heme Fe is taken up by a not yet discovered transporter and released from the heme molecule within the enterocyte. Fe may be stored within the enterocyte (in the Fe storage protein, ferritin) or transferred across the basolateral membrane to the plasma by the transport protein Ireg1 (McKie et al. 2000) (known also as ferroportin1 and MTP1). This latter process requires oxidation of Fe(II) to Fe(III) by hephaesti. Once Fe has entered the circulation, there are no significant physiologic mechanisms for Fe loss other than menstruation (Valko et al. 2005).

About 65% of Fe is bound to hemoglobin, 10% is a constituent of myoglobin, cytochromes, and iron-containing enzymes, and 25% is bound to the Fe storage proteins, ferritin and hemosiderin. Ferritin is a high-capacity and low-affinity storage protein with the storage capacity of about 4500 atoms of Fe per molecule. It is thought that only trace amounts of the metal remain free as non-chelated or loosely chelated Fe. To prevent the potential health disturbances produced by both Fe deficiency and Fe overload, mammals have evolved with numerous, integrated mechanisms regulating Fe metabolism (Valko et al. 2005).

Absorbed Fe is bound to circulating transferrin and passes initially through the portal system of the liver, which is the major site of Fe storage. Hepatocytes take up transferring-bound Fe via the classical transferrin receptor (TfR1) but likely in greater amounts by the recently identified homologous protein, TfR2 (Kawabata et al. 1999). The major site of Fe utilization is the bone marrow, where Fe is taken up via TfRs on erythrocyte precursors for use in heme synthesis (Figure 1).

The crypt cells obtain information about body Fe needs from two postulated regulators, the "stores regulator," which responds to body iron stores, and the "erythropoietic regulator," which responds to the body's requirement for erythropoiesis (Cavill, 1999). The capacity of the stores regulator to change Fe absorption is low relative to the erythropoietic regulator. Regardless of this, it plays an essential role in meeting increased Fe needs and in preventing excess (Valko et al. 2005).

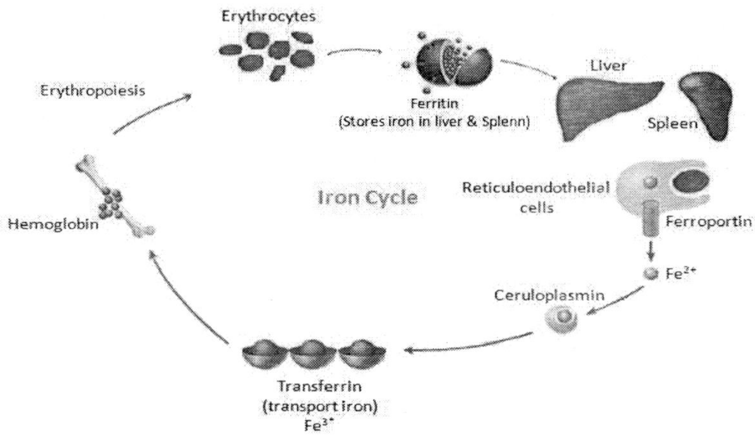

Modified from Madsen and Gitlin, 2007.

Figure 1. Iron cycle. Systemic iron cycle allows for rapid utilization of iron and is dependent on ceruloplasmin to establish a rate of iron oxidation sufficient for mobilization from reticuloendothelium.

Fe in the diet exists in the oxidized Fe(III) form. The HCI in the stomach achieves solubilization and dietary vitamin C (ascorbic acid, a reducing agent) reduces some of the iron to the Fe(II) state and facilitates its absorption. Fe taken up by the gut enters the plasma protein transferrin, which functions as a carrier molecule. Transferrin is a glycoprotein and each molecule has two separate binding sites to which Fe(III) attaches extremely tightly. The binding is assisted by the presence at each site of an anion, usually HCO_3^- or CO_3^{2-}. Under normal conditions the transferrin present in the bloodstream is only about 30% loaded with Fe on average, so that the amount of free Fe salts available in the blood plasma would be expected to be virtually zero, a result confirmed by experimentation (Gutteridge et al. 1981). A similar protein to transferrin, known as lactoferrin, is found in several body fluids and in milk and is produced by phagocytic cells (Reiter, 1979). Lactoferrin also binds two molecules of Fe(III)/mol of protein.

In mammals, the Fe balance is primarily regulated at the level of duodenal absorption of dietary Fe (Fleming and Sly, 2001). Disorders of Fe homeostasis, resulting in Fe deficiency or overload, are very common worldwide. Normal Fe homeostasis depends on a close link between dietary Fe absorption and body Fe needs. Over the past few years, an important body of information concerning the proteins involved in Fe absorption and in the regulation of Fe homeostasis has arisen from the study of inherited defects, both in humans and mice, leading to distinct Fe disorders (Fleming et al. 1997).

10.4. Health Effects

The efficiency of Fe(II) as an electron donor and of Fe(III) as an electron acceptor, with a redox potential compatible with the constrains of the cellular environment, is a fundamental feature for many biochemical reactions and renders Fe to an essential mineral and nutrient. However, this very property turns Fe into a potential biohazard, because under aerobic conditions, Fe can readily catalyze the generation of noxious radicals (Papanikolaou and Pantopoulos, 2005). Iron's toxicity is largely based on Fenton and Haber–Weiss chemistry (see below), where catalytic amounts of Fe are sufficient to yield hydroxyl radicals ($OH^{\bullet}$) from anion superoxide ($O_2^{\bullet-}$) and hydrogen peroxide (H_2O_2), collectively known as "reactive oxygen species" (ROS) (Halliwell and Gutteridge, 1990). Importantly, ROS are inevitable byproducts of aerobic respiration and emerge by incomplete reduction of dioxygen in mitochondria. ROS can also be generated during enzymatic reactions in other subcellular compartments, such as in peroxisomes, the endoplasmic reticulum or the cytoplasm (Papanikolaou and Pantopoulos, 2005).

ROS are also produced by the membrane-bound NADPH oxidase complex (Hampton et al. 1998), a multisubunit enzyme primarily expressed in phagocytic neutrophils and macrophages, but also in other cell types. NADPH oxidase is an important tool for the antimicrobial defense of the organism. The enzyme complex assembles upon infection and generates high levels of superoxide in a "respiratory burst", which is enzymatically and spontaneously dismutated to H_2O_2. The reaction products give rise to more potent oxidants such as peroxynitrite ($ONOO^-$) and hypochlorite (OCl^-), which amplify the bactericidal (and cytotoxic) capacity of phagocytic cells and constitute major toxic species *in vivo* (Ischiropoulos and Beckman, 2003). The former is generated by the spontaneous reaction of $O_2^{\bullet-}$ with Nitric Oxide (NO), while the latter is synthesized from H_2O_2 and chloride in a reaction catalyzed by myeloperoxidase (Papanikolaou and Pantopoulos, 2005).

For many years it has been well accepted that accumulations of Fe in organs such as the liver and heart can cause disease. In disorders such as genetic hemochromatosis and thalassemia, hepatic Fe overload causes cirrhosis; cardiac Fe overload leads to heart failure (Sheth and Brittenham, 2000). Accumulations of Fe are also frequently observed in areas of the brain that degenerate in disorders such as Parkinson and Alzheimer diseases (Rouault, 2001).

10.5. Iron in the Brain

Fe is required as a cofactor in Central Nervous System (CNS) metabolic processes including oxidative phosphorylation, neurotransmitter production, nitric oxide metabolism, and oxygen transport (Ponka, 1999). The observation that brain Fe content is increased in patients with Parkinson disease and other neurodegenerative disorders has created significant interest in the possibility that disturbances of brain Fe homeostasis may contribute to the pathogenesis of these diseases (Thomas and Jankovic, 2004).

Although numerous clinical and experimental studies have attempted to address this issue, a causative role for impaired Fe homeostasis has yet to be established in these more common neurodegenerative diseases. Nevertheless, interest in this neuropathology has focused increased attention on elucidating the mechanisms of brain Fe homeostasis and on defining iron's role in neurologic disease (Madsen and Gitlin, 2007).

Fe is distributed heterogenously in brain exhibiting higher concentrations in gray matter compared with white matter. In human brain, especially high levels are found in the globus pallidus (GP), putamen, and substantia nigra (SN). The regional variability in Fe levels has also been noted for rat brain, but in less proportion. Large variations in the concentration of Fe (μg/mg protein) existed between brain regions in comparable subcellular compartments as well as between subcellular compartments within a given region. Hallgren and Sourander (1958) have described the history of studies of Fe in brain; these authors found that Fe was concentrated primarily in GP, red nucleus, SN, putamen, dentate nucleus, and caudate nucleus (in decreasing order) with cerebral cortex, cerebellar cortex and medulla containing much less Fe. Connor and Menzies (1995) have observed that the amount of Fe in basal ganglia is similar to that in liver on per weight basis.

One of the methodological obstacles that must be considered when brain Fe is analyzed is blood contamination because of hemoglobin. For example, Kofod (1970) reported an average brain Fe value of 12.7 μg/g wet wt for rats that were perfused before analysis. In the same study a value of 18.2 μg/g was reported for unperfused specimens. Thus, Fe in brain can be divided into two roughly equivalent pools, one containing heme Fe and the other nonheme Fe (Prohaska, 1987).

The concentration of brain Fe (heme and nonheme) is influenced by several factors including species, age, and the environmental supply of Fe during development. The concentration of Fe in brain is higher for humans than for experimental rodents and for domestic animals (Kofod, 1970). The concentration of Fe is higher in adults than in infants. This has been shown for rats (Kofod, 1970) as well as for humans (Markesbery et al. 1984). Most of the increase occurs early postnatally. Nutritional deficiency of Fe during development leads to lowered brain (nonheme Fe) levels in several regions (Youdim et al. 1980).

10.5.1. Iron Transport to Brain

Among the processes involved in brain Fe regulation, the entrance of Fe to the cerebral compartment represents a critical step in controlling cerebral Fe homeostasis. Receptor-mediated endocytosis of Fe–transferrin is the most prevalent and important transport for physiological delivery of Fe into the brain (Moos et al. 2006). In the plasma, Fe circulates predominantly as Fe (III) complexed with transferrin, a glycoprotein with an approximate molecular weight of 80 KDa. Transport of Fe across the luminal membrane of the capillary endothelium occurs by receptor-mediated endocytosis of ferric transferring (Zheng et al. 2003). The receptor is a disulfide-linked, integral membrane glycoprotein with a molecular weight of approximately 180 kDa. Its affinity for apotransferrin is approximately two orders of magnitude lower than its affinity for diferric transferrin (K_d 2–7 nM) (Smith et al. 1997). It has been generally accepted that transferrin-bound Fe gains access to brain via transferring receptor (TfR) on brain capillary endothelial cells (Klausner et al. 1993).

Additional mechanisms for Fe delivery into the CNS may occur via a nonspecific metal transporter, divalent metal transporter 1 (DMT1). DMT1 has been cloned from the rat and belongs to the Nramp2 (natural-resistance-associated macrophage protein) family. Its cDNA encodes 562 amino acids with 12 putative membrane domains (Zheng et al. 2003). As a proton-coupled metal transporter originally discovered in duodenal enterocytes, DMT1 mediates the active transport of many divalent metal ions other than Fe, including zinc, manganese, copper, cobalt, cadmium, and nickel (Bannon et al. 2002; Garrick et al. 2003). A defective DMT1 allele encodes a protein that is rapidly degraded and has little or no activity in Fe uptake assays (Su et al. 1998).

Fe is also selectively taken up by the choroid plexus following an intravenous injection (Morris et al. 1992). With histochemical staining, a substantial amount of Fe can be seen in neuroglial cells and in the choroid plexus epithelial cells (Moos and Mollgard, 1993). In human subjects with a calcified choroid plexus, Fe concentration in the choroid plexus is about five times greater than in other brain tissues (Michotte et al. 1977).

The choroid plexus and the oligodendrocytes are the only two cell types in the CNS capable of producing transferrin. Thus, a role of the choroid plexus in brain Fe regulation has been suggested (Crowe and Morgan, 1992; Zheng et al. 1999). Although some studies with antibody against TfR failed to localize the receptors to the choroid plexus (Kissel et al. 1998), others have confirmed their presence (Moos, 1996; Zheng et al. 1999). Thus, it appears likely that the choroid plexus plays a pivotal role in co-regulating cerebral Fe homeostasis. The degree to which the blood cerebrospinal fluid contributes to overall brain Fe regulation, as compared to the role of the blood brain barrier (BBB), has yet to be established (Zheng et al. 2003) (Figure 2).

It has been reported that the most common cell type to stain for Fe under normal conditions is the oligodendrocyte (Connor and Menzies, 1995). However, neurons and microglia, as well as oligodendrocytes, express ferritin, indicating that all of these cell types have the capacity to store Fe. Interestingly, the relative abundance of H- and L-ferritin depends on the cell type (Connor et al. 1994) — neurons express mostly H-ferritin, microglia express mostly L-ferritin and oligodendrocytes express similar amounts of both subunits. Overall, the levels of H- and L-ferritin in neurons are much lower than in oligodendrocytes (Moos and Morgan, 2004; Zecca et al. 2004a). Very little ferritin expression is seen in astrocytes, indicating that these cells provide little Fe storage (Zecca et al. 2004b).

Fe requirements in the brain are much greater than the observed rate of Fe uptake into this tissue (Bradbury, 1997). This finding suggests most brain Fe used each day is derived from recycling behind the BBB analogous to what occurs in the periphery (Madsen and Gitlin, 2007). This mechanism would protect the brain from the effects of systemic Fe overload or deficiency and is supported by observations that disturbances of systemic Fe homeostasis exhibit minimal effects on CNS Fe content or metabolism (Moos and Morgan, 2004). The rate of Fe uptake is greatest during fetal life and postnatal Fe repletion is unproductive to correct the cognitive defects arising from Fe deficiency *in utero*; this concept indicates that following birth recycling, rather than uptake from the circulation, is the major Fe source for brain function (Lozoff et al. 2006). Existence of this brain Fe cycle has critical implications for interpreting any finding of brain Fe accumulation in disease. The inherited disorders aceruloplasminemia and neuroferritinopathy support the concept of a brain Fe cycle and demonstrate that dysregulation of brain Fe homeostasis can be a primary cause of neurodegeneration (Ponka, 2004).

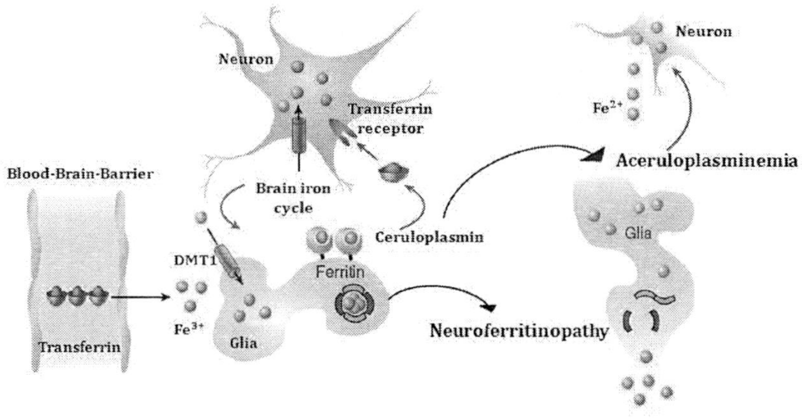

Modified from Masden and Gitlin, 2007.

Figure 2. Iron crosses the blood brain barrier via transferrin receptor pathway on endothelium. The brain iron cycle consists of glia and neurons, where gpi-linked ceruloplasmin functions in a similar role as in the periphery. In aceruloplasminemia, excess iron accumulation damages glia, and neurons are subsequently injured from loss of glia-derived factors, iron-deficiency, or accumulation of nontransferrin-bound iron. A similar pathogenesis is proposed for neuroferritinopathy.

10.6. Iron Neurotoxicity

Curtis et al. (2001) first describe a previously unrecognized, adult-onset neurodegenerative disease that affects the basal ganglia and is associated with Fe accumulation. They then identify a dominantly inherited gene that is mutated in the affected individuals, who are from northern England. All of the patients have the same mutation—insertion of an adenine within the portion of the gene that encodes the carboxy terminus of the L chain of ferritin, an iron-storage protein. The mutation leads to synthesis of a unique 22-amino-acid carboxy terminus. They showed that affected individuals have a mutation in the Ferritin Light Polipeptide gene (FTL), which encodes one of the subunits of the Fe storage protein ferritin. They suggest that this mutation interferes with the stability of the ferritin molecule, causing Fe to be inappropriately released into the cell and transported along the axon.

In the brains of these patients, the GP shows abundant spherical inclusions that contain ferritin. Throughout the white matter, axonal swellings are immunoreactive for neurofilaments, ubiquitin and tau—a characteristic of neurodegenerative diseases. Interestingly, altough serum ferritin levels are remarkably low, the pancreas, liver and heart appear to function normally. So, although the abnormal ferritin is almost certainly ubiquitously expressed, it would seem that only neurons develop significant pathology. The authors propose that this disorder should be referred to as neuroferritinopathy (Curtis et al. 2001).

Another genetic disorder is the Hallervorden–Spatz syndrome, the effects on Fe metabolism are less direct. Zhou et al. (2001) showed that patients with this recessively inherited movement disorder have a mutation in a gene that they named PANK2. This gene encodes pantothenate kinase, an enzyme that catalyses the phosphorylation of vitamin B5. This reaction uses cysteine, so in cells where the enzyme is defective, cysteine tends to accumulate. Cysteine is known to be a chelator of Fe, so the accumulation of Fe might be a secondary consequence of raised cysteine levels.

So, both of these diseases seem to be caused by mutations that affect Fe storage, and in both cases, it is suggested that the Fe deposits cause oxidative damage to axons. These studies show how Fe accumulation could be a cause of neurodegeneration, rather than a secondary effect, and it will be interesting to find out whether this metal contributes to the etiology of any other neurodegenerative diseases (Wood, 2001).

The inherited diseases of Fe metabolism are more common and more numerous than those of other metals, and the genetic basis of many of these disorders has been characterized (Andrews, 2002). Although many of the proteins shown to be essential for systemic Fe homeostasis are expressed within the brain (Wu et al. 2004, Zecca et al. 2004b), genetic disorders resulting in loss of function of these proteins rarely result in either brain Fe overload or deficiency or neurologic disease. For example, although ferroportin is abundantly expressed in the brain microvasculature and is the only known cellular Fe exporter, patients with mutations that impair this protein's function have no evidence of brain Fe accumulation (Pietrangelo, 2006). These observations suggest the presence of the unique mechanisms regulating Fe metabolism within the CNS.

10.6.1. Mechanism of Fe-induced Neurotoxicity

Neurons are highly polarized, with proteins being synthesized in the cell body and then transported over potentially long distances to synapses. Ferritin probably sequesters Fe in the cell body, but transport of ferritin down the axon may allow net transport of Fe to synapses. Normally, the half-life of ferritin appears to be determined by its degradation in lysosomes (Radisky and Kaplan, 1998); in the $Irp2^{-/-}$ mice (Fe regulatory protein 2–deficient mice), the degradation of overexpressed ferritin would lead to the increased release of free Fe. This could occur within lysosomes present in distal axons (Overly and Hollenbeck, 1996). The positively charged Fe atoms could then bind to axonal proteins, including negatively charged components of neurofilaments, causing oxidation and loss of function. In people with neuroferritinopathy, perhaps the ferritin heteropolymers containing the aberrant L chain are inherently unstable; here, the release of Fe might also occur spontaneously as ferritin is transported down the axon. Thus, iron-dependent oxidative damage to the axon may be an early event that is common to some iron-dependent diseases. The highly polarized nature of the neuron, together with axonal trafficking of iron-laden ferritin, may explain why significant pathology is seen only in the nervous system (Wood, 2001).

10.6.2. Fe and Free Radicals

Free radical-mediated tissue damage is generally accepted as a major mechanism underlying the occurrence of certain chronic diseases. The alterations in cell structure and function caused by Fe overload seem to be fundamentally related to free radical mediated damage of cell components (Fraga and Oteiza, 2002).

Many *in vitro* experiments confirmed the production of the radical $^{\bullet}OH$ which can be explained in terms of the following reactions (Valko et al. 2004).

$$Fe(III) + O_2^{\bullet-} \leftrightarrow Fe(II) + O_2 \quad (1)$$

$$Fe(II) + H_2O_2 \rightarrow Fe(III) + {}^{\bullet}OH + OH^- \text{ (Fenton reaction)} \quad (2)$$

The overall reaction of the combined steps is called Haber-Weiss reaction (Valko et al. 2004)

$$O_2^{\bullet-} + H_2O_2 \leftrightarrow O_2 + {}^{\bullet}OH + OH^- \quad (3)$$

In addition to the above reactions, the following reactions may also take place

$$OH + H_2O_2 \rightarrow H_2O + H^+ + O_2^{\bullet-} \quad (4)$$

$$OH + Fe(II) \rightarrow Fe(III) + OH^- \quad (5)$$

$$LOOH + Fe(II) \rightarrow Fe(III) + LO- + {}^{\bullet}OH \quad (6)$$

The Fe (II)-dependent decomposition of H_2O_2 (reaction 2) is called the Fenton reaction and produce hydroxyl radical (Lloyd et al. 1997; Toyokuni, 2002) which can react at diffusion-limited rates with various biomolecules, including lipids, proteins, and DNA. This implies that protecting of Fe from molecular oxygen and the surrounding media is a critical event in preventing iron-mediated oxidative stress. The tight binding of low molecular chelators via coordinating ligands such as O, N, S to Fe blocks the iron's ability to catalyze redox reactions. Since the maximal coordination number of Fe is six, it is often argued that the hexadentate chelators can provide more consistently inert complexes due to their ability to completely saturate the coordination sphere of the Fe atom. Consequently, a chelator molecule that binds to all six sites of the Fe ion completely deactivates the free Fe. Such chelators are termed "hexidentate", of which desferrioxamine is an example. There are many Fe chelators that inhibit the reactions of Fe, oxygen, and their metabolites (Lovejoy and Richardson, 2003). For example, desferrioxamine mesylate markedly decreases the redox activity of Fe(III) and is a very effective antioxidant through its ability to bind Fe (Valko et al. 2005).

In contrast a number of Fe-chelating agents mediate toxicity by stimulating Fe-mediated oxygen radical generation. For example ligands like nitrilotriacetate (NTA) and ATP are known to bind Fe and cause oxidation of lipids and DNA damage (Valko et al. 2005). The hexadentate ligand, ethylenediaminetetraacetic acid (EDTA), can also induce Fenton-reaction mediated radical damage (Engelmann et al. 2003). It was observed that autoxidation of Fe(II) is enhanced by EDTA and also by NTA, but inhibited by o-phenanthroline and deferoxamine. NTA enables Fe(III) to react with H_2O_2 and produce $^{\bullet}OH$. The redox activity of the Fe–EDTA complex has been explained by the incomplete coordination environment of Fe by this ligand. This allows the temporary binding of water and cellular reductants to the incompletely occupied sixth coordination site. The ability of a Fe chelator to promote redox cycling of Fe and consequent production of ROS is a property that may play a role in its antiproliferative activity (Chaston et al. 2003).

Generally, the $^{\bullet}OH$ may react by (1) hydrogen abstraction, (2) electron transfer and (3) addition reactions. The reaction of the $^{\bullet}OH$ with a biomolecule will produce another radical, usually of lower reactivity. As a result of the high reactivity of $^{\bullet}OH$, it often abstracts carbon bound hydrogen atoms more or less non-selectively, e.g. from glucose. Production of $^{\bullet}OH$ close to an enzyme molecule present in excess in the cell, such as lactate dehydrogenase, might have no biological consequences. However, attack by $^{\bullet}OH$ on a membrane lipid can cause a series of radical chain reactions that can severely damage the membranes (Zastawny et al. 1995).

The participation of Fe in the reactions occurring in cells is essential in: (1) the production of $^{\bullet}OH$ that can subsequently initiate lipid oxidation or oxidize almost any molecule present in biological systems, (2) the propagation of free radical reactions by decomposing peroxides. The relevance of iron-catalysed reactions *in vivo* is definitely supported by findings that Fe is present inside the cell. In contrast, based on the findings of O'Halloran and his group (Rae et al. 1999), the concentration of copper inside the cell is restricted to one ion per cell, thus the occurrence of Fenton chemistry *in vivo* is restricted predominantly to the presence of Fe and possibly other trace metals (Valko et al. 2005).

Iron-induced oxidative stress has the following implications: (1) failure in redox regulation leading to DNA damage, lipid peroxidation and oxidative protein damage and (2) free radical-induced activation of signal transduction pathways (Valko et al. 2005).

10.7. Iron and Neurodegenerative Diseases

Several neurogenerative disorders have been associated with misregulation of Fe metabolism in the CNS (Ponka, 2004). High Fe concentrations are found in the brains of patients with frequently encountered conditions such as Parkinson disease (PD) and Alzheimer disease (AD), or more rare conditions such as Huntington disease and Hallervorden–Spatz syndrome (Papanikolaou and Pantopoulos, 2005). In these disorders, iron-induced oxidative stress, combined with defective antioxidant capacities, promotes neuronal death and neurodegeneration. However, it is still controversial whether the extensive brain Fe accumulation is the initial pathogenic event, or a secondary effect. In the last few years, it has been demonstrated that mutations in genes of Fe metabolism origin the pathogenesis of two rare neurodegenerative disorders—aceruloplasminemia and neuroferritinopathy.

Concentrations of non-hem iron (mostly ferritin) increase in the putamen, motor cortex, prefrontal cortex, sensory cortex and thalamus during the first 30–35 years of life, and variable changes are observed in older individuals (Hallgren and Sourander, 1958). Recent studies have shown that levels of H-ferritin in older individuals (67–88 years of age) were higher than in younger controls (27–66 years) in the frontal cortex, caudate nucleus, putamen, SN and GP. In the case of L-ferritin, this increase is observed only in the SN and GP (Connor et al. 1995; Zecca et al. 2001).

The increased Fe concentrations in certain brain regions could result from the altered vascularization that is observed during ageing and in neurodegenerative diseases (Faucheux et al. 1999; Snowdon, 2003). The role of vascular factors in AD and the common features shared by AD and vascular dementia has been discussed (Iadecola, 2004). The increase in Fe concentrations in neurons, astrocytes and microglia, which normally have low Fe contents up to middle age, is typically present in regions such as the cortex, hippocampus and SN, which are particularly susceptible to the neuropathological changes that characterize AD and PD (Jellinger et al. 1990). This Fe 'invasion' and 'perennial occupation' might directly damage these cells or perturb the cellular environment, making it more susceptible to toxins and activation processes of a pathogenic type (for example, inflammation, factor release, morphological change or apoptosis). Moreover, during brain ageing, Fe is partially converted from its stable and soluble form (ferritin) into hemosiderin and other oxyhydroxide derivatives that contain Fe at higher reactivity (for a review, see Crichton, 2001). So the prelude of the pathogenic role of Fe in brain ageing can be summarized in this triad: Fe accumulation, invasion and increased reactivity (Zecca et al. 2004b).

On the other hand, large amounts of Fe are sequestered in neuromelanin granules in the dopaminergic neurons of the SN and the noradrenergic neurons of the Locus Coeruleus (Zecca et al. 1996). Neuromelanin is synthesized by the oxidation of excess cytosolic catechols that are that are preferentially targeted in neurodegenerative diseases such as AD and PD (Zecca et al. 2004b).

10.7.1. Alzheimer Disease

Transition metals zinc, copper and Fe have a well-documented interaction with the biochemical substrates of AD and their modulation with age may initiate AD-related cascades/pathology. Conversely, the formation of many of the biochemical substrates of AD may relate to the regulation of metal ion homeostasis that may be corrupted in the disease (Adlard and Bush, 2006).

A number of the alterations in metal ions may be underscored by alterations in the different metal storage and transport proteins that are suggested to be dysregulated in the AD brain. Ceruloplasmin, for example, that transports and oxidizes Fe, is variably reported to be altered in AD, with reports suggesting that it is increased in the brain and CSF (Cerebrospinal fluid) of AD patients (for review see Moreira et al. 2005), although other studies suggest that it may be decreased in brain tissue (Connor et al. 1993).

Accumulation of Fe in the brain, particularly in cells that are associated with neuritic plaques, is a consistent observation in AD and has been extensively investigated (Levine et al. 2004). In the brains of AD patients, Fe accumulation occurs without the normal age-related increase in ferritin (Connor et al. 1992), thereby increasing the risk of oxidative stress (Thompson et al. 2003). Recent evidence indicates that ageing and AD are associated with a decline in myelin, and this could be related to the Fe content of the myelin (Bartzokis, 2004). Fe might also have a direct impact on plaque formation through its effects on amyloid precursor protein (APP) processing (Rogers et al. 2002). Furthermore, the ability of α-secretase to cleave APP can be modulated by Fe (Bodovitz et al. 1995). Fe seems to promote both deposition of amyloid-β (Aβ) and oxidative stress, which is associated with the neuritic plaques (Huang et al. 2000; Rottkamp et al. 2001), although some authors have argued that by binding Fe, Aβ might, in fact, protect the surrounding neurons from oxidative stress (Perry et al. 2002). Therefore, understanding the cellular mechanisms for regulating Fe could be fundamentally important for understanding the biological basis of AD. Chelators for metals such as Fe that have redox potential are being developed and considered for the treatment of AD (Ritchie et al. 2003; Ben Shachar et al. 2004; Youdim et al. 2004).

A link between congenital Fe overload (hemochromatosis, or gene HFE1) and AD has recently become apparent. The HFE protein is found in blood vessels in the brain and the choroid plexus, and is expressed by cells associated with neuritic plaques (Lee and Connor, 2005). HFE is also expressed by reactive astrocytes in the brains of patients with AD, as well as by neurons (Zecca et al. 2004b). The pattern of neuronal staining for HFE in AD might indicate that HFE is induced by stress, and neurons that stain for TAU are also HFE-positive (Connor et al. 2001). Mutations in the HFE gene that are associated with this iron-overload disorder (Feder et al. 1996)—typically C282Y and H63D — are found at a higher carrier frequency in people of European origin than phenylketonuria and cystic fibrosis combined (Zecca et al. 2004b). The presence of the HFE mutation in AD strongly supports the idea that Fe imbalance in the brain contributes to AD, and its prevalence indicates that it could be an important risk factor (Moalem et al. 2000). HFE mutations are also associated with increased oxidative stress and severity of disease, as assessed neuropathologically (Pulliam et al. 2003), providing further support for the idea that Fe underlies the increase in oxidative stress that promotes neurodegeneration in AD (Zecca et al. 2004b).

The transferrin subtype C2 was also found with increased frequency in patients with AD when compared with age-matched controls (Zambenedetti et al. 2003). Patients with AD who are homozygous for the apolipoprotein E isoform E4 (APOE4) allele are twice as likely to have a transferring C2 allele than those without, or with only a single copy of APOE4 (Hussain et al. 2002). The presence of the C2 variant plus an HFE mutation increased the risk of AD five-fold, and this risk was increased even more in the presence of APOE4 (Robson et al. 2002). There is no significant difference between transferring C2 and other transferrin variants in their ability to bind Fe (Van Landeghem et al. 1998), so the impact of the transferring C2 mutation on AD is probably not mediated through the ability of transferrin to distribute Fe to the brain or to neurons. The combined data on transferrin and HFE mutations clearly indicate that genetic alterations specific to iron-management proteins can increase the risk of AD, and they provide strong evidence that Fe mismanagement in the brain can contribute to AD. To date, there is an ongoing investigation into the relationship between HFE mutations and AD, as some studies have found an association with AD and others have not (Zecca et al. 2004b).

10.7.2. Parkinson Disease

Some authors have reported an increase in total Fe concentration in the SN in the most severe cases of PD, but no changes in less severe cases (Riederer et al. 1989; Hirsch et al. 1991; Götz et al. 2004). Other studies have reported no increase in total Fe concentration in the SN of PD patients, probably because of methodology issues and the different disease stages of the patients (Uitti et al. 1989; Gałazka-Friedman et al. 1996). An increase in Fe content in the lateral GP in comparison to the medial GP was found, and this might be indicative of retrograde degeneration of dopaminergic neurons in PD (Griffiths et al. 1999). There is also a significant inverse relationship between DA concentration and Fe concentrations in the putamen, but not in the SN, of these patients, supporting the concept of a retrograde degenerative process (Gerlach et al. 1994).

Fe(III) deposits were found in microglia, oligodendrocytes, astrocytes located close to neurons, pigmented neurons and in the perimeter of Lewy bodies in the SN pars compacta of patients with PD. A similar picture of Fe accumulation in glia and neurons was found in the putamen and GP of the same patients. The number of ferritin-loaded microglia is increased in the SN of PD, and reactive microglia are often associated with degenerating and neuromelanin-loaded dopaminergic cells (Jellinger et al. 1990). In control patients with presymptomatic PD, which is indicated neuropathologically by neuronal loss and the presence of Lewy bodies, there was no evidence of total Fe increase (Owen et al. 1997). However, even in this case, Fe mismanagement might have occurred through intracellular (for example, from cytosol to mitochondria) and intercellular (for example, from oligodendrocytes to neurons) translocation (Zecca et al. 2004b).

An increase in redox-active Fe associated with neuromelanin occurs in SN neurons in PD patients. This increase is higher in patients that have the most severe neuronal loss, and it is not found in the immediate vicinity of melanized neurons, providing further evidence for a central role for neuromelanin in modulating Fe reactivity (Faucheux et al. 2003). PD is also associated with increased hemo oxygenase-1 immunoreactivity in the neuropil of the SN, and intense immunostaining in Lewy bodies in affected dopaminergic neurons (Schipper et al. 1998). In PD patients, the expression of the divalent cation transporter DCT1 is also increased in SN dopaminergic neurons (Qian and Wang, 1998).

The mechanism that underlies increased Fe concentrations in the SN is unknown. It seems to be independent of the transferrin/TfR system, as there is a significant decrease in TfR density on melanized neurons of the SN of PD patients (Faucheux et al. 1997). The main iron-storage protein, ferritin, shows a significantly increased Fe load in PD (Griffiths et al. 1999). Other studies showed an increase of lactoferrin receptors on SN neurons and microvessels, and an increase of lactoferrin content in neurons of PD patients (Leveugle et al. 1996). However, this generalized increase of the lactoferrin system could be secondary to Fe accumulation in the SN (Zecca et al. 2004b).

The mechanisms that might underlie iron-induced cell damage in PD are the mutations in α-synuclein, which cause a form of familial PD. Abnormal filamentous aggregates of misfolded α-synuclein protein are the main components of Lewy bodies with Fe deposits in the rim, perhaps indicating a pathogenic role for α-synuclein in PD (Zecca et al. 2004b). So far, little is known about the importance of α-synuclein in the nigral–dopaminergic pathway in either normal or pathological situations. In mice (Vila et al. 2000) and in non-human primates (Purisai et al. 2005), α-synuclein is highly expressed in the nigrostriatal pathway and SN, and together with Fe, is upregulated following the treatment with the neurotoxin MPTP (N-methyl-4-phenyl-1,2,3,6-tetrahydropyridine). The consequence of increased cellular Fe is degradation of Irp2 through the ubiquitin pathway (Mandel et al. 2004).

Radical scavengers, such as R-apomorphine, and Fe chelators deferoxamine mesylate and epigallocatechin gallate, induce neuroprotection and reverse these effects (Mandel et al. 2004).

As we mentioned above, Fe contributes to the enhanced generation of ROS, and increases oxidative stress and protein aggregation, including the aggregation of α-synuclein, and aggregations that lead to the formation of advanced glycation end products (Ostrerova-Golts et al. 2000; Uversky et al. 2001). The studies of Hashimoto et al. (1999) indicate that iron-catalysed oxidative reactions mediated by cytochrome c/H_2O_2 might be crucially involved in promoting α-synuclein aggregation, a process that is inhibited by desferrioxamine. In the presence of Fe and free-radical generators, such as dopamine or H_2O_2, BE-M17 human neuroblastoma cells that overexpress wild-type, A53T or A30P α-synuclein produce intracellular aggregates that contain α-synuclein and ubiquitin (Ostrerova-Golts et al. 2000). This is prevented by the action of desferrioxamine. Such aggregates disturb the cytosolic environment and interact with vesicles and their dopamine transporters and intraneuronal mitochondria, and these disturbances might result in activation of cell-death cascades. However, the question of whether α-synuclein is neurotoxic or neuroprotective is still open, as the formation of Lewy bodies and accumulation of α-synuclein could also be a compensatory process to enable the neuron to protect itself (Zecca et al. 2004b).

10.7.3. Other Neurological Diseases

Fe, particularly its accumulation, has been implicated in a series of other neurological diseases (Ke and Ming, 2003) and some of these are discussed briefly here.

10.7.3.1. Aceruloplasminemia

Aceruloplasminemia is an autosomal recessive disorder of Fe homeostasis caused by loss-of-function mutations in the ceruloplasmin gene (Yoshida et al. 1995). Patients have absent serum ceruloplasmin, decreased serum Fe, elevated serum ferritin, anemia, and insulin-dependent diabetes mellitus (Nittis and Gitlin, 2002). Despite these systemic features, most patients present with progressive dementia, dysarthria, and dystonia secondary to basal ganglia Fe accumulation (Logan et al. 1994; Morita et al. 1995). The neurologic disease in aceruloplasminemia is always associated with increased brain Fe as detected by magnetic resonance imaging or autopsy (Kono and Miyajima, 2006).

Histological findings from affected brain regions include neuronal cell loss, abnormal astrocyte architecture, and excess Fe deposition in glia and neurons (Kaneko et al. 2002, Morita et al. 1995; Oide et al. 2006).

The absence of ceruloplasmin results in the slow accumulation of Fe in the reticuloendothelial cells where this metal is normally stored and then mobilized for recycling (Patel et al. 2002) (Figure 2).

The absence of brain Fe accumulation in other diseases with increased nontransferrin-bound plasma Fe supports the idea that brain Fe accumulation in aceruloplasminemia results directly from impaired Fe homeostasis within the CNS. Consistent with this concept, ceruloplasmin is synthesized in astrocytes (Klomp et al. 1996, Klomp and Gitlin 1996) as a glycophosphatidylinositol-linked isoform (Jeong and David, 2003), suggesting that this membrane-bound ferroxidase facilitates the rate of Fe release from storage cells within the CNS. Ceruloplasmin is one of very few proteins established as playing a critical role in brain Fe homeostasis (Madsen and Gitlin, 2007).

10.7.3.2. Neuroferritinopathy

Neuroferritinopathy, an autosomal dominant extrapyramidal disease resulting from mutations in the gene encoding the light chain of ferritin, has revealed a primary role for ferritin in brain Fe homeostasis (Curtis et al. 2001; Maciel et al. 2005; Mancuso et al. 2005). Pathological examination of affected patients reveals cavitary degeneration in the basal ganglia nuclei, neuronal loss, and Fe and ferritin in both extracellular and cytoplasmic inclusion bodies of microglia. The autosomal dominant inheritance and the multimeric structure of ferritin suggest that these mutations impair ferritin assembly resulting in loss of Fe storage capacity within brain cells and subsequent iron-mediated cell injury (Levi et al. 2005). Although there are some phenotype-genotype correlations specific to each light chain mutation, all patients with neuroferritinopathy evidence dystonia in association with basal ganglia Fe accumulation (Chinnery et al. 2007).

Although the mechanism of neurodegeneration in neuroferritinopathy is unknown, the similar pathology to aceruloplasminemia suggests a role for ferritin in the brain Fe cycle and implies a common iron-dependent mechanism of neurodegeneration in these two diseases (Figure 2). Recent studies in aceruloplasminemia suggest that neuronal loss in this disease arises from Fe deficiency secondary to impaired Fe movement from astrocytes within the brain Fe cycle (Jeong and David, 2006). Support for a similar mechanism in neuroferritinopathy comes from a murine model of Irp2 deficiency that impairs ferritin regulation, results in Fe accumulation within oligodendrocytes, and leads to neurodegeneration most likely from secondary neuronal Fe deficiency (LaVaute et al. 2001).

The discovery of aceruloplasminemia and neuroferritinopathy demonstrates that dysregulation of brain Fe homeostasis can be a primary cause of neurodegeneration. These inherited disorders provide a platform for further mechanistic investigations that should reveal new insights into brain Fe homeostasis. Nevertheless, since there is a lack in understanding the molecular mechanisms of Fe homeostasis in the CNS, investigators currently cannot interpret the significance of Fe accumulation in the pathogenesis of other neurologic diseases (Lee et al. 2006; Zecca et al. 2004b).

10.7.3.3. Friedreich Ataxia

Friedreich ataxia is the most common of the early onset inherited ataxias, and it is characterized by degeneration of the large sensory neurons and spinocerebellar tracts, and cardiomyopathy (Patel and Isaya, 2001). The disease is caused by a substantial reduction in the concentration of the mitochondrial protein frataxin, which is provoked by a large GAA triplet-repeat expansion in the first intron of the frataxin gene, resulting in a reduction in frataxin expression by transcription inhibition (Zoghbi and Orr, 2000). This leads to increased mitochondrial Fe content, which seems to reflect the vital role of frataxin in iron–sulphur cluster biosynthesis (Muhlenhoff et al. 2002). Some success has been achieved in retarding the disease-associated cardiomyopathy using the ubiquinone analogue 2,3-dimethoxy-5-methyl-6-(10-hydroxydecyl)-1,4-benzoquinone (Muhlenhoff et al. 2002).

10.8. Neurodegeneration with Brain Iron Accumulation (NBIA)

Mutations in the gene that codes for pantothenate kinase 2 (PANK2) have been shown to be the main genetic defects associated with neurodegeneration with brain Fe accumulation (NBIA —formerly known as Hallervorden–Spatz syndrome) (Zhou et al. 2001; Hayflick, 2003). This autosomal recessive disease, which is characterized by dystonia and pigmentary retinopathy in children and speech or neuropsychiatric defects in adults, has a characteristic Magnetic resonance imaging (MRI) pattern in the GP, known as 'the eye of the tiger' because of its appearance (Hayflick, 2003). Pantothenate kinase is necessary for coenzyme-A biosynthesis, and it is targeted to mitochondria. It is proposed that accumulation of cysteine, which chelates Fe, causes oxidative stress and leads to the accumulation of Fe in the basal ganglia (Zecca et al. 2004b).

10.8.1. Huntington Disease and Multiple Sclerosis

Huntington disease and supranuclear palsy show significantly elevated levels of ferritin-associated Fe in the basal ganglia (Dobson, 2001) but the increases are present from the onset of the diseases and the relationship to the origin of these illnesses is uncertain. Multiple sclerosis (MS) has been associated with an accumulation of brain Fe (Bakshi et al. 2002) although the significance of this finding may be only a nonspecific indicator of tissue destruction in MS.

Final Considerations

Both, Fe excess and deficiency in the nervous system may result in neurodegenerative disorders with motor system and neurobehavioral abnormalities that may have devastating or even progressively deteriorating effects upon normal brain function. The progressive disability seen in many of these disorders is at present mostly incurable, although the promise of advances in human genetics and molecular biology hold the best opportunity of identifying the cause and either the prevention of effective treatments of sporadic and inherited neurodegenerative diseases.

During ageing, the total Fe concentration increases in some brain regions that are targeted by degenerative diseases such as AD, PD and other diseases. This inability to regulate Fe homeostasis generates an excess of reactive Fe, which invades cells such as neurons, astrocytes and microglia. Understanding the timing of Fe mismanagement in relation to the progression of neuronal loss would provide important information on pathogenesis, and would raise the possibility of monitoring Fe changes as an indicator of the disease progression, and possibly even pre-clinical diagnosis in conditions where Fe misregulation is an early event. It is possible that Fe accumulation is purely a consequence of neuronal loss and substitution by cells with higher Fe content, or a breakdown of the BBB that allows more Fe to access the brain.

More data is required on the age trend of Fe concentrations in different brain regions and the cellular distribution of Fe molecules in the same individuals.

Additional information on proteins and other molecules that are involved in Fe metabolism is also needed. The development of non-toxic and more selective Fe chelators could allow the deletion of potentially harmful Fe deposits in older people. Although experimental and preclinical data suggest the therapeutic potential of these drugs and their clinical applicability will be the most important challenge for future research.

More *in vitro* and *in vivo* studies should be done to investigate how Fe misregulation/accumulation synergizes with common endogenous and environmental toxins, because this is probably the most frequent pathogenic mechanism in sporadic AD and PD. Fe misregulation/accumulation alone can destroy neurons only in genetic disorders where Fe imbalance occurs rapidly and is extensive.

References

Adlard, P.A. and Bush, A.I. (2006) Metals and Alzheimer's disease. *J Alz Dis*. 10: 145–163.

Aisen, P., Enns, C. and Wessling-Resnick, M. (2001) Chemistry and biology of eukaryotic iron metabolism. *Int J Biochem Cell Biol*. 33: 940–959.

Andrews, N.C. (2002) A genetic view of iron homeostasis. *Semin Hematol*. 39:227–234.

Bakshi, R., Benedict, R.H., Bermel, R.A., Caruthers, S.D., Puli, S.R., Tjoa, C.W., Fabiano, A.J., and Jacobs, L. (2002) T2 hypointensity in the deep gray matter of patients with multiple sclerosis: a quantitative magnetic resonance imaging study. *Arch Neurol*. 59: 62-68.

Bannon, D., Portnoy, M.E., Olivi, L., Lees, P.S., Culotta, V.C. and Bressler, J.P. (2002) Uptake of lead and iron by divalent metal transporter 1 in yeast and mammalian cells. *Biochem Biophys Res Commun*. 295: 978–984.

Bartzokis, G. (2004) Age-related myelin breakdown: a developmental model of cognitive decline and Alzheimer's disease. *Neurobiol Aging*. 25: 5–18.

Ben Shachar, D., Kahana, N., Kampel, V., Warshawsky, A. and Youdim, M.B. (2004) Neuroprotection by a novel brain permeable iron chelator, VK-28, against 6-hydroxydopamine lesion in rats. *Neuropharmacology*. 46: 254–263.

Berg, D., Gerlach, M., Youdim, M.B., Double, K.L., Zecca, L., Riederer, P. and Becker, G. (2001) Brain iron pathways and their relevance to Parkinson's disease. *J Neurochem.* 79:225-236.

Bodovitz, S., Falduto, M.T., Frail, D.E. and Klein, W.L. (1995) Iron levels modulate α-secretase cleavage of amyloid precursor protein. *J Neurochem.* 64: 307–315.

Bradbury, M.W. (1997) Transport of iron in the blood-brain-cerebrospinal fluid system. *J Neurochem.* 69: 443–454.

Cargnoni, A., Ceconi, C., Gaia, G., Agnoletti, L. and Ferrari, R. (2002) Cellular thiols redox status: a switch for NF-kappaB activation during myocardial post-ischaemic reperfusion. *J Mol Cell Cardiol.* 34: 997-1005.

Cavill, I. (1999) Iron status as measured by serum ferritin: the marker and its limitations. *Am J Kidney Dis.* 34:S12-S17.

Chaston, T.B., Lovejoy, D.B., Watts, R.N. and Richardson, D.R. (2003) Examination of the antiproliferative activity of iron chelators: multiple cellular targets and the different mechanism of action of triapine compared with desferrioxamine and the potent pyridoxal isonicotinoyl hydrazone analogue 311. *Clin Cancer Res.* 9: 402-414.

Chinnery, P.F., Crompton, D.E., Birchall, D., Jackson, M.J., Coulthard, A., Lombès, A., Quinn, N., Wills, A., Fletcher, N., Mottershead, J.P., Cooper, P., Kellett, M., Bates, D. and Burn, J. (2007) Clinical features and natural history of neuroferritinopathy caused by the FTL1 460InsA mutation. *Brain.* 130:110–119.

Connor, J.R., Boeshore, K.L. and Benkovic, S.A. (1994) Isoforms of ferritin have a specific cellular distribution in the brain. *J Neurosci Res.* 37: 461–465.

Connor, J.R. and Menzies, S.L. (1995) Cellular management of iron in the brain. *J Neurol Sci.* 134: 33–44.

Connor, J.R., Milward, E.A., Moalem, S, Sampietro, M., Boyer, P., Percy, M.E., Vergani, C., Scott, R.J. and Chorney, M. (2001) Is hemochromatosis a risk factor for Alzheimer's disease? *J Alzheimer's Dis.* 3: 471–477.

Connor, J.R., Snyder, B.S., Arosio, P., Loeffler, D.A. and LeWitt, P.A. (1995) quantitative analysis of isoferritins in select regions of aged, parkinsonian and Alzheimer Disease brains. *J Neurochem.* 65: 717–724.

Connor, J.R., Snyder, B.S., Beard, J.L., Fine, R.E. and Mufson, E.J. (1992) Regional distribution of iron and iron-regulatory proteins in the brain in aging and Alzheimer's disease. *J Neurosci Res.* 31: 327–335.

Connor, J.R., Tucker, P., Johnson, M. and Snyder, B. (1993) Ceruloplasmin levels in the human superior temporal gyrus in aging and Alzheimer's disease. *Neurosci Lett.* 159: 88–90.

Crichton, R.R. (2001) Inorganic Biochemistry of Iron Metabolism: From Molecular Mechanisms to Clinical Consequences (John Wiley & Sons, Chichester).

Crichton, R.R., Wilmet, S., Legssyer, R. and Ward, R.J. (2002) Molecular and cellular mechanisms of iron homeostasis and toxicity in mammalian cells. *J Inorg Biochem.* 91: 9-18.

Crowe, A. and Morgan, E.H. (1992) Iron and transferrin uptake by brain and cerebrospinal fluid in the rat. *Brain Res.* 592: 8–16.

Curtis, A.R., Fey, C., Morris, C.M., Bindoff, L.A., Ince, P.G., Chinnery, P.F., Coulthard, A., Jackson, M.J., Jackson, A.P., McHale, D.P., Hay, D, Barker, W.A., Markham, A.F., Bates, D., Curtis, A. and Burn, J. (2001) Mutation in the gene encoding ferritin light polypeptide causes dominant adult-onset basal ganglia disease. *Nat Genet*. 28: 350-354.

Dobson, J. (2001) nanoscale biogenic iron oxides and neurodegenerative disease. *FEBS Lett*. 496: 1-5.

Engelmann, M.D., Bobier, R.T., Hiatt, T. and Cheng, I.F. (2003) Variability of the Fenton reaction characteristics of the EDTA, DTPA, and citrate complexes of iron. *Biometals*. 6: 519-527.

Faucheux, B.A., Bonnet, A.M., Agid, Y. and Hirsch, E.C. (1999) Blood vessels change in the mesencephalon of patients with Parkinson's disease. *Lancet*. 353: 981–982.

Faucheux, B.A., Hauw, J.J., Agid, Y. and Hirsch, E.C. (1997) The density of [125I]–transferrin binding sites on perikarya of melanized neurons of the substantia nigra is decreased in Parkinson's disease. *Brain Res*. 749: 170–174.

Faucheux, B.A., Martin, M.E., Beaumont, C., Hauw, J.J., Agid, Y. and Hirsch, E.C. (2003) Neuromelanin associated redox-active iron is increased in the substantia nigra of patients with Parkinson's disease. *J Neurochem*. 86: 1142-1148.

Feder, J.N., Gnirke, A., Thomas, W., Tsuchihashi, Z., Ruddy, D.A., Basava, A., Dormishian, F., Domingo, R. Jr., Ellis, M.C., Fullan, A, Hinton, L.M., Jones, N.L., Kimmel, B.E., Kronmal, G.S., Lauer, P., Lee, V.K., Loeb, D.B., Mapa, F.A., McClelland, E., Meyer, N.C., Mintier, G.A., Moeller, N., Moore, T., Morikang, E., Prass, C.E., Quintana, L., Starnes, S.M., Schatzman, R.C., Brunke, K.J., Drayna, D.T., Risch, N.J., Bacon, B.R. and Wolff, R.K. (1996) A novel MHC class I-like gene is mutated in patients with hereditary haemochromatosis. *Nature Genet*. 13: 399–408.

Fleming, R.E. and Sly, W.S. (2001) Hepcidin: a putative iron-regulatory hormone relevant to hereditary hemochromatosis and the anemia of chronic disease. *Proc Natl Acad Sci USA*, 98: 8780-8785.

Fleming, M.D., Trenor, C.C. 3rd, Su, M.A., Foernzler, D., Beier, D.R., Dietrich, W.F. and Andrews, N.C. (1997) Microcytic anaemia mice have a mutation in Nramp2, a candidate iron transporter gene. *Nat Genet*. 16: 383-386.

Fraga, C.G. and Oteiza, P.I. (2002) Iron toxicity and antioxidant nutrients. *Toxicology*, 180: 23-32.

Galazka-Friedman, J., Bauminger, E.R., Friedman, A., Barcikowska, M., Hechel, D. and Nowik, I. (1996) Iron in parkinsonian and control substantia nigra: a Mossbauer spectroscopy study. *Mov Disord*. 11: 8–16.

Garrick, M.D., Dolan, K.G., Horbinski, C., Chio, A.J., Higgins, D., Porubcin, M., Moore, E.G., Hainsworth, L.N., Umbreit, J.N., Conrad, M.E., Feng, L., Lis, A., Roth, J.A., Singleton, S. and Garrick, L.M. (2003) DMT1: a mammalian transporter for multiple metals. *Biometals*, 16: 41–54.

Gerlach, M., Ben-Shachar, D., Riederer, P., and Youdim, M.B. (1994) Altered brain metabolism of iron as a cause of neurodegenerative diseases? *J Neurochem*. 63: 793–807.

Götz, M.E., Double, K., Gerlach, M., Youdim, M.B. and Riederer, P. (2004) The relevance of iron in the pathogenesis of Parkinson's disease. *Ann NY Acad Sci*. 1012: 193–208.

Griffiths, P.D., Dobson, B.R., Jones, G.R. and Clarke, D.T. (1999) Iron in the basal ganglia in Parkinson's disease: an in vitro study using extended X-ray absorption fine strucutre and cryo-electron microscopy. *Brain*, 122: 667–673.

Gutteridge, J.M. (1981) Thiobarbituric acid-reactivity following iron-dependent free-radical damage to amino acids and carbohydrates. *FEBS Lett.* 128: 343-346.

Hallgren, B. and Sourander, P. (1958) The effect of age on the non-haemin iron in the human brain. *J Neurochem*. 3:41–51.

Halliwell, B. and Gutteridge, J.M.C. (1990) The role of free radicals and catalytic metal ions in human disease: an overview. *Methods Enzymol.* 186: 1 – 85.

Hampton, M.B., Kettle, A.J. and Winterbourn, C.C. (1998) Inside the neutrophil phagosome: oxidants, myeloperoxidase, and bacterial killing. *Blood*, 92: 3007–3017.

Hashimoto, M., Takeda, A., Hsu, L.J., Takenouchi, T. and Masliah, E. (1999) Role of cytochrome c as a stimulator of α-synuclein aggregation in Lewy body disease. *J Biol Chem*. 274: 28849–28852.

Hayflick, S.J. (2003) Unravelling the Hallervorden–Spatz syndrome: pantothenate kinase-associated neurodegeneration is the name. *Curr Opin Pediatr.* 15: 572–577.

Hirsch, E.C., Brandel, J.P., Galle, P., Javoy-Agid, F. and Agid, Y. (1991) Iron and aluminum increase in the substantia nigra of patients with Parkinson's disease: an X-ray microanalysis. *J Neurochem.* 56: 446–451.

Huang, X., Cuajungco, M.P., Atwood, C.S., Moir, R.D., Tanzi, R.E. and Bush, A.I. (2000) Alzheimer's disease, β-amyloid protein and zinc. *J Nutr.* 130: 1488S–1492S.

Hussain, R.I., Ballard, C.G., Edwardson, J.A. and Morris, C.M. (2002) Transferrin gene polymorphism in Alzheimer's disease and dementia with Lewy bodies in humans. *Neurosci Lett*. 317: 13–16.

Iadecola, C. (2004) Neurovascular regulation in the normal brain and in Alzheimer's disease. *Nature Rev Neurosci*. 5: 347–360.

Ischiropoulos, H. and Beckman, J.S. (2003) Oxidative stress and nitration in neurodegeneration: cause, effect, or association? *J Clin Invest.* 111: 163– 169.

Jellinger, K., Paulus, W., Grundke-Iqbal, I., Riederer, P. and Youdim, M.B. (1990) Brain iron and ferritin in Parkinson's and Alzheimer's disease. *J Neural Transm Park Dis Dement Sect.* 2, 327–340.

Jeong, S.Y. and David, S. (2003) Glycosylphosphatidylinositol-anchored ceruloplasmin is required for iron efflux from cells in the central nervous system. *J Biol Chem.* 278: 27144–27148.

Kaneko, K., Yoshida, K., Arima, K., Ohara, S., Miyajima, H., Kato, T., Ohta, M. and Ikeda, S.I. (2002) Astrocytic deformity and globular structures are characteristic of the brains of patients with aceruloplasminemia. *J Neuropathol Exp Neurol*. 61:1069–1077.

Kawabata, H., Yang, R, Hirama, T., Vuong, P.T., Kawano, S., Gombart, A.F. and Koeffler, H.P. (1999) Molecular cloning of transferrin receptor 2. A new member of the transferrin receptor-like family. *J Biol Chem.* 274: 20826-20832.

Ke, Y. and Ming Qian, Z. (2003) Iron misregulation in the brain: a primary cause of neurodegenerative disorders. *Lancet Neurol*. 2: 246–253.

Kissel, K., Hamm, S., Schulz, M., Vecchi, A., Garlanda, C. and Engelhardt, B. (1998) Immunohistochemical localization of the murine transferrin receptor (TfR) on blood–

tissue barriers using a novel anti-TfR monoclonal antibody. *Histochem Cell Biol.* 110: 63–72.

Klausner, R.D., Rouault, T.A. and Harford, J.B. (1993) Regulating the fate of mRNA: the control of cellular iron metabolism. *Cell*, 72, 19–28.

Klomp, L.W., Farhangrazi, Z.S., Dugan, L.L. and Gitlin, J.D. (1996). Ceruloplasmin gene expression in the murine central nervous system. *J Clin Invest.* 98:207–215.

Klomp, L.W. and Gitlin, J.D. (1996) Expression of the ceruloplasmin gene in the human retina and brain: implications for a pathogenic model in aceruloplasminemia. *Hum Mol Genet.* 5:1989–1996.

Kofod, B. (1970) Iron, copper, and zinc in rat brain. *Eur J Pharmacol.* 13: 40-45.

Kono, S. and Miyajima, H. (2006) Molecular and pathological basis of aceruloplasmineinia. *Biol Res*. 39:15–23.

Lao, T.T. and Ho, L.F. (2004) Impact of iron deficiency anemia on prevalence of gestational diabetes mellitus. *Diabetes Care*, 27: 650-656.

LaVaute, T., Smith, S., Cooperman, S., Iwai, K., Land, W., Meyron-Holtz, E., Drake, S.K., Miller, G., Abu-Asab, M., Tsokos, M., Switzer, R. 3rd, Grinberg, A., Love, P., Tresser, N. and Rouault, T.A. (2001) Targeted deletion of the gene encoding iron regulatory protein-2 causes misregulation of iron metabolism and neurodegenerative disease in mice. *Nature Genet.* 27: 209– 214.

Lee, D.W., Andersen, J.K. and Kaur, D. (2006) Iron dysregulation and neurodegeneration: the molecular connection. *Mol Interv*. 6:89–97.

Lee, S.Y. and Connor, J.R. (2005) Regulation of Hfe by stress factors in BV-2 cells. *Neurobiol Aging*, 26: 803-812.

Leveugle, B., Faucheux, B.A., Bouras, C., Nillesse, N., Spik, G., Hirsch, E.C., Agid Y. and Hof, P.R. (1996) Cellular distribution of the iron-binding protein lactotransferrin in the mesencephalon of Parkinson's disease cases. *Acta Neuropathol* (Berl.) 91: 566–572.

Levi, S., Cozzi, A. and Arosio, P. (2005) Neuroferritinopathy: a neurodegenerative disorder associated with L-ferritin mutation. *Best Pract Res Clin Haematol.* 18:265–276.

Levine, S., Connor, J. and Schipper, H. (2004) Redox-active metals in neurological disorders. *Ann NY Acad Sci.* 1012.

Lloyd, R.V., Hanna, P.M. and Mason, R.P. (1997) The origin of the hydroxyl radical oxygen in the Fenton reaction. *Free Radic Biol Med.* 22: 885-888.

Logan, J.I., Harveyson, K.B.,Wisdom, G.B., Hughes, A.E. and Archbold, G.P. (1994) Hereditary caeruloplasmin deficiency, dementia and diabetes mellitus. *Q J Med.* 87:663–670.

Lovejoy, D.B. and Richardson, D.R. (2003) Iron chelators as anti-neoplastic agents: current developments and promise of the PIH class of chelators. *Curr Med Chem.* 10: 1035-1049.

Lozoff, B., Beard, J., Connor, J., Barbara, F., Georgieff, M. and Schallert, T. (2006) Long-lasting neural and behavioral effects of iron deficiency in infancy. *Nutr Rev.* 64: S34–43.

Maciel, P., Cruz, V.T., Constante, M., Iniesta, I., Costa, M.C., Gallati, S., Sousa, N., Sequeiros, J., Coutinho, P. and Santos, M.M. (2005) Neuroferritinopathy: missense mutation in FTL causing early-onset bilateral pallidal involvement. *Neurology*, 65:603–605.

Madsen, E. and Gitlin, J.D. (2007) Copper and Iron Disorders of the Brain. *Annu Rev Neurosci*. 30: 317-337.

Mancuso, M., Davidzon, G., Kurlan, R.M., Tawil, R., Bonilla, E., Di Mauro, S. and Powers, J.M. (2005) Hereditary ferritinopathy: a novel mutation, its cellular pathology, and pathogenetic insights. *J Neuropathol Exp Neurol.* 64:280–294.

McKie, A.T., Marciani, P., Rolfs, A., Brennan, K., Wehr, K., Barrow, D., Miret, S., Bomford, A., Peters, T.J., Farzaneh, F., Hediger, M.A., Hentze, M.W. and Simpson, R.J. (2000) A novel duodenal iron-egulated transporter, IREG1, implicated in the basolateral transfer of iron to the circulation. *Mol Cell.* 5: 299-309.

Mandel, S., Maor, G. and Youdim, M.B.H. (2004) Iron and α-synuclein in the substantia nigra of MPTP treated mice; effect of neuroprotective drugs R-apomorphine and green tea polyphenol epipgallocatechin-3-gallate. *J Mol Neurosci.* 24: 401-416.

Markesbery, W.R., Ehmann, W.D., Alauddin, M., Loomis, T.C., Yee, G. and Stahl, W.L. (1984) Brain trace element concentrations in aging. *Neurobiol Aging*. 5: 19-28.

Michotte, Y., Massart, D.L., Lowenthal, A., Knaepen, L., Pelsmaekers, J. and Collard, M. (1977) A morphological and chemical study of calcification of the choroid plexus. *J Neurol.* 216: 127–133.

Moalem, S., Percy, M.E., Andrews, D.F., Kruck, T.P., Wong, S., Dalton, A.J., Mehta, P., Fedor, B. and Warren, A.C. (2000) Are hereditary hemochromatosis mutations involved in Alzheimer disease? *Am J Med Genet.* 93: 58–66.

Moos, T. (1996) Immunohistochemical localization of intraneuronal transferrin receptor immunoreactivity in the adult mouse central nervous system. *J Comp Neurol*. 375: 675–692.

Moos, T. and Mollgard, K. (1993) A sensitive post-DAB enhancement technique for demonstration of iron in the central nervous system. *Histochemistry*, 99: 471–475.

Moos, T. and Morgan, E.H. (2004) The metabolism of neuronal iron and its pathogenic role in neurological disease: review. *Ann NY Acad Sci.* 1012: 14–26.

Moos, T., Skjoerringe, T., Gosk, S. and Morgan, E.H. (2006) Brain capillary endothelial cells mediate iron transport into the brain by segregating iron from transferrin without the involvement of divalent metal transporter 1. *J Neurochem.* 98: 1946–1958.

Moreira, P.I., Siedlak, S.L., Aliev, G., Zhu, X., Cash, A.D., Smith, M.A. and Perry, G. (2005) Oxidative stress mechanisms and potential therapeutics in Alzheimer disease. *J Neural Transm*. 112: 921-932.

Morita, H., Ikeda, S., Yamamoto, K., Morita, S., Yoshida, K., Nomoto, S., Kato, M. and Yanagisawa, N. (1995) Hereditary ceruloplasmin deficiency with hemosiderosis: a clinicopathological study of a Japanese family. *Ann Neurol.* 37:646–656.

Morris, C.M., Keith, A.B., Edwardson, J.A. and Pullen, R.G. (1992) Uptake and distribution of iron and transferrin in the adult rat brain. *J Neurochem*. 59: 300–306.

Muhlenhoff, U., Richhardt, N., Ristow, M., Kispal, G. and Lill, R. (2002) The yeast frataxin homolog Yfh1p plays a specific role in the maturation of cellular Fe/S proteins. *Hum Mol Genet.* 11: 2025–2036.

National Research Council. (1979) Iron. Baltimore, MD, University Park Press.

National Research Council. (1989) Recommended dietary allowances, 10th Ed. Washington, DC, National Academy Press.

Nemeth, E., Tuttle, M.S., Powelson, J., Vaughn, M.B., Donovan, A., Ward, D.M., Ganz T. and Kaplan, J. (2004) Hepcidin regulates cellular iron efflux by binding to ferroportin and inducing its internalization. *Science*, 306:2090–2093.

Nittis, T. Gitlin, J.D. (2002) The copper-iron connection: hereditary aceruloplasminemia. *Semin Hematol.* 39:282–289.

Ostrerova-Golts, N., Petrucelli, L., Hardy, J., Lee, J.M., Farer, M. and Wolozin, B. (2000) The A53T α-synuclein mutation increases iron-dependent aggregation and toxicity. *J Neurosci.* 20: 6048–6054.

Oide, T., Yoshida, K., Kaneko, K., Ohta, M. and Arima, K. (2006) Iron overload and antioxidative role of perivascular astrocytes in aceruloplasminemia. *Neuropathol Appl Neurobiol.* 32:170–176.

Overly, C.C. and Hollenbeck, P.J. (1996) Dynamic organization of endocytic pathways in axons of cultured sympathetic neurons. *J Neurosci.* 16:6056-6064.

Owen, A.D., Schapira, A.H.V., Jenner, P. and Marsden, C.D. (1997) Indices of oxidative stress in Parkinson's disease, Alzheimer's disease, and dementia with Lewy bodies. *J Neural Transm.* 51: 167–173.

Papanikolaou, G. and Pantopoulos, K. (2005) Iron metabolism and toxicity. *Toxicol Appl Pharmacol.* 202: 199-211.

Patel, B.N., Dunn, R.J., Jeong, S.Y., Zhu, Q., Julien, J.P. and David, S. (2002) Ceruloplasmin regulates iron levels in the CNS and prevents free radical injury. *J Neurosci.* 22: 6578–6586.

Patel, P.I. and Isaya, G. (2001) Friedreich ataxia: from GAA tripletrepeat expansion to frataxin deficiency. *Am J Hum Genet.* 69: 15–24.

Perry, G., Sayre, L.M., Atwood, C.S., Castellani, R.J., Cash, A.D., Rottkamp, C.A. and Smith, M.A. (2002) The role of iron and copper in the aetiology of neurodegenerative disorders: therapeutic implications. *CNS Drugs*, 16: 339–352.

Pietrangelo, A. (2006) Hereditary hemochromatosis. *Biochim Biophys Acta*, 1763:700–710.

Ponka, P. (1999) Cellular iron metabolism. *Kidney Int.* 69: S2–S11.

Ponka, P. (2004) Hereditary causes of disturbed iron homeostasis in the central nervous system. *Ann NY Acad Sci.* 1012: 267–281.

Prohaska, J.R. (1987) Functions of Trace Elements in Brain Metabolism. *Physiol Rev.* 67: 858-901.

Pulliam, J.F., Jennings, C.D., Kryscio, R.J., Davis, D.G., Wilson, D., Montine, T.J., Schmitt, F.A. and Markesbery, W.R. (2003) Association of HFE mutations with neurodegeneration and oxidative stress in Alzheimer's disease and correlation with APOE. *Am J Med Genet.* 119B: 48–53.

Purisai, M.G., McCormack, A.L., Langston, W.J., Johnston, L.C. and Di Monte, D.A. (2005) Alpha-synuclein expression in the substantia nigra of MPTP-lesioned non-human primates. *Neurobiol Dis.* 20: 898-906.

Qian, Z.M. and Wang, Q. (1998) Expression of iron transport proteins and excessive iron accumulation in the brain in neurodegenerative disorders. *Brain Res Rev.* 27: 257–267.

Radisky, D.C. and Kaplan, J. (1998) Iron in cytosolic ferritin can be recycled through lysosomal degradation in human fibroblasts. *Biochem J.* 336: 201-205.

Rae, T.D., Schmidt, P.J., Pufahl, R.A., Culotta, V.C. and O'Halloran, T.V. (1999) Undetectable intracellular free copper: the requirement of a copper chaperone for superoxide dismutase. *Science*, 284: 805-808.

Rasmussen, M.L., Folsom, A.R., Catellier, D.J., Tsai, M.Y., Garg, U. and Eckfeldt, J.H. (2001) A prospective study of coronary heart disease and the hemochromatosis gene (HFE) C282Y mutation: the Atherosclerosis Risk in Communities (ARIC) study. *Atherosclerosis*, 154:739-746.

Reiter, B. (1979) The lactoperoxidase-thiocyanate-hydrogen peroxide antibacterium system in oxygen free radicals and tissue damage. *CIBA Found Symp*. 65: 285-294.

Riederer, P, Sofic, E., Rausch, W.D., Schmidt, B., Reynolds, G.P., Jellinger, K. and Youdim, M.B. (1989) Transition metals, ferritin, glutathione, and ascorbic acid in parkinsonian brains. *J Neurochem*. 52: 515-520.

Ritchie, C.W., Bush, A.I., Mackinnon, A., Macfarlane, S., Mastwyk, M., MacGregor, L., Kiers, L., Cherny, R., Li, Q.X., Tammer, A., Carrington, D., Mavros, C., Volitakis, I., Xilinas, M., Ames, D., Davis, S., Beyreuther, K., Tanzi, R.E. and Masters, C.L. (2003) Metal-protein attenuation with iodochlorhydroxyquin (clioquinol) targeting Aβ amyloid deposition and toxicity in Alzheimer disease: a pilot phase 2 clinical trial. *Arch Neurol*. 60: 1685–1691.

Robson, K.J., Lehmann, D.J., Wimhurst, V.L., Livesey, K.J., Combrinck, M., Merryweather-Clarke, A.T., Warden, D.R. and Smith, A.D. (2002) Synergy between the C2 allele of transferrin and the C282Y allele of the haemochromatosis gene (Hfe) as risk factors for developing Alzheimer's Disease. *J Med Genet*. 41: 261–265.

Rogers, J.T., Randall, J.D., Cahill, C.M., Eder, P.S., Huang, X., Gunshin, H., Leiter, L., McPhee, J., Sarang, S.S., Utsuki, T., Greig, N.H., Lahiri, D.K., Tanzi, R.E., Bush, A.I., Giordano, T.and Gullans, S.R. (2002) An iron-responsive element type II in the 5'-untranslated region of the Alzheimer's amyloid precursor protein transcript. *J Biol Chem*. 277: 45518–45528.

Rottkamp, C.A., Raina, A.K., Zhu, X., Gaier, E., Bush, A.I., Atwood, C.S., Chevion, M., Perry, G. and Smith, M.A. (2001) Redox-active iron mediates amyloid-β toxicity. *Free Radic Biol Med*. 30: 447–450.

Rouault, T.A. (2001) Iron on the brain. *Nat Gen*. 28: 299 – 300.

Schipper, H.M., Liberman, A. and Stopa, E.G. (1998) Neural heme oxygenase-1 expression in idiopathic Parkinson's disease. *Exp Neurol*. 150: 60–68.

Schümann, K., Moret, R., Künzle, H. and Kühn, L.C. (1999) Iron regulatory protein as an endogenous sensor of iron in rat intestinal mucosa. Possible implications for the regulation of iron absorption. *Eur J Biochem*. 260: 362-72.

Sheth, S. and Brittenham, G.M. (2000) Genetic disorders affecting proteins of iron metabolism: clinical implications. *Annu Rev Med*. 51: 443-464.

Smith, Q.R., Rabin, O. and Chikhale, E.G. (1997) Delivery of metals to brain and the role of the blood–brain barrier, in: Connor, J.R. (Ed.), Metals and Oxidative Damage in Neurological Disorders, Plenum, New York, pp. 113–130.

Snowdon, D.A. (2003) Nun Study. Healthy aging and dementia: findings from the Nun Study. *Ann Intern Med*. 139: 450–454.

Su, M.A., Trenor, C.C., Fleming, J.C., Fleming, M.D., and Andrews, N.C. (1998) The G185R mutation disrupts function of the iron transporter Nramp-2. *Blood*, 92: 2157–2163.

Thomas, M. and Jankovic, J. (2004) Neurodegenerative disease and iron storage in the brain. *Curr Opin Neurol*. 17:437–42.

Thompson, K., Menzies, S., Muckenthaler, M., Torti, F.M., Wood, T., Torti, S.V., Hentze, M.W., Beard, J. and Connor, J. (2003) Mouse brains deficient in H-ferritin have normal iron concentration but a protein profile of iron deficiency and increased evidence of oxidative stress. *J Neurosci Res*. 71: 46–63.

Toyokuni, S. (2002) Iron and carcinogenesis: from Fenton reaction to target genes. *Redox Rep*. 7: 189-197.

Uitti, R.J., Rajput, A.H., Rozdilsky, B., Bickis, M., Wollin, T. and Yuen, W.K. (1989) Regional metal concentrations in Parkinson's disease, other chronic neurological diseases, and control brains. *Can J Neurol Sci*. 16: 310-314.

Uversky, V.N., Li, J. and Fink, A.L. (2001) Metal-triggered structural transformations, aggregation, and fibrillation of human α-synuclein. A possible molecular link between Parkinson's disease and heavy metal exposure. *J Biol Chem*. 276: 44284–44296.

Valko, M., Izakovic, M., Mazur, M., Rhodes, C.J. and Telser, J. (2004) Role of oxygen radicals in DNA damage and cancer incidence. *Mol Cell Biochem*. 266: 37-56.

Valko, M., Morris, H. and Cronin, M.T. (2005) Metals, toxicity and oxidative stress. *Curr Med Chem*. 12: 1161-208.

Valko, M., Morris, H., Mazur, M., Rapta, P. and Bilton, R.F. (2001) Oxygen free radical generating mechanisms in the colon: do the semiquinones of vitamin K play a role in the aetiology of colon cancer? *Biochim Biophys Acta*. 1527: 161-166.

Van Landeghem, G., Sikstrom, C., Beckman, L., Adolfsson, R. and Beckman, R. (1998) Transferrin C2, metal binding and Alzheimer's Disease. *Neuroreport*, 9: 177–179.

Vila, M., Vukosavic, S., Jackson-Lewis, V., Neystat, M., Jakowec, M. and Przedborski, S. (2000) α-synuclein up-regulation in substantia nigra dopaminergic neurons following administration of the parkinsonian toxin MPTP. *J Neurochem*. 74: 721–729.

Walker, E.M. Jr. and Walker, S.M. (2000) Effects of iron overload on the immune system. *Ann Clin Lab Sci*. 30: 354-365.

WHO. (1996) Iron in Drinking-water 2nd ed. Vol. 2. Health criteria and other supporting information. World Health Organization, Geneva.

Wood, H. (2001) Neurodegeneration: Ironing out disease. *Nat Rev Neurosci*. 2: 605.

Wu, L.J., Leenders, A.G., Cooperman, S., Meyron-Holtz, E., Smith, S., Tsai, R.Y., Berger, U.V., Sheng, Z.H. and Rouault, T.A. (2004) Expression of the iron transporter ferroportin in synaptic vesicles and the blood-brain barrier. *Brain Res*. 1001:108–117.

Yoshida, K., Furihata, K., Takeda, S., Nakamura, A., Yamamoto, K., Morita, H., Hiyamuta, S., Ikeda, S., Shimizu, N. and Yanagisawa, N. (1995) A mutation in the ceruloplasmin gene is associated with systemic hemosiderosis in humans. *Nat Genet*. 9:267–272.

Youdim, M.B.H., Fridkin, M. and Zheng, H. (2004) Novel bifunctional drugs targeting monoamine oxidase inhibition and iron chelation as an approach to neuroprotection in Parkinson's disease and other neurodegenerative diseases. *J Neural Transm*. 111: 455-471.

Youdim, M.B.H., Green, A.R., Bloomfield, M.R., Mitchell, B.D., Heal, D.J.D. and Grahamesmith, G. (1980) The effects of iron deficiency on brain biogenic monoamine biochemistry and function in rats. *Neuropharmacol.* 19: 259-267.

Zambenedetti, P., De Bellis, G., Biunno, I., Musicco, M. and Zatta, P. (2003) Transferrin C2 variant does confer a risk for Alzheimer's disease. *J Alzheimers Dis.* 5: 423–427.

Zastawny, T.H., Altman, S.A., Randers-Eichhorn, L., Madurawe, R., Lumpkin, J.A., Dizdaroglu, M. and Rao, G. (1995) DNA base modifications and membrane damage in cultured mammalian cells treated with iron ions. *Free Radic Biol Med.* 18: 1013-1022.

Zecca, L., Gallorini, M., Schünemann, V., Trautwein, A.X., Gerlach, M., Riederer, P., Vezzoni, P. and Tampellini, D. (2001) Iron, neuromelanin and ferritin content in the substantia nigra of normal subjects at different ages: consequences for iron storage and neurodegenerative processes. *J Neurochem.* 76: 1766-1773.

Zecca, L., Shima, T., Stroppolo, A., Goj, C., Battiston, G.A., Gerbasi, R., Sarna, T. and Swartz, H.M. (1996) Interaction of neuromelanin and iron in substantia nigra and other areas of human brain. *Neuroscience*, 73: 407-415.

Zecca, L., Stroppolo, A., Gatti, A., Tampellini, D., Toscani, M., Gallorini, M., Giaveri, G., Arosio, P., Santambrogio, P., Fariello, R.G., Karatekin, E., Kleinman, M.H., Turro, N., Hornykiewicz, O. and Zucca, F.A. (2004a) The role of iron and copper molecules in the neuronal vulnerability of locus coeruleus and substantia nigra during aging. *Proc Natl Acad Sci U S A*, 101: 9843-9848.

Zecca, L., Youdim, M.B., Riederer, P., Connor, J.R. and Crichton, R.R. (2004b) Iron, brain ageing and neurodegenerative disorders. *Nat Rev Neurosci.* 5:863–873.

Zheng, W., Aschner, M. and Ghersi-Egea, J.F. (2003) Brain barrier systems: a new frontier in metal neurotoxicological research. *Toxicol Appl Pharmacol.* 192: 1–11.

Zheng, W., Zhao, Q., Slavkovich, V., Aschner, M. and Graziano, J.H. (1999) Alteration of iron homeostasis following chronic exposure to manganese in rats. *Brain Res.* 833: 125–132.

Zhou, B., Westaway, S.K., Levinson, B., Johnson, M.A., Gitschier, J. and Hayflick, S.J. (2001) A novel pantothenate kinase gene (PANK2) is defective in Hallervorden-Spatz syndrome. *Nat Genet.* 28: 345-349.

Zoghbi, H.Y. and Orr, H.T. (2000) Glutamine repeats and neurodegeneration. *Ann Rev Neurosci.* 23: 217–247.

In: Progress in Neurodegeneration
Editors: M.R. Avila-Costa and V.A. Martinez
ISBN 978-1-60741-317-2

Chapter 11

Mercury

Maria Luisa Juárez-Seres, Ana Luisa Gutierrez Valdéz and Maria Rosa Avila-Costa
Department of Neuroscience, Neuromorphology Lab
UNAM, Mexico

11.1. General Description

Mercury (Hg) is the 80th element of the Periodic table of elements. Hg is unique in that it is found in nature in several chemical and physical forms. At room temperature, elemental (or metallic) Hg exists as a liquid with a high vapor pressure and consequently is released into the environment as Hg vapor. Hg also exists as a cation with an oxidation state of +1 (mercurous) or 2+ (mercuric). Of the organic forms of Hg, methylmercury (MeHg) is the most frequently encountered compound in the environment. It is formed mainly as the result of methylation of inorganic (mercuric) forms of Hg by microorganisms in soil and water (Valko et al. 2005).

Hg is a naturally occurring element, which despite the fact that is uncommon in the earth's crust is made available for human contact by the processes of erosion and volcanism (Gochfeld, 2003). Hg exposure is the second-most common cause of toxic metal poisoning. Eventually it also enters the atmosphere through anthropogenic processes such as the combustion of fossil fuels. Once oxidized to ionic form in the atmosphere, it is washed by precipitation into bodies of fresh or saltwater (Dalton, 2004), here aquatic bacteria act on Hg and convert it to MeHg, which enters the aquatic food chain via plankton in the water. Fish then consume the plankton, which are at the bottom of the aquatic food chain, and the concentration of Hg within organisms increases as it moves through successive trophic levels. This biomagnification results in extraordinarily high concentrations in fish at the top of the food chain–as much as a million times the concentration in the surrounding waters (Dalton, 2004). Also, the elemental Hg content of dental amalgams has long been a topic of political and medical debate (Patrick, 2002).

According to the Agency for Toxic Substances and Disease Registry (ATSDR) of the U.S. Department of Health and Human Services, Hg is listed as the third-most frequently found (lead and arsenic are first and second), and the most toxic substance around the world (ATSDR, 2001).

Hg is found in the environment in three basic states: elemental Hg or Hg vapor, inorganic Hg, and organic Hg (ethyl-, methyl-, alkyl-, or phenylmercury). Each form has an individual toxicological profile and metabolic fate. On an individual exposure basis, the estimated intake and retention of elemental Hg vapor (from dental amalgams and atmospheric pollution) in non-occupationally exposed individuals has a much broader range (3.9-21.0μg /day) than either inorganic (4.3 μg /day) or MeHg (1-6 μg/day) exposure (ATSDR, 2001).

11.2. Sources of Mercury Exposure

Organic Hg is considered the most toxic and most frequent form of Hg exposure, organic Hg is found in fish, poultry that has been led fishmeal, pesticides, fungicides, insecticides, and thimerosal-containing vaccines. Thimerosal, which is 49.6% MeHg (a form of organic Hg), has been used as a preservative in vaccinations since the 1930s (Patrick, 2002).

Inorganic Hg (mercury salts) is found in cosmetic products, laxatives, teething powders, diuretics, and antiseptics (Ozuah, 2000). Inorganic Hg can be formed from the metabolism of elemental Hg vapor or MeHg (Clarkson, 2002). Although inorganic Hg does not normally reach the placenta or cross the blood–brain barrier (BBB), it has been found in the neonatal brain due to the absence of a fully formed BBB (Patrick, 2002).

11.2.1. Air

As a natural element Hg is ubiquitous in the environment, approximately 10,000 tons originates from degassing of earth's crust, to this amount approximately 20,000 tons/year is added by anthropogenic activity (Hansen and Dasher, 1997). Hg emissions from the coal smoke are the main source of anthropogenic discharge and Hg pollution in atmosphere. It is estimated that the Hg emissions will increase at a rate of 5% a year (Zhang et al. 2002).

Other important contributors to regional emissions included municipal waste combustion (5.6%), mercury-cell chlor-alkali plants and hazardous-waste incinerators (4% each), stationary internal combustion engines (3.5%), industrial, commercial and institutional boilers (3.3%) and lime manufacturing (3.0%) and medical waste incineration (1%) (Murray and Holmes, 2004).

11.2.2. Water

Hg in air eventually passes into rivers, lakes and oceans after traveling long distances together with wind. With Hg contaminating rain (Domagalski et al. 2004), ground and seawater, no one is safe (Beldowski and Pempkowiak, 2003).

Hg cloud water concentrations ranged from 7.5 to 71.8 ng 1 (−1), with a mean of 24.8 ng 1 (−1). Liquid water content explained about 60% of the variability in Hg cloud concentrations (Malcolm et al. 2003).

There are also linkages between acidic deposition and fish Hg contamination and eutrophication of estuaries (Driscoll et al. 2003). Numerous factories that directly pump untreated effluents pollute groundwater (Zahir et al. 2005).

11.2.3. Food

11.2.3.1. Food of Animal Origin

The emitted Hg both natural and anthropogenic is in an inorganic form predominantly metallic vapor, which is carried off to great distances by winds and eventually falls in water bodies. In aquatic environments, inorganic Hg is microbiologically transformed into lipophilic organic compound, MeHg. This transformation makes Hg more prone to biomagnification in food chains. Consequently, populations with traditionally high dietary intake of food originating from fresh or marine environment have highest dietary exposure to Hg (Hansen and Dasher, 1997)

11.2.3.2. Food of Plant Origin

Emissions of Hg from the province of Guizhou in Southwestern China to the global atmosphere have been estimated to be approximately 12% of the world total anthropogenic emissions primarily due to mining, chemical discharge and electricity production (Zahir et al. 2005). Even though the major source of Hg is inorganic, it was observed that active transformation of inorganic Hg to organic Hg species, MeHg takes place in water, sediments and soils. It has been reported that the concentration of Hg in rice grains can reach up to 569μg/kg of total Hg of which 145μg/kg is in MeHg form (Horvat et al. 2003).

11.2.4. Pharmaceuticals and Utility Products

Hg has always been a popular choice for dental amalgams. Skin whitening creams and soaps from developing countries is a recognized source of chronic Hg poisoning (Harada et al. 2001). Hg level of almost 2000 times above the allowable limit was found in blood of Indonesian domestic workers (Soo et al. 2003).

11.3. Absorption, Distribution, Metabolism, and Excretion

Elemental Hg, found in thermometers, thermostats, dental amalgams, and Hg added to latex paint, eventually enters a vaporized state. Eighty percent of inhaled elementary Hg vapor is absorbed and can cross the BBB or reach the placenta (Ozuah, 2000). Hg vapor in the gastrointestinal tract is converted to mercuric sulfide and excreted in the feces (National

Research Council, 2000). Hg vapor in the kidneys, however, the main repository for elemental Hg, is carried to all parts of the central nervous system (CNS) as a lipid-soluble gas. Hg vapor can also be oxidized to inorganic Hg by catalase and can attach to the thiol groups in most proteins--enzymes, glutathione, or almost any structural protein (Clarkson, 2002). Elemental Hg can also be methylated by microorganisms in soil and water and potentially the human gastrointestinal track (Yannai, 1991) where it can then be transformed into organic MeHg, the form found in fish, fungicides, and pesticides.

MeHg is absorbed in the gastrointestinal tract where 95% of that which is ingested is absorbed and subsequently distributed to all tissues within about 36h (Ozuah, 2000; Clarkson, 2002). MeHg is known to cross the placental barrier and has been detected in cord blood in concentrations greater than those found in maternal blood. It has also been detected in the fetal brain at concentrations 5–7-fold higher than that found in the maternal blood (Clarkson, 2002). MeHg also affects the permeability of the BBB (Charleston et al. 1996). Once MeHg permeates the BBB, it preferentially collects in astrocytes, which have the ability to modulate neurotoxicity (Aschner, 1996).

MeHg is present in the body as a water-soluble complex, mainly with the sulfur atom of thiol ligands (Clarkson, 2002), and crosses the BBB complexed with L-cysteine in a molecule resembling methionine (Patrick, 2002) (Figure 1).

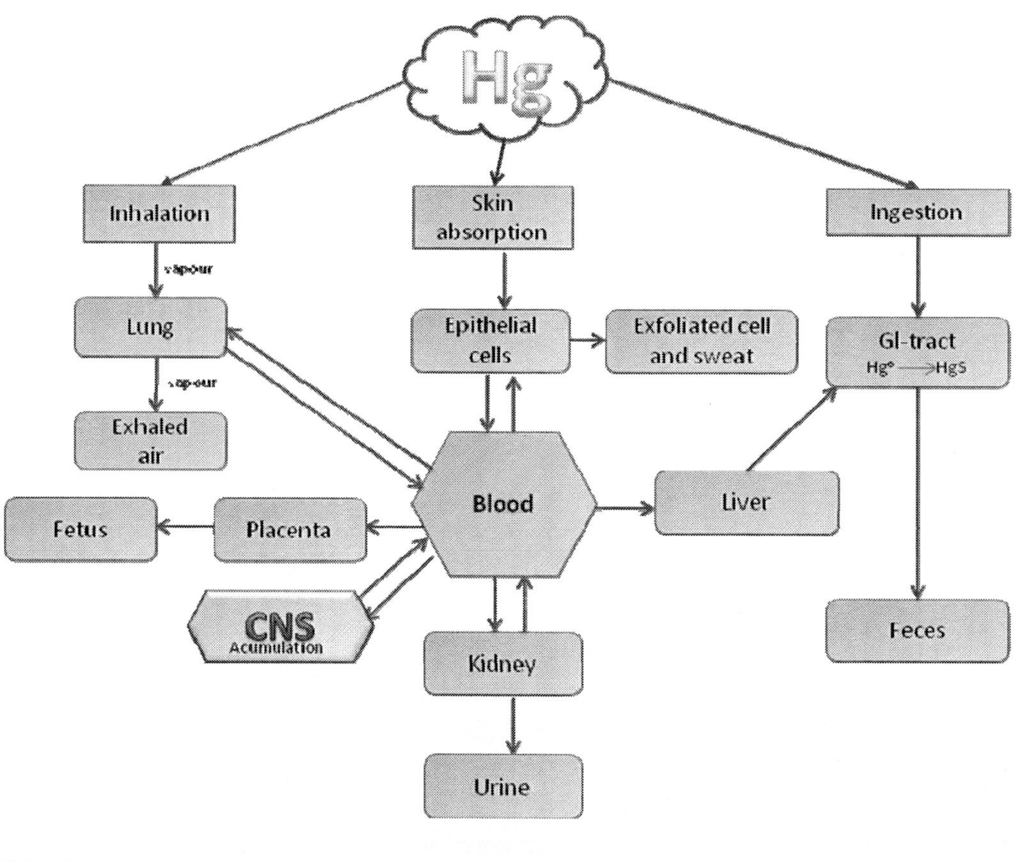

Figure 1. Mercury absorption, distribution and excretion.

11.4. Toxicity

Elemental Hg and its metabolites have the toxic effect of denaturing biological proteins, inhibiting enzymes, and interrupting membrane transport and the uptake and release of neurotransmitters (Clarkson, 2002). Chronic exposure most commonly manifests as a triad of increased excitability and irritability, tremors, and gingivitis (Ozuah, 2000). Less commonly, chronic exposure causes central and peripheral nervous system damage, manifesting as a characteristic fine tremor of the extremities and facial muscles, emotional lability, and irritability. Rarely, significant exposure can cause acrodynia or "pink disease," involving a pink rash on the extremities, pruritis, paresthesias, and pain (Tunnessen et al. 1987).

The toxicity of MeHg in occupational settings has been recognized dating back to the 19th century (Clarkson et al. 2003). In recent years concern was directed toward the consequences of environmental exposures. The most significant incidents resulting in neurotoxic effects in humans occurred in Japan and Iraq. In Japan the ingestion of contaminated fish resulted in human poisoning (Eto, 2000). The symptoms of this disease include sensory disturbances in the extremities, ataxia, disequilibrium, constriction of the visual field, impairment of gait and speech, muscle weakness, tremor, abnormal eye movement, and hearing impairment (Eto, 1997). It is also clear that the developing fetus is far more sensitive to MeHg exposure than the adult (WHO, 2003), and fetal exposure resulted in cerebral palsy (Eto, 1997).

In Iraq in 1972, bread made from grain treated with a methylmercurial fungicide poisoned 6530 individuals (Bakir et al. 1973). The relationship between symptoms and exposure was dose dependent (Marsh et al. 1987). South American gold mining operations are yet another site of human exposure. In these settings children were exposed both to elemental Hg vapor from Hg amalgam used in the extraction process, as well as to MeHg via the consumption of contaminated fish in the region. Counter et al. (2002) found that exposed children in the vicinity of an Andean goldmine had blood Hg levels that ranged from 4 to 89 μg L-1. Note that the current US EPA reference dose (RfD) for MeHg is 5.8 μg L-1 (CDC, 2004). Some of these children also exhibited neurological complaints consistent with Hg poisoning.

11.4.1. Mechanisms of Mercury Toxicity

Hg can cause biochemical damage to tissues and genes through diverse mechanisms, such as interrupting intracellular calcium (Ca^{2+}) homeostasis, disrupting membrane potential, altering protein synthesis, and interrupting excitatory amino acid pathways in the CNS (Yee and Choi, 1996). Mitochondrial damage, lipid peroxidation, microtubule destruction (National Research Council, 2000) and the neurotoxic accumulation of serotonin, aspartate, and glutamate are all mechanisms of MeHg neurotoxicity (Yee and Choi, 1996).

Glutathione, as both a carrier of Hg and an antioxidant, has three specific roles in protecting the body from Hg toxicity. First, glutathione, specifically binding with MeHg, forms a complex that prevents Hg from binding to cellular proteins and causing damage to both enzymes and tissue (Kromidas et al. 1990). Glutathione-mercury complexes also reduce intracellular damage by preventing Hg froth entering tissue cells and becoming an intracellular toxin (Patrick, 2002).

Second, glutathione-mercury complexes have been found in the liver, kidney, and brain, and appear to be the primary form in which Hg is transported and eliminated from the body (Zalups, 2000). The transport mechanism is unclear, but complexes of glutathione and Hg are the predominant form of Hg in both the bile and the urine (Ballatori and Clarkson, 1984). Glutathione and cysteine, acting as carriers of Hg, actually appear to control the rate of Hg efflux into bile; the rate of Hg secretion in bile appears to be independent of actual bile flow (Ballatori and Clarkson, 1983).

Cells of the BBB (brain capillary endothelial cells) release Hg in a glutathione complex. Inhibiting glutathione production in these cells inhibits their ability to release Hg (Kerper et al. 1996). Hg accumulates in the CNS primarily in astrocytes, the cells that provide the first line of defense for the CNS against toxic compounds (Tiffany-Castiglion and Qian, 2001). Astrocytes are the first cells in brain tissue to encounter metals crossing the BBB. They also contain high levels of metallothionein and glutathione, both carriers for heavy metals. It is hypothesized that astrocytes are the main depository of Hg in the brain (Cookson and Pentreath, 1996). In studies with astrocytes, the addition of glutathione, glutathione stimulators, or glutathione precursors significantly enhances the release of Hg from these cells in a complex with glutathione. Fujiyama et al. (1994) also suggest that conjugation with glutathione is the major pathway for Hg efflux from astrocytes. Glutathione also increases Hg elimination from renal tissue. Studies in mammalian renal cells reveal glutathione is 50% as effective as the chelating agent DMSA (2,3-dimercaptosuccinic acid) in preventing inorganic Hg accumulation in renal cells (Endo and Sakata, 1995).

Third, glutathione increases the antioxidant capacity of the cell, providing a defense against hydrogen peroxide, singlet oxygen, hydroxyl radicals, and lipid peroxides produced by Hg (Patrick, 2002). The addition of glutathione to cell cultures exposed to MeHg also prevented the reduction of cellular levels of glutathione peroxidase, a crucial antioxidant enzyme necessary for protection against the damaging effects of lipid peroxidation (Kromidas et al. 1990).

11.5. Mercury in the Brain

The nervous system provides an important site for the deposition of Hg whose neurotoxicity begins with its facility to cross the BBB and enter cells. The pivotal chemical interaction is methylmercury's ability to bind to sulfhydryl groups (Aschner and Clarkson, 1988). The complex thus formed makes use of translocator molecules within the cell membrane, thus enabling metallic influx into cells (Clarkson, 1993; Ballatori, 2002). Brain uptake is specifically linked to binding to L-cysteine and transfer via the neutral amino acid carrier site (Ballatori, 2002; Aschner et al. 1990). This phenomenon is possible as a result of molecular mimicry, which in this context is defined as encompassing those situations in which "a toxic metal forms a complex with an endogenous ligand such that the resulting compound mimics the behavior of a normal substrate". In this case the L-cysteine-methylmercury complex was found to mimic L-methionine (Clarkson, 1993). This author speculated that the mimicry facilitates membrane transport because certain carriers do not have highly specific structural requirements for their substrates, as do enzymes and their receptors. It was also demonstrated that administration of L-methionine was inhibitory to the uptake of MeHg in the rat brain (Aschner and Clarkson, 1988). Since the MeHg-cysteine complex mimics methionine, this observation is consistent with the notion of competition for a membrane transporter. To further confirm this relationship, it was shown that the methylmercury-L-cysteine complex inhibited the uptake of radiolabeled L-methionine (Aschner and Clarkson, 1988). Similar results have been demonstrated in brain capillary endothelial cells *in vitro* (Aschner and Clarkson, 1989). Moreover, Kajiwara et al. (1996) reported results, which support this same transport on the neutral amino acid carrier as the mode by which MeHg crosses the placental barrier from mother to fetus. More recently two specific L-type amino acid transporter proteins, LAT-1 and LAT-2 were identified (Simmons-Willis et al. 2002). Both were found in brain tissue and LAT-1 was expressed at high levels in the endothelial cells of the BBB (Simmons-Willis et al. 2002).

11.5.1. Mercury Transport to Brain

In comparative studies with different Hg species, MeHg has been revealed as the most toxic form (Toimela and Tähti, 2004; Silva-Pereira et al. 2005). This fact can be associated to MeHg ability to easily cross through the BBB, thus accumulating in different brain areas as the cerebellum, brain cortex and retina (Erie et al. 2005).

The microenvironment of the brain parenchyma is separated from fluctuations in ion and metabolite concentrations in the blood by two barrier structures. The barrier, which separates the systemic blood circulation from the interstitial fluid, is defined as the BBB, while the one that separates the systemic circulation from the cerebrospinal-fluid (CSF) compartment is known as the blood–CSF barrier (BCB). During the last few decades, the role of brain barrier systems in controlling CNS homeostasis has received substantial attention (Zheng et al. 2003).

Because of its high lipophilicity, MeHg by itself is capable of diffusing through the cell membrane without any specific carrier system; however, given its reactivity toward sulfhydryl groups, the amount of free MeHg in biological fluids at any given time is likely to be infinitesimal. An active process at the BBB has been proposed to transport MeHg into the brain. In blood, this highly neurotoxic metal is bound almost exclusively to proteins and low-molecular-weight sulfhydryl-containing compounds, such as cysteine. The increased uptake of MeHg following its conjugation with L-cysteine is inhibited by co-injection of other large neutral amino acids (such as methionine), but not by injection of acidic amino acids (such as L-aspartic acid). These characteristics are consistent with L system-mediated transport. Thus, a MeHg complex is formed that mimics the behavior of normal endogenous substrates, utilizing transport systems inherent to the BBB to gain access to the CNS (Zheng et al. 2003).

In general, inorganic mercurial salts are less neurotoxic than organic mercurial compounds, most likely because their rate of transport into the CNS parenchyma is lower than organic mercurials. Nevertheless some inorganic Hg compounds, such as mercuric chloride ($HgCl_2$), can act as direct barrier toxicants. In a cat study by Peterson and Cardoso

(1983), an intracarotid injection of $HgCl_2$ induced BBB damage, as manifested by fluorescent display of Evans blue staining in brain parenchyma. Direct application of Hg containing tissue fixative to the brain tissue corroborated diffusion of intravenously administered Evans blue into brain parenchyma, suggesting an immediate breakdown of the BBB (Marlin et al. 1980). By tracing the ultrastructural localization of cerebral alkaline phosphatase, a cerebro-microvascular marker primarily located on the luminal side of cerebral endothelial cells, Albrecht et al. (1994) observed a disappearance of alkaline phosphatase reactivity from luminal endothelial cell membranes in $HgCl_2$-exposed animals. This was accompanied by ultrastructural changes typical of the formation of "leaky" microvessels.

11.5.2. Mercury Neurotoxicity

Vulnerable periods during the development of the nervous system are sensitive to environmental insults because they are dependent on the temporal and regional emergence of critical developmental processes (i.e. proliferation, migration, differentiation, synaptogenesis, myelination, and apoptosis). Evidence from numerous sources demonstrates that neural development extends from the embryonic period through adolescence. Different behavioral domains (e.g. sensory, motor and various cognitive functions) are subserved by diverse brain areas (Rice and Barone, 2000).

Low concentrations of some metals, including Hg can directly induce α–synuclein fibril formation, which are the major constituent of intracellular protein inclusions (Lewy bodies and Lewy neurites) in dopaminergic neurons of the substantia nigra leading to Parkinson disease (Uversky et al. 2001). Moreover, low concentrations of cobalt and Hg are able to induce oxidative stress, cell cytotoxicity and increase the secretion of β–amyloid 1–40 and 1–42 which may lead to neurodegenerative alterations, such as Alzheimer and Parkinson diseases (Olivieri et al. 2002) (Figure 2).

The effects of Hg exposure at levels around 0.05 mg/m^3 or lower have been of concern, consist of increased complaints of tiredness, memory disturbance, subclinical finger tremor, abnormal EEG by computerized analysis and impaired performance in neurobehavioral or neuropsychological tests (Satoh, 2000). Neuropsychological effects in Hg vapor exposed male chloralkali workers with low concentrations of urinary Hg mean U-Hg 5.9 nmol/mmol creatinine indicated lowering of visuomotor/psychomotor speed and attention, and immediate visual memory (Ellingsen et al. 2001). Depression and impairment of short-term auditory memory was found in workers exposed to low levels of Hg (Soleo et al. 1990).

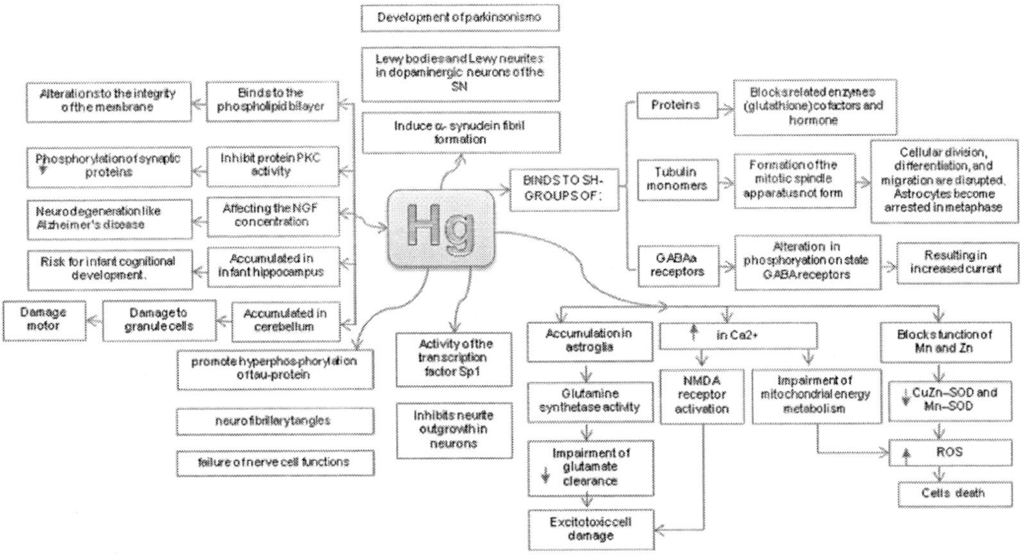

Figure 2. Effects of Hg exposure in the CNS. Abbreviations: Sp1, activity of specificity protein 1 (a transcription factor involved in the regulation of the *APP* gene); GABA, gamma-aminobutyric acid, BBB, blood–brain barrier; SH sulfhydryl group; NGF, nerve growth factor; SN Sustantia nigra

Following exposure, Hg accumulates in the BBB and BCB. Autopsy data from a Minamata Bay accident victim show that total Hg remained high in the brain as long as 26 years after exposure. Hg deposition was histochemically localized to microglial cells and Bergmann's glial cells, in neurons of specific brain areas, and in epithelial cells of the choroid plexus (Takeuchi et al. 1979). Studies in Hg-treated animals also demonstrate significant cerebral edema, vacuolar change, spongy degeneration, and the loss of parenchyma (Choi et al. 1988).

Concerning the neurotoxic effects induced by MeHg, it has been reported that pre- and early postnatal stages in the development of the CNS are very sensitive to the toxic effects of MeHg. The influence of MeHg on the level of nerve growth factor (NGF) during the development of CNS was studied by Lärkfors et al. (1991). They analyzed the level of NGF in cortical areas and in the septum with a sensitive enzyme immunoassay. The pups exposed to MeHg exhibited a 50% elevation in the level of NGF in the hippocampus on postnatal day 25 and postnatal day 50 compared to control animals. Concomitantly, the level of NGF decreased by 30% in the septum on P 25 and P 50, suggesting that the retrograde transport of NGF from hippocampus to septum could be affected by the exposure of MeHg. The exact mechanism by which the low level of Hg is affecting the NGF concentration in the developing brain is yet unknown. The increase of NGF in the hippocampus and the decrease of NGF measured in the septum could reflect altered conditions for neurotrophic support in these areas of the brain as a result of the exposure to heavy metal. Thus, this finding might indicate a connection between exposure of heavy metals and neurodegeneration, such as that found in the basal forebrain in Alzheimer disease (Lärkfors et al. 1991).

Detectable subtle effects on brain function in domains of language, memory and motor appeared at prenatal MeHg exposure particularly during second trimester. Neurobehavioral dysfunction was reported even if maternal hair Hg is 6 μg/g; corresponding value for blood is approximately 24 μg/L (Grandjean et al. 1994).

On the other hand, Autism is a disorder that can lead to life-long disability. Though not proved, there is potential link between Hg toxicity and autism in children (Lee et al. 2003). Subtle neurological disorders in children over Hg exposure have been widely reported (Counter and Buchanan, 2004; Johnson, 2004).

It is well known that organic and inorganic mercurials cause neurobehavioral changes in both animals and humans, mainly during early phases of development, leading to loss of cognitive and motor functions in the childhood (Roegge et al. 2004).

Consumption of high MeHg levels during pregnancy may lead to encephalopathy in the offspring (Davis et al. 1994). There are distinct differences in the distribution of pathological changes in young compared to adult brain upon MeHg exposure (Takeuchi et al. 1979). CNS damage following MeHg exposure in adults is primarily in specific areas, such as the granule layer of cerebellum and the visual cortex of cerebrum. When exposures to MeHg occur *in utero* or at an early age, the damage to the CNS is ubiquitous. Generally, the earlier the exposure, the more generalized the damage (Choi, 1989). It has, thus, been hypothesized that, in addition to a higher susceptibility of immature or differentiating neural cells, these differences are also caused by an immature BBB, leading to a more generalized distribution of MeHg in the developing brain. After adult exposure, however, MeHg is found throughout the brain, and the localization does not correlate with pathological changes, suggesting distinct vulnerability of various regions to this metal (Zheng et al. 2003).

Furthermore, it has been speculated that the symptoms even may appear later in life as a neurodegenerative disorder such as Parkinson or Alzheimer diseases. Moreover, decreasing levels of NGF could hypothetically cause degeneration amongst the affected cells, since NGF supports the cholinergic neurons in the basal forebrain, known to degenerate in Alzheimer disease. The exact mechanism by which MeHg is affecting the developing brain is not clear. However, MeHg and other organic mercurials are hydrophobic and subsequently penetrate easily the BBB thereby causing severe damage (Aschner and Aschner, 1990). MeHg has a high affinity for binding to sulfhydryl and chloride ligands and thereby interferes with many metabolic reactions (Magos, 1987).

Some clinical signs in motor impairment have been observed in adult humans after environmental exposure to MeHg (Auger et al. 2005; Zahir et al. 2005). Moreover, the inhibitory effects of MeHg in rodent locomotion have been extensively reported (Dietrich et al. 2005). Thus, two functional tests related to the locomotor activity in open field (rearing and ambulation) were selected to assess the neurological changes, and apply as potential predictive biomarkers of Hg neurotoxic effects (dos santos et al. 2007).

Studies with MeHg *in vitro* have shown that one initial site of damage appears to be on the neuritic membranes (Choi, 1989); the disruption and degeneration of neuritic processes have been characterized by the disappearance of microtubules, necessary for structural support and for axoplasmic transport. Thus, MeHg binds to SH-groups on the tubulin monomers to disrupt the assembly processes (Margolis and Wilson, 1981). In addition, it causes depolymerization of microtubules (Abe et al. 1975).

Some epidemiological studies reported that the MeHg related neuropsychological deficits were mainly found in the domains of cognitional parts, such as language attention, memory, and so forth (Grandjean et al. 1997), the deficits of inorganic Hg to children is seldom reported. In a study Feng et al. (2004) found that Hg was mainly accumulated in infant hippocampus and cerebellum, whereas for maternal rats, Hg was almost average stored in all brain regions. These results suggested that the toxicity of such Hg exposure between mothers and their offspring is different. Since hippocampus and cerebellum are closely related to the function of learning and memory in brain regions, therefore, they assumed that prenatal and weaning exposure to low dose of inorganic Hg would have a risk for infant cognitional development.

Measurement of tremor has been used in several occupational studies of workers with long-term exposure to Hg vapor (Hg^0). Tremor is defined as an involuntary rhythmic movement of a body part, and may originate from mechanical oscillations produced in sensorimotor loops, or from normal and pathological oscillators in central neuronal networks (Elble, 1996). As reviewed by Beuter and de Geoffroy (1996), and in some more recent studies (Ellingsen et al. 2001; McCullough et al. 2001; Lucchini et al. 2002), as well as in studies of previously exposed workers (Powell, 2000; Frumkin et al. 2001; Bast-Pettersen et al. 2005). Increased tremor has been observed at urinary Hg levels about 30-μg/g creatinine (WHO, 2003), but some studies indicate a possible effect at even lower exposure levels (Chapman et al. 1990; Langworth et al. 1992).

Franco et al. (2007) examined the effects of $HgCl_2$ exposure exclusively through maternal milk on biochemical parameters related to oxidative stress (glutathione and thiobarbituric acid reactive substances levels, glutathione peroxidase and glutathione reductase activities) in the cerebellum of weanling mice. These parameters were also evaluated in the cerebellum of mothers, which were subjected to intraperitoneal injections of $HgCl_2$ (0, 0.5 and 1.5 mg/Kg, once a day) during the lactational period. Considering the relationship between cerebellar function and motor activity, the presence of motor impairment was also evaluated in the offspring exposed to $HgCl_2$ during lactation. After treatments, pups lactationally exposed to inorganic Hg showed high levels of this metal in the cerebellar tissue, as well as significant impairment in motor performance in the rotarod task and decreased locomotor activity in the open field. Offspring and dams did not show changes in cerebellar glutathione levels or glutathione peroxidase activity. In pups, lactational exposure to inorganic Hg significantly increased cerebellar lipoperoxidation, as well as the activity of cerebellar glutathione reductase. However, these phenomena were not observed in dams. These results indicate that inorganic Hg exposure through maternal milk is capable of inducing biochemical changes in the cerebellum of weanling mice, as well as motor deficit and these phenomena appear to be related to the pro-oxidative properties of $HgCl_2$ (Franco et al. 2007).

Finally, it has been suggested that concentrations of 2.5–10 μM of MeHg in the brain (estimated from human blood and hair mercury levels) have been associated with delayed psychomotor development in children and adults with minimal signs of MeHg poisoning (Carta et al. 2003; Pinheiro et al. 2006). The results of Crespo-López et al. (2007) suggest that exposure to levels ten times lower than those referred may lead to genotoxic consequences for the developing CNS, pointing to the necessity of reviewing the tolerance values published in 2003 by the World Health Organization (WHO-IPCS, 2003).

11.6. Mechanism of Mercury-Induced Neurotoxicity

As we mentioned earlier, Hg binds to sulfhydryl groups of proteins and disulfide groups in amino acids resulting in inactivation of sulfur and blocks related enzymes, cofactors hormones (Markovich and James, 1999). Besides this, it also alters permeability of cellular membrane by binding to sulfhydryl radical (Bapu et al. 1994). Blocked or inhibited sulfur oxidation at cellular levels has been found in many chronic neurodegenerative disorders, including Parkinson disease (PD), Alzheimer disease (AD), Amiotrophic Lateral sclerosis (ALS), Lupus, Rheumatoid arthritis, Autism, etc. (Wilkinson and Waring, 2002) (Figure 2).

Hg neurotoxicity to cerebellum at higher doses has been related to impairment of motor function (Marcelo et al. 2005) and its genotoxicity to neuronal cells in fetal state may result in abnormal offspring or fetal deaths but its exact mode of activity at low doses, particularly at environmentally relevant concentrations which lead to subtle delays in neurodevelopment remain unexplored. Basically it blocks essential functional groups in biomolecules and also displaces essential metal ions from them. Mercuric ion is known as one of the strongest thiol-binding agents. Intracellular Hg therefore attaches itself to thiol residues of proteins particularly glutathione and cysteine resulting in inactivation of sulfur and blocks related enzymes, cofactors and hormones (Mathieson, 1995). Its molecular interactions with sulfhydryl groups in molecules of albumin, metallothionein, glutathione, and cysteine have been implicated in mechanisms involved in renal (Zalups, 2000) and neuronal toxicity (Fonnum and Lock, 2004; James et al. 2005).

Examination of victims of Minamata disease revealed CNS damage involving cell lysis, especially in neuroglia and granule cells (Segall and Wood, 1974). At the molecular level, MeHg binds to the phospholipid bilayer, which results in alterations to the integrity of the membrane (Girault et al. 1997). Recent work using *Drosophila* CNS cell lines points to MeHg interference with Notch, a transmembrane protein integral to an important signaling pathway. MeHg thereby activates a gene [E(spl)] which represses differentiation of neuroectodermal precursor cells and inhibits neurite outgrowth in neurons (Bland and Rand, 2006). A more circuitous route observed in human astrocytoma D384 cells results in membrane damage from lipoperoxidation following lysosomal rupture which releases proteolytic enzymes (Dare et al. 2001). It is noteworthy that Roberg et al. (1999) also observed that rupture of the lysosome occurs as a result of oxidative stress, and that this destabilization is an early event in the sequence of toxic events within the cell (Reardon and Bhat, 2007).

MeHg disrupts intracellular Ca^{2+} homeostasis (Shafer and Atchison, 1991; Shafer et al. 2002; Faustman et al. 2002). Shafer and Atchison (1991) were first to report that exposure to micromolar concentrations of MeHg disrupted Ca^{2+} currents in rat pheochromocytoma (PC12) cells in a concentration-dependent manner. Furthermore, Shafer et al. (2002) made the important discovery that prolonged low-dose exposure (i.e. in the nanomolar range) also significantly reduced Ca^{2+} currents in PC12 cells. Other disruptions in Ca^{2+} homeostasis may involve alteration of plasma membrane permeability as well as the release of intracellular Ca^{2+} stores (Faustman et al. 2002). MeHg also reduces voltage-activated Ca^{2+} channel currents in rat dorsal root ganglion neurons, most likely by binding to the outside of the channel pore (Leonhardt et al. 1996). Increased intracellular Ca^{2+}results in inhibition of mitochondrial phosphorylation and, thus prevents cells from getting the needed high-energy phosphate (Braughler et al. 1985). This increased intracellular Ca^{2+} ion may also exacerbate oxidative free radical injury as well as activate degradative enzymes such as phopholipases and proteases (Zhang et al. 2003; Shanker and Aschner, 2003; Kang et al. 2006). Ultimately increased intracellular Ca^{2+} can activate signal transduction pathways, which may result in cell death (Marty and Atchison, 1998).

Depending on the dosage, MeHg exposure may result in either necrosis or apoptosis in affected cells (Castoldi et al. 2000). There is evidence that MeHg induces cell cycle arrest in the G2/M phase leading to apoptosis (Miura et al. 1999). Ou et al. (1999) observed that MeHg led to the induction of the p21 cell cycle regulatory gene, which is implicated in G1, and G2 phases of cell cycle arrest. Transcriptional profiling revealed the activation of genes related to intracellular signaling, cell cycling and apoptosis (Wilke et al. 2003). The role of p53, another cell cycle regulatory protein, is also of interest. Recently Gribble et al. (2005) observed in murine fibroblasts both p53-dependent and independent pathways for cell cycle arrest while cell death itself was dependent on p53.

On the other hand it has been reported that fetus is especially vulnerable to MeHg since developing fetal brain processes, such as cellular division, differentiation, and migration are disrupted by binding of Hg to thiol groups of tubulin, the principal protein constituent of neuronal microtubules (Clarkson, 1992). The chromosomal genotoxicity of Hg salts could be due to interaction of Hg^{2+} with the motor protein kinesin mediating cellular transport processes (Zahir et al. 2005). Genotoxicity of mercurials could have far reaching consequences ranging from birth of abnormal offspring to neurodegenerative disorders. Recently in a major breakthrough, rise in apolipoprotein-E ε4 genotype has been proposed as a biomarker for low-dose Hg toxicity (Godfrey et al. 2003) and rise in apolipoprotein-E due to Hg has been advocated as a pathogenic factor for AD (Godfrey et al. 2003).

11.6.1. Oxidative Stress

Reactive oxygen species (ROS) include superoxide anion, (O_2), hydrogen peroxide (H_2O_2) and hydroxyl radical ($HO^{\bullet}$) (Boelsterli, 2003). Though ROS are produced at a basal level by the mitochondrial respiratory chain and in cytochrome p450 (CYP)-mediated oxidation processes (Boelsterli, 2003), cells have endogenous antioxidants (e.g., cytochrome oxidase, superoxide dismutases, catalases, peroxidases, glutathione, to name a few) to control ROS (Buonocore et al. 2001). Certain disease states as well as pathologies resulting from toxic substances, including MeHg, produce ROS in numbers that can devastate the cellular defense mechanisms and result in oxidative stress. ROS can damage vital cellular molecules such as DNA, proteins, and lipids. Since ROS can also play a role via redox signaling in the regulation of signal transduction (Suzuki et al. 1997), excess production of these substances may prove lethal to the cell.

At the cellular level, Hg interferes with mitochondrial respiration, causing oxidative stress (Konigsberg et al. 2001). Hg exposure produces increased levels of ROS in the motor neurons of mice and in cultured astrocytes and in various regions of the brains of adult male Sprague-Dawley rats (Hussain et al. 1997; Brawer et al. 1998; Pamphlett et al. 1998).

Exposure of CNS cells to MeHg results in the preferential accumulation of this compound in astrocytes (Aschner, 1996) and in an increase in ROS (Atchison and Hare, 1994). The cell's endogenous defenses against ROS include metallothioneins, heat shock proteins, heme oxygenase, and glutathione (Aschner et al. 1997; Sweet and Zelikoff, 2001). Ample evidence has accumulated to support the involvement of ROS in MeHg neurotoxicity. Though MeHg's ability to interfere with protein synthesis as well as with RNA synthesis was demonstrated (Reardon and Bhat, 2007), even greater cellular toxicity, beyond that for which RNA and protein synthesis disruption can account, appears to be associated with the action of ROS. Lipoperoxidation was identified as one manifestation of MeHg toxicity (Sarafian and Verity, 1991), but appears again not to be the primary cause of cellular toxicity. This is demonstrated by the fact that administration of α-tocopherol, which scavenges for free radicals at the plasma membrane, to cerebellar granule cells (CGCs) decreased MeHg-induced lipoperoxidation without markedly improving cell viability (Sarafian and Verity, 1991).

A persistent question concerns the relationship between oxidative stress and the disruption of Ca^{2+} homeostasis. Still unknown is the exact sequence of events by which MeHg exposure culminates in cell death. Marty and Atchison (1998) showed in rat CGCs results consistent with previous studies by finding that MeHg exposures of 0.5 and 1.0 mM resulted in a two-phase increase in intracellular Ca^{2+} concentration. The first phase was a result of release of Ca^{2+} from intracellular Ca^{2+} stores while the second phase was due to influx of Ca^{2+} across the cell membrane. They noted that the increase of Ca^{2+} concentration after MeHg exposure did not immediately produce cell death, but did decrease cell viability in a concentration-dependent manner after 3.5 h (Marty and Atchison, 1998). They also found that two voltage-dependent Ca^{2+} channel blockers (nifedipine and ω-conotoxin-MVIIC) delayed but did not eliminate both phases of increased Ca^{2+} in the cell (Marty and Atchison, 1998). Cell viability was observed to be as in controls at 3.5 h. Use of 1, 2-bis (2-aminophenoxy) ethane-N,N, N',N',-tetraacetic acid a Ca^{2+}-specific chelator, also protected cells initially. However, this protective effect was limited and after 24 h cell death occurred (Marty and Atchison, 1998). Subsequent work by Castoldi et al. (2000) on rat CGCs also showed that after 18 h of 0.5–10 mM MeHg exposure, there was no increased survival of cells observed after pre-treatment of cells with an NMDA receptor antagonist, a Ca^{2+} channel blocker, or two Ca^{2+} chelators. These results leave unresolved the question of whether Ca^{2+} changes precede oxidative stress or are a consequence of it (Reardon and Bhat, 2007).

Earlier studies using cultured CNS cells reported that the mitochondria are a principal site of MeHg accumulation as well as the most likely site for ROS generation (Yee and Choi, 1996). Investigations using rat C6 cells, which are a glial model, pointed to oxidation in both mitochondria and nuclei following MeHg exposure (Belletti et al. 2002). These authors, using MeHg concentrations in the 10^{-5} to 10^{-8} M range, used both visual and biochemical analysis in a time course assessment to present evidence of concentration-dependent mitochondrial dysfunction resulting from changes in the mitochondrial membrane potential. Nuclear involvement was evidenced by the generation of 8-hydroxy- 20-deoxyguanosine (8-OHdG), a marker of oxidative DNA damage. These authors proposed a sequence of events based on their observations: ROS generation leading to oxidative DNA damage which parallels a simultaneous change in the mitochondrial membrane potential due to a permeability transition. Of particular interest is the fact that the production of ROS was noted within 5 min of exposure to MeHg, and at 30 min there was a significant doubling of the fluorescence from the marker chloromethyldihydrodichlorofluorescein diacetate (or CM-H2DCFDA) for ROS. These results are compared to the earlier findings of Yee and Choi (1996) that inhibition of oxygen uptake into CNS cell cultures (astrocytes, oligodendrocytes, cerebral cortical neurons, and cerebellar granule neurons) occurred rapidly within seconds following MeHg exposure. Their findings also implicated the cytochrome C reduction–oxidation process as the locus of adverse effects of the MeHg (Yee and Choi, 1996). Fonfria et al. (2002) provided further details of the mitochondrial involvement by determining that MeHg exposure in CGCs resulted in the translocation of the apoptosis-inducing factor from the mitochondrial intermembrane space to the nucleus. This transition resulted in chromatin condensation and DNA degradation (Reardon and Bhat, 2007).

The recent work of Garg and Chang (2006) with murine microglia demonstrated that MeHg exposure led to the ROS-induced inactivation of mitochondrial aconitase. Aconitase, an essential enzyme in the Krebs cycle, is an iron-regulatory protein, which is sensitive to oxidative stress. Inactivation of this enzyme resulted in a large iron influx into the cells, which may underlie at least in part the resulting cytotoxicity. This work was also significant in its finding for the first time of ROS production in microglia as well as increased IL-6 production following MeHg exposure (Garg and Chang, 2006).

11.6.2. Neurotransmission

Hg obstructs neurotransmission by acting as a strong competitive inhibitor of muscarinic cholinergic receptors (Coccini et al. 2000). Hg also affects the GABAergic system. In the 1980s it was reported that treatment of rats with MeHg for either 3 or 20 days increased the total number of benzodiazepine-binding sites on GABAergic cells, although GABA binding did not change (Corda et al. 1981). This increase in benzodiazepine-binding sites probably resulted in increasing the frequency of channel opening, without altering the conductance or duration of its opening. Such results would partially explain why primary cultures from rat dorsal root ganglion neurons treated with 1–10 AM $HgCl_2$ resulted in an increase in GABA currents (Arakawa et al. 1991).

A suggested mechanism for the Hg-induced modulation of GABA currents comes from several experiments; Hg is capable of interacting with the sulfhydryl groups on GABA A receptors, resulting in increased current (Fonfria et al. 2001). Another possible mechanism of action involves altering the phosphorylation of the GABA A receptor complex. Using rat dorsal root ganglion cells, it was determined that a protein kinase A (PKA) inhibitor, but not a protein kinase C (PKC) inhibitor, could block the potentiation of the GABA-induced current (Huang and Narahashi, 1997a). Other complementary data suggest the involvement of second messenger systems using the Gi/ Go proteins (Huang and Narahashi, 1997b). Thus, the data suggest that Hg affects the GABAergic system by altering the properties of various GABA receptors, rather than affecting the synthesis or sequestering and storage of the neurotransmitter (Fitsanakis and Aschner 2005).

Concerning glutamate (Glu), it has been mentioned that treatment with Hg in its various forms results in excitotoxicity. For example, when 10 μM and 100 μM Hg was dialyzed into the frontal cortex in conscious animals, there was a significant increase in the concentration of Glu in the dialysis fluid (Juarez et al. 2002). This increase is not due to alterations in the basal metabolism of the amino acid, but due to perturbations in the uptake/clearance of the Glu. Rats treated with a single dose of 8mg MeHg/Kg and sacrificed on either postnatal day 21 or 60 did not show any difference in the total amount of Glu or aspartate in the hippocampus (Zanoli et al. 2001). It is noteworthy that Hg exposure early in the animal's life is sufficient to alter the glutamatergic system throughout the lifespan. Rats receiving 10 mg MeHg/Kg for 7 consecutive days, starting at postnatal day 16, showed severe damage in the occipital lobe. When MK-801, a non-competitive antagonist of NMDA receptors, was administered, the MeHg-induced damage was ameliorated (Miyamoto et al. 2001), implicating the NMDA receptors in the observed damage. Others have also shown altered NMDA receptor properties in neonatal and adult rodents (Rajanna et al. 1997), and similar results corroborating increases in extracellular Glu have also been observed in Aplysia neurons (Gyori et al. 1994). Thus, the question concerning Hg and Glu relates more to the mechanism by which excess concentrations of Glu accumulate extracellularly.

Earlier reports examining extracellular Glu concentrations suggested that Hg exerted an inhibitory effect on the ability of astrocytes to clear synaptically released Glu from the extracellular space (Aschner, 1996; Aschner et al. 2000). Such data further suggest that astrocytes are an important target of Hg, and that the subsequent excitotoxicity is a secondary effect of Hg caused by astrocytic dysfunction (Fitsanakis and Aschner 2005).

$HgCl_2$ causes a decrease in the activity of glutamine synthetase, the astrocytic enzyme responsible for the conversion of Glu to glutamine (Allen et al. 2001). This is important because it is this astrocytic glutamine that neurons use to synthesize pools of Glu for release during synaptic transmission. As such, inhibition of glutamine synthetase could result in an inability of astrocytes to clear extracellular Glu, to influence the Glu catabolism in neurons, and it may potentially affect ammonia metabolism, since the latter is metabolized in the process of glutamine to Glu conversion (Fitsanakis and Aschner, 2005).

Fitsanakis and Aschner (2005) conclude that the role of GABA and Glu in the neurotoxicity of Hg is not disputed. It appears that Hg can affect the GABA A receptor, leading to an increase in GABA-induced current by binding to an extracellular site of the receptor. This is in contrast to its effect on the glutamatergic system, which seems to be a secondary consequence of astrocytic inability to clear Glu from the synapse. Taken together with the inhibition of glutamine synthetase, these data suggest a modulation of amino acid metabolism, catabolism and regulation that results in the neurotoxicity of Hg.

On the other hand, Vidal et al. (2007) also have proposed that organic and inorganic Hg enhanced striatal dopamine (DA) release by different mechanisms; MeHg might act through a transporter-dependent mechanism (Faro et al. 2002a,b), whereas $HgCl_2$ seems to act by an exocytotic mechanism dependent on vesicular stores and external Ca^{2+} to increase DA release (Vidal et al. 2007).

The dopaminergic terminals of the nigrostriatal pathway in the striatum contain AMPA, NMDA and kainate glutamatergic ionotropic receptors types (Grenamyre et al. 1995). Moreover, this brain area also exhibits an elevated nitric oxide synthase (NOS) activity (Bredt et al. 1991), which seems to be modulated by activation of Glu ionotropic receptors. Thus, the activation of NMDA receptors by endogenous Glu increases Ca^{2+} conductance leading to a rise in intracellular Ca^{2+} concentration that binds calmodulin and this complex activates NOS producing an increase in NO-concentration (Garthway and Boulton, 1995). Faro et al. (2002a,b) have demonstrated that the effect of MeHg on striatal DA release could be produced, at least in part, by overstimulation of NMDA receptors and the subsequent increase in NO production. These data suggest that overactivation of Glu receptors could be involved in neurotoxic effects of MeHg. These results indicate that neuronal DA release induced by $HgCl_2$ in rat striatum could be mediated by the activation of NMDA Glu receptors, whilst a role of AMPA/Kainate receptors seems to be discarded (Vidal et al. 2007).

Moreover, MeHg and $HgCl_2$ increase the extracellular concentrations of Glu by stimulating its release and/or inhibiting its uptake, disturbing Glu homeostasis (Aschner et al. 2000; Fonfria et al. 2005). The increase in extracellular Glu levels could overstimulate NMDA receptors on dopaminergic terminals (Shuman and Madison, 1994), so then the protective effects observed with MK 801 and AP5 pre-treatment could explain that NMDA activation by Hg^{2+} induces DA release (Vidal et al. 2007).

Since DA release in rat striatum is under regulation of excitatory aminoacids, and activation of NMDA receptors by endogenous Glu can induce DA release (Zigmond et al. 1998), then the infusion of MK 801 and AP5 would block the DA release induced by Glu. Thus, the results of Vidal et al. (2007) show that inhibition of NMDA receptors reduces $HgCl_2$-induced DA release, reason by which it seems that $HgCl_2$ could act through an over-stimulation of NMDA receptors and the subsequent activation of NO production.

Finally, Hg and other toxic metals also inhibit binding of opioid receptor agonists to opioid receptors, while magnesium stimulates binding to opioid receptors (Tejwani and Hanissian, 1990).

11.7. Mercury and Neurodegenerative Diseases

It is well known that Hg is a potent neurotoxin that can delay neurological development in neonates, and has been proposed to be an environmental risk factor for several neurodegenerative conditions. The mechanisms by which environmental factors may influence the propagation of neurodegenerative diseases are not yet well understood. However, it is known that neurons require trophic factor support for maintenance and survival following traumatic physical and toxic insults (Monroe and Halvorsen, 2006). These authors found that $HgCl_2$ inhibited ciliary neurotrophic factor and interferon-γ receptor mediated Janus tyrosine kinase (Jak)/signal transducers and activators of transcription (STAT) pathway activation in SK-N-BE(2)-C neuroblastoma cell cultures, but did not inhibit the fibroblast growth factor receptor tyrosine kinase.

Results of dichlorofluorescein experiments showed increased levels of oxidative stress in $HgCl_2$-treated cells that was similar in magnitude to that caused by treatment with H_2O_2. The antioxidant agents, glutathione, N-acetylcysteine, and sodium ascorbate each protected neurons against $HgCl_2$-induced inhibition of STAT activation. $HgCl_2$ also inhibited Jak-STAT signaling in cultures of chick retina neurons, but did not affect signaling in nonneuronal HepG2 cells and chick skeletal myotubes. The specific inhibition of growth factor– mediated Jak-STAT signaling pathways in neurons by $HgCl_2$- induced oxidative stress offers a new mechanism by which Hg may produce neurotoxic symptoms in the developing nervous system, promote neurodegeneration in mature neurons, and inhibit recovery following neurotrauma (Monroe and Halvorsen, 2006).

There are several shared characteristics between Hg neurotoxicity and some neurodegenerative diseases, including increased intracellular levels of oxidative stress and dysfunctions in neurotrophic signaling pathways. The finding that Jak-STAT signaling is specifically inhibited in neurons by several pro-oxidant compounds, including heavy metals, suggests that this pathway could be a common denominator explaining the shared pathologies of metal neurotoxicity and diseases such as AD, PD and ALS (Monroe and Halvorsen, 2006; Monnet-Tschudi et al. 2006).

The effects of Hg were also tested in aggregating brain-cell cultures of fetal rat telencephalon, a three-dimensional brain-cell culture system. The continuous application for 10 to 50 days of non-cytotoxic concentrations of heavy metals resulted in their accumulation in brain cells and the occurrence of delayed toxic effects. When applied at non-toxic concentrations, MeHg becomes neurotoxic under pro-oxidant conditions. Furthermore, Hg induces glial cell reactivity, a hallmark of brain inflammation. It seems that Hg increases the expression of the amyloid precursor protein; Hg also stimulates the formation of insoluble beta-amyloid, which plays a crucial role in the pathogenesis of AD and causes oxidative stress and neurotoxicity *in vitro* (Monnet-Tschudi et al. 2006).

Leong et al. (2001) demonstrated axon degeneration and formation of neurofibrillary tangles in animal neuronal cell cultures within minutes and with lowest amounts of inorganic Hg (2 μl 100 nM in 2 ml neuronal cell culture nourishing solution). This neurodegenerative effect was not demonstrable with other metals like aluminum, lead, cadmium, or manganese (Leong et al. 2001). In neuronal stem cells, inorganic Hg of 2 and 5 μg/ml impaired tubulin functions for 48 hours. It caused apoptosis of nerve cells, and induced expression of heat shock proteins (Cedrola et al. 2003).

11.7.1. Alzheimer Disease

The etiology of most cases of AD is as yet unknown. The central pathogenetic mechanism is neurodegeneration and inflammatory processes, which in turn produce oxidative stress that accelerates neuron damage. The neurodegeneration starts with a hyperphosphorylation of the tau-protein due to as yet unknown reasons. These in turn leads to a breakdown of microtubules, which form the cytoskeleton of the neuron and are essential for the neuron's metabolism and functioning. The vital processes of the neuron are disturbed and finally neuron death ensues (Mutter et al. 2004).

Studies on migration suggest that exogenous factors might be responsible for triggering this pathological positive feedback circle; also, epidemiological studies suggest that environmental factors may be involved beside genetic risk factors (Grant, 1999; Haley, 2007). Some studies have shown higher Hg concentrations in *post mortem* brains and in blood of living patients with AD. In the past 20 years, a number of studies were published suggesting a potential pathogenic role of inorganic Hg in AD (Mutter et al. 2004; 2007; Haley, 2007).

It was shown that both organic (Falconer et al. 1994) and inorganic Hg (Duhr et al. 1993) cause the biochemical changes in tubulin, the major protein component of the neuronal cytoskeleton (Duhr et al. 1993). In healthy human brain tissue cultures, only Hg, even in lower concentrations, but not aluminum, lead, zinc or iron were able to inhibit binding to guanosine-tri-phosphate (GTP), which is necessary for tubulin synthesis and thus for neuron function (Duhr et al. 1993). Hg inhibits ADP-ribolyzation of tubulin and actin (Palkiewicz et al. 1994). This process leads to an inhibition of polymerization of tubulin to microtubulin. As a result, neurofibrillary tangles and senile plaques are formed. Living rats exposed to Hg vapor (250+300μg/m^3) four times a day exhibit the same molecular changes in their brain tissue as those caused in human brain cell cultures after 14 days (Pendergrass et al. 1997). These changes are similar to those found *post mortem* in brains of AD patients (Pendergrass and Haley, 1996). Tubulin is assumed to be the most vulnerable protein for Hg, because administration of very low doses of Hg^{2+} does not inhibit other GTP- or ATP-binding proteins (Pendergrass and Haley, 1995). Tubulin has at least 14 sulfhydryle groups which bind Hg with high affinity resulting in functional losses of tubulin and formation of neurofibrillary tangles. Since human nerve cells do not regenerate, any blocking of neurotubulin is particularly serious (Mutter et al. 2004).

In this way, administration of very low doses of Hg^{2+} (0.18μM) has been shown to promote hyperphosphorylation of tau-protein in neuronal cell cultures within 24 hours (Olivieri et al. 2000). Hyperphosphorylation of tau is the first biochemical change to be observed in the development of AD and results in formation of neurofibrillary tangles and failure of nerve cell functions. Administration of Hg to nerve cells provokes also production of β-amyloid 40 and 42 (Olivieri et al. 2000).

As we mentioned earlier, the main human sources for Hg are fish consumption (MeHg) and dental amalgam (Hg vapor). Moreover, amalgam consists of approx. 50% of elementary Hg which is constantly being vaporized and absorbed by the organism. Hg levels in brain tissues are 2 - 10 fold higher in individuals with dental amalgam (Mutter et al. 2007). Persons showing a genetically determined subgroup of transportation protein for fats (apolipoprotein E4) have an increased AD risk. Apolipoprotein E (APO E) is found in high concentrations in the CNS. The increased AD risk through APO E4 might be caused by its reduced ability to bind heavy metals (Mutter et al. 2004). Latest therapeutic approaches to the treatment of AD accept pharmaceuticals, which remove or bind metals from the brain. Preliminary success has been documented with chelation of synergistic toxic metals (iron, aluminum, zinc, copper) and therefore also Hg. The available data does not answer the question, whether Hg is a relevant risk factor in AD distinctively.

In sum, the findings from epidemiological and demographical studies, the frequency of amalgam application in industrialized countries, clinical studies, experimental studies and the dental state of Alzheimer patients in comparison to controls suggest a decisive role for Hg^{2+} in the etiology of AD. Other factors currently discussed as causes (e. g. other metals, inflammations, dietetic factors, vitamin deficiency, oxidative distress, and metabolic impairments) may act as co-factors (Mutter et al. 2007).

11.7.2. Parkinson Disease

PD is characterized by hypokinesia, rigidity, tremor, and in some cases dementia. Tremor is the most characteristic, and often the first symptom in PD. A still more incapacitating symptom is akinesia, which for the patients with the disorder results in augmenting difficulties at every movement (Nadeau, 1997).

The etiology is known in 25% of the cases of (medicaments, poisonings, cerebrospinal meningitis, etc.), and in 75% of the cases the etiology is unknown. Cases of unknown etiology are named idiopatic Parkinson disease. PD has probably a multifactorial etiology involving genetic, environmental, trauma and possibly other factors (Rybicki et al. 1993).

The deficiency of neurotransmitters, such as dopamine, is well established in the etiology of PD. Studies of Parkinson patients have demonstrated low levels of monoamine transmitters encountered in the basal ganglia, decreased values of homovanilic acid (HVA) and 5-HIFAA in the cerebrospinal fluid, and loss of the dark melanin pigment in the dopaminergic substantia nigra of the basal ganglia. A failure of the neurons in the substantia nigra result in decreased production of dopamine and leads secondarily to a loss of function in the corpus striatum. The consequence of this process is the clinical picture of PD (Nadeau, 1997).

Tremor is a classical symptom among victims of Hg^{2+} poisoning, as well as among MeHg poisoning victims. "Tremor Mercurialis" has been known since antiquity (Bjørklund, 1995a). The tremor of MeHg poisoning is different from physiological tremor and other pathological tremors in frequency and amplitude (Yamanaga, 1983).

Heavy metals, like Hg and copper, can produce lesions of the basal ganglia, with symptoms like hyperkinesia and tremor (Bjørklund, 1995a). According to Komulainen and Tuomisto (1981) Hg has a significant action on adrenergic neurons. Researchers at the Henry Ford Hospital in Detroit, Michigan have studied PD with respect to heavy metal exposure (Rybicki et al. 1993). They have calculated mortality rates for PD in Michigan counties for 1986-1988 with respect to exposure to iron, zinc, copper, mercury, magnesium, and manganese from industry based on recent census data. The death rates are statistically significantly higher in counties with an industry in the chemical and metal related-industrial categories than counties without these industries. The authors concluded: "These ecologic findings suggest a geographic association between PD mortality and the industrial use of heavy metals."

Ngim and Devathasan (1989) have done a case-control study among the multiethnic population of Singapore. They tested the hypothesis that a high level of body burden Hg is associated with an increased risk of PD. In 54 cases of idiopathic PD and 95 hospital-based controls, detailed interviews were completed (Clarkson et al. 1988). The two groups were matched for age, sex and ethnicity, between July 1985 and July 1987. The researchers found that there was a clear monotonic dose-response association between blood Hg levels and PD.

The result was adjusted for potential confounding factors, including dietary fish intake, medications, smoking and alcohol consumption. Ngim and Devathasan (1989) listed the following factors that could contribute to the body burden of Hg: dietary fish intake, ethnic over-the-counter medications, occupational exposures and dental amalgam fillings.

According to Störtebecker (1988) a possible exposure to Hg should be considered in the etiology of PD. He asks: "... why shouldn't a daily release of small amounts of mercury from dental amalgam fillings be capable of producing similar neurological symptoms." Dental amalgams are the predominant source of Hg^{2+} and Hg vapor in the general population (Nylander et al. 1987). There is a direct correlation between the number and surfaces of dental amalgam fillings and the amount of Hg in the brain.

More recently Iwata et al. (2006) investigated the neuromotor effects of occupational exposure to Hg vapor, by measuring hand tremor and postural sway in 27 miners and smelters (i.e., exposed workers) and 52 unexposed subjects. They found that total tremor intensity and frequency-specific tremor intensities at 1–6 and 10–14 Hz were significantly larger in the exposed workers than in the unexposed subjects. Their findings suggest that postural control, as well as hand tremor, may be affected by elemental Hg vapor exposure, suggesting a possibility for PD susceptibility.

11.7.3. Amyotrophic Lateral Sclerosis

Amyotrophic lateral sclerosis (ALS) is a progressive neurodegenerative disease involving primarily the motor neurons of the cerebral cortex, brainstem and spinal cord. It is one of the most common neurodegenerative disorders of adult onset. The incidence of ALS in the general population ranges about 1.5 new cases per year per 100,000 residents. Usually sporadic, this pathology is encountered in all ages of adult life. At this time, even in cases with mutations in Superoxide Dismutase 1 gene, the pathogenesis remains unclear. Several facts suggest the involvement of environmental factors as the existence of conjugal ALS or cluster of ALS cases (Corcia et al. 2003). Heavy metals have been suspected because of their proved neurotoxicity. Among all metals implicated in environmental or occupational intoxication, Hg appears as one of the most neurotoxic (Clarkson, 2002). Neurological clinical features suggestive of Hg intoxication comprise peripheral and CNS signs as erethism, tremor, peripheral neuropathy or cortical blindness (Clarkson et al. 2003). ALS cases related to Hg intoxication (Kantarjian, 1961; Barber, 1978; Adams et al. 1983; Schwarz et al. 1996) and professional exposure have been reported (Currier and Haerer, 1968; Felmus et al. 1976).

To identify antecedent events contributing to the development of ALS, Felmus et al. (1976) studied 25 ALS patients in whom they tabulated the incidence of factors previously associated with motor neuron disease and compared the incidences with those found in 25 hospitalized patients and 25 normal people. More ALS patients reported exposure to lead and Hg. Thus, they assumed that the exposure to lead and Hg, seems to be possible risk factors that may predispose to the development of ALS.

Recently, Praline et al. (2007) reported a case study of an 81-year-old woman in whom clinical signs and features of electromyographic activity patterns were consistent with ALS.

Increased blood level and massive urinary excretion of Hg proved Hg intoxication. Despite a chelation treatment with Meso 2–3 dimercaptosuccininc acid, she died after 17 months. In this way, motor neuron disorders following Hg intoxication have already been reported (Kantarjian, 1961; Barber, 1978; Adams et al. 1983). Sometimes, ALS followed accidental mercury injection (Schwarz et al. 1996). These case reports led to consider occupational exposure to Hg as a risk factor for ALS (Pralin et al. 2007). This relationship between Hg intoxication and motor neuron disease was also underscored through neuropathological examination that showed accumulation of Hg within cortical and spinal motor neurons in a patient after a suicide attempt with Hg injection (Pamphlett and Waley, 1996a). Experimental studies based on autometallography analysis demonstrated an accumulation of metallic Hg in motor neurons in mice or rats after blood (Su et al. 1998) or peritoneal (Pamphlett and Waley, 1996b) injection of Hg or after exposure to vapors of Hg (Pamphlett et al. 1998). Another experience showed that Hg is transported retrogradely from muscular nerves terminals to spinal and brainstem motor neurons in rats or mice following intramuscular injection of Hg^{2+} (Schiønning, 1993).

According to Praline et al. (2007) to date, the exact mechanism of Hg neurotoxicity related to ALS remains unknown. However, it has been demonstrated that this metal reacts and depletes free sulfhydryl groups and causes a reduction in the activity of superoxide dismutase (SOD) (Stohs and Bagchi, 1995). Thus the presence of Hg into motor neurons may lead to the formation of an oxidative stress, which is implicated in the pathophysiology of ALS (Shaw, 2005). To date, mutations of the SOD1 gene are the most frequent cause of familial ALS and some susceptibility genes such as SMN1 have been described in sporadic ALS (Gros-Louis et al. 2006). In mice, acute Hg vapor exposure led to atrophy of large myelinated motor axons and irregularity in axon shape which may be caused by damage of cytoskeletal components (Stankovic, 2006). Abnormal axonal transport is another suspected mechanism implicated in pathophysiology of ALS. Indeed, accumulation and abnormal assembly of neurofilaments are common pathological hallmarks of ALS (Shaw, 2005). Thus, it is important to underlie that Hg exposure should be suspected in patients with motor neuron disease. Biological metal analysis should be performed individually when exposition is suspected.

11.7.4. Other neurological Diseases

11.7.4.1. Multiple Sclerosis

Multiple sclerosis (MS) is a disorder of the CNS with an estimated incidence rate of 100 per 100 000 population (Dean, 1994). At this time, its etiology is poorly understood.

Dental amalgam is one of the suspected risk factors for MS (Ingalls, 1983). Dental amalgams containing Hg have been in use since 1818 (Bangsi et al. 1998). Ingalls (1983) analyzed data from areas where clusters of MS were reported, and suggested that this correlation was due to Hg in dental fillings, postulating its possible role in the etiology of MS.

In this way, Aminzadeh and Etminan (2007) explored and quantify the association between amalgam restorations and MS. These authors have conducted a systematic review and meta-analysis of the literature. The search was conducted in Medline (from 1966 to April 2006), EMBASE (2006, Week 16), and the Cochrane library (Issue 2, 2006) for English-language articles meeting specific definitions of MS and amalgam exposure. The pooled for the risk of MS among amalgam users was consistent, with a slight, nonstatistically significant increase between amalgam use and risk of MS.

Aminzadeh and Etminan (2007) concluded that future studies that take into consideration the amalgam restoration size and surface area along with the duration of exposure are needed in order to definitively rule out any link between amalgam and MS.

11.7.4.2. Acrodynia

Acrodynia is principally a syndrome of chronic Hg poisoning. The almost forgotten disease, mostly affecting infants and young children, is probably the best studied effect of human Hg poisoning.

Acrodynia is characterized by a) profound changes in temperament, b) skin alterations, c) neurologic symptoms, d) tachycardia, and e) stomatitis (Gorlin et al. 1976).

The syndrome is also known as Pink Disease, Swift Disease, Feer Disease, Selter Disease, Vegetativ neurose, and Vegetativ encphalit. These terms describe different aspects of the syndrome (Bjørklund, 1995b).

The etiology of Acrodynia was somewhat unclear, and it was various theories at different times (Obura, 1965). The following theories were most popular: avitaminosis, chronic infections, combination of dietary deficiency and infection, suprarenal insufficiency, derangement of sympathetic nervous system, and Hg poisoning.

Acrodynia has been described as due to unusual sensitivity or idiosyncrasy to Hg (Warkany and Hubbard, 1953). Clinical manifestations may include several of the following symptoms which, in the well established cases, are so distinctive that there is practically no differential diagnosis (Gorlin et al. 1976): pink hands and feet, scarlet tip of nose and cheeks, extreme irritability and restlessness alternating with periods of apathy, insomnia, anorexia, pain in extremities, profuse perspiration, generalized skin rashes, photophobia, desquamation, itching, salivation, loss of teeth, hypotonia which permits the child to assume many different and bizarre positions. There may be albuminuria but no blood or CSF changes, and no characteristic urinary findings except an abnormally raised level of Hg (Obura, 1965). The incidence and mortality rate of acrodynia have fallen dramatically since mercury-containing teething powders were withdrawn from the market (Bjørklund, 1995b).

11.7.4.3. Autism

Autism spectrum disorders (ASDs) are prevalent neurodevelopmental disorders that, based on a recent survey, affect not less than 1 in 150 children during the early 1990s (Geier et al. 2009). ASD diagnoses are characterized by impairments in social relatedness and communication, repetitive behaviors, abnormal movement patterns, and sensory dysfunction (Eigsti and Shapiro, 2003). Further, common co-morbidity conditions often associated with an ASD diagnosis include gastrointestinal disease and dysbiosis (White, 2003), autoimmune disease (Sweeten et al. 2003), and mental retardation (Bolte and Poustka, 2002).

In attempting to understand the underlying pathogenesis in those with ASDs diagnosis, a considerable body of research has been conducted to evaluate potential candidate causal genes. Genetic studies, to date, have not uncovered genes of strong effect. It has recently been postulated that increasing rates and less than 100% monozygotic concordance support a more inclusive reframing of ASDs as a multi-system disorder with genetic influence and environmental contributors (Herbert et al. 2006). Research into the metabolic basis for ASDs has been relatively underutilized compared to other approaches. In considering potential environmental contributors to ASDs, some studies have reported that exposure to Hg can cause immune, sensory, neurological, motor, and behavioral dysfunctions similar to traits defining or associated with autistic disorders, and that these similarities extend to neuroanatomy, neurotransmitters, and biochemistry (Mutter et al. 2005; Kern and Jones, 2006). In addition, investigators from the US National Institute of Environmental Health Sciences (1999) and the National Institute for Occupation Safety and Health of the Centers for Disease Control and Prevention (Nelson, 1991) have described a role for Hg exposure in the pathogenesis of autism. Hg poisoning has also sometimes been presumptively diagnosed as autism of unknown etiology until Hg poisoning has been uncovered (Chrysochoou, et al. 2003). Further, investigators reported on the effects of Hg on neuronal development: "…mercury exposure altered cell number and cell division; these impacts have been postulated as modes of action for the observed adverse effects in neuronal development. The potential implications of such observations are evident when evaluated in context with research showing that altered cell proliferation and focal neuropathologic effects have been linked with specific neurobehavioral deficits (e.g., autism)" (Faustman et al. 2000). Finally, the Collaborative on Health and the Environment's Learning and Developmental Disabilities recently published a consensus statement reporting that there is no doubt Hg exposure may produce autism spectrum disorders (2008).

It may be hypothesized that autism results from a combination of genetic and biochemical susceptibilities in the form of a reduced ability to excrete Hg and/or increased environmental exposure at key times in development. This would mean that individuals exposed to relatively high Hg could be affected even if their bodies were innately efficient eliminators. In contrast, only if an exposed fetus or infant also had genetic and/or biochemical susceptibilities, which decrease one's ability to remove Hg, would a lesser level of Hg exposure leads to problems (Geier et al. 2009).

Final Considerations

Modern medicine has been unable to determine the etiology of most important neurological diseases such as AD, autism, ALS, MS, and PD. This may be due to the hesitant to look at the possibility that iatrogenic Hg toxicity may be the main causal factor.

The incidence of neurodegenerative disease like PD and AD increases severely with age; only a small percentage is directly related to familial forms. The etiology of the most abundant, sporadic forms is complex and multifactorial, involving both genetic and environmental factors. Several environmental pollutants have been associated with neurodegenerative disorders. Hg for example, has been shown to interfere with a multitude of intracellular targets, thereby contributing to several pathogenic processes typical of neurodegenerative disorders, including oxidative stress, mitochondrial dysfunction, deregulation of protein turnover, and brain inflammation. Exposure to heavy metals such as Hg early in development can precondition the brain for developing a neurodegenerative disease later in life. On the other hand, Hg can exert its harmful effects through acute neurotoxicity or through slow accumulation during prolonged periods of life. The pro-oxidant effects of Hg can aggravate the age-related increase in oxidative stress that is related to the decline of the antioxidant defense systems. Brain inflammatory reactions also generate oxidative stress. Chronic inflammation can contribute to the formation of the senile plaques that are typical for AD. Hg can also directly induce α–synuclein fibril formation, which is the main constituent of Lewy bodies in dopaminergic neurons of the substantia nigra leading to PD.

Taken together, a considerable body of evidence suggests that the heavy metals such as mercury contribute to the etiology of neurodegenerative diseases and underline the importance of taking preventive measures in this regard.

References

Abe, T., Haga, T. and Kurokawa, M. (1975) Blockage of axoplasmic transport and depolymerization of reassembled microtubules by methylmercury. *Brain Res.* 86: 504-508.

Adams, C.R., Ziegler, D.K. and Lin, J.T. (1983) Mercury intoxication simulating amyotrophic lateral sclerosis. *JAMA*, 250: 642-643.

Agency for Toxic Substances and Disease Registry (ATSDR). 2001 CERCLA Priority List of Hazardous Substances. Atlanta, GA: U.S. Department of Health and Human Services, Public Health Service.

Albrecht, J., Szumanska, G., Gadamski, R., and Gajkowska, B. (1994) Changes of activity and ultrastructural localization of alkaline phosphates in cerebral cortical microvessels of rat after single intraperitoneal administration of mercuric chloride. *Neurotoxicology*, 15: 897–902.

Allen, J.W., Mutkus, L.A. and Aschner, M. (2001) Mercuric chloride, but not methylmercury, inhibits glutamine synthetase activity in primary cultures of cortical astrocytes. *Brain Res.* 891: 148– 157.

Aminzadeh, K.K. and Etminan, M. (2007) Dental amalgam and multiple sclerosis: a systematic review and meta-analysis. *J Public Health Dent.* 67:64-66.

Arakawa, O., Nakahiro, M. and Narahashi, T. (1991) Mercury modulation of GABA-activated chloride channels and non-specific cation channels in rat dorsal root ganglion neurons. *Brain Res.* 551: 58–63.

Aschner, M. (1996) Astrocytes as modulators of mercury-induced neurotoxicity. *Neurotoxicology*, 17: 663–670.

Aschner, M. and Aschner, J.L. (1990) Mercury neurotoxicity: mechanisms of blood-brain barrier transport. *Neurosci Biobehav Rev.* 14: 169-176.

Aschner, M., Cherian, M.G., Klaassen CD, Palmiter RD, Erickson JC, and Bush, A.I. (1997) Metallothioneins in brain–the role in physiology and pathology. *Toxicol Appl Pharmacol.* 142:229–242.

Aschner, M. and Clarkson, T.W. (1988) Uptake of methylmercury in the rat brain: effects of amino acids. *Brain Res.* 462: 31–39.

Aschner, M. and Clarkson, T.W. (1989) Methyl mercury uptake across bovine brain capillary endothelial cells in vitro: The role of amino acids. *Pharmacol Toxicol.* 64:293–299.

Aschner, M., Eberle, N.B., Goderie, S. and Kimelberg, H.K. (1990) Methylmercury uptake in rat primary astrocyte cultures: The role of the neutral amino acid transport system. *Brain Res.* 521:221–228.

Aschner, M., Yao, C.P., Allen, J.W. and Tan, K.H. (2000) Methylmercury alters glutamate transport in astrocytes. *Neurochem Int.* 37: 199–206.

Atchison, W.D. and Hare, M.F. (1994) Mechanisms of methylmercury-induced neurotoxicity. *FASEB J.* 8:622–629.

Auger, N., Kofman, O., Kosatsky, T. and Armstrong, B. (2005) Low-level methylmercury exposure as a risk factor for neurologic abnormalities in adults. *Neurotoxicology*, 26: 149–157.

Bakir, F., Damluji, S.F., Amin-Zaki L, Murtadha M, Khalidi A, Al-Rawi NY, Tikriti S, Dhahir HI, Clarkson, T.W., Smith, J.C. and Doherty, R.A. (1973). Methylmercury poisoning in Iraq. *Science*, 181:230–241.

Bapu, T.C., Purohit, R.C., and Sood, P.P. (1994) Fluctuation of trace elements during methylmercury toxication, and chelation therapy. *Hum Exp Toxicol.* 13: 815–823.

Bast-Pettersen, R., Ellingsen DG, Efskind J, Jordskogen, R. and Thomassen Y. (2005) A neurobehaviorial study of chloralkali workers after the cessation of exposure to mercury vapor. *Neurotoxicology*, 26: 427–437.

Ballatori, N. (2002) Transport of toxic metals by molecular mimicry. *Environ Health Persp.* 110: 689–694.

Ballatori, N. and Clarkson, T.W. (1983) Biliary transport of glutathione and methylmercury. *Am J Physiol.* 244: G435-G441.

Ballatori, N. and Clarkson, T.W. Dependence of biliary secretion of inorganic mercury on the biliary transport of glutathione. *Biochem Pharmacol.* 33:1093-1098.

Bangsi, D., Ghadirian, P., Ducic, S., Morisset, R., Ciccocioppo, S., McMullen, E. and Krewski, D. (1998) Dental amalgam and multiple sclerosis: a case-control study in Montreal, Canada. *Int J Epidemiol.* 27: 667-671.

Barber, T.E. (1978) Inorganic mercury intoxication reminiscent of amyotrophic lateral sclerosis. *J Occup Med*, 20: 667–669.

Beldowski, J. and Pempkowiak, J. (2003) Horizontal and vertical variabilities of mercury concentration and speciation in sediments of the Gdansk Basin, Southern Baltic Sea. *Chemosphere*, 52: 645–654.

Belletti, S., Orlandini, G., Vettori, M.V., Mutti A, Uggeri J, Scandroglio R, Alinova, R. and Gatti, R. (2002) Time course assessment of methylmercury effects on C6 glioma cells: Submicromolar concentrations induce oxidative DNA damage and apoptosis. *J Neurosci Res*. 70:703–711.

Beuter, A. and Geoffroy, A. (1996) Can tremor be used to measure the effects of chronic mercury exposure in human subjects? *Neurotoxicology*, 17: 213–228.

Bjørklund, G. (1995a) Parkinson's Disease and Mercury. *J of Orthomol Med.* 10: 147.

Bjørklund, G. (1995b) Mercury and Acrodynia. *J of Orthomol Med.* 10: 145-146.

Bland, C. and Rand, M.D. (2006) Methylmercury induces activation of NOTCH signaling. *Neurotoxicology*, 27:982–991.

Boelsterli, U.A. (2003) Mechanistic toxicology: The molecular basis of how chemicals disrupt biological targets. 1st ed. London and New York: Taylor and Francis.

Bolte, S. and Poustka, F. (2002) The relation between general cognitive level and adaptive behavior domains in individuals with autism with and without co-morbid mental retardation. *Child Psychiatry Hum Dev*. 33: 165–172.

Braughler, J.M., Duncan, L.A. and Goodman, T. (1985) Calcium enhances in vitro free radical-induced damage to brain synaptosomes, mitochondria, and cultured spinal cord neurons. *J Neurochem*. 45:1288–1293.

Brawer, J.R., McCarthy, G.F., Gornitsky M, Frankel D, Mehindate K, and Schipper, H.M. (1998) Mercuric chloride induces a stress response in cultured astrocytes characterized by mitochondrial uptake of iron. *Neurotoxicology*, 19: 767–776.

Bredt, D.S., Glatt, C.E., Hwang, P.M., Fotuhi, M., Dawson, T. and Snyder, H.S. (1991) Nitric oxide synthase protein and mRNA are discretely localized in neuronal populations if the mammalian CNS together with NADPH-diaphorase. *Neuron*, 7: 615–624.

Buonocore, G., Perrone, S. and Bracci, R. (2001) Free radicals and brain damage in the newborn. *Biol Neonate*, 79:180–186.

Carta, P., Flore, C, Alinovi R, Ibba A, Tocco MG, Aru G, Carta R, Girei E, Mutti A, Lucchini, R. and Randaccio, F.S.. (2003) Sub-clinical neurobehavioral abnormalities associated with low level of mercury exposure through fish consumption. *Neurotoxicology*, 24:617–623.

Castoldi, A.F., Barni, S., Turin, I., Gandini, C. and Manzo, L. (2000) Early acute necrosis, delayed apoptosis and cytoskeletal breakdown in cultured cerebellar granule neurons exposed to methylmercury. *J Neurosci Res.* 59:775–787.

Cedrola, S., Guzzi, G., Ferrari D, Gritti A, Vescovi AL, Pendergrass, J.C., and La Porta, C.A. (2003) Inorganic mercury changes the fate of murine CNS stem cells. FASEB J. 17: 869–871.

CDC. (2004) Blood mercury levels in young children and childbearing-aged women–United States, 1999–2002. *Mortal Morbid Weekly Rep*. 53:1018–1020.

Chapman, L.J., Sauter, S.L., Henning RA, Dodson, V.N., Reddan, W.G. and Matthews, C.G. (1990) Differences in frequency of finger tremor in otherwise asymptomatic mercury workers, *Br J Ind Med.* 47: 838–843.

Charleston, J.S., Body, R.L., Bolender RP, Mottet NK, Vahter, M.E. and Burbacher, T.M. (1996) Changes in the number of astrocytes and microglia in the thalamus of the monkey Macaca fascicularis following long-term subclinical methylmercury exposure. *Neurotoxicology*, 17:127–138.

Choi, B.H. (1989) The effects of methylmercury on the developing brain. *Prog Neurobiol.* 32: 447–470.

Choi, B.H., Kim, R.C. and Peckham, N.H. (1988) Hydrocephalus following prenatal methylmercury poisoning. *Acta Neuropathol.* 75; 325–330.

Chrysochoou, C., Rutishauser, C., Rauber-Lüthy C, Neuhaus T, Boltshauser E. and Superti-Furga, A. (2003) An 11-month-old boy with psychomotor regression and auto-aggressive behaviour. *Eur J Pediatr.* 162:559-561.

Clarkson, T.W. (2002) The three modern faces of mercury. *Environ Health Persp Suppl.*1: 11–23.

Clarkson. T/W. (1992) Mercury: major issues in environmental health. *Environ Health Perspect.* 100: 31–38.

Clarkson, .TW. (1993). Molecular and ionic mimicry of toxic metals. *Ann Rev Pharmacol Toxicol.* 32:545–571.

Clarkson, T.W., Magos, L. and Myers, G. (2003). The toxicology of mercury–current exposures and clinical manifestations. *New Eng J Med.* 349:1731–1737.

Clarkson, T.W., Friberg, L., Nordberg, G.F. and Sager, P.R. editors. (1988) Biological monitoring of toxic metals. New York: Plenum Press.

Coccini, T., Randine, G., Candura, S.M., Nappi, R.E., Prockop, L.D. and Manzo, L. (2000) Low level exposure to methylmercury modifies muscarinic cholinergic receptor binding characterstics in rat brain and lymphocytes: physiologic implications and new opportunities in biologic monitoring. *Environ Health Perspect.* 108: 29–33.

Cookson, M.R. and Pentreath, V.W. (1996) Protective roles of glutathione in the toxicity of mercury and cadmium compounds to C6 glioma cells. *Toxicol In Vitro*, 10: 257-264.

Collaborative on Health and the Environment's Learning and Developmental Disabilities Initiative. (2008) LDDI Scientific consensus statement on environmental agents associated with neurodevelopmental disorders, pp. 1–35.

Corcia, P., Jafari-Schluep, H.F., Lardillier, D, Mazyad, H., Giraud, P., Clavelou, P., Pouget, J. and Camu, W. (2003) A clustering of conjugal amyotrophic lateral sclerosis in southeastern France. *Arch Neurol.* 60:553-557.

Corda, M.G., Concas, A., Rossetti, Z., Guarneri, P., Corongiu, F.P. and Biggio, G. (1981) Methyl mercury enhances [3H]-diazepam binding in different areas of the rat brain. *Brain Res.* 229: 264– 269.

Counter, S.A. and Buchanan, L.H. (2004) Mercury exposure in children: a review. *Toxicol Appl Pharmacol.* 198: 209–230.

Counter, S.A., Buchanan, L.H., Ortega, F. and Goran, L. (2002). Elevated blood mercury and neuro-otological observations in children of the Ecuadorian gold mines. *J Toxicol Environ Health Part* A, 65:149–163.

Crespo-Lopéz, M.E., Lima de Sa, A., Manoel, H.A., Rodriguez, B.R. and Martins do Nascimiento, J.L. (2007) Methylmercury genotoxicity: A novel effect in human cell lines of the central nervous system. *Environ Int.* 33:141–146.

Currier, R.D. and Haerer, A.F. (1968) Amyotrophic lateral sclerosis and metallic toxins. *Arch Environ Health*, 17: 712-719.

Dalton, L.W. (2004). Methylmercury toxicity probed. *Chem. Eng. News*, 82:70–71.

Dare, E., Li, W., Zhivotovsky, B., Yuan, X. and Ceccatelli, S. (2001) Methylmercury and H_2O_2 provoke lysosomal damage in human astrocytoma D384 cells followed by apoptosis. *Free Radical Biol Med.* 30:1347–1356.

Davis, L.E., Kornfeld M, Mooney H.S., Fiedler K.J., Haaland K.Y., Orrison W.W., Cernichiari, E. and Clarkson T.W. (1994) Methylmercury poisoning: long-term clinical, radiological, toxicological, and pathological studies of an affected family. *Ann Neurol.* 35: 680–688.

Dean, G. (1994) How many people in the world have multiple sclerosis? *Neuroepidemiology*, 13:1-7.

Dietrich, M.O., Mantese, C.E., Dos Anjos G, Souza, D.O. and Farina, M. (2005) Motor impairment induced by oral exposure to methylmercury in adult mice. *Environ Toxicol Pharmacol.*19: 169–175.

Domagalski, J.L., Alpers, C.N., Slotton, D.G., Suchanek, T.H. and Ayers, S.M. (2004) Mercury and methylmercury concentrations and loads in the Cache Creek watershed, California. *Sci Total Environ.* 327: 215–237.

dos santos, A.P.M., Mateus, M.L., Carvalho, C.M.L., Camila, M. and Batoréu, C. (2007) Biomarkers of exposure and effect as indicators of the interference of selenomethionine on methylmercury toxicity. *Toxicol Lett.* 169: 121–128

Driscoll, C.T., Driscoll, K.M., Mitchell, M.J. and Raynal, DJ. (2003) Effects of acidic deposition on forest and aquatic ecosystems in New York State. *Environ Pollut.* 123: 327–336.

Duhr, E.F., Pendergrass, J.C., Slevin, J.T. and Haley, B.E. (1993) HgEDTA complex inhibits GTP interactions with the E-site of brain beta-tubulin. *Toxicol Appl Pharmacol.* 122: 273–280.

Eigsti, I.M. and Shapiro, T. (2003) A systems neuroscience approach to autism: biological, cognitive, and clinical perspectives. *Ment Retard Dev Disabil Res Rev.* 9: 205-215.

Elble, R.J. (1996) Central mechanisms of tremor. *J Clin Neurophysiol.* 13:133–144.

Ellingsen, D.G., Bast-Pettersen, R., Efskind, J. and Thomassen, Y. (2001) Neuropsychological effects of low mercury vapor exposure in chloralkali workers. *Neurotoxicology*, 22: 249–258.

Endo T, Sakata M. (1995) Effects of sulfhydryl compounds on the accumulation, removal and cytotoxicity of inorganic mercury by primary cultures of rat renal cortical epithelial cells. *Pharmacol Toxicol.* 76: 190-195.

Erie, J.C., Butz, J.A. and Good, J.A. (2005) Erie EA, Burritt MF, Cameron JD. Heavy metal concentrations in human eyes. *Am J Ophthalmol.* 139:888–93.

Eto, K. (1997) Pathology of Minamata disease. *Toxicol Pathol.* 25:614–623.

Eto, K. (2000) Minamata disease. *Neuropathology Suppl.* 1:14–19.

Falconer, M.M., Vaillant A, Reuhml ,K.R., Laferrierem, N. and Brown, D.L. (1994) The molecular basis of microtubule stability in neurons. *Neurotoxicology*, 15:109–122.

Faro, L.R.F., do Nascimento, J.L.M., Alfonso, M. and Duran, R. (2002a). Mechanism of action of methylmercury on *in vivo* striatal dopamine release. Possible involvement of dopamine transporter. *Neurochem Int.* 40: 455–465.

Faro, L.R.F., do Nascimento, J.L.M., Alfonso, M. and Duran, R. (2002b) Protection of methylmercury effects on the *in vivo* dopamine release by NMDA receptor antagonists an nitric oxide synthase inhibitors. *Neuropharmacology*, 42: 612–618.

Faustman, E.M., Ponce, R.A., Ou, Y.C., Mendoza, A.C., Lewandowski, T. and Kavanagh, T. (2002) Investigations of methylmercury-induced alterations in neurogenesis. *Environ Health Persp Suppl.* 5: 859–864.

Faustman, E.M., Silbernagel, S.M., Fenske, R.A., Burbacher, T.M. and Ponce, R.A. (2000) Mechanisms underlying Children's susceptibility to environmental toxicants. *Environ Health Perspect.* 108: 13-21.

Felmus, M.T., Patten, B.M. and Swanke, L. (1976) Antecedent events in amyotrophic lateral sclerosis. *Neurology*, 26:167-172.

Feng, W., Wang, M., Li, B., Liu, J., Chai, Z., Zhao, J. and Deng, G. (2004) Mercury and trace element distribution in organic tissues and regional brain of fetal rat after in utero and weaning exposure to low dose of inorganic mercury. *Toxicol Lett.* 152: 223–234.

Fitsanakis, A.V. and Aschner, M. (2005) The importance of glutamate, glycine, and g-aminobutyric acid transport and regulation in manganese, mercury and lead neurotoxicity. *Toxicol App Pharmacol.* 204: 343– 354.

Fonfria, E., Dare, E., Benelli, M., Sunol, C. and Ceccatelli, S. (2002) Translocation of apoptosis-inducing factor in cerebellar granule cells exposed to neurotoxic agents inducing oxidative stress. *Eur J Neurosci.* 16:2013–2016.

Fonfria, E., Rodriguez-Farre, E. and Sunol, C. (2001) Mercury interaction with the GABA(A) receptor modulates the benzodiazepine binding site in primary cultures of mouse cerebellar granule cells. *Neuropharmacology*, 41, 819– 833.

Fonfria, E., Vilaro, M.T., Babot, Z., Rodriguez-Farre, E. and Suñol, C. (2005) Mercury compounds disrupt neuronal glutamate transport in cultured mouse cerebellar granule cells. *J Neurosci Res.* 79: 545–553.

Fonnum, F. and Lock, E.A. (2004) The contributions of excitotoxicity, glutathione depletion and DNA repair in chemically induced unjury to neurones: exemplefied with toxic effects on cerebellar granule cells. *J Neurochem.* 88: 513–531.

Franco, J.L., Braga, H.C., Nunes, K.C.C., Ribas MC, Stringari J, Silva, P.A., Garcia Pomblum, S.C., Moro, A.M., Bohrerm, D., Santos, R.S.A., Dafre, A.L. and Farina, M. (2007) Lactational exposure to inorganic mercury: Evidence of neurotoxic effects. *Neurotoxicol Teratol.* 29: 360–367.

Frumkin, H., Letz, R., Williams, P.L., Gerr, F., Pierce, M., Sanders, A., Elon, L., Manning, C.C., Woods, J.S., Hertzberg, V.S., Mueller, P. and Taylor, B.B. (2001) Health effects of long-term mercury exposure among chloralkali plant workers. *Am J Ind Med.* 39: 1–18.

Fujiyama, J., Hirayama, K. and Yasutake, A. (1994) Mechanism of methylmercury efflux from cultured astrocytes. *Biochem Pharmacol.* 47: 1525-1530.

Garg, T.K. and Chang, J.Y. (2006) Methylmercury causes oxidative stress and cytotoxicity in microglia: Attenuation by 15-deoxy-delta 12, 14-prostaglandin J2. *J Neuroimmunol.* 171:17–28.

Garthway, J. and Boulton, C.L. (1995) Nitric oxide signaling in the central nervous system. *Annu Rev Physiol.* 57: 683–706.

Geier, D.A., Kern, J.K., Garver, C.R., Adams, J.B., Audhya, T., Nataf, R. and Geier, M.R. (2009) Biomarkers of environmental toxicity and susceptibility in autism. *J Neurol Sci.* 280: 101-108.

Girault, L., Boudou, A. and Dufourc, E.J. (1997) Methyl mercury interactions with phospholipid membranes as reported by fluorescence, 31P and 199Hg NMR. *Biochim Biophys Acta (BBA)–Biomembranes*, 1325:250–262.

Gochfeld, M. (2003) Cases of mercury exposure, bioavailability, and absorption. *Ecotoxicol Environ Safety*, 56:174–179.

Godfrey, M., Wojeik, D. and Cheryl, K. (2003) Apolipoprotein-E as a potential biomarker for mercury neurotoxicity. *J Alz Dis.* 5: 189–195.

Gorlin, R.J., Pindborg, J.J. and Cohen, M.M. Jr. (1976) Syndromes of the head and neck. McGraw-Hill Book Company, New York, 1976.

Grandjean, P., Weihe, P. and Nielson, J.B. (1994) Significance of intrauterine and postnatal exposures. *Clin Chem.* 40: 1935–2000.

Grandjean, P., Weihe P., White, R.F., Debes, F., Araki, S., Yokoyama, K., Murata, K., Sorensen, N., Dahl, R. and Jorgensen, P.J. (1997) Cognitive deficit in 7-year-old children with prenatal exposure to methylmercury. *Neurotoxicol Teratol.* 19: 417–428.

Grant, W.B. (1999) Dietary Links to Alzheimer's Disease: 1999 Update. *J Alz Dis.* 1:197–201.

Grenamyre, J.T., Olson, J.M.M., Penney, J.B. and Young, A.B. (1995) Autoradiographic characterization of *N*-methyl-D-aspartate-, quisqualate-, and Kainate-sensitive glutamate binding sites. *J Pharmacol Exp Ther.* 223: 254–263.

Gribble, E.J., Hong, S.W. and Faustman, E.M. (2005) The magnitude of methylmercury-induced cytotoxicity and cell cycle arrest is p53-dependent. *Birth Defects Res Part* A, 73:29–38.

Gros-Louis, F., Gaspar, C. and Rouleau, G.A. (2006) Genetics of familial and sporadic amyotrophic lateral sclerosis. *Biochim Biophys Acta*, 1762: 956-972.

Gyori, J., Fejtl, M., Carpenter, D.O. and Salanki, J. (1994) Effect of $HgCl_2$ on acetylcholine, carbachol, and glutamate currents of Aplysia neurons. *Cell Mol Neurobiol.* 14: 653–654.

Haley, B. (2007) The relationship of the toxic effects of mercury to exacerbation of the medical condition classified as Alzheimer's disease. *Medical Veritas*, 4, 1510–1524.

Hansen, J.C. and Dasher, G. (1997) Organic mercury: an environmental threat to the health of dietary exposed societies? *Rev Environ Health*, 12: 107–116.

Harada, M., Nakachi, S., Tasaka, K., Sakashita, S., Muta K, Yanagida K, Doi R, Kizaki T. and Ohno, H. (2001) Wide use of skin-lightening soap may cause mercury poisoning in Kenya. *Sci Total Environ.* 269: 183–187.

Herbert, M.R., Russo, J.P., Yang, S., Roohi, J., Blaxill, M., Kahler, S.G., Cremer, L. and Hatchwell, E. (2006) Autism and environmental genomics. *Neurotoxicology*, 27: 671-684.

Horvat, M., Nolde, N., Fajon, V., Jereb, V., Logar, M., Lojen, S., Jacimovic, R., Falnoga, I., Liya, Q., Faganeli, J. and Drobne, D. (2003) Total mercury, methylmercury and selenium in mercury polluted areas in the province Guizhou, China. *Sci Total Environ*, 304: 231–256.

Huang, C.S. and Narahashi, T. (1997a) The role of phosphorylation in the activity and mercury modulation of GABA-induced currents in rat neurons. *Neuropharmacology*, 36: 1631– 1640.

Huang, C.S. and Narahashi, T. (1997b) The role of G proteins in the activity and mercury modulation of GABA-induced currents in rat neurons. *Neuropharmacology*, 36: 1619–1630.

Hussain, S., Rodgers, D.A., Duhart, H.M. and Ali, S.F. (1997) Mercuric chloride-induced reactive oxygen species and its effect on antioxidant enzymes in different regions of rat brain. *J Environ Sci Health* B, 32: 395–409.

Ingalls, H.T. (1983) Epidemiology, etiology and prevention of multiple sclerosis, hypothesis and fact. *Am Forensic Med Pathol.* 4:55-61.

Iwata, T., Sakamoto, M., Feng, X., Yoshida, M., Liu, X.J., Dakeishi, M., Li, P., Qiu, G., Jiang, H., Nakamura, M. and Murata, K. (2006) Effects of mercury vapor exposure on neuromotor function in Chinese miners and smelters. *Int Arch Occup Environ Health*, 80: 381-387.

James, S.J., Slikker, III W., Melnyk, S., New, E., Pogribna, M. and Jernigan, S. (2005) Thimerosal neurotoxicity is associated with glutathione depletion: protection with glutathione precursors. *Neurotoxicology*, 26: 1–8.

Johnson, C.L. (2004) In the environment: sources, toxicities, and prevention of exposure. *Pediatr Ann.* 33: 437–442.

Juarez, B.I., Martiez, M.L., Montante, M., Dufour, L., Garcia, E. and Jimenez-Capdeville, M.E. (2002) Methylmercury increases glutamate extracellular levels in frontal cortex of awake rats. *Neurotoxicol Teratol.* 24: 767–771.

Kajiwara, Y., Yasutake, A., Adachi, T. and Hirayama, K. (1996) Methylmercury transport across the placenta via neutral amino acid carrier. *Arch Toxicol.* 70: 310–314.

Kang, M.S., Jeong, J.Y., Seo, J.H., Jeon, H.J., Jung, K.M., Chin, M., Moon, C., Bonventre, J.V., Jung, S.Y. and Kim, D.K. (2006). Methylmercury-induced toxicity is mediated by enhanced intracellular calcium through activation of phosphatidylcholine-specific phospholipase C. *Tox Appl Pharmacol.* 216:206–215.

Kantarjian, A.D. (1961) A syndrome clinically resembling amyotrophic lateral sclerosis following chronic mercurialism. *Neurology*, 11: 639–644.

Kern, J.K. and Jonesm, A.M. (2006) Evidence of toxicity, oxidative stress, and neuronal insult in autism. *J Toxicol Environ Health B Crit Rev.* 9: 485-499.

Kerper, L.E., Mokrzan, E.M., Clarkson, TW. and Ballatori, N. (1996) Methylmercury efflux from brain capillary endothelial cells is modulated by intracellular glutathione but not ATP. *Toxicol Appl Pharmacol.* 141: 526-531.

Konigsberg, M., Lopez-Diazguerrero, N.E., Bucio, L., Gutierrez-Ruiz, M.C. (2001) Uncoupling effect of mercuric chloride on mitochondria isolated from an hepatic cell line. *J Appl Toxicol.* 21: 323-329.

Kromidas, L., Trombetta, L.D. and Jamall, I.S. (1990) The protective effects of glutathione against methylmercury cytotoxicity. *Toxicol Lett.* 51: 67-80.

Komulainen, H. and Tuomisto, J. (1983) Effects of heavy metals on dopamine, noradrenaline and serotonin uptake and release in rat brain synaptosomas. *Acta Pharmacol Toxicol.* 48: 199-204.

Langworth, S., Almkvist, O., Söderman, E. and Wikström, B.O. (1992) Effects of occupational exposure to mercury vapour on the central nervous system. *Br J Ind Med.* 49: 545–555.

Lärkfors, L., Oskarsson, A., Sundberg, J. and Ebendal, T. (1991) Methylmercury induced alterations in the nerve growth factor level in the developing brain. *Dev Brain Res.* 62: 287-291.

Lee, D.A., Lopez-Alberola, R. and Bhattacharjee, M. (2003) Childhood autism: a circuit syndrome? *Neurologist*, 9: 99–109.

Leong, C.C., Syed, N.I. and Lorscheider, F.L. (2001) Retrograde degeneration of neurite membrane structural integrity of nerve growth cones following in vitro exposure to mercury. *Neuroreport*, 12:733–737.

Leonhardt, R., Pekel, M., Platt, B., Haas, H.L. and Busselberg, D. (1996) Voltage-activated calcium channel currents of rat DRG neurons are reduced by mercuric chloride and methylmercury. *Neurotoxicology*, 17:85–92.

Lucchini, R., Cortesi, I., Facco, P., Benedetti L, Camerino D, Carta, P., Urbano, M.L., Zaccheo, A. and Alessio, L. (2002) Neurotoxic effect of exposure to low doses of mercury. *Med Lav.* 93: 202–214.

Magos, L. (1987) Absorption, distribution and excretion of methylmercury. In C.U. Eccles and Z. Annau (Eds.), *The Toxicity of Methyl Mercury,* Johns Hopkins University Press, Baltimore, pp. 24-44.

Malcolm, E.G., Keeler, G.J., Lawson, S.T. and Sherbatskoy, T.D. (2003) Mercury and trace elements in cloud water and precipitation collected on Mt. Mansfield, Vermont. *J Environ Monit.* 5: 584–590.

Marcelo, O., Dietricha, C.E., Mantesea, G.A., Diogo, O.S. and Marcelo, F. (2005) Motor impairment induced by oral exposure to methylmercury in adult mice. *Environ Toxicol Pharmacol.* 19: 169–175.

Margolis, R.L. and Wilson, L. (1981) Microtubule treadmills – possible molecular machinery, *Nature*, 293: 705-711.

Markovich, D. and James, K.M. (1999) Heavy metals mercury, cadmium, and chromium inhibit the activity of the mammalian liver and kidney sulfate transporter sat-1. *Toxicol Appl Pharmacol.* 154: 181–187.

Marlin, A.E., Brown, W.E., Huntington, H.W. and Epstein, F. (1980) Effect of the dural application of Zenker's solution on the feline brain. *Neurosurgery*, 6: 45–48.

Marsh, D.O., Clarkson, T.W., Cox, C., Myers, G.J., Amin-Zaki, L. and Al-Tikriti, S. (1987) Fetal methylmercury poisoning. Relationship between concentration in single strands of maternal hair and child effects. *Arch Neurol.* 44:1017–1022.

Marty, M.S. and Atchison, W.D. (1998) Elevations of intracellular Ca^{2+} as a probable contributor to decreased viability in cerebellar granule cells following acute exposure to methylmercury. *Toxicol Appl Pharmacol.* 150:98–105.

Mathieson, P.W. (1995) Mercury: god of TH2 cells. *Clin Exp Immunol.* 102: 229–230.

McCullough, J.E., Dick, R. and Rutchik, J. (2001) Chronic mercury exposure examined with a computer-based tremor system. *J Occup Environ Med.* 43: 295–300.

Miura, K., Koide, N., Himeno, S., Nakagawa, I. and Imura, N. (1999) The involvement of microtubular disruption in methylmercury-induced apoptosis in neuronal and nonneuronal cell lines. *Toxicol Appl Pharmacol.* 160: 279–288.

Miyamoto, K., Nakanishi, H., Moriguchi, S., Fukuyama, N., Eto, K., Wakamiya, J., Murao, K., Arimura, K. and Osame, M. (2001) Involvement of enhanced sensitivity of N-methyl-d-aspartate receptors in vulnerability of developing cortical neurons to methylmercury neurotoxicity. *Brain Res.* 901: 252–258.

Monnet-Tschudi, F., Zurich, M.G., Boschat, C., Corbaz, A. and Honegger, P. (2006) Involvement of environmental mercury and lead in the etiology of neurodegenerative diseases. *Rev Environ Health*, 21: 105-117.

Monroe, R.K. and Halvorsen, S.W. (2006) Cadmium blocks receptormediated Jak/STAT signaling in neurons by oxidative stress. *Free Radic Biol Med.* 41: 493–502.

Murray, M. and Holmes, S.A. (2004). Assessment of mercury emissions inventories for the Great Lakes states. *Environ Res*. 95: 282–297.

Mutter, J., Naumann, J., Sadaghiani, C., Schneider, R. and Walach, H. (2004) Alzheimer disease: mercury as pathogenetic factor and apolipoprotein E as a moderator. *Neuro Endocrinol Lett.* 25: 331-339.

Mutter, J., Naumann, J., Schneider, R., Walach, H. and Haley, B. (2005) Mercury and autism: accelerating evidence? *Neuro Endocrinol Lett.* 26: 439-446.

Mutter, J., Naumann, J., Schneider, R. and Walach, H. (2007) Mercury and Alzheimer's disease. *Fortschr Neurol Psychiatr*. 75:528-538.

Nadeau, S.E. (1997) Parkinson's disease. *J Am Geriatr Soc*. 45: 233-240.

National Institute of Environmental Health Sciences. (1999) A research-oriented framework for risk assessment and prevention of children's exposure to environmental toxicants. *Environ Health Perspect.* 107, p. 510.

National Research Council. (2000) Effects of Methylmercury. Washington, DC: National Academy Press. 33-35.

Nelson, B.K. (1991) Evidence for behavioral teratogenicity in humans. *J Appl Toxicol.* 11: 33–37.

Ngim, C.H. and Devathasan, G. (1989) Epidemiologic study on the association between body burden mercury level and idiopathic Parkinson's disease. *Neuroepidemiology*, 8: 128-141.

Nylander, M., Friberg, L. and Lind, N. (1987) Mercury consentrations in the human brain and kidneys in relation to exposure from dental amalgam fillings. *Swed Dent J.* 11:179-187.

Obura, C.W. (1965) Pink Disease. Report of a Case. *Br Dent. J.* 119: 273-274.

Olivieri, G., Brack, C., Müller-Spahn, F., Stähelin, H.B., Herrmann, M., Renard, P., Brockhaus, M. and Hock, C. (2000) Mercury induces cell cytotoxicity and oxidative stress and increases beta-amyloid secretion and tau phosphorylation in SHSY5Y neuroblastoma cells. *J Neurochem*. 74:231–236.

Olivieri, G., Novakovic, M., Savaskan, E., Meier, F., Baysang, G., Brockhaus, M. and Muller-Spahn, F. (2002) The effects of β–estradiol on SHSY5Y neuroblastoma cells during heavy metal induced oxidative stress, neurotoxicity and beta-amyloid secretion. *Neuroscience*, 113: 849–855.

Ou, Y.C., Thompson, S.A., Ponce, R.A., Schroeder, J., Kavanagh, T.J. and Faustman, E.M. (1999) Induction of the cell cycle regulatory gene p21 (Waf1, Cip1) following methylmercury exposure in vitro and in vivo. *Toxicol Appl Pharmacol*. 157: 203–212.

Ozuah, P.O. (2000) Mercury poisoning. *Curr Probl Pediatr*. 30: 91-99.

Palkiewicz, P., Zwiers, H. and Lorscheider, F.L. (1994) ADP-ribosylation of brain neuronal proteins is altered by in vitro and in vivo exposure to inorganic mercury. *J Neurochem.* 62: 2049–2052.

Pamphlett, R., Slater, M. and Thomas, S. (1998) Oxidative damage to nucleic acids in motor neurons containing mercury. *J Neurol Sci*. 159: 121–126.

Pamphlett, R. and Waley, P. (1996a) Uptake of inorganic mercury by the human brain. *Acta Neuropathol*. 92: 525-527.

Pamphlett, R. and Waley, P. (1996b) Motor neuron uptake of low dose inorganic mercury. *J Neurol Sci*. 135: 63–67.

Patrick, L. (2002) Mercury toxicity and antioxidants: Part 1: Role of glutathione and alpha-lipoic acid in the treatment of mercury toxicity. *Altern Med Rev.* 7: 456–471.

Pendergrass, J.C. and Haley, B.E. (1995) Mercury-EDTA Complex Specifically Blocks Brain-Tubulin-GTP Interactions: Similarity to Observations in Alzheimer's Disease. In: Friberg LT, Schrauzer GN, eds. Status Quo and Perspective of Amalgam and Other Dental Materials. International Symposium Proceedings. Thieme, Stuttgart- New-York, 98–105.

Pendergrass, J.C. and Haley, B.E. (1996) Inhibition of Brain Tubulin-Guanosine 5'-Triphosphate Interactions by Mercury: Similarity to Observations in Alzheimer's Diseased Brain. In: Sigel H, Sigel A, eds. Metal Ions in Biological Systems. V34. Marcel Dekker, New York. Pp. 461–78.

Pendergrass, J.C., Haley, B.E., Vimy, M.J., Winfield, S.A. and Lorscheider, F.L. (1997) Mercury vapor inhalation inhibits binding of GTP to tubulin in rat brain: similarity to a molecular lesion in Alzheimer diseased brain. *Neurotoxicology*, 18: 315–324.

Peterson, E.W. and Cardoso, E.R. (1983) The blood–brain barrier following experimental subarachnoid hemorrhage. part 2: Response to mercuric chloride infusion. *J Neurosurg.* 58: 345–351.

Pinheiro, M.C.N., Oikawa, T., Vieira, J.L.F., Gomes, M.S.V., Guimarães, G.A., Müller, R.C.S., Amoras, W.W., Ribeiro, D.R., Rodrigues, A.R., Côrtes, M.I. and Silveira, L.C. (2006) Comparative study of human exposure to mercury in riverside communities of Amazon. *Braz J Med Biol Res.* 39: 411–414.

Powell, T.J. (2000) Chronic neurobehavioral effects of mercury poisoning on a group of Zulu chemical workers. *Brain Inj.* 14:797–814.

Praline, J., Guennoc, A.M., Limousin, N., Hallak, H., de Toffol, B. and Corcia, P. (2007) ALS and mercury intoxication: a relationship? *Clin Neurol Neurosurg.* 109: 880-883.

Rajanna, B., Rajanna, S., Hall, E. and Yallapragada, P.R. (1997) In vitro metal inhibition of N-methyl-d-aspartate specific glutamate receptor binding in neonatal and adult rat brain. *Drug Chem Toxicol.* 20: 21– 29.

Reardon, M.A. and Bhat, H.K. (2007). Methylmercury neurotoxicity: Role of oxidative stress. *Toxicol Environ Chem.* 89: 535–554.

Rice, D. and Barone, Jr. S. (2000) Critical periods of vulnerability for the developing nervous system: evidence from humans and animal models. *Environ Health Perspect.* 108: 511–513.

Roberg, K., Johansson, U. and Ollinger, K. (1999). Lysosomal release of cathepsin D precedes relocation of cytochrome C and loss of mitochondrial transmembrane potential during apoptosis induced by oxidative stress. *Free Radical Biol Med.* 27:1228–1237.

Roegge, C.S., Wang, V.C., Powers, B.E., Klintsova, A.Y., Villareal S, Greenough, W.T. and Schantz, S.L. (2004) Motor impairment in rats exposed to PCBs and methylmercury during early development. *Toxicol Sci.* 77: 315–324.

Rybicki, B.A., Johnson, C.C., Uman, J. and Gorell, J.M. (1993) Parkinson's disease mortality and the industrial use of heavy metals in Michigan. *Mov Disord.* 8: 87-92.

Sarafian, T.A. and Verity, M.A. (1991) Oxidative mechanisms underlying methyl mercury neurotoxicity. *Int J Dev Neurosci.* 9:147–153.

Satoh, H. (2000) Occupational and environmental toxicology of mercury and its compounds. *Ind Health*, 38: 153–164.

Schiønning, J.D. (1993) Retrograde axonal transport of mercury in rat sciatic nerve. *Toxicol Appl Pharmacol*. 121: 43-49.

Schwarz, S., Husstedt, I., Bertram, H.P. and Kuchelmeister, K. (1996) Amyotrophic lateral sclerosis after accidental injection of mercury. *J Neurol Neurosurg Psychiatry*, 60: 698.

Segall, H.J. and Wood, J.M. (1974) Reaction of methylmercury with plasmalogens suggests a mechanism for neurotoxicity of metal-alkyls. *Nature*, 248:456–458.

Shafer, T.J. and Atchison, E.D. (1991) Methylmercury blocks N- and L- type Caþþ channels in nerve growth factordifferentiated pheochromocytoma (PC12) cells. *J Pharmacol Exp Ther*. 258:149–157.

Shafer, T.J., Meacham, C.A. and Barone, S. (2002) Effects of prolonged exposure to nanomolar concentrations of methylmercury on voltage-sensitive sodium and calcium currents in PC12 cells. *Dev Brain Res*. 136:151–164.

Shanker, G. and Aschner, M. (2003) Methylmercury-induced reactive oxygen species formation in neonatal cerebral astrocytic cultures is attenuated by antioxidants. *Mol Brain Res*. 110:85–91.

Shaw, P.J. (2005) Molecular and cellular pathways of neurodegeneration in motor neurone disease. *J Neurol Neurosurg Psychiatry*, 76: 1046-1057.

Shuman, E.M. and Madison, D.V. (1994) Nitric oxide and synaptic function. *Ann Rev Neurosci*. 17: 153–183.

Silva-Pereira, L.C., Cardoso, P.C.S., Leite, D.S., Bahia, M.O., Bastos, W.R., Smith, M.A.C. and Burbano, R.R. (2005) Cytotoxicity and genotoxicity of low doses of mercury chloride and methylmercury chloride on human lymphocytes in vitro. *Braz J Med Biol Res*. 38:901–907.

Simmons-Willis, T., Koh, A., Clarkson, T.W. and Ballatori, N. (2002) Transport of a neurotoxicant by molecular mimicry: The methylmercury-L-cysteine complex is a substrate for human L-type large neutral amino acid transporter LAT 1 and LAT 2. *Biochem J*. 367:239–246.

Soleo, L., Urbano, M.L., Petrera, V. and Ambrosi, L. (1990) Effects of low exposure to inorganic mercury on psychological performance. *Br J Ind Med*. 47: 105–109.

Soo, Y.O., Chow, K.M., Lam, C.W., Lai, F.M., Szeto, C.C., Chan, M.H. and Li, P.K. (2003) A whitened face woman with nephrotic syndrome. *Am J Kidney Dis*. 41: 250–253.

Stankovic, R. (2006) Atrophy of large myelinated motor axons and declining muscle grip strength following mercury vapor inhalation in mice. *Inhal Toxicol*. 18: 57-69.

Stohs, S.J. and Bagchi, D. (1995) Oxidative mechanisms in the toxicity of metal ions. *Free Radic Biol Med*. 18: 321–336.

Störtebecker, P. (1988) Neurology for barefoot doctors in all countries. Correct Diagnosis by simple methods. Täby/Stockholm: Störte becker Foundation for Research.

Su, M., Wakabayashi, K., Kakita, A., Ikuta, F. and Takahashi, H. (1998) Selective involvement of large motor neurons in the spinal cord of rats treated with methylmercury. *J Neurol Sci*. 156: 12-17.

Suzuki, Y.J., Forman, H.J. and Sevanian, A. (1997) Oxidants as stimulators of signal transduction. *Free Rad Biol Med*. 22:269–285.

Sweet, L.I. and Zelikoff, J.T. (2001) Toxicology and immunotoxicology of mercury: A comparative review in fish and humans. *J Toxicol Environ Health Part B*, 4:161–205.

Sweeten, T.L., Bowyer, S.L., Posey, D.J., Halberstadt, G.M. and McDougle, C.J. (2003) Increased prevalence of familial autoimmunity in probands with pervasive developmental disorders. *Pediatrics*, 112:e420.

Takeuchi, T., Eto, N. and Eto, K. (1979) Neuopathology of childhood cases of methylmercury poisoning (Minamata disease) with prolonged symptoms, with particular reference to the decortication syndrome. *Neurotoxicology*, 1: 1–20.

Tejwani, G.A. and Hanissian, S.H. (1990) Modulation of mu, delta, and kappa opioid receptors in rat brain by metal ions and histidine. *Neuropharmacology*, 29: 445-452.

Tiffany-Castiglion, E. and Qian, Y. (2001) Astroglia as metal depots: molecular mechanisms for metal accumulation, storage and release. *Neurotoxicology*, 22: 577-592.

Toimela, T. and Tähti, H. (2004) Mitochondrial viability and apoptosis induced by aluminium, mercuric mercury and methylmercury in cell lines of neural origin. *Arch Toxicol.* 78: 565–74.

Tunnessen, W.W., McMahon, K.J. and Baser, M. (1987) Acrodynia: exposure to mercury from fluorescent light bulbs. *Pediatrics*, 79: 786-789.

Uversky, V.N., Li, J. and Fink, A.L. (2001) Metal-triggered structural transformations, aggregation, and fibrillation of human α-synuclein. A possible molecular NK between Parkinson's disease and heavy metal exposure. *J Biol Chem.* 276: 44284– 44296.

Valko, M., Morris, H. and Cronin, M.T. (2005) Metals, toxicity and oxidative stress. *Curr Med Chem.* 12: 1161-208.

Vidal, L., Durán, R., Faro, L.F., Campos, F., Cervantes, R.C., and Alfonso, M. (2007) Protection from inorganic mercury effects on the *in vivo* dopamine release by ionotropic glutamate receptor antagonists and nitric oxide synthase inhibitors. *Toxicology*, 238: 140-146.

Warkany, J. and Hubbard, D.M. (1953) Acrodynia and mercury. *J Ped Gastroenterol.* 42: 365-386.

WHO. (2003) Elemental mercury and inorganic mercury compounds: human health aspects. World Health Organization, Geneva.

White, J.F. (2003) Intestinal pathophysiology in autism. *Exp Biol Med.* 228: 639–649.

Wilke, R.A., Kolbert, C.P., Rahimi, R.A. and Windebank, A.J. (2003) Methylmercury induces apoptosis in cultured rat dorsal root ganglion neurons. *Neurotoxicology*, 24:369–378.

Wilkinson, L.J. and Waring, K.H. (2002) Cysteine dioxygenase: modulation of expression in human cell lines by cytokines and control of sulphate production. *Toxicol In Vitro*, 16: 481–483.

Yamanaga, H. (1983) Quantitative analysis of tremor in Minamata disease. *Tohukv J Exp Med.* 141: 13-22.

Yannai, S., Berdicevsky, I. and Duek, L. (1991) Transformations of inorganic mercury by Candida albicans and Saccharomyces cerevisiae. *Appl Environ Microbiol.* 57: 245-247.

Yee, S. and Choi, B.H. (1996) Oxidative stress in neurotoxic effects of methylmercury poisoning. *Neurotoxicology*, 17: 17-26.

Zahir, F., Rizwi, J.S., Haq, K.S. and Khan, H.R. (2005) Low dose mercury toxicity and human health. *Environ Toxicol Pharmacol.* 20: 351–360.

Zalups, R.K. (2000) Molecular interactions with mercury in the kidney. *Pharmacol Rev.* 52: 113-143.

Zanoli, P., Cannazza, G. and Baraldi, M. (2001) Prenatal exposure to methyl mercury in rats: Focus on changes in kynurenine pathway. *Brain Res Bull*, 55: 235–238.

Zhang, J., Miyamoto, K., Hashioki, S., Hao, H.P., Murao, K., Saido, T.C. and Nakanishi, H. (2003) Activation of mu-calpain in developing cortical neurons following methylmercury treatment. *Dev Brain Res.* 142:105–110.

Zhang, M.Q., Zhu, Y.C. and Deng, R.W. (2002) Evaluation of mercury emissions to the atmosphere from coal combustion. *China Ambio*, 31: 482–484.

Zheng, W., Aschner, M. and Ghersi-Egea, J. (2003) Brain barrier systems: a new frontier in metal neurotoxicological research. *Toxicol App Pharmacol.* 192: 1–11.

Zigmond, M.J., Castro, S.L., Keefe, K.A., Abercrombie, E.D. and Sved, A.F. (1998) Role of excitatory amino acids in the regulation of dopamine synthesis and release in neostriatum. *Amino acids*, 14: 57–62.

In: Progress in Neurodegeneration
Editors: M.R. Avila-Costa and V.A. Martinez

ISBN 978-1-60741-317-2

Chapter 12

Copper

Verónica Anaya-Martínez and Maria Rosa Avila-Costa
Department of Neuroscience, Neuromorphology Lab
UNAM, Mexico

12.1. General Description

Copper (Cu) is a transition metal well known for its distinctive reddish brown color. Cu takes on a bright metallic luster, and is malleable, ductile, and is the second better conductor of heat and electricity (silver is the first) for that the electrical industry is one of the greatest users of Cu (Schäfer et al. 1999).

Cu has been known from ancient times, the early use of Cu possibly resulted from the natural incidence in native form. The "Copper Age" followed the "Stone Age". In compounds found in the environment it usually has a valence of 2 but can exist in the metallic, +1 and +3 valence states. Cu is found naturally in a wide variety of mineral salts and organic compounds, and in the metallic form (Kim et al. 2008).

Cu is present in relatively high levels in brain and is distributed heterogeneously in two oxidation states (cupric Cu^{2+} and cuprous Cu^{+}). The distribution is dependent on brain region, subcellular location, and other variables such as age, species, environmental, and genetic factors. Neural distribution of Cu correlates well with known neural functions of Cu (Prohaska, 1987).

Cu is required for cellular respiration, iron oxidation, pigment formation, neurotransmitter biosynthesis, antioxidant defense, peptide amidation, and connective tissue formation (Pena et al. 1999). This metal is essential for the Central Nervous System (CNS) development, and disruption of Cu homeostasis during fetal life leads to perinatal mortality, severe growth retardation, and neurodegeneration (Keen et al. 1998). Experiments in mice reveal that the developmental timing of perinatal Cu deficiency influences the severity of neurological outcome, suggesting a critical period for brain Cu acquisition (Prohaska and Brokate 2002). Acquired Cu deficiency in adults results in myelopathy with lower limb spasticity and sensory ataxia due to ascending sensory tract dysfunction and neurodegeneration of the dorsal column (Kumar et al. 2004; Prodan et al. 2006).

12.2. Sources of Copper Exposure

It is well know that the main natural sources of Cu include windblown dust, volcanoes, decaying vegetation, forest fires and sea spray. The humans are responsible for emissions in the smelters, iron foundries, power stations and combustion sources such as municipal incinerators (WHO, 1998). Other sources of Cu in air may be tobacco smoke and stack emissions of coal burning power plants (Lide, 1998). In the soil the major release of Cu is from tailings and overburdens from Cu mines and sewage sludge. Agricultural use of Cu products accounts for 2% of Cu released to soil (WHO, 1998).

Occurrence (all forms) in earth crust: 70 ppm, in seawater: 0.001-0.02 ppm. Found in nature in its native state; also in combined form in several minerals. Cu is present in concentrations averaging about 4 ppm in limestone, 55 ppm in igneous rocks, 50 ppm in sandstones, and 45 ppm in shales (Lide, 1998).

The concentration of Cu in the continental crust, generally estimated at 50 ppm, tends to be highest in the ferromagnesium minerals, such as the basalts pyropene and biotite, where it averages 140 ppm. Sandstones contain 10-40 ppm, shales 30-150 ppm, and marine black shales 20-300 ppm (Schäfer et al. 1999).

12.2.1. Air

Exposure to Cu by inhalation is determined by air concentrations, particulate size and the respiratory rate. Concentrations of Cu determined in Canada over the period 1984-1993 averaged 0.014 $\mu g/m^3$. The maximum value, detected in 66% of samples, was 0.418 μg Cu/m^3 (Dann, 1994). In the USA, air levels of Cu vary between 96 and 138 ng/m^3 in urban samples and 76 and 176 ng/m^3 in non-urban settings, though levels as high as 4629 ng/m^3 have also been recorded (Lide, 1998; WHO, 1998).

Inhalation of dusts and fumes of metallic Cu causes irritation of the upper respiratory tract, congestion of nasal mucous membranes, ulceration and perforation of the nasal septum, and pharyngeal congestion. Inhalation of Cu fumes may give rise to metal fume fever (high temperature, metallic taste, nausea, coughing, general weakness, muscle aches, and exhaustion) (Kim et al. 2008).

12.2.2. Water

Because the concentration of Cu in drinking water is highly variable, means are of limited significance. A combination of low pH and soft water passing through Cu pipes and fittings may produce high Cu levels in drinking water; however, only a little over 1% of USA drinking water exceeds the drinking water standard of 1 mg/l, with the average Cu concentration in drinking water reported as approx 0.13 mg/l (Schäfer et al. 1999).

12.2.3. Food

Clinical studies in humans indicate that most adults ingest about 1mg of Cu daily in the diet (NRC, 1989). This needs further assessment and suffers from a lack of sensitive, reliable and easy measures for detecting marginal deficiency as well as excess Cu accumulation. Nevertheless, the available evidences suggest that levels of intake between 1 and 3mg are indeed safe and most probably are adequate for the normal human adult (NRC 1989; Linder, 2001).

Cu ingestion causes nausea, vomiting, abdominal pain, metallic taste, and diarrhea. Ingestion of large doses may cause stomach and intestine ulceration, jaundice, and kidney and liver damage (Owen 1981; Madsen and Gitlin, 2007).

12.3. Absorption, Distribution, Metabolism, and Excretion

Cu is readily available in the diet and, following absorption through the stomach and duodenum, probably as complexes with amino acids, such as histidine, or small peptides (Pickart and Thaler, 1980), is rapidly removed from the portal circulation by hepatocytes in the liver. These Cu complexes enter the blood and most of the Cu binds tightly to serum albumin in equilibrium with a small 'pool' of Cu complexes. In the liver Cu is taken up and incorporated into the glycoprotein ceruloplasmin (CP), which is then released into the circulation (normal plasma content is 200-400mg/l). About 95% of total plasma Cu is found in this protein, the rest being attached to albumin or amino acids as mentioned above (Harrison and Hoare, 1980). CP contains 6 or 7mol of copper/mol of protein. It does not exchange Cu readily nor will it tightly bind extra Cu, so it does not have the characteristics of a Cu transport protein (by contrast transferring can easily bind Fe(III) and release it again within cells). It seems that cells must take up and degrade CP in order to obtain Cu from it (Gutteridge, 1981).

Biliary excretion is the only physiological mechanism of Cu elimination, and at steady state the amount of Cu excreted into the bile is equivalent to that absorbed from the intestine (Gitlin, 2003). The rate of Cu excretion into the bile increases promptly in response to an increase in dietary Cu, and excess of this metal does not occur in the absence of an underlying metabolic defect (Gollan and Deller, 1973). This aspect of Cu homeostasis is critical to the interpretation of any study that implicates increases in dietary Cu content with neurologic disease (Sparks and Schreurs, 2003).

12.4. Toxicity

The fumes and dust cause irritation of the upper respiratory tract. Inhalation of Cu fume results in the irritation of the upper respiratory tract. Contact with Cu fumes will also cause irritation of the eyes, nose and throat (Lide, 1998).

Cu has a contraceptive effect when present in the uterus. It is added to some intrauterine contraceptive devices permitting reduction in their size with concomitant reduction in the associated side effects such as pain and bleeding (Lide, 1998).

Gastrointestinal irritation, rarely serious, can result following the drinking of carbonated water or citrus fruit juices, which have been in contact with Cu vessels, pipes, tubing or valves. Such beverages are acidic enough to dissolve irritant quantities of Cu (Schäfer et al. 1999).

Most animals share certain pathological characteristics of Cu deficiency; in man a lack of Cu is infrequently (Owen, 1981). Many enzymes are affected by Cu deficiency, in rat CP cinoxidase is depressed and succinic dehydrogenase is increased (Gallagher et al. 1956), the cytochrome oxidase is reduced in some organs (Dallman, 1967), Cu-Zn superoxide dismutase falls in the erythrocytes (Bohnenkamp and Weser, 1976), tyrosine hydroxylase in the brain is drop (Morgan and O'Dell, 1977), and lysyl oxidase is reduced (Sandberg et al. 1969).

The decrease of this last enzyme was relating by the phatological process: lathyrism, in which the experimental animals show degeneration of the aorta, skin fragility, tendinous loosening, and epiphyseal and periosteal changes (Rojkind et al. 1968); symptoms related by the alterations in collagen and elastin (Dasler et al. 1961).

The human Cu deficiency was showed in premature infants (Graham, 1971), in infants with severe diarrhea (Cordano and Graham, 1966), or malnutrition (Yuen et al. 1979). Cu deficiency leads to anemia, anorexia, dermatitis, diarrhea, osteoporosis, neurotrophenia, pallor, hepatomegalia and slowed growth (Hambidge, 1977). Cu deficiency is a hallmark of the hereditary disorder called Menkes disease (Madsen and Gitlin, 2007) in which the infants suffer hypothermia, hypopigmentation, abnormal hair, tortuous arteries, intractable seizures, and failure to thrive, all the mentioned taken place by the lack of defects in Cu transports across the placenta, brain, and gastrointestinal tract due to loss-of-function mutations in the gene Atp7a (Chelly et al. 1993; Mercer et al. 1993; Vulpe et al. 1993).

About the hereditary disorders related with Cu, Wilson disease is an autosomal recessive disorder presented in the second-third decade of life and resulted in hepatic cirrhosis, dystonia, dysarthria and progressive basal ganglia degeneration due to loss-of-function mutations in the gene encoding the copper-transporter Atp7b (Yamaguchi et al. 1993; Madsen and Gliti, 2007).

In humans are more common the cases of toxicity for Cu over exposition by accidental ingest, suicide or industrial poisoning. Workers in Cu smelters and in brass foundries had symptoms as headache, sweating, nausea and exhaustion after they were exposed to Cu (Lyle et al. 1976).

The occurrence of lung cancer is bigger in Cu smelter workers and in women in nearby cities (Newman et al. 1976). Cu contamination is also a problem in the area of lead smelters; the Cu content of farm dust, soil and plants was increased about 6-fold (Dorn et al. 1975).

Prolonged or repeated exposure to Cu can discolor skin and hair and irritate the skin; also may cause mild dermatitis, runny nose, and irritation of the mucous membranes. Repeated ingestion may damage the liver and kidneys. Repeated inhalation can cause chronic respiratory disease (Madsen and Gitlin, 2007).

Persons with pre-existing skin disorders or impaired liver, kidney, or pulmonary function or pre-existing Wilson disease may be more susceptible to the effects of this metal (Schäfer et al. 1999).

12.4.1. Mechanisms of Copper Toxicity

The mechanism of Cu toxicity is not fully understood. In addition to gastrointestinal irritation, the target organ for Cu is the liver, as observed in the congenital Wilson disease or the anthropogenic Cu overload in childhood cirrhosis. Cu is transported from the plasma to the liver cells by albumin. In the liver cells, Cu ions are transferred to metallothionein and to CP, a process of limited capacity. At the same time, Cu ions are secreted with the bile. In cases of a deteriorated elimination capacity (Wilson disease, childhood cirrhosis) and an exhaustion of intracellular binding proteins, free Cu ions are likely to become available. The cytotoxicity of free Cu ions is well known. It may result in hepatic cirrhosis, as is observed in Cu overload diseases (Schäfer et al. 1999).

12.5. Copper in the Brain

Many metals such as Cu, magnesium, manganese, iron, zinc, cobalt, and molybdenum, are essential and required for optimal CNS function. They play important roles in brain function as catalysts, second messengers, and gene expression regulators. Being essential cofactors for functional expressions of many proteins, trace elements are needed to activate and stabilize enzymes, such as superoxide dismutase, metalloproteases, protein kinases, and transcriptional factors containing zinc finger proteins. Apparently, metals must be supplied to the CNS at an optimal level, because both deficiency and excess can result in aberrant CNS function. Thus, transport of metal ions across the blood-brain barrier (BBB) is the first step in regulating their CNS levels. A series of active or receptor-mediated transport systems inherent to the BBB vasculature serve to control the transport of these metals into the brain, maintaining their optimal concentrations (Zheng et al. 2003).

It has been known for some time that brain contains a number of Cu proteins. Some of these proteins have identified catalytic properties, whereas the function of others remains unknown. Some brain enzyme activities change after dietary Cu deficiency, but these are indirect changes and not Cu dependent in a strict sense (Prohaska, 1987).

One of the earliest enzyme activities known to be Cu dependent was the catalysis of cytochrome-c oxidation. Shortly after the essentiality of Cu was established, Cohen and Elvehjem (1934) showed that in brain of copper-deficient rats, the oxidase activity and a component of heme were reduced. After a rather argumentative period, it was shown that cytochrome oxidase did contain Cu (Wainio et al. 1959).

In brain, cytochrome oxidase is an important enzyme in oxidative metabolism. Its distribution within neurons reflects neuronal activity exclusive to that pathway (Kageyama and Wong-Riley, 1982). When an activity of 10 U cytochrome oxidase/nmol Cu is assumed, the amount of Cu in young adult rat brain associated with cytochrome oxidase is estimated at -20% of the total brain Cu, which indicates that this protein is one of the major cuproproteins of brain (Prohaska, 1987).

On the other hand, eucaryotic cells, including neurons and glia, contain two distinct proteins that catalyze the disproportionation of superoxide ($O_2^{\bullet -}$), the univalent reduction product of dioxygen (Sayre et al. 2005). Both are called superoxide dismutases (SOD). One is a homodimer of 32,000 mol wt. Each subunit contains 1 mol Cu and zinc (CuZn-SOD). The other SOD contains one manganese (MnSOD) per subunit and is a tetramer of 80,000 mol wt. The subcellular distribution of CuZn-SOD and MnSOD is unique with MnSOD located in the matrix of mitochondria, whereas CuZn-SOD is located primarily within the cytoplasm (Prohaska, 1987).

CuZn-SOD is a major cuproprotein of brain and may account for an estimated 25% of the total Cu in rat brain (Prohaska and Brokate, 2002). The CNS is a highly aerobic organ consuming, in the resting state, about one-fifth of the total oxygen debt. The CNS is also sensitive of oxygen toxicity. Thus SOD activity has been assumed to be important in protecting brain from elevation of $O_2^{\bullet -}$ during normal conditions and under various neuropathological states (Prohaska, 1987).

A number of neuropathological diseases in which brain lipid peroxidation has been implicated have been studied, and the activity of CuZn-SOD has been measured. These include multiple sclerosis, Down syndrome, neuronal ceroid-lipofuscinosis and chronic alcoholism. No changes in CuZn-SOD in brain were found, except in Down syndrome. In this condition, trisomy 21, genetic amplification of CuZn-SOD results in a 50% increase in enzyme and a corresponding increase in lipid peroxidation (as measured by malondialdehyde formation *in vitro*) (Brooksbank and Balazs, 1984). This occurs, as suggested by the authors, because of an increase in H_2O_2 levels from excessive dismutation of $O_2^{\bullet -}$. In brain, no compensatory rise in glutathione peroxidase (GSH) occurred (Brooksbank and Balazs, 1984). Thus, Prohaska (1987) assume that Cu, through the CuZn-SOD molecule, exerts one of its biological functions in superoxide metabolism.

A third copper-dependent enzyme is dopamine-β-monooxygenase (DBM). This large (mol wt 290,000) glycoprotein tetramer has interested neuropharmacologists greatly, since the enzyme is responsible for synthesis of the neurotransmitter norepinephrine (Prohaska, 1987). Regional distribution of DBM correlates with noradrenergic neurons. Cell bodies of noradrenergic neurons are enriched in the locus coeruleus, brain stem, and posterior hypothalamus, and in these regions DBM is also enriched as determined enzymatically (Coyle and Axelrod, 1972) or immunocytochemically (Cimarusti et al. 1979). The histochemical method has also shown DBM to be present in distal axons where, presumably, it is located near the norepinephrine storage granules. Thus, in contrast to cytochrome oxidase and CuZn-SOD, DBM has a unique and highly specialized distribution (Prohaska, 1987).

Finally, the tripeptide glutathione (GSH) is present at high concentration in a number of tissues (Deneke and Fanburg, 1989), and has been shown to bind Cu and to play a role in its biliary excretion (Houwen et al. 1990). In addition, GSH can transfer Cu to cuproproteins including metallothioneins (Ferreira et al. 1993), a family of proteins that play an important role in metal detoxification. It has been known for some time that intake of high quantities of zinc increases the level of metallothioneins and reduces the absorption of Cu (Turnlund, 1999). It has been proposed that sequestration of Cu by metallothioneins in mucosal cells may account for the decrease in Cu absorption (Fischer et al. 1983). Bertinato and L'Abbé (2004) assumed that GSH and metallothioneins play an important role for intracellular Cu metabolism that extends beyond their role as metal detoxification proteins.

12.5.1. Copper Transport to Brain

Uncertainty still exists regarding transport of Cu from blood to brain. Some believe CP (Hellman and Gitlin, 2002), the plasma cuproprotein, is responsible; others suggest that copper-amino acid complexes are taken up. Neural Cu can exist in a pool with low-molecular-weight ligands, as shown by some (Hunt, 1980).

Recently, it has been proposed that cooper transporter 1 (Ctrl) is responsible for Cu-transpor to brain; Ctrl is a plasma membrane protein essential for early embryonic development and intestinal Cu uptake (Kuo et al. 2001, Lee et al. 2001, Nose et al. 2006) (Figure 1). Ctrl is present on endothelial cells of the BBB, and expression at this site increases following perinatal Cu deficiency (Kuo et al. 2006). These observations suggest that Cu is transported from the plasma into the brain via Ctr1. Intracellular Cu metabolism is dependent on the Cu transport Atpases, Atp7a and Atp7b. One of these Atpases resides in the late Golgi of every cell, delivering Cu to the secretory pathway for incorporation into cuproenzymes and excretion (Madsen and Gitlin, 2007). In the brain, Atp7a is expressed in endothelial cells of the BBB and facilitates Cu movement across the basolateral membrane into the extravascular space of the brain. Atp7a is also expressed within specific populations of neurons in several brain regions including the cerebellum and hippocampus (Schlief et al. 2005). Atp7b is expressed in hepatocytes and is required for Cu excretion from these cells into the bile. Cu-dependent trafficking of these Atpases serves as the primary mechanism determining intracellular Cu homeostasis (Lutsenko and Petris 2003) (Figure 1).

Intracellular trafficking of Cu also requires proteins termed metallochaperones that target this metal to specific cellular pathways (O'Halloran and Culotta 2000). These chaperones include Atox1, which delivers Cu to the copper-transporting Atpases in the late Golgi (Hamza et al. 2001, 2003), Cu chaperone for superoxide dismutase (CCS), which is required for Cu incorporation into cytoplasmic Cu/Zn-SOD (Culotta et al. 2006; Wong et al. 2000), and Cox17, Sco1, and Sco2, which deliver Cu to mitochondrial cytochrome c oxidase (Hamza and Gitlin, 2002) (see figure 1). Sco1 and Sco2 also function in a pathway of mitochondrial Cu homeostasis (Leary et al. 2007). Metallothionein is a cysteine-rich cytoplasmic protein that chelates Cu and is essential to protect against the toxicity caused by excess Cu (Kelly and Palmiterm, 1996).

Cu is distributed throughout most regions of the brain and is most abundant in the basal ganglia (Madsen and Gitlin, 2007). Studies have detected this metal in the cell bodies of cortical pyramidal and cerebellar granular neurons, in the neuropil of the cerebral cortex, in the hippocampus and cerebellum, and in synaptic membranes of afferent nerves (Kozma et al. 1981, Sato et al. 1994, Trombley and Shepherd, 1996). In some neurons, Cu is released at the synapse (Hartter and Barnea, 1988) reaching micromolar concentrations (Kardos et al. 1989) that can abrogate long-term potentiation in the hippocampus (Doreulee et al. 1997). Cu is an antagonist at N-methyl-D-aspartate (NMDA) receptors in cultured hippocampal neurons (Vlachova et al. 1996), modulating activation of calcium-dependent cascades that contribute to synaptic plasticity (Lu et al. 2001). Synaptic NMDA receptor activation results in rapid and reversible trafficking of Atp7a in cultured hippocampal neurons in association with Cu release, suggesting the presence of a mechanism linking Cu homeostasis and neuronal activation within the brain (Schlief et al. 2005).

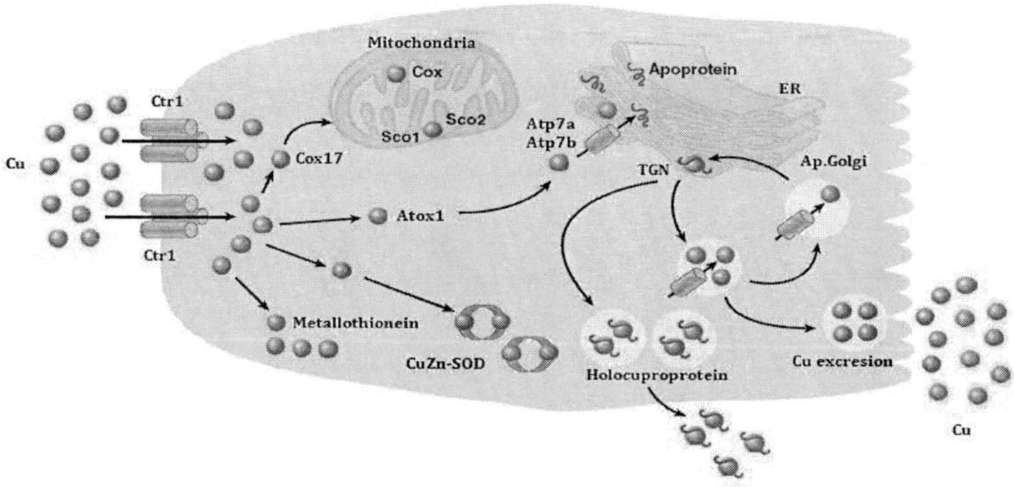

Modified from Madsen and Gitlin 2007.

Figure 1. Cellular Cu homeostasis. Model of Cu trafficking in polarized cell reveals Cu entry via Ctr1 followed by distribution to the Cu chaperones. The Cu chaperone for superoxide dismutase (CCS) delivers copper to Cu/Zn superoxide dismutase (SOD1), Atox1 delivers Cu to one of the Atpases (Atp7a/Atp7b) in the late Golgi and Cox17, and Sco1 and Sco2 are involved in the pathway of copper trafficking to mitochondria and cytochrome oxidase (Cox). The Atpases transport Cu into the secretory pathway for incorporation into newly synthesized cuproproteins and for export from the cell. Metallothionein serves to chelate most available copper and is critical for cell survival in Cu excess.

12.5.2. Copper Neurotoxicity

There is substantial interest in the role of Cu, and other trace redox-active transition metals in the neuropathology of neurodegenerative disorders such as Parkinson disease (PD), Alzheimer disease (AD) and amyotrophic lateral sclerosis (ALS). These metals are essential in most biological reactions (e.g. in the synthesis of DNA, RNA and proteins), and as cofactors of numerous enzymes, particular those involved in respiration; thus, their deficiency can lead to disturbances in CNS and other organ function. However, excessive tissue accumulation of redox-active transition metals can be cytotoxic, in particular because perturbations in metal homeostasis result in an array of cellular disorders characterized by oxidative stress and increased free radical production. Oxidative stress, defined as the imbalance between biochemical processes leading to production of reactive oxygen species (ROS) and the cellular antioxidant cascade, causes molecular damage that can lead to a critical failure of biological functions and ultimately cell death (Sayre et al. 2005).

Wilson disease and Menkes disease are the known inherited disorders of Cu metabolism in humans (Madsen and Gitlin, 2007). The essential role of Cu in the developing CNS is evidenced by Menkes disease, during which impaired Cu transport into and within the developing brain results in demyelination and neurodegeneration (Kodama et al. 1999). Brain Cu accumulation in Wilson disease results in dystonia, dysarthria, and other parkinsonian symptoms, as well as psychiatric symptoms of depression, cognitive deterioration, personality change, psychosis, and schizophrenia (Ferenci, 2004). Although the signs and symptoms of Menkes and Wilson diseases are distinct, each disorder results from inherited loss-of-function mutations in genes encoding homologous copper-transporting P-type adenosine triphosphatases (Atpases) Atp7a and Atp7b (Culotta and Gitlin, 2001).

12.6. Mechanism of Copper-Induced Neurotoxicity

The overexpression of human CuZnSOD in mice, resulting in a tenfold higher level in both myocytes and endothelial cells, was able to extinguish a burst of superoxide (in electron paramagnetic resonance detection) and reduce functional damage following 30 minutes global ischemia (Wang et al. 1988). These results suggest that superoxide is an important factor in protecting against postischemic injury and it is therefore unexpected that CuZnSOD knockout mice show little if any neurodegenerative phenotype (Bruijn et al. 1998). Nonetheless, it is apparent that reductions in CuZnSOD activity can lead to a perturbation of cellular antioxidant defense mechanisms to promote a pro-oxidant condition. So, while absences of CuZnSOD have little effect, enzyme inactivation by metal-catalyzed oxidation promotes oxidative damage (Kwon et al. 1998).

Ceruloplasmin, an important Cu storage protein, is one of the key proteins that responds to oxidative stress. CP is considered the major iron(II)-oxidizing enzyme in the CNS, and an inherited metabolic disorder called aceruloplasminemia is associated with an impairment in iron homeostasis and consequent neurodegeneration (Harris et al. 1988). Interestingly, while CP is increased in brain tissue and cerebrospinal fluid in AD, PD and Huntington disease (HD) patients (Loeffler et al. 1996), neuronal levels of CP remain unaffected (Castellani et al. 1999). Therefore, while increased CP may indicate a compensatory response to augmented oxidative stress in AD, its failure to do so in neurons may play an important role in metal-catalyzed damage (Castellani et al. 1999). In fact, studies directed at clarifying the relationship between oxidative stress and tissue metal-ion levels indicate that both the ratio of Cu to Zn and the levels of CP are significantly higher with increasing age, and higher yet in cases with neurodegeneration (Mezzetti et al. 1998). Since the Cu-Zn ratio is significantly correlated with systemic oxidative stress, (i.e. lipid peroxidation), these findings suggest that increased oxidative stress burden in aging and neurodegeneration may reflect, in large part, copper-mediated ROS production (Sayre et al. 2005). In this case, redox-inert Zn may serve as an antioxidant by preventing binding of pro-oxidant Cu at tissue sites.

12.6.1. Oxidative Stress

The principal ROS cause of oxidative stress is the hydroxyl radical, which inflicts damage to biomacromolecules at diffusion-controlled rates. Although peroxynitrite also appears capable of hydroxyl radical-like activity, most hydroxyl radicals reflect the Fenton reaction between reduced transition metals (usually iron[II] or copper[I]) and H_2O_2. Re-reduction of the resulting oxidized transition metal ions (iron[III] or copper[II]) can be effected by superoxide or by other available cellular reductants such as vitamin C. In addition to their redox role, transition metals, along with redox-inactive metal ions, may contribute to neurodegeneration through their toxic effects on protein and peptide structure, such as a pathological aggregation phenomenon. In these cases, transition metals can sometimes exert dual neurotoxic properties (Sayre et al. 2005).

Cu, in excess of cellular needs, mediates free radical production and direct oxidation of lipids, proteins, and DNA. Therefore the balance between intracellular and extracellular contents of Cu is driven by cellular transport systems that regulate uptake, export and intracellular compartmentalization (Harris, 1991). The balance between Cu requirement and toxicity is achieved both at the cellular level and at the tissue and organ levels (Bull and Cox, 1994).

One of the most accepted explanations for copper-induced cellular toxicity comes from the assumption that Cu ions are prone to participate in the formation of ROS (Gaetke and Chow, 2003). Cupric and cuprous Cu ions can act in oxidation and reduction reactions. The cupric ion (Cu(II)), in the presence of biological reductants such as ascorbic acid or GSH, can be reduced to cuprous ion (Cu(I)) which is capable of catalyzing the formation of reactive hydroxyl radicals through the decomposition of H_2O_2 via Fenton reaction (Lloyd et al. 1997).

The hydroxyl radical is particularly reactive and can further react with practically any biological molecules in the near vicinity via hydrogen abstraction leaving behind carbon-centered radical e.g. form a lipid radical from unsaturated fatty acids (Buettner and Jurkiewicz, 1996). Experimental studies confirmed that Cu is also capable of inducing DNA strand breaks and oxidation of bases via oxygen free radicals (Brezova et al. 2003). Cu in both oxidation states (cupric or cuprous) was more active that iron in enhancing DNA breakage induced by the genotoxic benzene metabolite 1,2,4-benzenetriol. DNA damage occurred mainly by a site-specific Fenton reaction (Brezova et al. 2003).

When considering Cu as a metal participating in Fenton reaction it must be contemplate the discovery that upper limit of "free" pools of Cu is far less than a single ion per cell (Rae et al. 1999). This suggests a significant overcapacity for chelation of Cu in the cell.

Opposing oxidant damage is an array of antioxidant defenses that, in mammalian cells, include both enzymatic and nonenzymatic entities. Cytosolic CuZnSOD and mitochondrial Mn-SOD, which convert superoxide to O_2, and H_2O_2, are important in enzymatic defenses. H_2O_2 is detached by catalase and peroxidases, which have ubiquitous tissue distribution.

However, several articles have reported that the overexpression of CuZn-SOD causes oxidative damage to cells. A mechanism by which an excess of SODs accelerates oxidative stress was investigated (Lombardi et al. 1998). The presence of CuZn-SOD, Mn-SOD or Mn(II) enhanced the frequency of DNA damage induced by H_2O_2 and Cu(II), and altered the site specificity of the latter: H_2O_2 induced Cu(II)-dependent DNA damage with a high frequency at the 5'-guanine of poly G sequences; when SODs were added, the frequency of cleavages at the thymine and cytosine residues increased. SODs also enhanced the formation of 8-oxo-7,8-dihydro-2'-deoxyguanosine by H_2O_2 and Cu(II). It was concluded that overexpression of SODs may increase carcinogenic risks (Valko et al. 2005).

12.6.2. Neurotransmission

Cu, at high concentrations, may influence uptake and efflux of serotonin, dopamine and norepinephrine from presynaptic terminals (Komulainen and Tuomisto, 1982). Also copper-amino acid complexes may influence secretion of releasing hormones from hypothalamic granules (Barnea and Cho, 1984), further suggesting a role for Cu in neuroendocrine homeostasis. This final effect is mediated via prostaglandin E2 (Barnea et al. 1985). Cu may have neural functions distinct from the cuproenzymes; however, because of strong reactivity, cupric ions at nonphysiological doses can exert marked effects on biomolecules (Prohaska, 1987).

Some investigators found that Cu is accumulated in synaptic vesicles (Rajan et al. 1976) and might be released from axon terminals upon depolarization in a calcium-dependent manner (Hartter and Barnea, 1988). Moreover, it was shown that Cu influences other neurotransmitter systems, e.g., opiate or $GABA_A$ receptors (Sadée et al. 1982; Ma and Narahashi, 1993).

12.7. Copper and Neurodegenerative Diseases

12.7.1. Alzheimer Disease

AD is characterized pathologically by the presence of neurofibrillary tangles (NFT), senile plaques (SP), neuropil threads, amyloid-β (Aβ) deposition, and a selective loss of neurons. Much enthusiasm in recent years has come from the discovery of a multiplicity of mutations in the genes encoding the β-protein precursor (βPP) and/or the presenilins, which account for the bulk of the familial cases of AD on the basis of overproduction of βPP and/or altered βPP proteolytic processing, both leading to increased Aβ (Sayre et al. 2005). Aβ is a high-affinity Cu protein and amyloid precursor protein (APP) and is known to be a chaperone involved in Cu trafficking across the plasma membrane (Barnham et al. 2003; Bayer et al. 2003; Bellingham et al. 2004a; b).

Of the various hypotheses that have been suggested for the etiology of sporadic AD, the one receiving most attention is the role of oxidative stress. Several studies have indicated imbalances of trace elements, including aluminum, silicon, lead, mercury, zinc, Cu and iron.

A disruption of homeostasis of Cu and iron is particularly significant, recent evidence for increases in oxidative stress parameters such as lipid peroxidation, as well as increases in markers of oxidative damage to both the protein constituents of the pathological hallmark structures of AD (NFT and SP) (Sayre et al. 1997; Smith et al. 1998) and nucleic acids (Nunomura et al. 1999).

By means of microparticle-induced X-ray emission, it was found that Cu(II) and other trace metals are significantly elevated in AD neuropil and that these metals are significantly further concentrated within the core and periphery of senile plaques (Lovell et al. 1998).

In histochemical studies, Sayre et al. (2005) found that redox activity of the lesions in AD can be detected directly, can be inhibited by prior exposure of tissue sections to copper- and iron-selective chelators, and can be reinstated following reexposure of the chelator-treated sections to either Cu or iron salts. Their studies indicated the presence of redox-active iron and Cu in AD pathology suggesting that these metals are the main producers of ROS and are responsible not only for the numerous oxidative stress markers that appear on NFT and SP, but also for the more global oxidative stress parameters observed in AD.

Other studies of oxidative stress in AD have focused on the inducible mitochondrial Mn-SOD and the constitutive cytoplasmic CuZnSOD enzymes. The CuZnSOD gene is associated with AD neuropathology, and levels of both Mn-SOD mRNA and CuZnSOD were found to be increased in AD, whereas the total antioxidant status was decreased (De Leo et al. 1998). However, since SOD enzymes are key components of the cellular antioxidant armamentarium, any pro-oxidant mechanism linked to SOD must derive from the balance in the local concentrations of superoxide and H_2O_2, which together can produce the hydroxyl radical by the Haber-Weiss process (Sayre et al. 2005).

Zappasodi et al. (2008) studied the correlation between 'free' Cu and EEG rhythms alterations in AD, a step further by checking the hypothesis that this relationship might be different in carriers and non-carriers of the APOE-ε4 allele. They found that the correlations between EEG spectral abnormalities typical of the AD brain and higherthan-normal levels of serum 'free' Cu are stronger in APOE-ε4 carriers than in non-carriers. In particular, their AD patients, whose alpha rhythm amplitudes in parietal, occipital, and temporal areas were markedly lower than controls' amplitudes, showed higher-than-normal levels of serum 'free' Cu correlating with alpha power abnormalities in an APOE-ε4 dependent manner. Their results clearly indicate that the APOE-ε4 allele modulates the effect of Cu on EEG 'slowing' in AD, suggesting that the modulation of Cu dysfunction may be one of the mechanisms that make APOE-ε4 a risk factor for AD.

12.7.2. Parkinson Disease

Recent evidence implicates Cu (II) and Fe(III) in the overproduction of free radicals as a possible causal factor in the death of nigral cells associated to PD (Bush, 2000). Iron deposits selectively involving neuromelanin in substantia nigra neurons of PD patients have been found; in addition, contents of nigral Cu are decreased, whereas its concentration in cerebrospinal fluid is raised. Augmented oxidation markers and impaired mitochondrial electron transport mechanisms appear to be closely related to interactions between iron and neuromelanin, which result in accumulation of iron and a continuous production of cytotoxic species leading to neuronal death (Jellinger, 1999). In familial PD, a mutation of the α-synuclein gene has been identified, which is important because α-synuclein is a component of Lewy bodies, which typify the neuropathology. A report of Paik et al. (1999) proposes abnormal interaction of α-synuclein with Cu(II) in the formation of Lewy bodies.

12.7.3. Amyotrophic Lateral Sclerosis

A crucial breakthrough of ALS comes from the finding that many of the familial cases (FALS) are associated with one or another mutation in the CuZnSOD gene (Carrí et al. 2003). The protein products of these mutations retain nearly identical SOD activity, but take on altered properties linked to oxidative stress, possibly involving a gain-of-function peroxidase activity (Wiedau-Pazos et al. 1996). In this regard, transgenic mice overproducing a human FALS CuZnSOD mutant demonstrate increases in protein carbonyls suggestive of increased hydroxyl radical production or lipoxidation-derived radicals (Andrus et al. 1998). Also, using *in vivo* microdialysis, increased hydroxyl radical production in the striatum, as determined by conversion of 4-hydroxybenzoic acid to 3,4-dihydroxybenzoic acid, was seen for mice overexpressing mutant CuZnSOD relative to mice overexpressing the wild type human enzyme (Bogdanov et al. 1998). The hypothesis that mutant SOD-induced neurodegeneration is associated with disturbances in neuronal free-radical homeostasis is further supported by observations made on several neuronal cell cultures expressing the mutant SOD (Ghadge et al. 1997); however, the mechanism underlying the link between the SOD mutations and oxidative stress indicators appears not to be simply increased production of hydroxyl radicals, since no increases in $HO^{\bullet}$ adducts are seen *in vitro* for the Gly93→Ala and Ala4→Val mutants relative to wild type enzyme (Singh et al. 1998).

For the predominant, sporadic form of ALS, an imbalance in trace metal ions, possibly attached to increased oxidative stress, has been considered for some time. Some studies provide evidence for decreases in Cu in cerebrospinal fluid and serum, and increases in manganese in serum relative to age-matched controls (Hart et al. 1998).

12.7.4. Other Neurological Diseases

12.7.4.1. Wilson Disease

Wilson disease is an autosomal recessive disorder resulting in hepatic cirrhosis and progressive basal ganglia degeneration due to loss-of-function mutations in the gene encoding the copper-transporter Atp7b (Bull et al. 1993; Yamaguchi et al. 1993). The resulting impairment in biliary Cu excretion leads to hepatocyte Cu accumulation, copper-mediated liver damage, activation of cell-death pathways, leakage of Cu into the plasma, and eventual Cu overload in all tissues (Gitlin 2003; Tao and Glitin, 2003). Although Atp7b is expressed in some regions of the brain, in Wilson disease Cu overload in extrahepatic tissues is due to excess accumulation from the plasma following liver injury because this is entirely reversed following liver transplantation (Emre et al. 2001, Schumacher et al. 2001).

CP is an essential ferroxidase that contains greater than 95% of the Cu present in plasma. This protein is synthesized in hepatocytes and secreted into the plasma following the incorporation of six Cu atoms in the late secretory pathway (Madsen and Gitlin, 2007). In Wilson disease, loss of function of Atp7b results in synthesis of apoceruloplasmin that is rapidly degraded in the plasma (Figure 2). As a result, the serum CP concentration is a valuable diagnostic indicator of Wilson disease (Hellman and Gitlin, 2002).

Almost half of all patients with Wilson disease present signs and symptoms of neuropsychiatric illness (Gollan and Gollan, 1998). Although such neurological features may initially be subtle, without chelation, patients will progress to severe parkinsonian symptoms, consistent with the neuropathologic findings of basal ganglia, Cu accumulation and neurodegeneration (Oder et al. 1991).

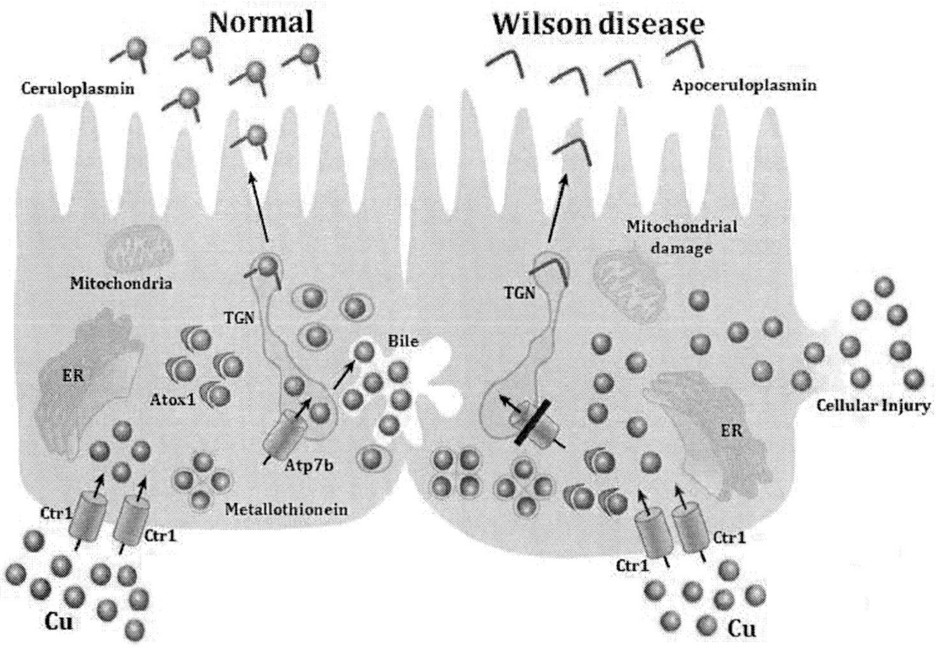

Modified from Madsen and Gitlin 2007.

Figure 2. Pathogenesis of Wilson disease. Model of the proposed pathways and proteins relevant to Cu metabolism in the human hepatocyte. Cu transport to the trans-Golgi network (TGN) is shown as the process mediating intracellular homeostasis by Atp7b. Dysfunction results in cytosolic Cu accumulation with associated cellular damage.

Neurodegeneration in Wilson disease results directly from Cu accumulation, but the precise mechanisms of cellular injury are unknown. Psychiatric symptoms are also common and range from behavioral problems to psychosis (Dening, 1991). These abnormalities, although without neuropathologic correlates, are clearly the result of brain Cu accumulation because rapid improvement is observed upon treatment with oral chelating agents to restore Cu homeostasis (Madsen and Gitlin, 2007). The reversible nature of many of these psychiatric symptoms is consistent with the concept that Cu can modulate synaptic function (Schlief and Gitlin, 2006) and suggests innovative strategies for investigation in common psychiatric disorders.

12.7.4.2. Menkes Disease

Menkes disease is an X-linked disorder characterized by growth failure, fragile hair, hypopigmentation, arterial tortuosity, and neuronal degeneration due to loss-of-function mutations in the gene encoding Atp7a (Chelly et al. 1993; Vulpe et al. 1993). The pleiotropic features of this disease are the result of impaired activity of specific cuproenzymes resulting from impaired Atp7a function (Figure 1). The neurologic features are present in early infancy, revealing a critical role for Atp7a and Cu in neuronal development (Mercer, 1998). Magnetic resonance imaging of the brain reveals defective myelination with cerebellar and cerebral atrophy (Leventer et al. 1997), and neuropathologic examination demonstrates focal degeneration of the gray matter and neuronal loss most prominent in the hippocampus and cerebellum (Barnard et al. 1978; Okeda et al. 1991).

Although several of the known cuproenzymes play critical roles in brain biochemistry, the mechanisms of neurodegeneration in Menkes disease are unknown and not explained by impaired activity of any of these enzymes. Studies in a murine model of Menkes disease suggest a role for Atp7a and Cu in axon extension and synaptogenesis during development (El Meskini et al. 2005). Atp7a mediates the availability of an NMDA receptor–dependent, releasable pool of copper in hippocampal neurons (Schlief et al. 2005), and the absence of Atp7a activity in Menkes disease markedly accentuates NMDA receptor–mediated excitotoxicity in these neurons (Schlief et al. 2006). These data suggest a model where loss of Atp7a contributes to both seizures and neuronal degeneration in affected patients and raise the possibility of therapeutic approaches based on NMDA receptor blockade (Hardingham and Bading, 2003; Schlief and Gitlin 2006).

Systemic Cu treatment is not successful in patients with Menkes disease because Cu transport into the brain is dependent on Atp7a (Madsen and Gitlin, 2007). Occasional patients with some residual Atp7a activity have less CNS pathology despite significant systemic disease (Møller et al. 2000) and when treated with Cu manifest neurodevelopmental improvement (Christodoulou et al. 1998). This clinical observation reveals a hierarchy of Cu distribution preferential to the brain under circumstances of limited Cu availability (Madsen and Gitlin, 2007), a concept supported by recent observations in a zebrafish model of Menkes disease (Mendelsohn et al. 2006). The mechanisms of this hierarchy are unclear but this process illustrates copper's fundamental role in brain development.

12.7.4.3. Prion Diseases

The growing interest in prion diseases, (e.g. Creutzfeldt-Jakob disease), apparently caused by infection with an altered prion protein (PrPC) conformation that is transferred to endogenous protein, rises the question as to the normal role of CP in healthy individuals. In this regard, several lines of evidence indicate that the PrPC binds Cu(II) and thus, as with CP, may serve a cytoprotective role (Brown et al. 1997). One possible scenario, consistent with the finding in cell cultures that Cu stimulates endocytosis of PrPC from the cell surface, is that prion protein controls Cu metabolism by helping as a recycling receptor for uptake of extracellular Cu (Pauly and Harris, 1998).

There is extensive data that suggest a link between Cu and the PrPC. A mutant form of the protein PrPSc is the culprit in a number of neurodegenerative disorders. PrPC is a glycosylphosphatidylinositol-anchored protein that is expressed in the CNS as well as peripheral tissues (Bertinato and L'Abbé, 2004). Thus far, no known function has been found for PrPC. It has been shown that Cu can bind the PrPC *in vivo* (Brown et al. 1997) and can rapidly stimulate endocytosis of PrPC from the cell surface in a process that is reversible (Pauly and Harris, 1998). Furthermore, Cu binding at the outer plasma membrane has been shown to be related to the expression level of PrPC (Rachidi et al. 2003). Whether PrPC plays a significant role in the uptake or intracellular trafficking of Cu remains to be determined.

Final Considerations

Recent insights into the molecular physiology of Cu metabolism have led to a better understanding of the role of Cu in normal brain development and function. Afterward, Cu has been related to some devastating neurological diseases, and a common theme that emerges is the possibility of Cu-induced free radical production leading to neuronal damage. For example, the amyloid precursor protein involved in AD, is a copper-binding molecule, and can reduce Cu(II) to Cu(I), which can form reactive hydroxyl radicals. The prion protein involved in Creutzfeldt–Jakob disease is a copper-binding protein and might have a function in Cu uptake in neurons and transport Cu to specific target proteins. However the link between this role and the pathogenesis of the disease remains unclear. PD appears to be associated with abnormalities of iron and Cu. Patients have been found to have lower levels of CP in their cerebrospinal fluid and this might be related to the raised iron levels in the basal ganglia. ALS is a neurodegenerative disease affecting motor neurons, which leads to progressive muscular weakness, atrophy and death. An autosomal dominant form is originated by mutations in the Cu-Zn SOD. It appears that the disease results from a gain of function of Cu-Zn SOD, associated with abnormal forms of the protein in which Cu is bound in a manner that allows the generation of ROS.

The inherited disorders of Cu metabolism reveal copper's essential role in brain development and neuropsychiatric disease. The mechanisms of psychiatric disease observed in patients with Wilson disease are of particular concern. Chelation is an effective therapy in such individuals, indicating Cu direct role in pathogenesis and suggesting strategies for study that may be broadly applicable to brain function and mental health. Although elucidation of the genetic basis of Menkes and Wilson diseases has revealed much about the cell biology of Cu metabolism, much more needs to be learned about the normal mechanisms of Cu homeostasis in the brain. In particular, the role of the basal ganglia in Cu storage and distribution needs further study as do the mechanisms of this brain region's vulnerability to excess of Cu.

However, caution is necessary because more knowledge of Cu homeostasis within the brain and copper's role in specific neurologic functions is required before any such therapeutic approaches can be considerately undertaken or interpreted.

Cu diseases illustrate the equally devastating results of deficiency and excess and the consequences of breakdown of homeostatic mechanisms. The list of possible diseases involving this essential trace element continues to increase. More research into the molecular basis of Cu homeostasis will have an impact not just on the uncommon genetic disorders but also on frequent and serious progressive neurological diseases.

References

Andrus, P.K., Fleck, T.J., Gurney, M.E., and Hall, E.D. (1998) Protein oxidative damage in a transgenic mouse model of familial amyotrophic lateral sclerosis. *J Neurochem*. 71: 2041-2048.

Barnard, R.O., Best, P.V. and Erdohazi, M. (1978) Neuropathology of Menkes' disease. *Dev Med Child Neurol.* 20: 586–597.

Barnea, A. and Cho, G. (1984) Evidence that copper-amino acid complexes are potent stimulators of the release of luteinizing hormone-releasing hormone from isolated hypothalamic granules. *Endocrinology*. 115: 936-943.

Barnea, A., Cho, G. and Colombani-Vidal, M. (1985). A role for extracellular copper in modulating prostaglandin E2 (PGE2) action: facilitation of PGE2 stimulation of the release of gonadotropin-releasing hormone (LHRH) from median eminence explants. *Endocrinology*. 117: 415-417.

Barnham, K.J., McKinstry, W.J., Multhaup, G., Galatis, D., Morton, C.J., Curtain, C.C., Williamson, N.A., White, A.R., Hinds, M.G., Norton, R.S., Beyreuther, K., Masters, C.L., Parker, M.W. and Cappai, R. (2003) Structure of the Alzheimer's disease amyloid precursor protein copper binding domain. A regulator of neuronal copper homeostasis. *J Biol Chem*. 278: 17401–17407.

Bayer, T.A., Schäfer, S., Simons, A., Kemmling A, Kamer T, Tepest R, Eckert A, Schussel K, Eikenberg, O., Sturchler-Pierrat, C., Abramowski, D., Staufenbiel, M. and Multhaup, G. (2003) Dietary Cu stabilizes brain superoxide dismutase 1 activity and reduces amyloid Abeta production in APP23 transgenic mice. *Proc Natl Acad Sci USA*. 100: 14187–14192.

Bellingham, S.A., Ciccotosto, G.D., Needham, B.E., Fodero, L.R., White, A.R., Masters, C.L., Cappai, R. and Camakaris, J. (2004a) Gene knockout of amyloid precursor protein and amyloid precursor-like protein-2 increases cellular copper levels in primary mouse cortical neurons and embryonic fibroblasts. *J Neurochem*. 91: 423–428.

Bellingham, S.A., Lahiri, D.K., Maloney, B., La Fontaine, S., Multhaup, G. and Camakaris, J. (2004b) Copper depletion down-regulates expression of the Alzheimer's disease amyloid-beta precursor protein gene. *J Biol Chem.* 279: 20378–20386.

Bertinato, J. and L'Abbé, M.R. (2004) Maintaining copper homeostasis: regulation of copper-trafficking proteins in response to copper deficiency or overload. *J Nutr Biochem.* 15: 316-322.

Bogdanov, M.B., Ramos, L.E., Xu, Z. and Beal, M.F. (1998) Elevated "hydroxyl radical" generation in vivo in an animal model of amyotrophic lateral sclerosis. *J Neurochem*. 71: 1321-1324.

Bohnenkamp, W. and Weser, U. (1976) Copper deficiency and erythrocuprein (2Cu, 2Zn-superoxide dismutase). *Biochim Biophys Acta*. 444: 396-406.

Brezova, V., Valko, M., Breza, M., Morris, H., Telser, J., Dvoranova, D., Kaiserova, K., Varecka, L., Mazur, M. and Leibfritz, D. (2003) Role of radicals and singlet oxygen in photoactivated DNA cleavage by the anticancer drug camptothecin: An electron paramagnetic resonance study. *J Phys Chem B,* 107: 2415-2425.

Brooksbank, B.W. and Balazs, R. (1984) Superoxide dismutase, glutathione peroxidase and lipoperoxidation in Down's syndrome fetal brain. *Brain Res.* 318: 37-44.

Brown, D.R., Qin K, Herms, J.W., Madlung, A., Manson, J., Strome, R., Fraser, P.E., Kruck, T., von Bohlen, A., Schulz-Schaeffer, W., Giese, A., Westaway, D. and Kretzschmar, H. (1997) The cellular prion protein binds copper in vivo. *Nature*, 390: 684-687.

Bruijn, L.I., Houseweari, M.K., Kato, S., Anderson, K.L., Anderson, S.D., Ohama, E., Reaume, A.G., Scott, R.W. and Cleveland, D.W. (1998) Aggregation and motor neuron toxicity of an ALS-linked SOD1 mutant independent from wild-type 5001. *Science*, 281:1851- 1854.

Buettner, G.R. and Jurkiewicz, B.A. (1996) Catalytic metals, ascorbate and free radicals: combinations to avoid. *Rad Res.* 145: 532-541.

Bull, P.C. and Cox, D.W. (1994) Wilson disease and Menkes disease: new handles on heavy-metal transport. *Trends Genet.* 10: 246-252.

Bull, P.C., Thomas, G.R., Rommens, J.M., Forbes, J.R. and Cox, D.W. (1993) The Wilson disease gene is a putative copper transporting P-type Atpase similar to the Menkes gene. *Nat Genet.* 5: 327–337.

Bush, A.I. (2000) Metals and neuroscience. *Curr Opinion Chem Biol.* 4:184–191.

Carrí, M.T., Ferri, A., Cozzolino, M., Calabrese, L. and Rotilio, G. (2003) Neurodegeneration in amyotrophic lateral sclerosis: the role of oxidative stress and altered homeostasis of metals. *Brain Res Bull.* 61: 365-374.

Castellani, R.J., Smith, M.A., Nunomura A., Harris, P.L.R. and Perry, G. (1999) Is increased redox-active iron in Alzheimer disease a failure of the copperbinding protein ceruloplasmin? *Free Radic Biol Med.* 26: 1508-1512.

Chelly, J., Tumer, Z., Tonnesen, T., Petterson, A., Ishikawa-Brush, Y., Tommerup, N., Horn, N. and Monaco, A.P. (1993) Isolation of a candidate gene for Menkes disease that encodes a potential heavy metal binding protein. *Nat Genet.* 3:14-19.

Christodoulou, J., Danks, D.M., Sarkar, B., Baerlocher, K.E., Casey, R., Horn, N., Tümer, Z. and Clarke, J.T. (1998) Early treatment of Menkes disease with parenteral copper-histidine: long-term follow-up of four treated patients. *Am J Med Genet.* 76: 154–164.

Cimarusti, D.L., Saito, K., Vaughn, J.E., Barber, R., Roberts, E. and Thomas, P.E. (1979). Immunocytochemical localization of dopamine-beta-hydroxylase in rat locus coeruleus and hypothalamus. *Brain Res.* 162: 55-67.

Cohen, E. and Elvehjem, C.A. (1934) The relation of iron and copper to the cytochrome and oxidase content of animal tissues. *J BioL Chem.* 107: 7-105.

Cordano, A. and Graham, G.G. (1966) Copper deficiency complicating severe chronic intestinal malabsorption. *Pediatrics*, 38: 596-604.

Coyle, J.T. and Axelrod, J. (1972) Dopamine-β-hydroxylase in the rat brain: developmental characteristics. *J Neurochem.* 19: 449-459.

Culotta, V.C. and Gitlin, J.D. (2001) Disorders of copper transport. In The Molecular and Metabolic Basis of Inherited Disease, ed. CR Scriver, AL Beaudet, WS Sly, D Valle, pp. 3105–36. New York: McGraw-Hill.

Culotta, V.C., Yang, M. and O'Halloran, T.V. (2006) Activation of superoxide dismutases: putting the metal to the pedal. *Biochim Biophys Acta*, 1763: 747–758.

Dallman, P.R. (1967) Cytochrome oxidase repair during treatment of copper deficiency: relation to mitochondrial turnover. *J Clin Invest.* 46: 1819-1827.

Dann, T. (1994) PM10 and PM2.5: Concentrations at Canadian urban sites, 1984-1993. Ottawa, Environmental Protection Service, Environment Canada, Environmental Technology Centre, (Report series No. PMD 94-3).

Dasler, W., Stoner, R.E. and Milliser, R.V. (1961) Effect of osteolathyrism on soluble collagen fractions of rat connective tissue. *Metabolism*, 10: 883-887.

De Leo, M.E., Borrello, S., Passantino, M., Palazzotti, B., Mordente, A., Daniele, A., Filippini, V., Galeotti, T. and Masullo, C. (1998) Oxidative stress and overexpression of manganese superoxide dismutase in patients with Alzheimer's disease. *Neurosci Lett.* 250: 173-176.

Dening, T.R. (1991) The neuropsychiatry of Wilson's disease: a review. *Int J Psychiatr Med.* 21: 135–148.

Deneke, S.M. and Fanburg, B.L. (1989) Regulation of cellular glutathione. *Am J Physiol.* 257: 163–173.

Doreulee, N., Yanovsky, Y. and Haas, H.L. (1997) Suppression of long-term potentiation in hippocampal slices by copper. *Hippocampus*, 7: 666–669.

Dorn, C.R., Pierce, J.O., Chase, G.R. and Phillips, P.E. (1975) Environmental contamination by lead, cadmium, zinc, and copper in a new lead-producing area. *Environ Res.* 9: 159-172.

El Meskini, R., Cline, L.B., Eipper, B.A. and Ronnett, G.V. (2005) The developmentally regulated expression of Menkes protein Atp7A suggests a role in axon extension and synaptogenesis. *Dev Neurosci.* 27: 333–348.

Emre, S., Atillasoy, E.O., Ozdemir, S., Schilsky, M. and Rathna Varma, C.V. (2001) Orthotopic liver transplantation forWilson's disease: a single-center experience. *Transplantation,* 72:1232–36.

Ferenci, P. (2004) Pathophysiology and clinical features of Wilson disease. *Metab. Brain Dis.* 19:229–239.

Ferreira, A.M., Ciriolo, M.R., Marcocci, L., and Rotilio, G. (1993) Copper(I) transfer into metallothionein mediated by glutathione. *Biochem J.* 292:673–676.

Fischer, P.W., Giroux, A. and L'Abbé, M.R. (1983) Effects of zinc on mucosal copper binding and on the kinetics of copper absorption. *J Nutr.* 113: 462–469.

Gaetke, L.M. and Chow, C.K. (2003) Copper toxicity, oxidative stress, and antioxidant nutrients. *Toxicology*, 189: 147-163.

Gallagher, C.H., Judah, J.D. and Rees, K.R. (1956) The biochemistry of copper deficiency. I. Enzymological disturbances, blood chemistry and excretion of amino-acids. *Proc R Soc Lond B Biol Sci.* 144: 134-150.

Ghadge, G.D., Lee, J.P., Bindokas, V.P., Jordan, J., Ma, L., Miller, R.J. and Roos, R.P. (1997) Mutant superoxide dismutase-1-linked familial amyotrophic lateral sclerosis: molecular mechanisms of neuronal death and protection. *J Neurosci.* 17: 8756-87566.

Gitlin, J.D. (2003) Wilson disease. *Gastroenterology*, 125:1868–1877.

Gollan, J.L. and Deller, D.J. (1973) Studies on the nature and excretion of biliary copper in man. *Clin Sci.* 44:9–15.

Gollan, J.L. and Gollan, T.J. (1998) Wilson disease in 1998: genetic, diagnostic and therapeutic aspects. *J Hepatol.* 28: 28–36.

Graham, G.G. (1971) Human copper deficiency. *N Engl J Med.* 285: 857-858.

Gutteridge, J.M. (1981) Thiobarbituric acid-reactivity following iron-dependent free-radical damage to amino acids and carbohydrates. *FEBS Lett.* 128: 343-346.

Hambidge, K.M. (1977) The role of zinc and other trace metals in pediatric nutrition and health. *Pediatr Clin North Am.* 24: 95-106.

Hamza, I., Faisst, A., Prohaska, J., Chen, J., Gruss, P. and Gitlin, J.D. (2001) The metallochaperone Atox1 plays a critical role in perinatal copper homeostasis. *Proc Natl Acad Sci USA*, 98: 6848–6852.

Hamza, I. and Gitlin, J.D. (2002) Copper chaperones for cytochrome c oxidase and human disease. *J Bioenerg Biomembr.* 34: 381–388.

Hamza, I., Prohaska, J. and Gitlin, J.D. (2003) Essential role for Atox1 in the copper-mediated intracellular trafficking of the Menkes Atpase. *Proc Natl Acad Sci USA*, 100: 1215–1220.

Hardingham, G.E. and Bading, H. (2003) The Yin and Yang of NMDA receptor signalling. *Trends Neurosci.* 26: 81–89.

Harris, E.D. (1991) Copper transport: an overview. *Proc Soc Exp Biol Med.* 196: 130-140.

Harris, Z.L., Klomp, L.W. and Gitlin, J.D. (1988) Aceruloplasminemia: an inherited neumdegenerative disease with impairment of iron homeostasis. *Am J Clin Nutr.* 87: S972-S977.

Harrison, P.M. and Hoare, R.J. (1980) Metals in Biochemistry, Chapman and Hall, London.

Hart, P.J., Liu, H., Pellegrini, M., Nersissian, A.M., Gralla, E.B., Valentine, J.S. and Eisenberg, D. (1998) Subunit asymmetry in the three-dimensional structure of a human CuZnSOD mutant found in familial amyotrophic lateral sclerosis. *Protein Sci.* 7: 545-555.

Hartter, D.E. and Barnea, A. (1988) Evidence for release of copper in the brain: depolarization-induced release of newly taken-up ^{67}copper. *Synapse*, 2: 412–415.

Hellman, N.E. and Gitlin, J.D. (2002) Ceruloplasmin metabolism and function. *Annu Rev Nutr.* 22: 439–458.

Houwen, R., Dijkstra, M., Kuipers, F., Smit, E.P., Havinga, R. and Vonk, R.J. (1990) Two pathways for biliary copper excretion in the rat. The role of glutathione. *Biochem Pharmacol.* 39: 1039–1044.

Hunt, D.M. (1980) Copper and neurological function. In: Biological Roles of Copper, edited by D. Evered and G. Lawrenson. Amsterdam: *Excerpta Med.* p. 247-266.

Jellinger, K.A. (1999) The role of iron in neurodegeneration: prospects for pharmacotherapy of Parkinson's disease. *Drugs Aging.* 14: 115-140.

Kageyama, G.H. and Wong-Riley, M.T. (1982) Histochemical localization of cytochrome oxidase in the hippocampus: correlation with specific neuronal types and afferent pathways. *Neuroscience*, 7: 2337-2361.

Kardos, J., Kovacs, I., Hajos, F., Kalman, M. and Simonyi, M. (1989) Nerve endings from rat brain tissue release copper upon depolarization. A possible role in regulating neuronal excitability. *Neurosci Lett.* 103: 139–144.

Keen, C.L., Uriu-Hare, J.Y., Hawk, S.N., Jankowski, M.A., Daston, G.P., Kwik-Uribe, C.L. and Rucker, R.B. (1998) Effect of copper deficiency on prenatal development and pregnancy outcome. *Am J Clin Nutr.* 67:1003S–100311.

Kelly, E.J. and Palmiterm R.D. (1996) Amurine model of Menkes disease reveals a physiological function of metallothionein. *Nat Genet.* 13: 219–222.

Kim, B.E., Nevitt, T. and Thiele, D.J. (2008) Mechanisms for copper acquisition, distribution and regulation. *Nat Chem Biol.* 4: 176–185.

Kodama, H., Murata, Y. and Kobayashi, M. (1999) Clinical manifestations and treatment of Menkes disease and its variants. *Pediatr Int.* 41:423–429.

Komulainen, H. and Tuomisto, J. (1982) Effect of copper on the uptake and release of monoamines in rat brain synaptosomes. *Acta Pharmacol Toxicol.* 48: 205-213.

Kozma, M., Szerdahelyi, P. and Kasa, P. (1981) Histochemical detection of zinc and copper in various neurons of the central nervous system. *Acta Histochem.* 69: 12–17.

Kumar, N., Gross, J.B.J. and Ahlskog, J.E. (2004) Copper deficiency myelopathy produces a clinical picture like subacute combined degeneration. *Neurology*, 63:33–39.

Kuo, Y.M., Gybina, A.A., Pyatskowit, J.W., Gitschier, J. and Prohaska, JR. (2006) Copper transport protein (Ctr1) levels in mice are tissue specific and dependent on copper status. *J Nutr*. 136: 21–26.

Kuo, Y.M., Zhou, B., Cosco D. and Gitschier, J. (2001) The copper transporter CTR1 provides an essential function in mammalian embryonic development. *Proc Natl Acad Sci USA*, 98: 6836–6841.

Kwon, O.J., Lee, S.M., Floyd, R.A. and Park, J.W. (1998) Thiol-dependent metalcatalyzed oxidation of copper, zinc superoxide dismutase. *Biochim Biophys Acta*, 1387:249-256.

Lide, D.R. (1998) Handbook of chemistry and physics. 78TH Edition, CRC, USA.

Leary, S.C., Cobine, P.A., Kaufman, B.A., Guercin, G.H., Mattman, A., Palaty, J., Lockitch, G., Winge, D.R., Rustin, P., Horvath, R. and Shoubridge, E.A. (2007) The human cytochrome c oxidase assembly factors SCO1 and SCO2 have regulatory roles in the maintenance of cellular copper homeostasis. *Cell Metab.* 5: 9–20.

Lee, J., Prohaska, J.R. and Thiele, D.J. (2001) Essential role for mammalian copper transporter Ctrl in copper homeostasis and embryonic development. *Proc Natl Acad Sci USA*, 98: 6842–6847.

Leventer, R.J., Kornberg, A.J., Phelan, E.M. and Kean, M.J. (1997) Early magnetic resonance imaging findings in Menkes' disease. *J Child Neurol.* 12: 222–224.

Linder, M.C. (2001) Copper and genomic stability in mammals. *Mutation Res.* 475:141–152.

Lloyd, R.V., Hanna, P.M. and Mason, R.P. (1997) The origin of the hydroxyl radical oxygen in the Fenton reaction. *Free Radic Biol Med.* 22: 885-888.

Loeffler, D.A., LeWitl, P.A., Juneau, P.L., Sima, A.A., Nguyen, H.U., DeMaggio, A.J., Bridunan, C.M., Brewer, G.J., Dick, R.D., Troyer, M.D. and Kanaley, L. (1996)

Increased regional brain concentrations of ceruloplasmin in neurodegenerative disorders. *Brain Res.* 738: 265-274.

Lombardi, V., Valko, L., Stolc, S., Valko, M., Ondrejicková, O., Horáková, L., Placek, J. and Troncone A. (1998) Free radicals in rabbit spinal cord ischemia: electron spin resonance spectroscopy and correlation with SOD activity. *Cell Mol Neurobiol.* 18: 399-412.

Lovell, M.A., Robertson, J.D., Teesdale, W.J., Campbell, J.L. and Markesbery, W.R. (1998) Copper, iron and zinc in Alzheimer's disease senile plaques. *J Neurol Sci.* 158: 47-52.

Lu, W., Man, H., Ju, W., Trimble, W.S., MacDonald, J.F. and Wang, Y.T. (2001) Activation of synaptic NMDA receptors induces membrane insertion of new AMPA receptors and LTP in cultured hippocampal neurons. *Neuron.* 29: 243–254.

Lutsenko, S. and Petris, M.J. (2003) Function and regulation of the mammalian copper-transporting Atpases: insights from biochemical and cell biological approaches. *J Membr Biol.* 191: 1–12.

Lyle, W.H., Payton, J.E. and Hui, M. (1976) Haemodialysis and copper fever. *Lancet,* 1: 1324-1325.

Ma, J.Y., Narahashi, T. (1993) Differential modulation of $GABA_A$ receptor-channel complex by polyvalent cations in rat dorsal root ganglion neurons. *Brain Res.* 607: 222-232.

Madsen, E. and Gitlin, J.D. (2007) Copper and Iron Disorders of the Brain. *Annu Rev Neurosci.* 30: 317-337.

Mendelsohn, B.A., Yin, C., Johnson, S.L., Wilm, T.P., Solnica-Krezel, L. and Gitlin, J.D. (2006) Atp7a determines a hierarchy of copper metabolism essential for notochord development. *Cell Metab.* 4: 155–162.

Mercer, J.F. (1998) Menkes syndrome and animal models. *Am J Clin Nutr.* 67: 1022S–1028.

Mercer, J.F., Livingston, J., Hall, B., Paynter, J.A., Begy, C., Chandrasekharappa, S., Lockhart, P., Grimes, A., Bhave, M., Siemieniak, D. and Glover, T.W. (1993) Isolation of a partial candidate gene for Menkes disease by positional cloning. *Nat Genetic.* 3:20–25.

Mezzetti, A., Pierdomenico, S.D., Costantini, F., Romano, F., De Cesare, D., Cuccurullo, F., Imbastaro, T., Riario-Sforza, G., Di Giacomo, F., Zuliani, G. and Fellin, R. (1998) Copper/zinc ratio and systemic oxidant load: effect of aging and aging-related degenerative diseases. *Free Radic Biol Med.* 25:676-681.

Møller, L.B., Tümer, Z., Lund, C., Petersen, C., Cole, T., Hanusch, R., Seidel, J., Jensen, L.R. and Horn, N. (2000) Similar splice-site mutations of the Atp7A gene lead to different phenotypes: classical Menkes disease or occipital horn syndrome. *Am J Hum Genet.* 66: 1211–1220.

Morgan, R.F. and O'Dell, B.L. (1977) Effect of copper deficiency on the concentrations of catecholamines and related enzyme activities in the rat brain. *J Neurochem.* 28: 207-213.

Newman, J.A., Archer, V.E, Saccomanno, G., Kuschner, M., Auerbach, O., Grondahl, R.D. and Wilson, J.C. (1976) Histologic types of bronchogenic carcinoma among members of copper-mining and smelting communities. *Ann N Y Acad Sci.* 271:260-268.

Nose, Y., Kim, B.E. and Thiele, D.J. (2006) Ctrl drives intestinal copper absorption and is essential for growth, iron metabolism, and neonatal cardiac function. *Cell Metab.* 4: 235–244.

NRC (1989) Recommended dietary allowances. Food and Nutrition Board, National Research Council and National Academy of Sciences, Washington, D.C.

Nunomura, A., Perry, G., Pappolla, M.A., Wade, R., Hirai, K., Chiba, S. and Smith, M.A. (1999) RNA oxidation is a prominent feature of vulnerable neurons in Alzheimer's disease. *J Neurosci.* 19: 1959-1964.

Oder, W., Grimm, G., Kollegger, H., Ferenci, P., Schneider, B. and Deecke, L. (1991) Neurological and neuropsychiatric spectrum ofWilson's disease: a prospective study of 45 cases. *J Neurol.* 238: 281–287.

O'Halloran, T.V. and Culotta, V.C. (2000) Metallochaperones, an intracellular shuttle service for metal ions. *J Biol Chem*. 275: 25057–25060.

Okeda, R., Gei, S., Chen, I., Okaniwa, M., Shinomiya, M. and Matsubara, O. (1991) Menkes' kinky hair disease: morphological and immunohistochemical comparison of two autopsied patients. *Acta Neuropathol* (Berl.). 81: 450–457.

Owen, C.A. (1981) Cooper deficiency and toxicity. Acquired and inherited, in plants, animals, and man. Noyes Publications, NJ, USA.

Paik, S.R., Shin, H.J., Lee, J.H., Chang, C.S. and Kim, J. (1999) Copper(II)-induced self-oligomerization of alpha-synuclein. *Biochem J.* 340:821-828.

Pauly, P.C. and Harris, D.A. (1998) Copper stimulates endocytosis of the prion protein. *J Biol Chem*. 273: 33107-33110.

Pena, M.M., Lee, J. and Thiele, D.J. (1999) A delicate balance: homeostatic control of copper uptake and distribution. *J Nutr*. 129:1251–1260.

Pickart, L. and Thaler, M.M. (1980) Growth-modulating tripeptide (glycylhistidyllysine): association with copper and iron in plasma, and stimulation of adhesiveness and growth of hepatoma cells in culture by tripeptide-metal ion complexes. *J Cell Physiol.* 102: 129-139.

Prodan, C.I., Bottomley, S.S., Holland, N.R. and Lind, S.E. (2006) Relapsing hypocupraemic myelopathy requiring high-dose oral copper replacement. *J Neurol Neuros Psychiatr.* 77:1092–93.

Prohaska, J.R. (1987) Functions of trace elements in brain metabolism. *Physiol Rev*. 67: 858-901.

Prohaska, J.R. and Brokate, B. (2002) The timing of perinatal copper deficiency in mice influences offspring survival. *J Nutr*. 132: 3142–3145.

Rachidi, W., Vilette, D., Guiraud, P., Arlotto, M., Riondel, J., Laude, H., Lehmann, S. and Favier, A. (2003) Expression of prion protein increases cellular copper binding and antioxidant enzyme activities but not copper delivery. *J Biol Chem.* 278: 9064–9072.

Rae, T.D., Schmidt, P.J., Pufahl, R.A., Culotta, V.C. and O'Halloran, T.V. (1999) Undetectable intracellular free copper: the requirement of a copper chaperone for superoxide dismutase. *Science*, 284: 805-808.

Rajan, K.S., Colburn, R.W. and Davis, J.M. (1976) Distribution of metal ions in the subcellular fractions of several rat brain areas. *Life Sci.* 18: 423-431.

Rojkind, M., Rhi, L. and Aguirre, M. (1968) Biosynthesis of the intramolecular cross-links in rat skin collagen. *J Biol Chem*. 243: 2266-2272.

Sadée, W., Pfeiffer, A. and Herz, A. (1982) Opiate receptor: multiple effects of metal ions. *J Neurochem.* 39: 659-667.

Sandberg, L.B., Weissman, N. and Smith, D.W. (1969) The purification and partial characterization of a soluble elastin-like protein from copper-deficient porcine aorta. *Biochemistry*, 8: 2940-2945.

Sato, M., Ohtomo, K., Daimon, T., Sugiyama, T. and Iijima, K. (1994) Localization of copper to afferent terminals in rat locus ceruleus, in contrast to mitochondrial copper in cerebellum. *J Histochem Cytochem*. 42: 1585–1591.

Sayre, L.M., Moreira, P.I., Smith, M.A. and Perry, G. (2005) Metal ions and oxidative protein modification in neurological disease. *Ann Ist Super Sanità*. 41:143-164.

Sayre, L.M., Zelasko, D.A., Harris, P.L., Perry, G., Salomon, R.G. and Smith, M.A. (1997) 4-Hydroxynonenal-derived advanced lipid peroxidation end products are increased in Alzheimer's disease. *J Neurochem*. 68: 2092-2097.

Schäfer, S.G., Dawes, R.L.F., Elsenhans, B., Forth, W. and Schomann, K. (1999) Metals. In: Toxicology. Hans Marquardt, Siegfried G. Schäfer, Roger O. McClellan, and Frank Welsch (Eds). Academic Press, San Diego, Ca. Pp. 755-804.

Schlief, M.L., Craig, A.M. and Gitlin, J.D. (2005) NMDA receptor activation mediates copper homeostasis in hippocampal neurons. *J Neurosci*. 25: 239–246.

Schlief, M.L. and Gitlin, J.D. (2006) Copper homeostasis in the CNS: a novel link between the NMDA receptor and copper homeostasis in the hippocampus. *Mol Neurobiol*. 33: 81–90.

Schlief, M., West, T., Craig, A., Holtzman, D. and Gitlin, J. (2006) Role of the Menkes copper transporting Atpase in NMDA receptor-mediated neuronal toxicity. *Proc Natl Acad Sci USA*, 103: 14919–14924.

Schumacher, G., Platz, K.P., Mueller, A.R., Neuhaus, R., Luck, W., Langrehr, J.M., Settmacher, U., Steinmueller, T., Becker, M. and Neuhaus, P. (2001) Liver transplantation in neurologic Wilson's disease. *Transplant Proc*. 33: 1518–1519.

Singh, R.J., Karoui, H., Gunther, M.R., Beckman, J.S., Mason, R.P. and Kalyanaraman, B. (1998) Reexamination of the mechanism of hydroxyl radical adducts formed from the reaction between familial amyotrophic lateral sclerosis-associated Cu,Zn superoxide dismutase mutants and H2O2. *Proc Natl Acad Sci U S A*, 95: 6675-66780.

Smith, M.A., Sayre, L.M., Anderson, V.E., Harris, P.L., Beal, M.F., Kowall, N. and Perry, G. (1998) Cytochemical demonstration of oxidative damage in Alzheimer disease by immunochemical enhancement of the carbonyl reaction with 2,4-dinitrophenylhydrazine. *J Histochem Cytochem*. 46: 731-735.

Sparks, D.L. and Schreurs, B.G. (2003) Trace amounts of copper in water induce beta-amyloid plaques and learning deficits in a rabbit model of Alzheimer's disease. *Proc Natl Acad Sci USA*, 100: 11065–11069.

Tao, T.Y. and Gitlin, J.D. (2003) Hepatic copper metabolism: insights from genetic disease. *Hepatology*, 37: 1241–1247.

Trombley, P.Q. and Shepherd, G.M. (1996) Differential modulation by zinc and copper of amino acid receptors from rat olfactory bulb neurons. *J Neurophysiol*. 76: 2536–2546.

Turnlund, J.R. (1999) Copper. In: Shils ME, Olson JA, Shike M, Ross AC, editors. Modern Nutrition in Health and Disease 9th ed. Baltimore:Williams & Wilkins. pp. 241–52.

Valko, M., Morris, H. and Cronin, M.T.D. (2005) Metals, Toxicity and Oxidative Stress. *Curr Med Chem*. 12: 1161-1208.

Vlachova, V., Zemkova, H. and Vyklicky, L.J. (1996) Copper modulation ofNMDAresponses in mouse and rat cultured hippocampal neurons. *Eur J Neurosci.* 8: 2257–2264.

Vulpe, C., Levinson, B., Whitney, S., Packman, S. and Gitschier, J. (1993) Isolation of a candidate gene for Menkes disease and evidence that it encodes a copper-transporting Atpase. *Nat Genetic*. 3: 7-13.

Wainio, W.W., Vander Wende, C. and Shimp, N.F. (1959) Copper in cytochrome c oxidase. *J BioL Chem*. 234: 2433-2436.

Wang, P., Chen, H., Qin, H., Sankarapandi, S., Becher, M.W., Wong, P.C. and Zweier, J.L. (1988) Overexpression of human copper, zinc-superoxide dismutase (SOD1 1) prevents postischemic injury. *Proc Nat Acad SC USA*, 95: 4556-4560.

WHO (1998) Environmental health criteria for Copper, World Health Organization, Geneva.

Wiedau-Pazos, M., Goto, J.J., Rabizadeh, S., Gralla, E.B., Roe, J.A., Lee, M.K., Valentine, J.S. and Bredesen, D.E. (1996) Altered reactivity of superoxide dismutase in familial amyotrophic lateral sclerosis. *Science*, 271:515-518.

Wong, P.C., Waggoner, D., Subramaniam, J.R., Tessarollo, L., Bartnikas, T.B., Culotta, V.C., Price, D.L., Rothstein, J. and Gitlin, J.D. (2000) Copper chaperone for superoxide dismutase is essential to activate mammalian Cu/Zn superoxide dismutase. *Proc Natl Acad Sci USA*, 97: 2886–2891.

Yamaguchi, Y., Heiny, M.E. and Gitlin, J.D. (1993) Isolation and characterization of a human liver cDNA as a candidate gene for Wilson disease. *Biochem Biophys Res Commun*. 197: 271-277.

Yuen, P., Lin, H.J. and Hutchinson, J.H. (1979) Copper deficiency in a low birthweight infant. *Arch Dis Child*, 54: 553-555.

Zappasodi, F., Salustri, C., Babiloni, C., Cassetta, E., Del Percio, C., Ercolani, M., Rossini, P.M. and Squitti, R. (2008) An observational study on the influence of the APOE-epsilon4 allele on the correlation between 'free' copper toxicosis and EEG activity in Alzheimer disease. *Brain Res.* 1215: 183-189.

Zheng Wei, Aschner, M. and Ghersi-Egeac, J.F. (2003) Brain barrier systems: a new frontier in metal neurotoxicological research. *Toxicol and Applied Pharmacol.* 192: 1–11.

Index

B

C

D

E

F

H

M

N

O

Q

R

S

T

U

V

W

X

Y

Z